Probability Models and
Statistical Methods in Genetics

DATE DUE			

A WILEY PUBLICATION IN APPLIED STATISTICS

Probability Models and Statistical Methods in Genetics

REGINA C. ELANDT-JOHNSON

University of North Carolina, Chapel Hill

John Wiley & Sons, Inc.

New York · London · Sydney · Toronto

Library of Congress Catalogue Card Number: 75-140177

ISBN 0 471 23490 7

Printed in the United States of America.

10 9 8 7 6 5 4 3 2 1

Preface

The purpose of this book is to present the basic concepts of the theory of probability and statistics, with a rather extensive account of applications in the analysis of qualitative genetic traits. A knowledge of advanced mathematics is not assumed, but it is expected that readers will be mathematically minded and able to follow algebraic manipulations.

The book has been written mainly for biologists who wish to understand probability and statistics, and to learn about applications of this theory in genetics. As in any book which is on the border between two disciplines, there is some difficulty in meeting the needs of different types of readers, while at the same time keeping the book to a reasonable size. In the present case, readers who are mainly interested in learning probability may find the lack of detailed proofs a drawback; those mostly interested in applications, on the other hand, may complain that there are too many formal definitions, and sometimes insufficient discussion of complicated genetic problems. Perhaps both kinds of readers will find their understanding of the book improved by the following brief account of its structure and of the reasons which led me to present it in this form.

I always find it difficult to learn a new subject if the terminology is ambiguous and inconsistent. Neither statistics nor genetics is completely free from such defects. I have deliberately stated, in some detail, terms and definitions from both disciplines in the hope that this will clarify at least some of the discussion in the text.

The chapters reflect the nature of the book in that they fall into two classes: "statistical" (and "probabilistic") and "genetical."

Chapters 2, 5, 6, 8, 11, 12, and 13 constitute a regular introductory course in probability and statistical methodology, with emphasis on genetical applications. Although the more complicated proofs are omitted, the presentation is not just a collection of statistical "recipes" as care has been taken to explain the reasons for and the effects of using the techniques described. Major emphasis is on discrete random variables. Since quantitative genetics is not presented in this book, distributions of continuous random variables (apart from the normal distribution) are not discussed in detail.

The remaining twelve chapters are entirely devoted to the mathematical theory of the transmission of inheritance units in large populations and finite numbers of families of different sizes. Since the book is of a mathematical rather than biological nature, it cannot be claimed that a complete biological background is provided. But emphasis is laid on the importance of correct sampling procedures and the construction of experimental designs to discriminate among competing hypotheses (see Chapters 11, 13, and especially 16). It is also pointed out, in several places, that the mathematical models used are only approximations to the real situations, since they are derived under certain, usually simplified, assumptions.

There are separate chapters on special topics such as inbreeding (Chapter 9), selection (Chapter 10), blood groups (Chapter 14), analysis of family data (Chapters 17 and 18), and models for the genetics of tissue transplantation (Chapter 19). These chapters have been written so that they can be read more or less independently of each other. Each may be regarded as a small (although not exhaustive) monograph summarizing basic work in the field at present available in a number of different journals and books.

The references given at the end of each chapter are not always cited in the text. They have been included because they might be of some interest to the reader.

To save space, calculations have been skipped in some examples, but I have always tried to give the "key steps" so that the reader should be able to reconstruct them in detail.

Material marked ▼ is of relatively greater mathematical difficulty. It may be omitted at first reading but should not be very troublesome to master subsequently.

I should like to take this opportunity to acknowledge persons and institutes which, directly or indirectly, have aided me in writing this book. I think first of Poland, where I was born, and of all the people and universities concerned with my mathematical education. My first encounter with biological problems was in the Department of Agricultural Experimentation (later Plant Genetics) in Poznań Agricultural University, of which Professor S. Barbacki was (and is still) the chairman. I much appreciate what I learned from Professor Barbacki and my colleagues in regard to biological problems and the possibilities of using mathematical methods in their solutions.

My warm thanks go also to Professor B. G. Greenberg, chairman of the Department of Biostatistics at the University of North Carolina at Chapel Hill, where I am now working, for his encouragement and for the stimulating atmosphere in his department.

I am especially grateful to my husband, Dr. N. L. Johnson, who has thoroughly read the whole manuscript and made several useful comments on the statistical chapters. I should also like to thank Professor R. J. Walsh of the University of New South Wales for comments on the earlier version of Chapter 14, and the reviewers for suggestions and criticism leading to improvements both in theory and in the presentation of the text. In addition, thanks are due to Mmes. Betsy Sturdivant, Linda Williams, and Delores Gold for their careful typing and other assistance.

Finally, I should like to express my appreciation of the editorial work of the publisher, and in particular of Miss Beatrice Shube, for her encouragement and of Miss Marcia Heim for her careful editorial supervision.

REGINA C. ELANDT-JOHNSON

Chapel Hill, North Carolina
July 1970

Contents

Chapter 13.   Testing Genetic Hypotheses   345

Chapter 14.   Human Blood Groups   391

Chapter 15.   Autosomal Linkage in Experimental Populations   421

Chapter 16.   Statistically Equivalent Models of Inheritance   439

Chapter 17.   Segregation Ratios in Families. Simple Modes of
**   Inheritance   458**

ment $(0 < \pi \leqslant 1)$, with known π. Fixed numbers of ascertained families, 468 — 7. Maximum likelihood estimator of θ under complete ascertainment $(\pi = 1)$, 472 — 8. Evaluation of scores, amounts of information, and heterogeneity tests, 474 — 9. Approximate Weinberg's sib method, 479 — 10. An approximate method of Li and Mantel for estimating θ, 481 — 11. Efficiencies of the approximate estimators $\hat{\theta}_s$ and $\tilde{\theta}_s$, 482 — 12. Maximum likelihood estimator of θ under single ascertainment $(\pi \rightarrow 0)$, 484 — 13. Weinberg's "proband method," 486 — 14. Test of goodness of fit under single ascertainment, 486 — 15. The ascertainment probability, π, and its estimation, 487 — 16. The number of families under risk, 490 — 17. Ascertainment through the parents. Recessive trait, 492 — 18. Difficulties in analysis of family data, 492.

for one-way single-grafting designs, 540 — ▼ 14. A model including strong and weak loci for the double-grafting design $Pt_1 \rightarrow F_2 \leftarrow Pt_2$, 543 — 15. Histocompatibility systems in man. The *HL-A* system, 546 — 16. Histocompatibility models in randomly mating populations, 547 — 17. General remarks, 548.

SPECIAL NOTATION AND SYMBOLS

Greek Alphabet

A	α	alpha	I	ι	iota	P	ρ	rho
B	β	beta	K	κ $\varkappa$	kappa	Σ	σ	sigma
Γ	γ	gamma	Λ	λ	lambda	T	τ	tau
Δ	δ	delta	M	μ	mu	Υ	υ	upsilon
E	ϵ	epsilon	N	ν	nu	Φ	ϕ φ	phi
Z	ζ	zeta	Ξ	ξ	xi	X	χ	chi
H	η	eta	O	o	omicron	Ψ	ψ	psi
Θ	θ ϑ	theta	Π	π	pi	Ω	ω	omega

Some Conventional Symbols and Notations

$\equiv$ is identically equal to

$<$ is less than

$>$ is greater than

$\leqslant$ is less than or equal to (is not greater than)

$\geqslant$ is greater than or equal to (is not less than)

$\doteqdot$ (or $\simeq$ or $\approx$) is approximately equal to

$\neq$ is not equal to

∞ infinity

$\propto$ is proportional

$\rightarrow$ approaches

$\in$ belongs to

Σ summation sign (sum of)

Π multiplication sign (product of)

$\cup$ union (logical sum)

$\cap$ intersection (logical product)

$\frown$ is distributed as

$\stackrel{\cdot}{\frown}$ is approximately distributed as

$|x|$ modulus (absolute value, without respect to the sign) of x

(a, b) is equivalent to $a < x < b$, open interval

$\langle a, b \rangle$ is equivalent to $a \leqslant x \leqslant b$, closed interval

$(a, b \rangle$ is equivalent to $a < x \leqslant b$, left side, half-open interval

$|x| < a$ is equivalent to $-a < x < a$

$|x| > a$ is equivalent to $x < -a$ or $x > a$

$x = x_0 \,|\, y$ indicates that $x = x_0$ *given* y ($x = x_0$ when y is given)

$\{x\}$ is a set of values $x_1, x_2, \ldots$

Probability Models and
Statistical Methods in Genetics

Basic Terms and Definitions in Genetics

1.1. INTRODUCTION

Although this book is intended for research workers who should be familiar with basic genetics and genetic terminology, nevertheless it will be useful first to review briefly the basic concepts with which geneticists are concerned, so that the reader can later see more clearly how probability models can be applied to genetics. Of course, this review will not in any way amount to a regular course in genetics, and any mathematician who is really involved in research in statistical genetics will need to study at least one basic book on the principles of genetics. A few such books are included in the references at the end of this chapter.

Here we will give only the basic concepts and definitions necessary to understand the construction of simple probability models which can be applied to the Mendelian laws of segregation. Later, as more advanced probability and statistical theory is developed to meet genetic problems of greater complexity, the biological aspects and genetic terminology involved will be introduced at the appropriate points in the text.

1.2. CHROMOSOMES AND GENES

The hereditary information in a living organism is contained in a special set of genetic material called *chromosomes*. Chromosomes are located in the nucleus of the cell. Any species has its own definite number of different chromosomes. For example, the characteristic chromosome number for Drosophila is 4; for wheat, 7; and for man, 23. One set of such different chromosomes is called a *genome*. Organisms including more than two genomes are generally termed *polyploids*. Let k be the characteristic number of chromosomes in a genome. An individual with only one genome (i.e., one set of k chromosomes) is called a *monoploid;* with $2k$ chromosomes (i.e., two genomes), a *diploid;* with $3k$ chromosomes, a *triploid;* with $4k$, a

1

tetraploid; with 5k, a *pentaploid;* with 6k, a *hexaploid;* and so on. Most organisms are diploids, that is, they include in their somatic cells two sets of the same genome. The genomes are usually identical (i.e., are from the same species); such polyploids are called *autopolyploids.* But there can sometimes be polyploids with genomes from different species—these are called *allopolyploids.* The most frequent types of polyploids are autopolyploids with an even number of genomes, 2m–ploids, say.

The corresponding chromosomes in each of the genomes of a given organism are called *homologous* chromosomes. Thus, each diploid somatic cell includes k pairs of homologous chromosomes, a tetraploid somatic cell has k quadruplets of homologous chromosomes, and so on. The homologous chromosomes are identical, except for the sex chromosomes. Any chromosome other than a sex chromosome is called an *autosome.* Man, for example, has 22 homologous pairs of autosomes or 44 autosomes in all.

An organism can start life either as a cell (or a group of nonfusing cells) derived from a single parent—this is called an *asexual reproduction*—or by the union of two parental cells—this is called a *sexual reproduction.* In sexual reproduction two *germ* cells, called also *gametes,* join together in one cell termed a *zygote.* The zygote constitutes a new individual.

If an organism is, in general, 2m–ploid, each gamete includes only m genomes, but the zygote has 2m genomes. The germ cells (gametes) are called *haploid* cells. In particular, when the organism is a diploid, each haploid cell (gamete) includes only one genome; for tetraploids each haploid cell includes two genomes, but for triploids one haploid cell will have two and another only one genome. Thus in the crossing of tetraploids three kinds of zygotes are theoretically possible: diploids, triploids, and tetraploids (in fact, in most species only diploids and tetraploids are viable).

Let us consider bisexual diploid organisms. The sex chromosomes are denoted by X and Y. The sex with two identical sex chromosomes is called the *homogametic sex,* and the one with two different chromosomes the *heterogametic sex.* Usually XX (homogametic) is female, and XY (heterogametic) is male. But there are some species in which XY is female and XX male.

In some plants (e.g., most of the common vegetables), although a new life must start from the union of two gametes, two different sexes do not occur. In other words, there are no sex chromosomes.

Chromosomes are bearers of heredity units called *genes.* The term "gene" was first introduced by Johannsen (1911). He proposed to call a *gene* the smallest (indivisible) element of the genetic material which can be, as a whole, transmitted from the parent to the offspring and which determines character(s) or property(ies) or a physiological function of a living organism.

Each gene has an established position on the chromosome, termed its *locus*. A gene associated with a given locus can occur in different forms called *allelomorphs* or *alleles*. In a diploid organism only two alleles can be present simultaneously at the same locus; one in a given chromosome and another in its homologue at exactly the same position. Two individuals, within the same species, have the same characteristic set of chromosomes, with the same number of loci, but the alleles at each locus, in these two individuals, can be different. This gives rise to a genetic variation of traits controlled by the same locus (or loci) within a species.

In recent years, a tremendous amount of work on the biochemistry of genetic material has been done, and as a result the definition of the gene has been modified. It has been shown that the genetic material, present in chromosomes, is deoxyribonucleic acid, briefly, DNA. In modern terminology, a gene is defined as the segment of DNA which represents a *functional unit* of genetic material capable of producing a simple primary phenotypic effect or responsible for a certain cellular product. Using this definition, "gene" is no longer indivisible, and "locus" is no longer a point but rather a small *region* (segment) of a chromosome. This modern definition of a gene is a broader one and explains some facts which cannot be explained by the previous definition. For a gene defined in this way, the term *cistron* is often used.

Genes control phenotypes by determining which proteins have to be synthesized. It appears that each gene is responsible for the synthesis of a specific polypeptide chain. Thus, in cases when gene function can be determined by a biochemical reaction, molecular geneticists define the gene as a segment of DNA which controls a single polypeptide chain ("one gene, one polypeptide chain").

The reader interested in details of the biological and biochemical backgrounds of these definitions of the gene is referred to specialist books on genetics [see, for instance, Herskowitz, (1962), Chapters 22 and 31, or Srb et al., (1965), Chapter 5].

In higher organisms, a locus (and therefore an allele associated with it) is not always in one-to-one correspondence with a cistron or any other unit introduced by molecular biologists. This is particularly true for genetic traits whose phenotypic effects are the results of serological tests [e.g., blood groups (Chapter 14) or leucocyte antigens (Chapter 19)]. For this reason a term, *system*, has been approved by the World Health Organization Committee on Notation of Immunoglobulins (1965), which states: "The term *system* is used here for a unit of closely linked genetic information which cannot be separated into subunits at the level of resolution allowed by formal human genetics." Such a piece of information can be transmitted as some sort of entity to the offspring. [A slightly different definition of a system is given by Mi and Morton (1966).]

A system can exist in alternative forms, which will be called "alleles." The region on the chromosome assigned to a system will be termed a "locus." A "locus" or an "allele" defined in a system may include only one cistron but may also consist of two or more cistrons. For the purpose of this book we shall adopt, when necessary, the terms "locus" and "allele" as defined above, and for simplicity "gene" will be used sometimes for a unit defined in a system.

Alleles are usually denoted by one letter, but sometimes by two (or even three) letters. Let us call a certain locus A. At least two alleles exist at each locus. These can be denoted by A and a or A_1 and A_2. If more than two alleles exist for a given locus A, they are called *multiple* alleles and are usually denoted by $A_1, A_2, \ldots, A_s$, where s is the number of possible allelomorphs; and similarly for loci B, C, etc. The particular known loci are denoted by one or two letters indicating the character under consideration. For example, the locus which controls haptoglobins is called the Hp locus, and its alleles are denoted by Hp^1 and Hp^2.

Sudden heritable changes are called *mutations*. Mutations can occur spontaneously (under unknown natural forces) or can be induced by ionizing radiation or by other mutagenic agents. A mutation which occurs within a gene is called a *point mutation*. There is no rule that mutation must occur in the whole gene (cistron). It is usually assumed that a gene possesses several mutational sites at which the change can take place.

The smallest unit of genetic material which can undergo mutation is called a *muton*. Point mutation gives rise to a new allele.

In addition to point mutations, sudden changes can occur in the whole or part of a chromosome (chromosomal aberrations), or in the number of genomes. For instance, polyploids can be considered as genome mutations of diploids.

1.3. GENOTYPE AND PHENOTYPE. DOMINANCE AND RECESSIVENESS

The total genetic constitution of an organism is termed its *genotype*. In a narrower sense, we use this term with respect to one, two, etc., loci. If A and a are alleles at locus A, then the possible genotypes (with respect to this locus) in a diploid organism are AA, Aa, and aa.

An individual possessing identical alleles, at a given locus, on both homologous chromosomes is called a *homozygote;* if two different alleles are present at a given locus, the individual is a *heterozygote*. Thus, AA or aa are homozygotes and Aa is a heterozygote. When the germ cells (gametes) are formed during meiosis, the homologous chromosomes separate into different gametes. This process is known as the *law of segregation* or *Mendel's first law*. During fertilization, the two gametes join into a zygote and a new genotype is obtained. A new genotype can be different from both parental genotypes. For example, all the offspring of parental crosses $AA \times aa$ are Aa—they are different from the parental genotypes. The offspring of $AA \times aa$ are usually called the F_1 generation (or F_1 hybrids). From crosses $Aa \times Aa$, which are called *intercrosses*, we obtain the F_2 generation; some of the individuals in F_2 will be of a parental type, AA or aa, whereas others will be Aa.

The total physical expression of a genotype, influenced also by environment, is called its *phenotype*. In a narrower sense this term refers to the expression of some particular gene (or genes) at a given locus and under given environmental conditions. Genotype is the result of segregation; phenotype is the result of gene action.

The phenotypic expression of a heterozygote (with respect to a single locus) is usually one of these three types:

(*a*) The heterozygote shows only the properties of one of the two alleles. This is called *dominance*. For example, the color of pea seeds is controlled by a single locus. The yellow seeds are AA; the green seeds are aa. When a yellow pea AA is crossed with a green pea aa, the resulting heterozygotes Aa are all yellow. We say that the allele A is *dominant* over the allele a; and the allele a is called *recessive* with respect to the allele A.* In other words, the gene A acts in such a way on its allelic partner a that a cannot express itself in the heterozygote and the genotypes AA and Aa show the same phenotype; this can be written briefly as $A-$ (i.e., AA or Aa or aA). Dominance is a certain kind of interaction of allelic genes and can be due to several causes. For instance, a may just be inactive, so that the phenotypic effect of Aa is the same as that of AA; or the action of a is much weaker than that of A, so that A suppresses its expression. There are some characters for which, although A is dominant over a, some Aa express themselves as aa. We say, in such cases, that the heterozygotes Aa exhibit reduced (or incomplete) *penetrance*. Incomplete penetrance can also occur in homozygotes; AA may exhibit recessive character aa, and aa may appear as dominant AA (or Aa). Problems with incomplete penetrance will be considered in Chapter 16.

(*b*) The phenotypic expression of a heterozygote is *intermediate* between the two homozygotes. For example, let bb correspond to white, and BB to red, color of pea flowers. The heterozygotes Bb are pink. In fact,

---

* Dominance or recessiveness associated with a single locus is sometimes called *simple* dominance or *simple* recessiveness, respectively. Some other aspects of the term "dominance" as it is used in quantitative genetics will be discussed in section 6.8.

"intermediate" does not necessarily mean that the heterozygote must be halfway between *BB* and *bb*; the term indicates only that it is between *BB* and *bb*.

(*c*) The heterozygote shows the properties of both alleles. In such a case, the alleles are called *codominants*. Let us take as an example the blood group system *MN*. The heterozygote *MN* has the serological properties of both antigens—that controlled by the allele *M* and that controlled by *N*. But in the *ABO* blood group system the three main alleles, *A*, *B*, and *O*, say, have the following properties: *A* and *B* are dominants over *O*, but are codominants with respect to each other. There are six genotypes but only four phenotypes, that is,

$$\text{Genotype:} \quad OO \quad \underbrace{AA, AO} \quad \underbrace{BB, BO} \quad AB$$

$$\text{Phenotype:} \quad OO \quad A- \quad B- \quad AB.$$

Phenotypic effects resulting from gene actions are *traits* or *characters*. A trait may be associated with a single locus, but may result from the joint action of two or more loci. The interaction of non-allelic genes is called *epistasis*. Several kinds of epistatic traits will be discussed in Chapter 3.

Of special interest are so-called *quantitative* characters (e.g., height or weight). It is often assumed that they are controlled by more than one pair of genes from independent loci and that the effects of genes from each locus are additive (or, in the general case, cumulative). The term *polygenic* character is sometimes used in such situations.

On the other hand, one gene may influence more than one trait; such multiple effects of a single gene are called *pleiotropic*. For instance, in human beings a set of abnormal traits transmitted as a unit to the offspring is assumed to be a pleiotropic effect of a single deleterious gene; such a set of traits is called in medicine a *syndrome*.

1.4. AUTOSOMAL LINKAGE. CROSSING OVER

For any species, the number of chromosomes in a genome is limited. The number of genes in each species is considerably larger than the number of chromosomes. Thus, only genes located in different chromosomes can segregate independently (Mendel's second law). Others must be transmitted together; that is, they are *linked*. Each pair of homologous chromosomes belongs to a *linkage group*. There are as many linkage groups as there are chromosomes in a genome.

Let us consider two autosomal loci, *A* and *B*, each with two alleles, *A*, *a* and *B*, *b*, respectively. Suppose that these loci are located in the same chromosome; that is, they are linked. Let the genes A and B be in one chromosome, and the genes a and b in its homologue. We write the genotype in the form AB/ab. (Note that sometimes the notation $\dfrac{AB}{ab}$ is used.) The symbol $/$ (or, alternatively, $-$) indicates that the loci A and B are linked, and that A, B genes are in one chromosome and a, b genes in its homologue. The configuration AB/ab is called *coupling*, while the configuration Ab/aB is termed *repulsion*.

Suppose that we cross two homozygous populations $ABAB \times abab$, and assume that the loci are linked. The F_1 progeny will be AB/ab. If the linkage were complete, the individuals AB/ab would produce only gametes AB and ab. If we now cross the F_1 progeny AB/ab back to the parental population $abab$ (i.e., produce backcrosses), we should only obtain the parental types AB/ab and ab/ab. It has been demonstrated, in several species, that genotypes Ab/ab and aB/ab also occur, although in much smaller proportions than would be expected when the loci segregate independently. The formation of new genetic combinations different from the parental combinations of two linked loci is called *recombination;* the new genotypes are designated as *recombinants.* This phenomenon can be explained by supposing that the genetic material has been exchanged between homologous chromosomes, and it is called *crossing over.* The expected proportion of recombinants occurring in offspring due to crossing over is termed the *recombination fraction;* we denote it by λ. We see later that $0 \leqslant \lambda \leqslant \frac{1}{2}$. If $\lambda = 0$, loci are absolutely linked; if $\lambda = \frac{1}{2}$, we cannot distinguish linkage from the situation in which loci are on different chromosomes.

The smallest unit of genetic material which is able to undergo crossing over (i.e., is indivisible by recombination) is called a *recon.*

There have been a few theories on how crossing over occurs. The usually accepted theory is called the *four-strand* model. At a certain stage during meiosis two homologous chromosomes lie side by side in the nucleus; they are called *bivalents.* Each chromosome consists of two identical halves termed *chromatids,* so bivalents are also sometimes called *tetrads* [Fig. 1.1a]. At some later stage the chromosomes split up into two *chromatids* and some parts of chromatids from the homologous chromosomes exchange segments in X-like arrangements which are called *chiasmata* (crosses) [Fig. 1.1b]. After a certain time we have the configuration as in Fig. 1.1c.

As has been mentioned before, the recombination fraction for two given loci is constant; the bigger the distance between two loci the more often crossing over is expected to occur. Thus the recombination fraction can be used as a measure of the "distance" between two incompletely linked loci, in constructing *chromosome maps.* One per cent recombination is often called one *crossover unit* or a *centimorgan.* It should be noted that recombination

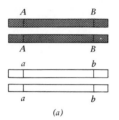 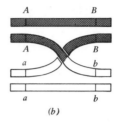 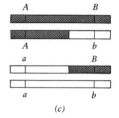

Figure 1.1

fractions in males and females are not always the same. For instance, in Drosophila no recombinants are observed in males (complete linkage); the linkage groups and chromosome mapping have been obtained using recombination fractions observed in females. Usually the heterogametic sex is the one which can exhibit complete linkage if such takes place.

In the early theory of genetics, the Mendelian unit of inheritance, called "gene," was equivalent with cistron, muton, and recon. Now, these terms are not always equivalent. Since a cistron, called here a gene, may consist of smaller, nonallelic subunits, recombination can occur *within* a locus (i.e., within a region on the chromosome assigned to the gene). In fact, the cistron was "discovered" by observing recombinations at a certain locus controlling the color of eyes in Drosophila. From recent investigations, it appears that recombinations can occur between two adjacent mutons, so that recon may be equivalent to muton. Recombinations within a locus are always very rare; sometimes thousands of observations are needed in order to find one recombinant. As yet, several known loci in plants, animals, and humans are still considered as "indivisible" units and genes or alleles can be detected from recombination. This might be because these loci are really "points," not "regions," or because the recombinations within a locus have not yet been observed. Therefore, for practical purposes, the probability models derived for analyzing genetic traits and their transmission are based on the premise that a Mendelian unit of inheritance, that can be transmitted as a whole and can undergo recombination as a unit is still a valid concept. However, the new definitions of cistron, muton, and recon are important concepts in bacterial genetics.

1.5. SEX-LINKED INHERITANCE

As was mentioned in section 1.2, the two sex chromosomes, X and Y, are different, and XX is usually a female, XY a male. The chromosomes X and Y have different shapes and are not homologous over their whole length. In several species the sex-linked genes are located in that part of the X chromosome which does not have a homologous segment in the Y chromosome. Traits controlled by these genes are called *X-linked*. For instance, hemophilia in man is due to a recessive X-linked gene. Female heterozygotes *Aa* are carriers of the hemophilic gene *a* and may transmit it to their sons. Thus males with the single gene *a* are hemophilic, whereas females must have *aa* to be hemophilic. The X-linked gene *a* (or *A*) is present in males only in a *single dose* in the X chromosome; in this case and similar ones the males are said to be *hemizygous* at the locus under consideration. There are but a few characters found in different species which are solely inherited by males, that is, are controlled by genes present in the segment of the Y chromosome which lacks a homologous part in X.

However, some segments of X and Y are homologous, and the genes present in the X chromosome may have alleles present in the Y chromosome. The type of sex linkage in which a gene can exist in both chromosomes is sometimes called *partial* or *incomplete sex linkage* (in earlier terminology, the X-linked inheritance was called *complete* sex-linked or, more briefly, sex-linked inheritance).

SUMMARY

Basic genetic terms such as chromosome, gene, allele, genotype, phenotype, and segregation have been defined. Interactions of allelic genes (dominance) and of nonallelic genes (epistasis) have been introduced. The mode of transmission of genes located in the same chromosomes (linked genes) and the mechanism of sex-linked inheritance have been briefly discussed.

REFERENCES

Bailey, N. T. J., *Introduction to the Mathematical Theory of Genetic Linkage*, Clarendon Press, Oxford, 1961.

Davenport, C. B., The imperfection of dominance and some of its consequences, *Amer. Nat.* **44** (1910), 129–35.

Herskowitz, I. H., *Genetics*, Little, Brown and Company, Boston, 1962.

Johannsen, W., The genotype conception of heredity, *Amer. Nat.* **45** (1911), 129–59.

King, R. C., *A Dictionary of Genetics*, Oxford University Press, New York, 1967.

Mi, P. M., and Morton, N. E., Blood factor association, *Vox Sang.* **11** (1966), 434–49.

Morgan, T. H., Chromosomes and heredity, *Amer. Nat.* **44** (1910), 449–96.

Morgan, T. H., The theory of the gene, *Amer. Nat.* **51** (1917), 545–59.

Muller, H. J., The mechanism of crossing over, *Amer. Nat.* **50** (1916), 284–305, 350–365, 421–433.

Serra, J. A., *Modern Genetics*, Vol. I, Academic Press, London, New York, 1965.

Srb, A. M., Owen, R. D., and Edgar, R. C., *General Genetics*, W. H. Freeman & Co., San Francisco, 1965.

Stern, C., *Principles of Human Genetics*, W. H. Freeman & Co., San Francisco, 1960.

Thomson, J. S., and Thomson, M. W., *Genetics in Medicine*, Philadelphia, 1966.

Watson, J. D., *Molecular Biology of the Gene*, W. A. Benjamin, Inc., New York, 1965.

Winchester, A. M., *Genetics. A Survey of the Principles of Heredity*, Houghton Mifflin Co. (third edition), New York, 1966.

World Health Organization, Notation for genetic factors of human immunoglobulins, *Bull. Wld. Hlth. Org.* **33** (1965), 721.

PART I

Population Genetics.
Probability Models

In a broad sense, the subject matter of *population genetics* is the analysis of genetic structures of populations originating from different outputs of genetic material and subject to several internal and environmental conditions. Thus we may consider randomly and nonrandomly mating populations under uniform conditions and also in circumstances where such forces as selection, mutation, or migration operate to change the structure of a population.

Biological populations are always finite, though they can be very large. A plant breeder who crosses heterozygous pea plants producing yellow seeds expects to obtain, in the next generation, plants producing yellow and green seeds in the ratio 3:1. In practice, the result will not be exact but will approach the ratio 3:1 when the number of such crosses becomes large. We then can speak about a *model* of the genetic structure of a population, the size of which becomes larger and larger, or we can say that the population is "infinite."

An *infinite* population is a hypothetical population which does not really exist. To think about such a population is convenient, however, since it can be often described by using well-known mathematical formulae of the theory of probability distributions.

Models for infinite populations can be of two types:

(*a*) *Deterministic* models describe the structure of the population at certain stages of its development, without taking into account what happens between these stages; such models are of discrete type and usually take one generation as a "jump" unit of time between two stages.

(*b*) *Stochastic* models, on the other hand, take into account dynamics of a population in time, and are continuous functions of a new variable which represents time.

Randomly mating natural populations, inbreeding, and selection should all be treated stochastically. However, in many cases, it is difficult to derive stochastic models, and quite often deterministic models with generation "jumps" give reasonably good approximations of the real situation.

In Chapters 2–10 we will present deterministic models which can be applied in studying the genetics of infinite populations.

Chapters 2, 5, 6, and 8 are devoted to the mathematical theory of probability, random variables, their distributions, and moments of random variables such as expected value, variance, and covariance.

Chapters 4, 7, 9, and 10 are concerned with applications of probability models to such problems as deriving equilibrium distributions for populations in which mating is "purely" random (Chapter 4), for populations practicing inbreeding (Chapters 7 and 9), and for populations subject to natural forces such as selection or mutation (Chapter 10).

CHAPTER 2

Probability Models.
General Concepts and Definitions

2.1. MODELS

The term "model" has several meanings. In everyday language it means something which is ideal, perfect, but, perhaps, can never be realized. "A model of an ideal happy population" or "a model of beauty" are examples.

"Model" also designates a copy (usually in miniature) of something which has to be constructed in reality. An engineer makes a plan or model of a new building, car, or airplane before construction is begun.

Utilizing experience with and theory regarding some phenomena, scientists construct models to account for them. A "gene," as a model of a hereditary unit, is an example of a scientific model.

Observing some mass phenomena, such as, for instance, sunrises and sunsets, or births and deaths, we notice that they occur with some regularity; we say that they obey natural laws. However, the time of sunset for each day can be closely determined, whereas the death of an individual cannot usually be precisely predicted, though it is more probable at some ages and less at others.

What do we mean by "more probable"? How shall we measure the *chance* that someone will survive up to a certain age, and what is the "probability model" for this natural law?

Situations in which there is only some chance that certain phenomena will occur are also frequently encountered in genetics. A plant breeder crosses two lines of wheat and wants to predict the structure of the population so obtained. Two normal parents who have a child affected by some mental disease worry that the next child might also be affected. What is the chance that it will be normal? Can we measure this chance, and what will be a model from which we can deduce the probability that the second child will not be affected?

In the last few problems we have introduced the term "probability" without precisely defining it, using it rather as a *measure of the chance* that an event will occur. To express this chance in a numerical form one has first to define probability in mathematical terms, and then to construct a probability model appropriate to the given problem.

The object of this book is to discuss several probability models which can be applied to the genetics of qualitative traits. These models will be constructed under some reasonable conditions or assumptions. It might happen that in a real situation the assumptions can be satisfied only approximately. Thus the model will fit the data only approximately, too. As has been said, a model is sometimes an ideal construction which fits exactly only in ideal, perhaps hypothetical, circumstances. Nevertheless, it is useful to construct such models and to analyze them in experiments.

2.2. SAMPLE SPACE, RANDOM EVENTS, PROBABILITY

2.2.1. Sample Space

The formal development of probability theory is not the main purpose of this book. However, it is useful to introduce a few concepts and definitions from probability theory and to show how they correspond to terms commonly used by geneticists. Let us consider a simple example.

Example 2.1. Let A be a locus with two alleles, A (dominant) and a (recessive). Suppose that the parental crosses are $Aa \times Aa$. Each parent can produce two kinds of gametes, A and a, and the genotype of an offspring is the result of the union of two parental gametes. A simple way to find all possible resultant genotypes is to use a checkerboard as follows:

TABLE 2.1   Possible genotypes

♀ ＼ ♂	A	a
A	AA	Aa
a	aA	aa

Table 2.1 shows four possible types of outcomes which can result from an experiment and distinguishes between the maternal (♀) and paternal (♂) gamete contributions. The experiment need not in this case, be actually performed. From long experience and cytological studies it is known that

the gametes are formed during meiosis and that, after fertilization, they join together randomly. Therefore, all possible outcomes are always associated with an experiment which is or could be performed.

Geneticists call Table 2.1 a checkerboard, but statisticians would call it a sample space. We now define the latter term more precisely.

DEFINITION 2.1.   *The collection (set) of all possible outcomes associated with an experiment is called the sample space for this experiment. We will use the symbol Ω to denote a sample space.*

DEFINITION 2.2.   *Each possible outcome is called a sample point or an elementary event; we will denote it by e.*

The sample space is also sometimes called the *fundamental set of elementary events* and might be denoted as $\Omega = \{e_1, e_2, \ldots\}$.

How can we construct the sample space? To answer this question we confine ourselves to the situation in which there is only a finite (though possibly very large) number of outcomes, that is, where the sample space is *finite* and *discrete*. The majority of genetics problems discussed in this book are of this kind. Some concepts of *continuous* sample space will be discussed briefly at the end of this chapter.

The outcomes given in Table 2.1 can also be written in the form of an array: *AA*, *Aa*, *aA*, *aa*. If one is not particularly interested in which is the maternal and which the paternal gamete, the genotypes *Aa* and *aA* are, in effect, the same. Then, only three genotypes (possible outcomes), *AA*, *Aa*, and *aa*, are distinct, and the sample space consists of three points. But in this case, of course, the information regarding the maternal and the paternal contributions is lost.

If we are interested only in phenotypes of the offspring, the sample space consists of just two sample points, *A*– (here the sign – stands for *A* or *a*) and *aa*. Now some information on the genotype as well as on the parental contributions to the offspring will be lost.

Which sample space should we use? This depends, of course, on the problem in which we are interested and on our knowledge about this problem. A reasonable general rule would be: use the sample space which is the most informative. Thus, in our Example 2.1 we would use the four-point sample space shown in Table 2.1.

2.2.2. Random Events

Let us consider our four-point sample space in Example 2.1 and ask this question: how, using our new terminology, can we express the fact that an offspring is a dominant? Three sample points, *AA*, *Aa*, and *aA*, refer to a dominant. These three points form a *subset* in the sample space and determine the event that an offspring is a dominant. Similarly, only one point, *aa*, determines a recessive. We now define an event in a sample space.

DEFINITION 2.3.   *Any subset, E, in the sample space* Ω *(or, in other words, any collection of elementary events) is called a random event.*

Since in probability theory we deal only with random events, the word "random" will usually be omitted and the term "event" will be used for "random event."

2.2.3.   Definition of Probability

Since, in most cases, events are predicted only with some chance to occur, we wish to define a "measure of the chance" that the event occurs. We start with a formal definition of probability.

DEFINITION 2.4.   *With every event E there is associated a real non-negative number,* $\Pr\{E\}$, *such that*

$$0 \leqslant \Pr\{E\} \leqslant 1, \tag{2.1}$$

which is called the probability of the event E.

How shall we calculate the probability of an event E defined in the sample space Ω?

Let us return to our Example 2.1 and first assume that all four possible types of offspring are *equally likely* to be formed. Let us assign to each of the elementary events, AA, Aa, aA, and aa, a *weight* equal to $\frac{1}{4}$. This will be called the *elementary probability* for each of these four elementary events. To the event $A-$ (i.e., an offspring is a dominant) correspond three elementary events, AA, Aa, and aA, so that the total weight assigned to the event $A-$ is $\frac{1}{4} + \frac{1}{4} + \frac{1}{4} = \frac{3}{4}$. We then have $\Pr\{A-\} = \frac{3}{4}$. For recessives we have $\Pr\{aa\} = \frac{1}{4}$. The phenotypic ratio Dominants: Recessives $= \frac{3}{4} : \frac{1}{4} = 3 : 1$.

However, it might happen that the gametes or zygotes are not equally viable, so that different genotypes are *not equally likely* to occur. In the general case, we may assign unequal weights, ω_1, ω_2, ω_3, and ω_4, say, subject to the condition $\omega_1 + \omega_2 + \omega_3 + \omega_4 = 1$, and such that $\Pr\{AA\} = \omega_1$, $\Pr\{Aa\} = \omega_2$, $\Pr\{aA\} = \omega_3$, and $\Pr\{aa\} = \omega_4$. The probability of the event that an offspring of the $Aa \times Aa$ parents will be a dominant is now $\Pr\{A-\} = \omega_1 + \omega_2 + \omega_3$.

How can we determine these weights and what will be their effective values?

If the causes and modes of gene action were known, we could possibly evaluate the weights by using theoretical arguments. But often this is not the case. The hypothetical phenotypic ratio Dominants: Recessives = 3:1 was first derived by Mendel, who based his argument on the results of his classical experiments on crossing a pure line of yellow pea seeds (AA) with a pure line of green pea seeds (aa). Crossing the F_1 hybrids, that is, $Aa \times Aa$, he observed in F_2 the ratio

$$\frac{\text{number of plants with yellow seeds}}{\text{number of plants in experiment}}.$$

Such a ratio is called a *relative frequency*. Repeating these experiments several times, Mendel remarked that the relative frequency appeared to be approaching a limit equal to $\frac{3}{4}$.

The concept of probability as a limit of a relative frequency is used by some authors. Suppose that in a series of n independent repetitions (called "trials") of an experiment, an event E occurs f_n times. The number f_n is called (absolute) frequency, and the ratio f_n/n is termed the relative frequency of an event E observed in n successive trials. Repeating a series of trials, with n becoming larger and larger ($n \to \infty$), we may observe that f_n/n approaches a certain limit. This limit is called the probability of the event E. We summarize this in the following definition—called sometimes the *statistical* definition—of probability.

DEFINITION 2.5. *If in a long series of independent trials the relative frequency, f_n/n, of the occurrence of the event E tends to some limit p, we will call this limit the probability of the event E, that is,*

$$\Pr\{E\} = \lim_{n \to \infty} \frac{f_n}{n} = p. \tag{2.2}$$

This definition has some vague aspects. For example, an infinite series of trials cannot, in fact, be carried out and it is not possible to evaluate the limit p. But, again, formula (2.2) is another "model" of probability and in several cases serves useful purposes.

Two special kinds of events should be mentioned. If an event E is *impossible*, $\Pr\{E\} = 0$. On the other hand, we notice that, using Definition 2.5 of probability, the relation

$$\lim_{n \to \infty} (f_n/n) = 0$$

does not mean that the event cannot ever occur. Events having probability 0 may occur, but only very rarely. Similarly, if the event E is *certain*, $\Pr\{E\} = 1$. But the statement

$$\lim_{n \to \infty} (f_n/n) = 1$$

does not necessarily imply that E always occurs.

We now summarize the results from the above discussions and examples. Using formal Definition 2.4, we can obtain the probabilities in a *discrete* sample space, Ω, as follows. To each elementary event $e_i (i = 1, 2, \ldots)$ in Ω, we assign a weight ω_i, called its probability, $\Pr\{e_i\} = \omega_i$, and satisfying the condition

$$\omega_1 + \omega_2 + \cdots = 1. \tag{2.3}$$

The probability that an event E is a subset of elementary events e's is the sum of the corresponding elementary probabilities ω's. Since the sample space exhausts all elementary events, we have the fundamental result

$$\Pr\{\Omega\} = 1. \tag{2.4}$$

In particular, if the sample space is finite of size N, and all elementary events are equally likely, we shall have $\omega_i = 1/N$, $i = 1, 2, \ldots, N$.

If the weights are not known *a priori*, one uses statistical Definition 2.5, employing a sequence of independent trials from which the relative frequencies for events can be calculated. Their limits (if they exist) are used as measures of the probabilities of corresponding events.

2.3. COMPOUND EVENTS

Events can be combined in several ways. Before we introduce some definitions of combined (or compound) events, we shall explain the meanings of two words, "or" and "and," which will be used in precisely defined ways in this book.

The word "or" has two meanings in everyday language. "I like green *or* blue color" usually means that I like green or blue or both. One can replace the word "or" by the phrase "at least one." "I want to use in this painting *at least one* of the two colors, green or blue" means that I want to use green or blue or both. On the other hand, the sentence "I will go to see this play on Saturday evening *or* Sunday afternoon" usually means that I will see the play only on one day, and not on both. When we use the word "or" in regard to events, it always has the first of these two meanings. It always means *one* or *another* or *both*, or, equivalently, *at least one*.

The word "and" always means *both*. Thus the statement "The pea seeds are yellow *and* round" means that they possess both characters.

Since events are defined as sets of points, the algebra of sets also applies to events. We shall introduce two useful concepts of compound events, union and intersection.

2.3.1. Union (Logical Sum) of Events

Let E_1 and E_2 be two events defined on the sample space Ω. The compound event E, consisting of E_1 *or* E_2 (i.e., E_1 or E_2 or both), is called the *union* (the *logical sum* or sometimes, briefly, the *sum*) of these events. To distinguish the logical sum from the arithmetic sum, it is customary to use the symbol $\cup$ (union) instead of the sign $+$, that is, to write

$$E = E_1 \cup E_2. \tag{2.5}$$

On the other hand, the symbol $\cup$ is rather less convenient than $+$, and therefore some authors prefer to write

$$E = E_1 + E_2. \tag{2.5a}$$

Since it is almost always clear from the context whether we are dealing with logical or arithmetic sums, no ambiguity should arise even if the symbol $+$ is used in the same problem for two different kinds of summation. In this book, we adopt the symbol $+$ for both union* and ordinary summation.

The concept of union can be generalized to m events. We have the following definition.

DEFINITION 2.6. *The union E of events $E_1, E_2, \ldots, E_m$ is the set of all the sample points belonging to E_1 or E_2 or $\cdots$ or E_m (and also to any pair $E_i E_j$, to any triplet $E_i E_j E_k, \ldots,$ etc.). We write*

$$E_1 + E_2 + \cdots + E_m = \sum_{i=1}^{m} E_i, \tag{2.6}$$

where Σ is a summation operator (here signifying "union").

The union of events can be presented diagramatically as shown, for example, in Fig. 2.1a or Fig. 2.1b.

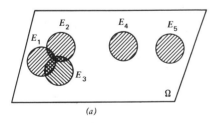 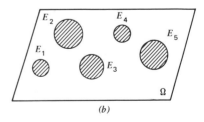

Figure 2.1

* We shall also use, for union, E_1 *or* E_2.

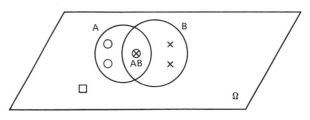

Figure 2.2

The occurrence of the union of events $E_1, E_2, \ldots, E_m$ is equivalent to the occurrence of at least one (but can be, of course, more than one) of these events.

Example 2.2.   Let us consider the *ABO* blood group system, with three "alleles," *A*, *B*, and *O*. We assume that *A* and *B* are each dominant to *O*, but codominants to each other. The six genotypes below represent a sample space (Ω) of genotypes associated with this locus:

Label (No.):	1	2	3	4	5	6
Genotype:	*AA*	*AO*	*BB*	*BO*	*AB*	*OO*
Symbol in Fig. 2.2:	○	○	×	×	⊗	□
Expected proportion:	P_1	P_2	P_3	P_4	P_5	P_6.

The recognition of a blood group (phenotype) of an individual is based on the serological reactions of red cells with special reagents called antisera. At this stage, we assume that two reagents, anti-A and anti-B, can identify the four groups: A, B, AB, and O. The red cells of *AA* or *AO* individuals react with anti-A, and of *BB* or *BO* with anti-B, but the red cells of *AB* react with both anti-A and anti-B, while *OO* react with neither of these sera (for more details, see Chapter 14).

Let E_1 denote the event that the red cells react with anti-A. The event E_1 consists of sample points labeled in Ω as 1, 2, 5 (and represented in Fig. 2.2 by ○). Similarly, let E_2 denote the event that the red cells react with anti-B (points labeled 3, 4, 5 and represented in Fig. 2.2 by ×). The event $E = E_1 + E_2$ means that the red cells react with anti-A or anti-B or both (points labeled 1, 2, 3, 4, 5). In other words, E is the event that an individual can have the blood group A or B or AB (but not O). (See Fig. 2.2.)

2.3.2.   Intersection (Logical Product) of Events

Suppose that on some occasions two events, E_1 and E_2, defined in Ω, can occur simultaneously, or, in other words, E_1 and E_2 have common points in Ω. The common subset of sample points belonging to both E_1 and

E_2 is called the *intersection* (or the *logical product*, or, briefly, the *product*) of these two events. The symbol $\cap$ is used to denote the intersection of events, to distinguish the logical product from the arithmetic product. If E' denotes the simultaneous occurrence of E_1 and E_2, we may write

$$E' = E_1 \cap E_2. \tag{2.7}$$

But the multiplication sign ($\cdot$) can also be used, since it is more convenient and usually the risk of ambiguity is rather small. In this book, we will use the symbols (or operators) of arithmetical multiplication also for intersection;* that is, instead of (2.7) we write

$$E' = E_1 \cdot E_2 = E_1 E_2. \tag{2.7a}$$

The concept of the intersection of more than two events is generalized in the following definition.

DEFINITION 2.7. *The intersection E' of the events $E_1, E_2, \ldots, E_m$ is the set of sample points belonging to E_1 and E_2 and $\cdots$ and E_m simultaneously. We write*

$$E' = E_1 E_2 \cdots E_m = \prod_{i=1}^{m} E_i, \tag{2.8}$$

where $\prod$ is a product operator (here signifying intersection).

In our Example 2.2, $E_1 E_2$ denotes individuals reacting with anti-A and anti-B, that is, having the blood group AB (see Fig. 2.2, point $\otimes$).

2.3.3. Mutually Exclusive and Exhaustive Events

Let $\varnothing$ denote an empty set, that is, a set with no points. The expression

$$E_1 E_2 = \varnothing \tag{2.9}$$

means that the intersection of E_1 and E_2 is an empty set, that is, that the occurrence of E_1 excludes the occurrence of E_2, and vice versa. Such events are called *mutually exclusive*. The definition can be extended to m events.

DEFINITION 2.8. *Events $E_1, E_2, \ldots, E_m$ are mutually exclusive if*

$$E_i E_j = \varnothing \tag{2.10}$$

for every pair (i, j) with $i \neq j$ (see Fig. 2.1).

When events are mutually exclusive, the occurrence of one of them excludes the occurrence of each of the others. For example, the three kinds

* We shall also sometimes use, for intersection, E_1 *and* E_2.

of genotypes *AA*, *Aa*, and *aa* represent three mutually exclusive events; if an individual is *AA*, it can be neither *Aa* nor *aa*.

DEFINITION 2.9.   *Let* Ω *be a sample space and E be an event defined on* Ω. *The event consisting of the sample points in* Ω *which do not belong to E is called the complement of E, and we denote it by* $\bar{E}$ *(non-E) in* Ω. *We have*

$$\bar{E} = \Omega - E. \tag{2.11}$$

For example, in the sample space of three genotypes *AA*, *Aa*, and *aa*, the events *A–* (dominant) and *aa* (recessive) are two complementary events.

DEFINITION 2.10.   *Events* $E_1, E_2, \ldots, E_m$ *such that*

$$E_1 + E_2 + \cdots + E_m = \Omega \tag{2.12}$$

are called exhaustive events (i.e., they include all possible outcomes).

DEFINITION 2.11.   *Events* $E_1, E_2, \ldots, E_m$ *satisfying both conditions* (2.10) *and* (2.12) *are called mutually exclusive and exhaustive events.*

For example, all possible genotypes (or phenotypes) associated with a given locus are mutually exclusive and exhaustive events on the sample space determined by genotypes at this locus.

2.4.   ADDITION LAW OF PROBABILITIES

Example 2.3.   Let us return to Example 2.2. Suppose that the genotypes given in Example 2.2 occur with some theoretical or "expected" relative frequencies $P_1, P_2, \ldots, P_6$, satisfying the condition

$$\sum_{j=1}^{6} P_i = 1.$$

These *P*'s play the role of weights, introduced in section 2.2.3, and represent the elementary probabilities in the sample space defined for genotypes at the *ABO* blood locus. If E_1, E_2, and $E = E_1 + E_2$ are the events as defined in Example 2.2, we shall have

$$\Pr\{E_1\} = P_1 + P_2 + P_5, \quad \Pr\{E_2\} = P_3 + P_4 + P_5,$$

$$\Pr\{E_1 E_2\} = P_5, \quad \Pr\{E_1 + E_2\} = P_1 + P_2 + P_3 + P_4 + P_5.$$

We notice that

$$\Pr\{E_1 + E_2\} = \Pr\{E_1\} + \Pr\{E_2\} - \Pr\{E_1 E_2\}. \tag{2.13}$$

Formula (2.13) is valid for any two events and represents the *addition law of probabilities for the union of two events. For three events, E_1, E_2, and

E_3, we have

$$\Pr\{E_1 + E_2 + E_3\} = \Pr\{E_1\} + \Pr\{E_2\} + \Pr\{E_3\} - \Pr\{E_1E_2\}$$
$$- \Pr\{E_1E_3\} - \Pr\{E_2E_3\} + \Pr\{E_1E_2E_3\}. \qquad (2.14)$$

The right-hand side of (2.14) can be easily obtained by substituting $E = E_1 + E_2$ into the left-hand side of (2.14) and applying formula (2.13). Formula (2.14) can be generalized to more than three events.

If the events E_1 and E_2 are exclusive, then

$$\Pr\{E_1E_2\} = 0, \qquad (2.15)$$

and formula (2.13) takes the form

$$\Pr\{E_1 + E_2\} = \Pr\{E_1\} + \Pr\{E_2\}. \qquad (2.16)$$

In general, if m events, $E_1, E_2, \ldots, E_m$, are mutually exclusive, we have

$$\Pr\{E_1 + E_2 + \cdots + E_m\} = \Pr\{E_1\} + \Pr\{E_2\} + \cdots + \Pr\{E_m\}$$
$$= \sum_{i=1}^{m} \Pr\{E_i\}, \qquad (2.17)$$

that is, the probability of the union (sum) of m mutually exclusive events is equal to the sum of their probabilities.

Let $E_1, E_2, \ldots, E_m$ be mutually exclusive and exhaustive events in Ω. Thus from (2.17), (2.12), and (2.4) we obtain

$$\Pr\left\{\sum_{i=1}^{m} E_i\right\} = \Pr\{\Omega\} = 1. \qquad (2.18)$$

2.5. PROBABILITY DISTRIBUTION OF EVENTS

DEFINITION 2.12. *If the sequence of events $E_1, E_2, \ldots, E_m$ satisfies conditions (2.10) and (2.12) (i.e., these events are mutually exclusive and exhaustive) in Ω, then the sequence of probabilities $\Pr\{E_1\}, \Pr\{E_2\}, \ldots, \Pr\{E_m\}$ satisfies condition (2.18). The sequence $\Pr\{E_1\}, \Pr\{E_2\}, \ldots, \Pr\{E_m\}$ is called the probability distribution of events $E_1, E_2, \ldots, E_m$ in the sample space Ω.*

Since we will be dealing only with probability distributions, the abbreviated form "distribution" for "probability distribution" will generally be used in this book.

DEFINITION 2.13. *The sample space together with the probability distribution is called the probability space.*

Example 2.4. The probabilities $P_1, \ldots, P_6$ in Example 2.3 represent the *genotype distribution* at the *ABO* blood locus. If we consider only phenotypes, that is, the blood groups A, B, AB, and O, we may denote $\Pr\{A\} = \Pr\{AA\} + \Pr\{AO\} = P_1 + P_2 = Q_1$; $\Pr\{B\} = \Pr\{BB\} + \Pr\{BO\} = P_3 + P_4 = Q_2$; $\Pr\{AB\} = P_5 = Q_3$; and $\Pr\{O\} = \Pr\{OO\} = P_6 = Q_4$. Thus the sequence Q_1, Q_2, Q_3, Q_4, with $\Sigma Q_j = 1$, represents the *phenotype distribution* of the ABO blood groups.

2.6. CONDITIONAL PROBABILITY AND MULTIPLICATION LAW

The concept of conditional probability is important in general applications. We will demonstrate its meaning, using again an example involving ABO blood groups.

Example 2.5. Let us return to Example 2.3 and ask this question: what is the "expected" proportion of individuals whose red cells react with anti-B among those reacting with anti-A? In other words we wish to reduce our sample space to genotypes which react with anti-A (i.e., to AA, AO, AB) and find the proportion of AB in this reduced sample space. Of course, now the original proportions, P_1, P_2, and P_5, do not add up to 1, but the proportions $P'_1 + P'_2 + P'_3$ as given below:

Genotype: AA AO AB

Expected
   proportion: $P'_1 = \dfrac{P_1}{P_1 + P_2 + P_5}$ $P'_2 = \dfrac{P_2}{P_1 + P_2 + P_5}$ $P'_3 = \dfrac{P_5}{P_1 + P_2 + P_5}$

do add up to 1.

Thus the expected proportion of individuals reacting with anti-B, restricted by the condition that they also react with anti-A, is $P'_3 = P_5/(P_1 + P_2 + P_5)$. Using the notation of Example 2.3, we find that $P'_3 = \Pr\{E_1 E_2\}/(\Pr\{E_1\})$.

This result can be generalized for any two events, E_1 and E_2, having nonzero unconditional probabilities, that is, $\Pr\{E_1\} > 0$, $\Pr\{E_2\} > 0$, denoting the "conditional probability of E_2 *given* that E_1 occurs" by $\Pr\{E_2 \mid E_1\}$.

DEFINITION 2.14. *The conditional probability of an event E_2, given that an event E_1 has occurred*, $\Pr\{E_2 \mid E_1\}$, *is*

$$\Pr\{E_2 \mid E_1\} = \frac{\Pr\{E_1 E_2\}}{\Pr\{E_1\}}, \qquad (2.19)$$

provided $\Pr\{E_1\} > 0$.

In a similar way we define

$$\Pr\{E_1 \mid E_2\} = \frac{\Pr\{E_1 E_2\}}{\Pr\{E_2\}}, \qquad (2.20)$$

provided that $\Pr\{E_2\} > 0$. Note that we have not defined $\Pr\{E_i \mid E_j\}$ for cases when $\Pr\{E_j\} = 0$, $i, j = 1, 2$ ($i \neq j$). We do not need such probabilities for the applications in this book.

From (2.19) and (2.20) we obtain

$$\Pr\{E_1 E_2\} = \Pr\{E_2 \mid E_1\}\Pr\{E_1\} = \Pr\{E_1 \mid E_2\}\Pr\{E_2\}. \tag{2.21}$$

We shall see later that in some problems it is difficult to find $\Pr\{E_1 E_2\}$, but is easy to evaluate $\Pr\{E_1 \mid E_2\}$ or $\Pr\{E_2 \mid E_1\}$. Then $\Pr\{E_1 E_2\}$ can be obtained from (2.21).

It can be shown that the probability of intersection of three events can be evaluated from the formula

$$\Pr\{E_1 E_2 E_3\} = \Pr\{E_1 \mid E_2 E_3\}\Pr\{E_2 \mid E_3\}\Pr\{E_3\}, \tag{2.22}$$

or from similar formulae, taking the E's in different orders. Extension to more than three events is straightforward.

2.7. INDEPENDENT EVENTS

The term "independence" has a very broad meaning in common language and is generally used to indicate the lack of any kind of relationship. In the theory of probability and statistics, independence (or dependence) of events is expressed solely in terms of relative frequencies or, more precisely, in terms of probabilities and is sometimes called *statistical independence*. Other kinds of relationships have different terminology, such as "interaction" or "correlation," and will be discussed later (see Chapters 3 and 6).

DEFINITION 2.15. *An event E_2 is (statistically) independent of E_1 if*

$$\Pr\{E_2 \mid E_1\} = \Pr\{E_2\}. \tag{2.23}$$

This means that the relative *frequency of occurrence* of E_2 is not affected by the fact that E_1 occurs or does not occur.

In a similar way we define E_1 as being independent of E_2 if

$$\Pr\{E_1 \mid E_2\} = \Pr\{E_1\}. \tag{2.24}$$

Substituting (2.23) [or (2.24)] into (2.21), we obtain

$$\Pr\{E_1 E_2\} = \Pr\{E_1\}\Pr\{E_2\}. \tag{2.25}$$

Provided $\Pr\{E_1\} > 0$, $\Pr\{E_2\} > 0$, we can deduce (2.23) and (2.24) from (2.25), and Definition 2.15 can be replaced by the following equivalent definition.

DEFINITION 2.15a.   *Two events E_1 and E_2 are independent if the probability of the simultaneous occurrence of E_1 and E_2 is the product of the probabilities of the occurrence of E_1 and E_2, that is,*

$$\Pr\{E_1 E_2\} = \Pr\{E_1\}\, P\{E_2\},$$

provided $\Pr\{E_i\} > 0$, $i = 1, 2$.

Since in practice we are usually interested in events with nonzero probabilities, the condition $\Pr\{E_i\} > 0$, $i = 1$, 2, is almost always satisfied. In this book we will assume that this is the case and will often drop it when using the Definition 2.15a of independence.

If the events E_1 and E_2 are statistically dependent, then

$$\Pr\{E_1 E_2\} \neq \Pr\{E_1\}\,\Pr\{E_2\}. \tag{2.26}$$

Three events, E_1, E_2, E_3 are *independent in a set* (i.e., E_i is independent of E_j and of E_k, and of $E_j E_k$, $i, j, k = 1, 2, 3$, $i \neq j \neq k$) if

$$\Pr\{E_1 E_2 E_3\} = \Pr\{E_1\}\Pr\{E_2\}\Pr\{E_3\}, \tag{2.27}$$

provided $\Pr\{E_i\} > 0$, $i = 1, 2, 3$.

It can be shown, by examples, that *pairwise* independence, alone that is,

$$\Pr\{E_i E_j\} = \Pr\{E_i\}\Pr\{E_j\}, \qquad i, j = 1, 2, 3, i \neq j,$$

is necessary but *not* sufficient for independence in a set of three events.

We now generalize the definition of independence for k events.

DEFINITION 2.16.   *The events, E_1, E_2, \ldots , E_k, are independent in a set if*

$$\Pr\{E_1 E_2 \cdots E_k\} = \Pr\{E_1\}\Pr\{E_2\} \cdots \Pr\{E_k\} \tag{2.28}$$

provided $\Pr\{E_i\} > 0$, $i = 1, 2, \ldots , k$.

We may notice that exclusive events and independent events are not the same things. For two exclusive events, E_1 and E_2, from (2.15) we have

$$\Pr\{E_1 E_2\} = 0.$$

For independent events, on the other hand, from (2.25) we have

$$\Pr\{E_1 E_2\} = \Pr\{E_1\}\Pr\{E_2\} > 0;$$

they cannot be exclusive.

2.8.   SOME APPLICATIONS OF THE PROBABILITY THEORY OF COMPOUND EVENTS

The results derived in section 2.3 are fundamental in probability theory and have broad applications in several fields of experimental science, including genetics. We give here three examples to demonstrate the practical use of these probability laws.

Example 2.6. *Mendel's laws* provide interesting examples of compound events.

1. *The principle of segregation* (Mendel's first law) states that, when an individual forms gametes, the allelic pairs separate from each other. This is an example of exclusive events. If a gamete contains the allele A, it cannot contain the allele a.

2. *The principle of independent assortment* (Mendel's second law) states that the allelic pairs located in different chromosomes segregate independently into gametes and join independently into zygotes.

Suppose that the genotypes of a parental population are Aa with respect to locus A, and Bb with respect to locus B, that is, the joint parental genotypes are $AaBb$. At locus A, the gamete a occurs with probability $\frac{1}{2}$, and the genotype aa in the offspring of $Aa \times Aa$ crosses with probability $\Pr\{aa\} = \Pr\{a\}\Pr\{a\} = \frac{1}{2} \cdot \frac{1}{2} = \frac{1}{4}$. The same is true for locus B, and we have $\Pr\{bb\} = \frac{1}{2} \cdot \frac{1}{2} = \frac{1}{4}$. If these loci are located in different chromosomes, segregation occurs independently and the genotype $aabb$ in the offspring of $AaBb \times AaBb$ crosses has the probability

$$\Pr\{aabb\} = \Pr\{aa\}\Pr\{bb\} = \frac{1}{4} \cdot \frac{1}{4} = \frac{1}{16}.$$

Similar rules apply to other genotypes.

Example 2.7. Let A and a be dominant and recessive alleles, respectively, at locus A.

(i) Suppose first that the heterozygotes Aa produce gametes A and a with equal frequencies. We then have $\Pr\{A\} = \Pr\{a\} = \frac{1}{2}$. Suppose that the matings $Aa \times Aa$ are performed. Since the gametes combine independently into zygotes, so that $\Pr\{AA\} = \Pr\{A\}\Pr\{A\} = \frac{1}{2} \cdot \frac{1}{2} = \frac{1}{4}$, $\Pr\{Aa\} = \Pr\{A\}\Pr\{a\} + \Pr\{a\}\Pr\{A\} = 2 \times \frac{1}{2} \cdot \frac{1}{2} = \frac{1}{2}$, and $\Pr\{aa\} = \Pr\{a\}\Pr\{a\} = \frac{1}{2} \cdot \frac{1}{2} = \frac{1}{4}$.

The genotype and phenotype distributions are, respectively,

Genotypes			Phenotypes	
AA	Aa	aa	$A-$	aa
$\frac{1}{4}$	$\frac{1}{2}$	$\frac{1}{4}$	$\frac{3}{4}$	$\frac{1}{4}$

(ii) Suppose that the gamete a is less viable than the gamete A, in the ratio $a : A = \alpha : 1$, where $0 < \alpha < 1$. Individuals with genotype Aa will produce gametes a and A with the following probabilities: $\Pr\{a\} = \frac{1}{2}\alpha$ and $\Pr\{A\} = 1 - \frac{1}{2}\alpha = \frac{1}{2}(2 - \alpha)$. Gametes combine independently into zygotes, so that now the genotype and phenotype distributions in the offspring of $Aa \times Aa$

matings are, respectively,

Genotypes			Phenotypes	
AA	Aa	aa	$A-$	aa
$\frac{1}{4}(2-\alpha)^2$	$\frac{2}{4}\alpha(2-\alpha)$	$\frac{1}{4}\alpha^2$	$\frac{1}{4}(2-\alpha)(2+\alpha)$	$\frac{1}{4}\alpha^2$.

Example 2.8. Let A_1 and A_2 be two codominant alleles at locus A. Under general assumptions of fully viable gametes and zygotes, the offspring of $A_1A_2 \times A_1A_2$ matings will have the genotype (and this will also be the phenotype) distribution shown in Example 2.7(i).

(i) Suppose that the genotypes A_1A_1 and A_1A_2 are less viable than A_2A_2 in the ratios $A_1A_1:A_1A_2:A_2A_2 = \gamma_1:\gamma_2:1$, where $0 < \gamma_1 < 1$, $0 < \gamma_2 < 1$. Since survival of a genotype and segregation are independent events, the genotypic ratios in an "infinite" population will be $A_1A_1:A_1A_2:A_2A_2 = \frac{1}{4}\gamma_1:\frac{1}{2}\gamma_2:\frac{1}{4} = \gamma_1:2\gamma_2:1$.

To find the distribution of the genotypes, we have to "normalize" these figures by making them add up to 1. Let $\Delta = \gamma_1 + 2\gamma_2 + 1$. The genotype distribution will be

	A_1A_1	A_1A_2	A_2A_2
	γ_1/Δ	$2\gamma_2/\Delta$	$1/\Delta$.

We notice that now these probabilities add up to 1.

(ii) Suppose that, in addition to the reduced viabilities of A_1A_1 and A_1A_2, the gene A_2 in the heterozygote has only the penetrance β [i.e., only the proportion β of A_1A_2 express themselves as A_1A_2, and the remaining proportion $(1-\beta)$ of A_1A_2 express themselves as A_1A_1]. We now have

$$\Pr\{A_1A_1\} = \frac{\gamma_1 + 2(1-\beta)\gamma_2}{\Delta}, \quad \Pr\{A_1A_2\} = \frac{2\gamma_2\beta}{\Delta}, \quad \Pr\{A_2A_2\} = \frac{1}{\Delta}.$$

These probabilities add up to 1, and their values represent the probability distribution of genotypes under the conditions stated in (ii).

It is worthwhile to notice that in all these cases the sample space is the same (all possible genotypes), but the probability spaces are different in each case.

2.9. PRODUCT OF SAMPLE SPACES. JOINT PROBABILITY OF COMPOUND EVENTS

Example 2.9. Let us return to Example 2.1. We may consider the maternal gamete output as a sample space, $\Omega^{(1)}$, say, consisting of two points, A and a. Similarly, the paternal gamete output may be represented by another sample space, $\Omega^{(2)}$, say, consisting also, in this case, of points A and a. The sample space, Ω, for the genotypes of the offspring consists of four points in two-dimensional Cartesian space as shown in Table 2.1. The sample space Ω is called the product of the sample spaces Ω_1 and Ω_2. We notice that, in this example, the sample points (A, A), (A, a), (a, A) and (a, a) correspond to the genotypes AA, Aa, aA, and aa, the symbols for which can be obtained by "formal multiplications" of gametic symbols (e.g., $A \times A = AA$). We generalize our result in the following definition.

DEFINITION 2.17.   *Let* $\Omega^{(1)} = \{e_1^{(1)}, e_1^{(1)}, \ldots\}$ *and* $\Omega^{(2)} = \{e_1^{(2)}, e_2^{(2)}, \ldots\}$ *be two sample spaces. The set of all ordered pairs*, $\{e_i^{(1)}, e_j^{(2)}\}$, *is called the product* (Ω) *of these two sample spaces. We write*

$$\Omega = \Omega^{(1)} \times \Omega^{(2)}. \tag{2.29}$$

Extension to the product of k sample spaces is straightforward.

It is worthwhile to notice that the two-dimensional sample space in Table 2.1 can be reduced to the one-dimensional sample space of genotypes AA, Aa, aa.

The concept of a product of sample spaces is especially useful if we want to emphasize two (or more) kinds of events, assigned to each sample space. In our Example 2.9 these two kinds of events were the maternal and the paternal gamete outputs.

Let E_i be an event defined in $\Omega^{(1)}$ and D_j an event defined in $\Omega^{(2)}$. The compound event including points of E_i and D_j is the product or intersection of E_i and D_j (briefly E_iD_j).

DEFINITION 2.18.   *The joint probability of the compound event* E_iD_j *in the sample space* $\Omega = \Omega^{(1)} \times \Omega^{(2)}$ *is a real, nonnegative number*, $\Pr\{E_iD_j\}$, *such that*

$$0 \leqslant \Pr\{E_iD_j\} \leqslant 1. \tag{2.30}$$

It can also be simply denoted as

$$\Pr\{E_iD_j\} = p_{ij}. \tag{2.31}$$

2.10.   JOINT PROBABILITY DISTRIBUTI

Let $E_1, E_2, \ldots, E_m$ be m mutually excl   exhaustive events in $\Omega^{(1)}$, that is,

$$\sum_{i=1}^{m} E_i = \Omega^{(1)}. \tag{2.32}$$

Similarly, let $D_1, D_2, \ldots, D_n$ be n mutually exclusive and exhaustive events in $\Omega^{(2)}$, that is,

$$\sum_{j=1}^{n} D_j = \Omega^{(2)}. \qquad (2.33)$$

It is easy to see that $m \times n$ compound events of type $(E_i D_j)$ represent a set of mutually exclusive and exhaustive events in $\Omega = \Omega^{(1)} \times \Omega^{(2)}$, that is,

$$\sum_{i=1}^{m} \sum_{j=1}^{n} (E_i D_j) = \Omega. \qquad (2.34)$$

From (2.34) and (2.31) we obtain

$$\sum_{i=1}^{m} \sum_{j=1}^{n} \Pr\{E_i D_j\} = \sum_{i=1}^{m} \sum_{j=1}^{n} p_{ij} = 1. \qquad (2.35)$$

DEFINITION 2.19. *If the set of compound events $(E_i D_j)$ defined in Ω satisfies condition (2.34), then the set of probabilities $\Pr\{E_i D_j\} = p_{ij}$ satisfying condition (2.35) is called the joint probability distribution of events $(E_i D_j)$ in Ω.*

The joint probability distribution in Ω is given in Table 2.2.

The sample space Ω together with the joint probability distribution is called the *joint probability space*.

Example 2.10. Let A and B be two loci with A, a and B, b alleles, respectively. The four possible gametes, AB, Ab, aB, and ab, may be considered as compound events in the two-dimensional sample space. Tables 2.3a and 2.3b show probability spaces for gametes produced by $ABab$

TABLE 2.2   Joint probability distribution

E \ D	D_1	D_2	$\cdots$	D_j	$\cdots$	D_n	Total
E_1	p_{11}	p_{12}	$\cdots$	p_{1j}	$\cdots$	p_{1n}	$p_{1.}$
E_2			$\cdots$	p_{2j}	$\cdots$	p_{2n}	$p_{2.}$
$\ldots$							$\ldots$
E_i			$\cdots$	p_{ij}	$\cdots$	p_{in}	$p_{i.}$
$\ldots$							$\ldots$
E_m			$\cdots$	p_{mj}	$\cdots$	p_{mn}	$p_{m.}$
Total	$p_{.1}$	$p_{.2}$	$\cdots$	$p_{.j}$	$\cdots$	$p_{.n}$	1

TABLE 2.3a Joint gamete distribution with respect to two independent loci		
Locus A \ Locus B	$\frac{1}{2}B$	$\frac{1}{2}b$
$\frac{1}{2}A$	$\frac{1}{4}AB$	$\frac{1}{4}Ab$
$\frac{1}{2}a$	$\frac{1}{4}aB$	$\frac{1}{4}ab$

TABLE 2.3b Joint gamete distribution with respect to two linked loci		
Locus A \ Locus B	$\frac{1}{2}B$	$\frac{1}{2}b$
$\frac{1}{2}A$	$\frac{1}{2}(1-\lambda)AB$	$\frac{1}{2}\lambda Ab$
$\frac{1}{2}a$	$\frac{1}{2}\lambda aB$	$\frac{1}{2}(1-\lambda)ab$

heterozygotes when loci are independent or linked, respectively. For linked loci, λ is the recombination fraction and the gametes are produced by heterozygotes AB/ab, in coupling. We notice that, although the "compound" gametes are obtained by "formal multiplication" (i.e., $AB = A \times B$), the joint probabilities in the case of linked loci are not the products of probabilities for gametes at single loci, since these events are not independent.

2.11. MARGINAL PROBABILITY DISTRIBUTIONS

Let us consider the event E_i in the product space Ω regardless of which of the events D occurs. We have

$$E_i = E_{i.} = \sum_{j=1}^{n}(E_iD_j). \qquad (2.36)$$

The notation $E_{i.}$ indicates that E_i is considered in Ω [not in $\Omega^{(1)}$] and has been obtained as a (logical) sum of events E_iD_j, summation being taken with respect to subscript j. Let

$$\Pr\{E_i\} = \Pr\{E_{i.}\} = \sum_{j=1}^{n}\Pr\{E_iD_j\} = \sum_{j=1}^{n}p_{ij} = p_{i.} \qquad (2.37)$$

be the probability of occurrence of the event E_i. Here again, the point in the subscript indicates that summation was with respect to the subscript j.

DEFINITION 2.20. *The sequence of probabilities* $p_{1.}, p_{2.}, \ldots, p_{m.}$, *defined in* (2.37) *for m mutually exclusive events* $E_1, E_2, \ldots, E_m$, *and satisfying the condition*

$$\sum_{i=1}^{m} p_{i.} = 1, \qquad (2.38)$$

is called the marginal probability distribution of these events determined in Ω.

Similarly, the sequence of probabilities

$$\Pr\{D_j\} = \Pr\{D_{.j}\} = \sum_{i=1}^{m} \Pr\{E_i D_j\} = \sum_{i=1}^{m} p_{ij} = p_{.j}, \qquad (2.39)$$

for $j = 1, 2, \ldots, m$, where $D_1, D_2, \ldots, D_n$ are mutually exclusive, and

$$\sum_{j=1}^{n} p_{.j} = 1, \qquad (2.40)$$

defines the marginal distribution of events $D_1, D_2, \ldots, D_n$ in Ω.

We also notice that

$$\sum_{i=1}^{m} \sum_{j=1}^{n} p_{ij} = \sum_{i=1}^{m} p_{i.} = \sum_{j=1}^{n} p_{.j} = 1. \qquad (2.41)$$

The corresponding marginal distributions are given in the last column and the last row of Table 2.2.

In our Example 2.10 the marginal distributions correspond to distributions of gametes at single loci A and B, respectively. We notice that they are the same, whether the loci are independent or linked.

2.12.   CONDITIONAL DISTRIBUTIONS.   INDEPENDENCE

From Table 2.2 we can easily find that the conditional probability of an event E_i, given that D_j has occurred, is

$$\Pr\{E_i \mid D_j\} = \frac{\Pr\{E_i D_j\}}{\Pr\{D_j\}} = \frac{p_{ij}}{p_{.j}}, \qquad (2.42)$$

provided that $\Pr\{D_j\} > 0$ for $i = 1, 2, \ldots, m$; $j = 1, 2, \ldots, n$. We obviously have

$$\sum_{i=1}^{m} \Pr\{E_i \mid D_j\} = \frac{\sum_{i=1}^{n} p_{ij}}{p_{.j}} = \frac{p_{.j}}{p_{.j}} = 1. \qquad (2.43)$$

DEFINITION 2.21.   *The sequence of probabilities defined in (2.42) and satisfying condition (2.43) is called the conditional probability distribution of events* $E_1, E_2, \ldots, E_m$, *given that D_j has occurred.*

There are n such conditional distributions, one for each D_j ($j = 1, 2, \ldots, n$). Similarly, from Table 2.2 we obtain

$$\Pr\{D_j \mid E_i\} = \frac{\Pr\{E_i D_j\}}{\Pr\{E_i\}}, \qquad (2.44)$$

provided $\Pr\{E_i\} > 0$, for $i = 1, 2, \ldots, m; j = 1, 2, \ldots, n$. We have

$$\sum_{j=1}^{n} \Pr\{D_j \mid E_i\} = \frac{\sum_{j=1}^{n} p_{ij}}{p_{i.}} = \frac{p_{i.}}{p_{i.}} = 1. \tag{2.45}$$

The probabilities (2.44) for all $j = 1, 2, \ldots, n$ represent the conditional distribution of events $D_1, \ldots, D_n$, given that E_i has occurred. There are m such distributions, one for each E_i ($i = 1, 2, \ldots, m$).

When E_i and D_j are independent, then

$$p_{ij} = p_{i.} \times p_{.j} \tag{2.46}$$

[see formula (2.25)].

Substituting (2.46) into (2.42), we obtain

$$\Pr\{E_i \mid D_j\} = \frac{p_{i.} \times p_{.j}}{p_{.j}} = p_{i.} = \Pr\{E_i\} \tag{2.47}$$

for $i = 1, 2, \ldots, m$; this is the same as (2.37).

Similarly, substituting (2.46) into (2.44), we obtain

$$\Pr\{D_j \mid E_i\} = \frac{p_{i.} \times p_{.j}}{p_{i.}} = p_{.j} = \Pr\{D_j\} \tag{2.48}$$

for $j = 1, 2, \ldots, n$; this is the same as (2.39).

We now define the independence of two sequences of events.

DEFINITION 2.22. *Let $E_1, \ldots, E_m$ be a sequence of mutually exclusive and exhaustive events in $\Omega^{(1)}$, and $D_1, \ldots, D_n$ be a sequence of mutually exclusive and exhaustive events in $\Omega^{(2)}$. These sequences are independent in $\Omega^{(1)} \times \Omega^{(2)}$ if their joint probability distribution can be expressed as the product of marginal distributions, that is,*

$$\Pr\{E_i D_j\} = \Pr\{E_i\}\Pr\{D_j\}, \qquad \Pr\{E_i\} > 0, \Pr\{D_j\} > 0 \tag{2.49}$$

or

$$p_{ij} = p_{i.} p_{.j}, \qquad p_{i.} > 0, \; p_{.j} > 0, \tag{2.49a}$$

for all $i = 1, 2, \ldots, m; j = 1 \; 2, \ldots, n$.

2.13. GENOTYPE AND PHENOTYPE DISTRIBUTIONS WITH RESPECT TO TWO LINKED LOCI

We now demonstrate application of the results derived in sections 2.9–2.12 in constructing genotype and phenotype distributions for two autosomal linked loci, each with two alleles.

TABLE 2.4a　The genotype distribution of the offspring of $AB/ab \times AB/ab$ mating (4×4 sample space)

♀ \ ♂	$\frac{1}{2}(1-\lambda)AB$	$\frac{1}{2}\lambda Ab$	$\frac{1}{2}\lambda aB$	$\frac{1}{2}(1-\lambda)ab$
$\frac{1}{2}(1-\lambda)AB$	$\frac{1}{4}(1-\lambda)^2$ $ABAB$	$\frac{1}{4}\lambda(1-\lambda)$ $ABAb$	$\frac{1}{4}\lambda(1-\lambda)$ $ABaB$	$\frac{1}{4}(1-\lambda)^2$ AB/ab
$\frac{1}{2}\lambda Ab$	$\frac{1}{4}\lambda(1-\lambda)$ $ABAb$	$\frac{1}{4}\lambda^2$ $AbAb$	$\frac{1}{4}\lambda^2$ Ab/aB	$\frac{1}{4}\lambda(1-\lambda)$ $Abab$
$\frac{1}{2}\lambda aB$	$\frac{1}{4}\lambda(1-\lambda)$ $ABaB$	$\frac{1}{4}\lambda^2$ AB/ab	$\frac{1}{4}\lambda^2$ $aBaB$	$\frac{1}{4}\lambda(1-\lambda)$ $aBab$
$\frac{1}{2}(1-\lambda)ab$	$\frac{1}{4}(1-\lambda)^2$ AB/ab	$\frac{1}{4}\lambda(1-\lambda)$ $Abab$	$\frac{1}{4}\lambda(1-\lambda)$ $aBab$	$\frac{1}{4}(1-\lambda)^2$ $abab$

The distribution of the gametes produced by Ab/ab individuals and given in Table 2.3b can be presented in the form of one-dimensional probability space as follows:

$$\begin{array}{lcccc} \text{Gamete:} & AB & Ab & aB & ab \\ \text{Probability:} & \tfrac{1}{2}(1-\lambda) & \tfrac{1}{2}\lambda & \tfrac{1}{2}\lambda & \tfrac{1}{2}(1-\lambda). \end{array} \qquad (2.50)$$

Since gametes join independently into zygotes, the simplest way of constructing the probability space for genotypes is to use formula (2.46) for the probabilities of independent events, as shown in Table 2.4a. The first column and the first row represent the maternal and the paternal gamete distributions (marginal distributions), respectively.

TABLE 2.4b　The genotype distribution of the offspring of $AB/ab \times AB/ab$ (3×3 sample space)

Locus A \ Locus B	$\frac{1}{4}BB$	$\frac{1}{2}Bb$	$\frac{1}{4}bb$
$\frac{1}{4}AA$	$\frac{1}{4}(1-\lambda)^2$ $AABB$	$\frac{1}{2}\lambda(1-\lambda)$ $AABb$	$\frac{1}{4}\lambda^2$ $AAbb$
$\frac{1}{2}Aa$	$\frac{1}{2}\lambda(1-\lambda)$ $AaBB$	$\frac{1}{2}[1-2\lambda(1-\lambda)]$ $AaBb$	$\frac{1}{2}\lambda(1-\lambda)$ $Aabb$
$\frac{1}{4}aa$	$\frac{1}{4}\lambda^2$ $aaBB$	$\frac{1}{2}\lambda(1-\lambda)$ $aaBb$	$\frac{1}{4}(1-\lambda)^2$ $aabb$

TABLE 2.4c The phenotype distribution of the offspring of
$AB/ab \times AB/ab$

Locus A \ Locus B	$\frac{3}{4}B-$	$\frac{1}{4}bb$
$\frac{3}{4}A-$	$\frac{1}{4}[2 + (1 - \lambda)^2]$ $A-B-$	$\frac{1}{4}[1 - (1 - \lambda)^2]$ $A-bb$
$\frac{1}{4}aa$	$\frac{1}{4}[1 - (1 - \lambda)^2]$ $aaB-$	$\frac{1}{4}(1 - \lambda)^2$ $aabb$

Since some of the genotypes shown in Table 2.4a are the same, we may
also use the 3×3 two-dimensional sample space, as in Table 2.4b,
which was obtained from Table 2.4a by combining the probabilities for
the appropriate genotypes. For example, $\Pr\{AaBb\} = 2\Pr\{AB/ab\} +$
$2\Pr\{Ab/aB\} = \frac{1}{2}(1 - \lambda)^2 + \frac{1}{2}\lambda^2 = \frac{1}{2}[1 - 2\lambda(1 - \lambda)]$, etc. The first column
represents the genotype distribution with respect to locus A, and the first
row the genotype distribution with respect to locus B. These are the marginal
distributions. But here the joint probabilities are *not* the products of marginal
probabilities, since the loci are not independent. However, if $\lambda = \frac{1}{2}$, that is,
if the loci are independent, the joint probabilities will be products of corresponding marginal probabilities in both presentations.

If A is dominant to a, and B is dominant to b, the phenotype distribution
will be obtained by adding appropriate terms. Thus,

$$\Pr\{A-B-\} = \frac{1}{4}(1 - \lambda)^2 + \frac{1}{2}\lambda(1 - \lambda) + \frac{1}{2}\lambda(1 - \lambda) + \frac{1}{2}[1 - 2\lambda(1 - \lambda)]$$
$$= \frac{1}{4}(3 - 2\lambda + \lambda^2) = \frac{1}{4}[2 + (1 - \lambda)^2].$$

In a similar way we obtain

$$\Pr\{A-bb\} = \Pr\{aaB-\} = \frac{1}{4}\lambda^2 + \frac{1}{2}\lambda(1 - \lambda) = \frac{1}{4}\lambda(2 - \lambda) = \frac{1}{4}[1 - (1 - \lambda)^2].$$

The phenotype distribution can be presented in the two-dimensional probability space, as in Table 2.4c. The marginal distributions correspond to the
phenotype distributions at loci A and B, respectively.

The genotype distribution shown in Table 2.4b can also be represented in a
one-dimensional space consisting of nine different points, and the phenotype
distribution from Table 2.4c in a one-dimensional space consisting of four
points.

2.14. CONTINUOUS SAMPLE SPACE

For most of the genetic problems considered in this book, the sample
space will be *discrete*, that is, will consist of a finite or an infinite but countable

number of outcomes, as the preceding discussion has assumed. But there are several situations in which the measured character is continuous, such as the height or weight of an individual. These problems of quantitative genetics will not be extensively treated in this book. It is useful, however, to indicate some differences between discrete and continuous sample spaces, since in some situations we will use approximations based on *normal* distribution, which is continuous and will be discussed in Chapter 8.

Consider a straight line, that is, a one-dimensional sample space. The number of points on the line is uncountable. The events which are defined in this sample space usually correspond to results of measurements. But we cannot measure points. A point on a straight line is dimensionless; its measure is zero. An interval, however, has a certain length. It can be as small as possible, but it must be an interval and not a solitary point or collection of isolated points.

Let x be a certain variable which defines events in a continuous sample space. By analogy to measures of points and intervals, we define the probability that x takes a certain value x_0 as being zero, that is, $\Pr\{x = x_0\} = 0$, and the probability that x is in a certain *interval* $(x_0 - \Delta x, x_0 + \Delta x)$, even if Δx is very small, as a positive measure of this event, that is,

$$\Pr\{x_0 - \Delta x < x < x_0 + \Delta x\} > 0.$$

Similarly, in two-dimensional continuous space we use small *areas* as elementary events, not points.

We will return to this problem in Chapter 5, where the concept of random variable will be introduced.

SUMMARY

In this chapter, we have introduced several concepts and definitions from probability theory, such as sample space, random event, union and intersection of events, probability, and probability distributions in one- and two-dimensional sample space. The simplest rules for calculating probabilities of union and intersection of events have been given. Conditional probabilities and (statistical) independence of events have been defined. Using simple examples from genetics, we have demonstrated how probabilistic models can be useful in genetic problems, emphasizing the correspondence of genetic terms and laws to those introduced in the theory of probability of random events.

REFERENCES

Cramer, H., *The Elements of Probability Theory and Some of Its Applications*, John Wiley & Sons, New York, 1955.

Gangolli, R. A., and Ylvisaker, D., *Discrete Probability*, Harcourt, Brace & World, New York, 1966.

Johnson, N. L., and Leone, F. C., *Statistics and Experimental Design*, Vol. I, John Wiley & Sons, New York, 1964.

Mosteller, F., Rourke, R. E. K., and Thomas, G. B., *Probability with Statistical Applications*, Addison-Wesley Publishing Co. Reading, Mass. 1961.

Srb, A. M., Owen, R. D., and Edgar, R. S., *General Genetics*, W. H. Freeman & Co., San Francisco, 1965.

Wilks, S. S., *Mathematical Statistics*, John Wiley & Sons, New York, 1962.

PROBLEMS

2.1. It is usually assumed that albinism is a simple recessive trait. Suppose that two nonalbino parents are $Aa \times Aa$.

(a) What is the probability that the first child will be albino?

(b) What is the probability that the second child will be albino?

(c) What is the probability that the second child will be albino, *given* that the first was albino? Did you expect the answer? Why?

(d) What is the probability, again assuming that the parents are $Aa \times Aa$, that in a family of three children all will be albinos? All will be nonalbinos?

2.2. Suppose that we observe genotypes in two experiments, each involving a different type of mating. In Experiment I the genotypes $AA \times aa$ were mating, in Experiment II the genotypes $Aa \times Aa$. Suppose that N offspring of mating $AA \times aa$ and N offspring of $Aa \times Aa$ are mixed together, and assume that $N \to \infty$.

(a) What is the probability that an individual selected from this population is Aa?

(b) Given that an offspring is Aa, what is the probability that he is the progeny of the parents from Experiment II? (See section 7.2.)

2.3. Answer questions (a) and (b) from Problem 2.2, assuming that the progeny were mixed in the following proportions: $\frac{1}{3}$ of the offspring from mating $AA \times aa$ and $\frac{2}{3}$ from the offspring of mating $Aa \times Aa$.

2.4. Suppose that the probability space for two mating populations is arranged in the 3×3 matrix shown below and that the pairs male-female mate randomly (i.e., the joint probability of selecting a particular pair of genotypes for a male-female pair is the product of the probabilities of selecting a male and a female with these genotypes; see also section 4.2).

♀ \ ♂	$d_2 AA$	$2h_2 Aa$	$r_2 aa$
$d_1 AA$			
$2h_1 Aa$			
$r_1 aa$			

(a) Insert the joint probabilities in this probability space.

(b) Find the proportions of offspring obtained from the following matings: $♀AA \times ♂Aa$, $♂AA \times ♀Aa$, $AA \times Aa$.

(c) What is the probability of obtaining an offspring AA, given that (i) the mother was AA? (ii) the father was AA? (iii) either the mother or the father (but not both) was AA?

2.5. It is assumed that the color of kernels in wheat is controlled by two independent loci with alleles R_1, r_1 at one locus and R_2, r_2 at another locus. The effects of the gene actions of both loci are similar, and the intensity of the red color depends only on the number of R genes present in the genotype. A genotype with four R genes ($R_1R_1R_2R_2$) gives dark red color; with three R genes ($R_1R_1R_2r_2$ or $R_1r_1R_2R_2$), medium dark red; with two R genes, medium red; with one R gene, light red; and with no R gene ($r_1r_1r_2r_2$), white kernels. Suppose that the crossing $R_1r_1R_2r_2 \times R_1r_1R_2r_2$ has been performed.

(a) Construct a probability model of inheritance for the color of kernels in wheat in the population resulting from these crosses. What kind of probability space do you think is the most appropriate?

(b) Which outcomes in the sample space correspond to four different shades of the red color and which to the white color? Define these events and find their probabilities.

(c) Find the probability of occurrence of red (without regard to shade) kernels.

(d) Which probability laws do you apply to answer each of these questions?

2.6. Answer questions (a), (b), and (c) in Problem 2.5 if the parental populations are $R_1R_1R_2R_2 \times r_1r_1r_2r_2$ (Dark red × White).

▼ **2.7.** Suppose that a certain trait is controlled by one locus with two alleles A, a, where A is dominant over a. Suppose that a is a lethal recessive and that the surviving ratios of genotypes are $AA:Aa:aa = 1:\alpha:0$, with $0 < \alpha < 1$. Suppose that the initial population consists of AA and Aa in proportions h_1 and h_2, respectively, with $h_1 + h_2 = 1$, and that the population is mating at random (for definition of *random mating* see Problem 2.4 and section 4.2).

(a) Find the probability distribution of the genotypes in the first generation.

(b) What is the probability of producing a zygote aa in the first generation?

(c) Suppose that the individuals which survive in the first generation are again subjected to random mating. Find the distribution of the genotypes in the second generation.

2.8. Let A and a be the dominant and recessive alleles at the locus A, and similarly B and b be the alleles at the locus B. We assume that these loci are linked with recombination fraction λ. Find the phenotype distribution in the offspring of the following crosses:

(a) $AB/ab \times Ab/aB$ (coupling × repulsion),

(b) $Ab/aB \times Ab/aB$ (repulsion × repulsion).

Genotypes and Phenotypes in Experimental Populations

3.1. SIMPLE COMBINATORIAL PROBLEMS

3.1.1. Permutations and Combinations

We first introduce some useful mathematical notation.

DEFINITION 3.1. *For any positive integer, n, the product of consecutive integers from 1 to n is defined as n factorial, briefly written as n!, and conventionally 0! = 1, that is,*

$$\left.\begin{array}{c} n! = n(n-1)(n-2)\cdots 3\cdot 2\cdot 1, \\ 0! = 1. \end{array}\right\} \tag{3.1}$$

We notice that

$$n! = n(n-1)! = n(n-1)(n-2)! = \cdots$$
$$= n(n-1)\cdots(n-k+1)(n-k)!, \qquad k \leqslant n. \tag{3.2}$$

For example,

$$5! = 5\cdot 4! = 5\cdot 4\cdot 3! = 5\cdot 4\cdot 3\cdot 2! = 5\cdot 4\cdot 3\cdot 2\cdot 1 = 120.$$

Suppose that we have n different objects, $a, b, c, d, \ldots$, etc., and we wish to arrange them in special orders.

DEFINITION 3.2. *Any arrangement of n objects in a definite order is called a permutation.*

How many different permutations of n different objects are possible? The first object can be selected in n ways from the remaining $n-1$ objects, the second can be selected in $n-1$ ways, and so on until only one object is left, which can be selected in just one way. If we denote by $P_n{}^n$ the number of permutations of n different objects, we obtain the formula

$$P_n{}^n = n(n-1)(n-2)\cdots 3\cdot 2\cdot 1 = n!. \tag{3.3}$$

For example, suppose that we have $n = 3$ objects, a, b, c. Then $P_3{}^3 = 3! = 3\cdot 2\cdot 1 = 6$, so there are six possible arrangements: abc, acb, bac, bca, cab, cba.

Suppose that among n objects, r_1 are of one kind (are indistinguishable), r_2 are of the second kind ... and r_k are of the kth kind with $r_1 + r_2 + \cdots + r_k = n$. For example, if the objects are $aaaaa$, bb, ccc, d, we have $n = 11$, $r_1 = 5$, $r_2 = 2$, $r_3 = 3$, $r_4 = 1$. If we interchange two a's in the first group, the arrangement will be the same since these a's are indistinguishable. Evidently, all 5! rearrangements of the a's produce no visible change in the permutation, and similarly for the remaining homogeneous groups. Therefore the number of *visually* different arrangements possible among 11! is $11!/(5!2!3!1!) = 27720$.

More generally, if we denote by $P_{r_1,r_2,\ldots,r_k}^n$, the number of permutations of n objects with k types of objects, where r_i is the number of indistinguishable objects in the ith type $(i = 1, 2, \ldots, k)$, we will have

$$P_{r_1,r_2,\ldots,r_k}^n = \frac{n!}{r_1!\, r_2! \cdots r_k!}. \tag{3.4}$$

In the special case, when there are only two kinds of objects: r objects of one kind and $n - r$ objects of another kind, we have

$$P_{r,n-r}^n = \frac{n!}{r!\,(n-r)!}. \tag{3.4a}$$

DEFINITION 3.3. *An r-combination from n distinct objects is a group of r selected from n objects without regard to their order. The number of possible such combinations, denoted often by $C_r{}^n$, is*

$$C_r{}^n = \frac{n!}{r!\,(n-r)!} = \binom{n}{r}. \tag{3.5}$$

This definition can be generalized.

DEFINITION 3.3a. *An $r_1, r_2, \ldots, r_k$ combination is an arrangement of n distinct objects into k mutually exclusive groups with r_i objects in the ith group $(i = 1, 2, \ldots, n)$ without regard to the order in any group and subject to the condition $r_1 + r_2 + \cdots + r_k = n$. Denoting the number of such possible combinations by $C_{r_1,r_2,\ldots,r_k}^n$, we have*

$$C_{r_1,r_2,\ldots,rk}^n = \frac{n!}{r_1!r_2! \cdots r_k!} = \binom{n}{r_1, r_2, \cdots, r_k}. \tag{3.6}$$

We note that the symbol $\binom{n}{r}$ is commonly used for $n!/[r_1!(n-r)!]$ and sometimes $\binom{n}{r_1, r_2, \cdots, r_k}$ for $n!/(r_1! \, r_2! \cdots r_k!)$. We also notice that (3.6) is identical with (3.4), and (3.5) with (3.4a).

It is easy to show that

$$\binom{n}{r} = \binom{n}{n-r}.$$

If we want to observe the order in each group, then the number of *ordered* selections of r objects will be

$$C_r{}^n r! = \frac{n!}{(n-r)!}. \tag{3.8}$$

Suppose, for example, that we have $n = 4$ objects, a, b, c, d, and wish to find the number of groups each having $r = 2$ objects. Thus the number of unordered groups is $C_2{}^4 = 4!/(2!2!) = 6$, and these are ab, ac, ad, bc, bd, cd. The number of ordered groups is $C_2{}^n \cdot 2! = 4!/2! = 12$, and these are:

$$ab, \ ac, \ ad, \ bc, \ bd, \ cd$$

$$ba, \ ca, \ da, \ cb, \ bd, \ cd.$$

3.1.2. Newton's Binomial Formula

It is well known that $(a + b)^2 = a^2 + 2ab + b^2$. Some people also remember the expansion of $(a + b)^3$; if not, it can be evaluated from $(a + b)^3 = (a + b)^2(a + b)$. The evaluation of $(a + b)^n$ from the recurrence formula $(a + b)^n = (a + b)^{n-1}(a + b)$, when n is large, will be rather tedious. But it can be found from *Newton's binomial formula*, that is,

$$(a + b)^n = \binom{n}{0}a^0 b^n + \binom{n}{1}a^1 b^{n-1} + \binom{n}{2}a^2 b^{n-2} + \cdots + \binom{n}{n}a^n b^0$$

$$= \sum_{r=0}^{n} \binom{n}{r}a^r b^{n-r} = \sum_{r=0}^{n} \binom{n}{r}a^{n-r}b^r. \tag{3.9}$$

There are $n + 1$ terms in the expansion in (3.9). The $(r + 1)$th term is

$$\binom{n}{r}a^r b^{n-r}. \tag{3.10}$$

The coefficients $\binom{n}{r}$ in the expansion $(a + b)^n$ can be found from formula

(3.5) or from *Pascal's triangle:*

$$
\begin{array}{ccccccccccccc}
 & & & & & 1 & & & & & & & n \\
 & & & & 1 & & 1 & & & & & & 0 \\
 & & & 1 & & 2 & & 1 & & & & & 1 \\
 & & 1 & & 3 & & 3 & & 1 & & & & 2 \\
 & 1 & & 4 & & 6 & & 4 & & 1 & & & 3 \\
1 & & 5 & & 10 & & 10 & & 5 & & 1 & & 4 \\
 & & & & & & & & & & & & 5
\end{array}
$$

Pascal's triangle is formed in the following way. In each row the first and the last numbers are always 1. The other numbers in the $(r + 1)$th row are the sums of the two adjacent numbers from the rth row. The second number in each row indicates the power of the binomial. For example, $(a + b)^4 = b^4 + 4ab^3 + 6a^2b^2 + 4a^3b + a^4$. In particular,

$$(1 + 1)^n = 2^n = \binom{n}{0} + \binom{n}{1} + \binom{n}{2} + \cdots \binom{n}{n} = \sum_{i=0}^{n} \binom{n}{i}. \qquad (3.11)$$

When n becomes larger and larger, even constructing the Pascal triangle is not easy, but tables of $n!$ are available.

The expression $(a_1 + a_2 + \cdots + a_k)^n$ is called a *multinomial.* It can be shown that the expansion of this multinomial is

$$(a_1 + a_2 + \cdots + a_k)^n = \sum \sum \cdots \sum \frac{n!}{r_1! \, r_2! \cdots r_k!} a_1^{r_1} a_2^{r_2} \cdots a_k^{r_k}, \qquad (3.12)$$

where r's can take any value $0, 1, \ldots, n$ subject to the condition

$$r_1 + r_2 + \cdots + r_k = \sum_{i=1}^{k} r_i = n. \qquad (3.13)$$

This brief review of some basic concepts from combinatorial mathematics will be useful throughout the book. Some other interesting and useful formulae for the reader to prove are given in the problems at the end of this chapter.

▼ 3.2. GENERATING FUNCTIONS

Let $a_0, a_1, a_2, \ldots$ be an infinite or a finite sequence of real numbers (we briefly denote such a sequence as $\{a_i\}$). Suppose that we can construct some function which contains these numbers as constant coefficients so that they can be found from this function. Any function which generates the sequence of numbers $a_0, a_1, a_2, \ldots$ in this manner is called a *generating function* of this sequence.

It is known from calculus that many functions can be expanded into power series (see Appendix II). If a power series in a variable quantity t,

$$a_0 + a_1t + a_2t^2 + \cdots + a_nt^n, \tag{3.14}$$

has the property that, when $n \to \infty$, the series given in (3.14) tends to a certain function $G(t)$ [converges to $G(t)$] for a certain range of values (even very small) of t, then we say that $G(t)$ can be expanded into a power series. Such a series is unique and can be used as a generating function.

DEFINITION 3.4. *If series (3.14) is such that it converges in some finite interval* $-t_0 < t < t_0$ *to the function* $G(t)$, *then the function*

$$G(t) = a_0 + a_1t + a_2t^2 + \cdots + a_nt^n + \cdots \tag{3.15}$$

is called the (power) generating function of the coefficients (numbers) a_0, $a_1, a_2, \ldots$.

Example 3.1. (*a*) It has been shown that the series $1 + t + t^2 + \cdots + t^n + \cdots$ approaches $1/(1-t)$ as n tends to infinity, if $-1 < t < 1$ [see Appendix II, formula (II.21)]. The sequence of coefficients, in this example, is $1, 1, \ldots$ (i.e., $a_i = 1$, for $i = 0, 1, \ldots$). Therefore the function

$$G(t) = \frac{1}{1-t} = 1 + t + t^2 + \cdots + t^n + \cdots$$

is the generating function of the sequence of numbers $1, 1, \ldots$.

(*b*) Suppose that the sequence is finite, that is, $a_i = 1$ for $i = 0, 1, \ldots, n$ and $a_i = 0$ for $i > n$. It is known [see formula (II.23)] that

$$1 + t + t^2 + \cdots + t^n = \frac{1 - t^{n+1}}{1 - t}$$

(this is a finite geometric series). Thus,

$$G(t) = \frac{1 - t^{n+1}}{1 - t} = 1 + t + t^2 + \cdots + t^n$$

is the generating function of the finite sequence $a_i = 1, i = 0, 1, \ldots, n$.

Example 3.2. Let

$$a_i = \binom{n}{i}, \qquad i = 0, 1, \ldots, n.$$

Expanding the function $(1 + t)^n$ as a polynomial in t, we obtain

$$G(t) = (1 + t)^n = \binom{n}{0} + \binom{n}{1}t + \binom{n}{2}t^2 + \cdots + \binom{n}{n}t^n.$$

Thus, $G(t) = (1 + t)^n$ is the generating function of the finite sequence of numbers defined as

$$a_i = \binom{n}{i}, \qquad i = 0, 1, \ldots, n.$$

The variable t in $G(t)$ has no special meaning; it is sometimes called a "dummy" variable. It can be replaced by some other symbol or label.

The advantage of using the generating function $G(t)$ is that often such a function is known and the sequence $\{a_i\}$ is unknown. From the expansion of $G(t)$ into a power series we can find the sequence $\{a_i\}$.

We shall now demonstrate how the results derived in sections 3.1 and 3.2 can be applied in some simple genetic problems. Several other applications will be shown throughout the book.

3.3. THE NUMBER OF GENOTYPES AND PHENOTYPES

3.3.1. A Single Locus with s Alleles

Let us first consider a single locus with s alleles, $A_1, A_2, \ldots, A_s$. There will be s possible homozygotes of type $A_i A_i$ $(i = 1, 2, \ldots, s)$ and $\binom{s}{2}$ possible heterozygotes of type $A_i A_j$ $(i, j = 1, 2, \ldots, s; i < j)$. Thus the number of possible genotypes with regard to this locus is

$$\binom{s}{2} + s = \frac{s(s-1)}{2} + s = \frac{(s+1)s}{2} = \binom{s+1}{2}. \tag{3.16}$$

In particular, for $s = 2$, the number of genotypes is $\binom{3}{2} = 3$, as is well known to all geneticists. The number of phenotypes depends on the mode of dominance. For example, in the *ABO* system with three alleles *A*, *B*, *O* there are six genotypes, but only four phenotypes, A, B, AB, and O (compare Example 2.2).

3.3.2. Several Loci

By genotype we usually understand the genetic constitution of an individual. The terms "genotype" and "zygote" are equivalent in the sense that both have the same "pattern" of symbols.

Let us consider n loci. Let s_t be the number of alleles at the tth locus, $t = 1, 2, \ldots, n$. Thus the number of genotypes (or "genetic patterns") with respect to these loci will be

$$\binom{s_1 + 1}{2}\binom{s_2 + 1}{2} \cdots \binom{s_n + 1}{2} = \prod_{t=1}^{n} \binom{s_t + 1}{2} = \frac{1}{2^n} \prod_{t=1}^{n} s_t(s_t + 1). \quad (3.17)$$

But if the term "genotype" describes not only the genic formula of an individual but also the way in which it breeds, the terms "zygote" and "genotype" may have a slightly different meaning. For two linked loci, individuals AB/ab and Ab/aB have the same genic patterns, but they are different with regard to frequency of gamete production. The individuals in coupling AB/ab produce gametes AB and ab with frequencies $\frac{1}{2}(1 - \lambda)$, and gametes Ab and aB with frequencies $\frac{1}{2}\lambda$, whereas for the individuals in repulsion, Ab/aB, these frequencies are reversed.

To distinguish between genotypes which, although having the same genetic patterns, represent different structures so far as the locations of genes on the chromosome are concerned, Fisher (1947) used the term *segregating genotype*. Thus, AB/ab and Ab/aB are two different segregating genotypes.

Let us now find the number of segregating genotypes of n linked loci, the tth locus having s_t alleles. Each of the two homologous chromosomes can be made in

$$S = s_1 \cdot s_2 \cdot \cdots \cdot s_n \quad (3.18)$$

ways. There will be $\binom{S}{2}$ "heterozygous" pairs of chromosomes (i.e., a set of genes in one chromosome differing by at least one gene from the set of genes in its homologue) and S "homozygous" pairs of chromosomes. Therefore, the total number of segregating genotypes will be

$$\binom{S}{2} + S = \binom{S + 1}{2} = \frac{S(S + 1)}{2}. \quad (3.19)$$

In particular, when $s_1 = s_2 = \cdots = s_n = 2$, we have $S = 2^n$. Thus, for $n = 2$ linked loci, each with two alleles, we find the number of segregating genotypes to be $4 \times 5/2 = 10$, while the number of different genetic patterns is $3 \times 3 = 9$.

▼ 3.3.3. Polyploids

The evaluation of genotypes and phenotypes becomes more complicated when the organisms are polyploids. At a single locus, the number of genotypes depends on what hypothesis regarding the mode of forming gametes is used. For $2m$-ploids, considering a limiting case in which a gamete is formed by selecting a set of m homologous chromosomes, associated with this locus, at random (random chromosome segregation), it can be shown [e.g., Elandt-Johnson (1969)] that the number of different genotypes is

$$\binom{s + 2m - 1}{2m} = \binom{s + 2m - 1}{s - 1},$$

(3.20)

and the number of gametes is

$$\binom{s + m - 1}{m} = \binom{s + m - 1}{s - 1}.$$

(3.21)

For instance, for tetraploids, $2m = 4$, with two alleles, A and a, we have $\binom{5}{4} = \binom{5}{1} = 5$ genotypes: $AAAA$, $AAAa$, $AAaa$, $Aaaa$, $aaaa$, while the number of gametes is $\binom{3}{1} = 3$, that is, AA, Aa, aa.

For n independent loci, the tth locus having s_t alleles, the number of genotypes will be

$$\binom{s_1 + 2m - 1}{2m}\binom{s_2 + 2m - 1}{2m} \cdots \binom{s_n + 2m - 1}{2m} = \prod_{t=1}^{n}\binom{s_t + 2m - 1}{2m}.$$

(3.22)

For linked loci, the number of genotypic patterns will be given by (3.22), while the number of segregating genotypes will be

$$\binom{S + 2m - 1}{2m},$$

(3.23)

with S given by (3.18).

Several other combinatorial problems, such as the number of different gametes produced by a given genotype or the number of genotypes having exactly q different genes, are enumerated in a paper by Elandt-Johnson (1969).

3.4. EVALUATION OF PHENOTYPIC RATIOS IN THE OFFSPRING OF INTERCROSSES AND BACKCROSSES, USING GENERATING FUNCTION

We shall consider loci each of which has two alleles, one allele being dominant and the other recessive.

In experimental populations the homozygous populations AA and aa or $AABB$ and $aabb$, and so on, are called the *parental populations*, Pt_1 and Pt_2*, say, respectively. From crosses $Pt_1 \times Pt_2$ we obtain F_1 (F_1 is called the

---

* Geneticists usually use symbols P_1 and P_2 for parental populations. Since we will use P for binomial and multinomial proportions, to avoid confusion we have introduced Pt_1 and Pt_2 for parental populations.

first generation) in which all individuals are heterozygotes *Aa* or *AaBb*, etc. The crosses $F_1 \times F_1$, that is, *Aa* × *Aa* or *AaBb* × *AaBb*, etc., are called *intercrosses*, and their offspring is denoted by F_2 (*second generation*). The crosses $F_1 \times Pt_1$ and $F_1 \times Pt_2$ are called *backcrosses*. However, in practical applications, most use is made of $F_1 \times Pt_2$, that is, *Aa* × *aa* or *AaBb* × *aabb*, etc., and the term "backcrosses" generally applies to these. Another term for backcrosses of heterozygotes × recessives is *testcrosses*.

Let us consider *n* independent loci, and find the phenotypic ratios in the population being an offspring of intercross and backcross populations respectively, using generating functions.

Intercrosses. Let *D* denote a dominant and *R* a recessive at a single locus. The F_2 generation (with respect to *n* loci) will consist of 2^n phenotypes of the form $D^{n-r}R^r$, which represents a recessive with respect to *r* loci and a dominant with respect to the remaining $n - r$ loci ($r = 0, 1, \ldots, n$). In particular, R^n denotes a *n*-tuple recessive and D^n a *n*-tuple dominant. It is easy to see that there will be $\binom{n}{n-r} = \binom{n}{r}$ different phenotypic expressions of the type $D^{n-r}R^r$.

Then the function

$$(D + R)^n = \sum_{r=0}^{n} \binom{n}{r} D^{n-r}R^r \tag{3.24}$$

is the generating function for the numbers of different phenotypes, each of the form $D^{n-r}R^r$ ($r = 0, 1, \ldots, n$).

We know that the probabilities of occurrence of a dominant and a recessive, at a single locus, are $\Pr\{D\} = \frac{3}{4}$ and $\Pr\{R\} = \frac{1}{4}$, respectively. Let us consider the generating function

$$(\tfrac{3}{4}D + \tfrac{1}{4}R)^n = \sum_{r=0}^{n} \binom{n}{r} (\tfrac{3}{4}D)^{n-r}(\tfrac{1}{4}R)^r. \tag{3.25}$$

The coefficient of $D^{n-r}R^r$, that is, $\binom{n}{r} \times (\tfrac{3}{4})^{n-r}(\tfrac{1}{4})^r$, consists of two terms: the first term, $\binom{n}{r}$, indicates how many phenotypes with *r* recessive and $n - r$ dominant loci are possible, and the second term, $(\tfrac{3}{4})^{n-r}(\tfrac{1}{4})^r$, gives the probability of occurrence of each phenotype of the form $D^{n-r}R^r$, assuming that the *n* loci are independent.

Example 3.3. Let $n = 4$. There will be $2^4 = 16$ different gametes, $3^4 = 81$ different genotypes, and $2^4 = 16$ different phenotypes. Expansion (3.25) gives $(\tfrac{3}{4}D + \tfrac{1}{4}R)^4 = \tfrac{81}{256}D^4 + 4 \cdot \tfrac{27}{256}D^3R + 6 \cdot \tfrac{9}{256}D^2R + 4 \cdot \tfrac{3}{256}DR^3 + \tfrac{1}{256}R^4$. This means that there is one phenotype of type D^4, with probability 81/256; four phenotypes of form D^3R, each with probability 27/256; six phenotypes of form D^2R^2, each with probability 9/256; and so on. The phenotypic ratios in the offspring of intercrosses $F_1 \times F_1$ (i.e., in F_2) is

$$81:27:27:27:27:9:9:9:9:9:9:3:3:3:3:1.$$

It can be appreciated that the preparation of the checkerboard will be, even for $n = 4$, rather laborious.

Backcrosses. When loci with dominant and recessive genes are concerned, the only distinguishable phenotypes will be from testcrosses heterozygotes × recessives, that is, $F_1 \times Pt_2$. The generating function analogous to (3.25) will be

$$(\tfrac{1}{2}D + \tfrac{1}{2}R)^n = \sum_{r=0}^{n} \binom{n}{r}(\tfrac{1}{2}D)^{n-r}(\tfrac{1}{2}R)^r. \qquad (3.26)$$

It is easy to see that the phenotypic ratios in the offspring of backcrosses $F_1 \times Pt_2$ will be $1:1:1: \cdots :1$.

3.5. INDEPENDENCE, DEPENDENCE AND INTERACTION

These words are used in a number of contexts, and quite often the same word has a different meaning in different contexts. It seems necessary to explain more precisely how these terms are used in probability and statistics.

3.5.1. Independent and Dependent Frequencies of Events

In Chapter 2, we defined two events, E_1 and E_2, as independent if

$$\Pr\{E_1E_2\} = \Pr\{E_1\}\Pr\{E_2\}, \quad \text{with} \quad \Pr\{E_1\} > 0, \Pr\{E_2\} > 0. \qquad (3.27)$$

These events were called dependent if

$$\Pr\{E_1E_2\} \neq \Pr\{E_1\}\Pr\{E_2\}. \qquad (3.28)$$

We recall that formula (3.27) defines (statistical) independence and states the *frequency of occurrence* of E_1 does not influence the frequency of occurrence of E_2, and vice versa.

Example 3.4. Two loci located in two different chromosomes are called *independent* in the sense that the frequency of transmission of the genetic units at locus A is not influenced by the frequency of transmission of the genetic units at locus B.

Two linked loci are called *dependent* since their location in the same chromosome affects the segregation process, so that genes from different loci cannot separate randomly.

3.5.2. Independent and Dependent Phenotypic Effects of Gene Action. Causal Dependence

The concepts of "gene" and "genotype" are, in some sense, abstract. For instance, the statement "An individual has a genotype $AaBb$" determines the event that the genes A, a, B, b are present in the cells of this individual.

The genes themselves are not observed, but "purely" genetic characters and traits are the results of *gene action*. Therefore, as long as we cannot determine a phenotypic expression for a given genotype, the term "genotype" does not mean much more than a "set of genes."

A geneticist, of course, is interested in what the genotype expresses, that is, what the phenotype is. Phenotypic expression is always a function of gene action; in other words, the phenotype always depends on the genotype, although the functional relationship is not always simple. Let us discuss a few examples.

Example 3.5.   (*a*) The heredity of flower color in the sweet pea is *intermediate*, that is, BB corresponds to red, Bb to pink, and bb to white color of the flowers. Phenotypic expression of pea flower color is a *function of the number of B alleles* present in the zygote.

(*b*) The yellow color of pea seeds is a dominant, and the green color is a recessive. If g (green) denotes the recessive and G (yellow) the dominant allele, then GG and Gg denote yellow and gg denotes green seeds. The phenotypic expression of seed color in the pea is a *function of the presence* (or absence) of the *G allele*.

Example 3.6.   Let G and g be the alleles for yellow and green color, respectively, and W and w the alleles (at a different locus) for round and wrinkled shape of seeds in the pea. It is known from classical (Mendel's) experiments that these loci are located in different chromosomes (are independent). Also the color of the seeds has no influence on their shape—the seeds can be wrinkled or round, independently of whether they are yellow or green.

Here we have a situation in which two *loci are independent and the phenotypic characters controlled by these loci are also independent.* We notice that the phenotypic ratios in F_2 are $9:3:3:1$.

Example 3.7.   Two independent loci, each with two alleles, R, r and P, p, respectively, control the shape of the comb in the chicken. From intercross

RrPp × *RrPp* the following results are obtained in F_2:

Phenotype:	*R–P–*	*R–pp*	*rrP–*	*rrpp*
Shape of the comb:	Walnut	Rose	Pea	Single
Ratio:	9 :	3 :	3 :	1.

The phenotypic ratios are still, as in Example 3.6, 9:3:3:1, but here only one trait (comb shape) appears and its appearance depends on what combinations of genes occur at both loci.

Now, the *loci are independent but the phenotype is the result of the joint action of both loci.*

Example 3.8. Hemophilia, in which the blood fails to clot normally, and red-green color blindness are two X-linked recessive traits, each controlled by a different locus. Since these loci are linked, the *transmission of these two characters is not an independent process, although the traits themselves are causally independent.*

Example 3.9. In Problem 2.5 we considered the intensity of the red color of kernels in wheat. It has been assumed that the intensity is controlled by two loci, each with two alleles, R_1, r_1 and R_2, r_2, respectively. The loci are located in different chromosomes, and the intensity of the red color is an *additive* function of the number of R genes. The color changes are from dark red $(R_1R_1R_2R_2)$ to white $(r_1r_1r_2r_2)$. Here, the *loci are independent in their transmission of heredity units, but the phenotypic expression is a function of the number of R genes at both loci.*

3.5.3. Dependence and Interaction

Phenotype is a result of, or depends on, gene *action* or gene *interaction* at one locus or several loci. In everyday language the words "dependence" and "interaction" are quite often used interchangeably. In statistics, however, "interaction" has a special meaning.

Let $G_1, G_2, \ldots, G_n$ represent n genes, not necessarily at the same locus. Suppose that the phenotypic effects of these genes can be measured in the same units. Let α_i be the value of the effect of the gene G_i. If the total effect of genes $G_1, G_2, \ldots, G_n$ is the sum of the effects of these genes, that is, can be expressed in the form

$$\alpha_1 + \alpha_2 + \cdots + \alpha_n \tag{3.29}$$

or even more generally in the form

$$k_1\alpha_1 + k_2\alpha_2 + \cdots + k_n\alpha_n, \tag{3.29a}$$

then we say that the effects of these genes are *additive*. In such cases we say that there is *no interaction*. In Example 3.5 we have described a situation in which the phenotypic expression was dependent on the number of allelic genes present in genotype (color of pea flowers), and in Example 3.8 color of wheat kernels was the function of the number of "color genes" present at two loci; in both cases the gene effects were additive—there was no interaction.

DEFINITION 3.5.   *If the action of two of more genes produces nonadditive effects, the departures from additivity are called interaction among these genes.*

In Example 3.6 the pea seed of genotypes *GG* and *Gg* were yellow, while of *gg*, green. One or two genes *G* in a genotype did not affect the phenotype. The interaction took place—we observed the effect of dominance.

DEFINITION 3.6.   *Dominance is a form of interaction of allelic genes.*

In Example 3.7 the phenotypic effect of nonallelic genes (from two loci) was such that the shapes of the combs of chicken were different depending on whether *R* and *P*, only *R*, only *P*, neither *R*, nor *P* were present in the genotype. We observed the interaction of genes from different loci.

DEFINITION 3.7.   *The interaction of nonallelic genes in the phenotypic expression of a genotype is called epistasis.*

When a gene at one locus suppresses or inhibits the expression of a gene at another locus, the first gene is said to be *epistatic* to the second.
Several forms of epistasis will be discussed in section 3.6.
From these examples and discussion we conclude:

1. Phenotype depends always on genotype, but the phenotypic expression is independent of the mode and frequency of transmission of the heredity units.

2. If the genes act in such a way that the phenotype is a result of the additive effects of genes, we say that there is no interaction. Interaction takes place when the gene effects are nonadditive. We have seen here that dominance is a phenotypic expression of the interaction of allelic genes, and epistasis an expression of the interaction of nonallelic genes.

3.5.4.   The Role of Genetic and Environmental Factors in Phenotypic Expression

So far, we have discussed only the relationship between genotype and phenotype, without taking into account the influence of environment. It is well known that almost always gene action is conditioned not only by internal but also by external circumstances. For instance, the production of chlorophyll in plants is genetically controlled, but the characteristic green color of chlorophyll can appear only in cooperation with sunlight. Several traits in plants and animals do not appear until a certain temperature is reached.

In such cases, genes (or genotypes) determine the potentialities of a living organism; collaboration (interaction?) between genetic and environmental factors is necessary for the realization of the phenotype.

It often is assumed that several continuous characters, such as weight or height, are results of additive effects of many genes from different loci. But two plants of the same genotypes, growing in relatively uniform conditions, may still show small differences in their heights and weights. In fact, several small events which are out of control can occur in the environment and cause small changes, called *random errors*, in the values of the characters.

Sometimes the external factors act in a similar way to the genetic ones. For example, cretinism can be due to the presence of certain genes or to a lack of iodine in the diet of a child (whatever his genotype). The term *phenocopy* has been used in such cases. It describes an individual whose phenotype, because of the action of nongenic factors, becomes like the phenotype normally caused by specific genes. It is sometimes difficult to distinguish a phenocopy from the genetically determined phenotype.

As can be seen from this brief discussion, environment can markedly change phenotypic expression. One might wonder, therefore, how the genetic properties of individuals can be studied and how sure we can be that our conclusions are correct.

Several traits, especially qualitative ones, are not sensitive to small changes in environmental conditions. Hemophilia, for example, will occur in genetically affected individuals irrespective of the conditions in which they live. On the other hand, insufficient observation of a trait may lead to incorrect conclusions. Hence well-planned experiments, combined with statistical analysis, are useful tools in the study of both genetic and nongenetic factors influencing the phenotype.

Since in this book we are concerned mostly with qualitative traits, genetic rather than environmental factors will be analyzed. By contrast, models in quantitative genetics must include environmental factors as well as genetic ones. The main aim of quantitative genetics is to establish what part of the variability observed is due to heredity and what part to changes in external conditions. In qualitative genetics, the mode of inheritance plays the more important role.

3.6. EPISTASIS

In section 3.5 we defined epistasis as an interaction of nonallelic genes in the phenotypic expression. We will give here a few examples [most of them taken from the book *General Genetics*, by Srb and Owen (1953)] of epistasis among two loci.

Example 3.10. Let C stands for color and c for lack of color of onion bulbs. Genotype CC or Cc will produce colored onion bulbs, and cc will produce white bulbs. Another pair of alleles, R and r, at a locus R independent of C, is responsible for red color (allele R) and for yellow color (allele r) of bulbs. R is dominant over r. The genes R, r can express themselves only in the presence of C.

The phenotypic ratios in F_2 from the intercrosses $CcRr \times CcRr$ are

$$C\text{–}R\text{–} \quad C\text{–}rr \quad \overbrace{ccR\text{–} \quad ccrr}$$

$$\text{Red} \quad \text{Yellow} \quad \text{White}$$
$$9 \quad : \quad 3 \quad : \quad 4 \quad .$$

This kind of epistasis is sometimes called *recessive epistasis*, since the recessive gene c, responsible for lack of color (white), suppresses the expression of the color genes, R and r, from another (independent) locus.

It is easy to show that the corresponding phenotypic ratios in the offspring of the backcrosses $CcRr \times ccrr$ are

$$\text{Red}:\text{Yellow}:\text{White} = 1:1:2.$$

Example 3.11. Another gene, at another independent locus, which inhibits the color in onion is a *dominant I* ("Inhibitor"). Thus in offspring of the intercrosses $IiRr \times IiRr$ we obtain

$$I\text{–}R\text{–} \quad I\text{–}rr \quad iiR\text{–} \quad iirr$$
$$\text{White (9)} \quad \text{White (3)} \quad \text{Red (3)} \quad \text{Yellow (1)},$$

that is, White:Red:Yellow = 12:3:1.

This is sometimes called *dominant epistasis*, since a dominant gene I suppresses the effect of genes R and r (here a red or yellow color) from another locus.

For the offspring of the backcrosses $IiRr \times iirr$, we have

$$\text{White}:\text{Red}:\text{Yellow} = 2:1:1.$$

It is worthwhile noticing that the phenotypic ratios in the offspring of backcrosses are the same whether recessive or dominant epistasis occurs.

Example 3.12. It is easy to show that in the offspring of the intercrosses $IiCc \times IiCc$, we obtain

$$I\text{–}C\text{–} \quad I\text{–}cc \quad iiC\text{–} \quad iicc$$
$$\text{White (9)} \quad \text{White (3)} \quad \text{Colored (3)} \quad \text{White (1)},$$

that is, the phenotypic ratio White:Colored = 13:1. This is sometimes called *mutual epistasis* since the dominant I inhibits the expression of the dominant C, and the recessive cc inhibits the expression of the recessive ii. In the offspring of the backcross of $IiCc \times iicc$, the phenotypic ratio is White:Colored $= 3:1$.

Example 3.13. There are some examples, especially in genetics of plants, in which a character does not (or, may be, does) occur if the individual is recessive with respect to one or another or both loci. In corn plants two dominant alleles, W_1 and W_2, must be both present at each of two independent loci for production of chlorophyll. Single recessives (recessive monohybrids) W_1-w_2w_2 or $w_1w_1W_2$-, as well as double recessive $w_1w_1w_2w_2$, often produce white seedlings. From the selfpollination of $W_1w_1W_2w_2$ (which is equivalent to the mating $W_1w_1W_2w_2 \times W_1w_1W_2w_2$) the resulting offspring is

$$W_1\text{-}W_2\text{-} \quad \underbrace{W_1\text{-}w_2w_2 \quad w_1w_1W_2\text{-} \quad w_1w_1w_2w_2}$$

$$\text{Green (9)} \qquad \qquad \text{White (7)}$$

or

$$\text{Green (9)} \qquad \text{Greenish (6)} \qquad \text{White (1)}.$$

The phenotypic ratio is Green:White $= 9:7$. This is sometimes called a *complementary* or *recessive duplicate* epistasis. Here the recessive w_2w_2 suppresses the effect of W_1 (or w_1), and vice versa, the recessive w_1w_1 suppresses the effect of W_2 (or w_2).

In fact the single dominants, W_1-w_2w_2 and W_2-w_1w_1, are slightly greenish. In other words, the full effect (green color) is produced when both dominants, W_1 and W_2, are present, that is by phenotypes W_1-W_2-. Genotypes with one dominant gene, W_1 or W_2, but not both, i.e., phenotypes W_1-w_2w_2 or W_2-w_1w_1, produce intermediate effect (greenish color), while double recessives, $w_1w_1w_2w_2$, produce no color. The phenotypic ratios are $9:6:1$; this is sometimes called *incomplete duplicate epistasis*.

Sometimes a character expresses itself only as a double recessive $aabb$. Therefore, in the offspring of $AaBb \times AaBb$ only two different phenotypes will be observed in the ratio $15:1$. This is called *duplicate dominant* epistasis. (A is epistatic to B or b, and B is epistatic to A or a.)

It will be shown in Chapter 16 that in naturally occurring populations (e.g., human populations), when the gene frequencies are unknown, the models of single and double recessive cannot be distinguished unless an especially designed experiment is performed.

SUMMARY

The subject matter of this chapter has been the evaluation of the number of different genotypes and phenotypes and their ratios in the offspring of intercross and backcross experimental (laboratory) populations. Basic definitions and formulae from combinatorial mathematics and the concept of generating functions have been introduced and applied to these problems. Also the meanings and definitions of terms: independence, dependence and interaction have been discussed, using several examples. A separate section has been devoted to the discussion on interaction of nonallelic genes called epistasis.

REFERENCES

Elandt-Johnson, Regina C., Application of occupancy and ordering theory in genetics of autopolyploids, *Zastosowania Matematyki* (1969), 193–204.

Feller, W., *An Introduction to Probability Theory and Its Applications*, John Wiley & Sons, New York, 1950.

Fisher, R. A., The theory of linkage in polysomic inheritance, *Phil. Trans. Roy. Soc. B* **233** (1947) 55–87.

Li, C. C., *Human Genetics*, McGraw-Hill Book Company, Inc., New York, 1961.

Srb, A. M., and Owen, R. D., *General Genetics*, W. H. Freeman & Co., San Francisco, 1953.

Stern, C., *Principles of Human Genetics*, W. H. Freeman & Co., San Francisco, 1960.

PROBLEMS

3.1. (a) Prove that

$$\binom{a}{0}\binom{b}{n} + \binom{a}{1}\binom{b}{n-1} + \binom{a}{2}\binom{b}{n-2} + \cdots + \binom{a}{n}\binom{b}{0} = \binom{a+b}{n}$$

for $n \leqslant \min(a, b)$.

(b) Using the results of (a), show that $\displaystyle\sum_{r=0}^{n}\binom{n}{r}^{2} = \binom{2n}{n}$;

(c) $\displaystyle\binom{n}{0} + \frac{1}{2}\binom{n}{1} + \frac{1}{3}\binom{n}{2} + \cdots + \frac{1}{n+1}\binom{n}{n} = \frac{2^{n+1} - 1}{n+1}$;

(d) $\displaystyle\binom{n}{n} + \binom{n+1}{n} + \binom{n+2}{n} + \cdots + \binom{n+m}{n} = \binom{n+m+1}{n+1}$.

3.2. Prove that

(a) $\displaystyle 1\binom{n}{1} + 2\binom{n}{2} + \cdots + n\binom{n}{n} = \sum_{r=1}^{n} r\binom{n}{r} = n2^{n-1}$;

(b) $\displaystyle 1^{2}\binom{n}{1} + 2^{2}\binom{n}{2} + \cdots + n^{2}\binom{n}{n} = \sum_{r=1}^{n} r^{2}\binom{n}{r} = n(n+1)2^{n-2}$.

3.3. It is usually assumed that albinism is a rare recessive trait. The albino offspring aa can be obtained from normal heterozygous parents, $Aa \times Aa$, from mating in which one parent is normal and the other albino, $Aa \times aa$, and from two albino parents $aa \times aa$. The matings $aa \times aa$ are very rare, so that analyses have been based mostly on the matings normal × normal and normal × albino.

It was observed, however, that in one English family with two albino parents all three children were nonalbino [Stern (1960), p. 106]. If we exclude mutation or the influence of environmental conditions, which of these three hypotheses appears to be most reasonable?

(*a*) Albinism is a simple dominant.

(*b*) Albinism is controlled by two independent loci with mutual epistasis.

(*c*) Albinism is controlled by two independent loci with complementary epistasis.

Could you distinguish between these hypotheses if certain data from a large number of observations in family studies were available? What segregation results could exclude the hypothesis (*b*)?

3.4. The colored (purple or red) grain in maize is due to the presence of aleurone and it has been suggested that it is controlled by four independent loci (Emerson, *Amer. Nat.* *46* (1912), p. 612). The dominant genes at each locus were defined as follows.

C is a gene for color; *C* must be present in order that any color develops;

R is a gene for production of red aleurone; the presence of *C* and *R* exhibits red color of grain;

P is a gene for purple, it is only effective in presence of *C* and *R*; thus *CRP* exhibits purple color, *CRp* gives red color, while other possible combinations exhibit white grain;

I is inhibitor of color development; this means that only *CPRi* gives purple and only *CRi* gives red.

(*a*) Show that the offspring of *CcRrPpIi* × *CcRrPpIi* (i.e., F_2 generation) has the phenotypic ratios 27 purple:9 red:220 white. Find the phenotypic ratios in the offspring of: (*b*) *CcRRPpii* × *CcRRPpii*; (*c*) *CcRRPpIi* × *CcRRPpIi*; (*d*) *CCRrPpIi* × *CCRrPpIi*.

3.5. Find the phenotypic ratios in the offspring of the backcrosses (testcrosses) when the trait is controlled by two loci with: (*a*) complementary (duplicate recessive) epistasis; (*b*) duplicate dominant epistasis; (*c*) incomplete duplicate epistasis.

3.6. Find the phenotypic ratios in F_2 generation for:

(*a*) complementary epistasis;

(*b*) mutual epistasis;

assuming that the loci are *linked* with recombination fraction λ and the parental matings are:

   (i) *AB/ab* × *AB/ab* (coupling × coupling),

   (ii) *AB/ab* × *Ab/aB* (coupling × repulsion),

   (iii) *Ab/aB* × *Ab/aB* (repulsion × repulsion).

(*Hint:* Use the results of Problem **2.8**.)

CHAPTER 4

Equilibrium Laws in Panmictic Populations

4.1. CONCEPT OF GENE FREQUENCY

We first restrict ourselves to genes located in the autosomes.

Let us consider a single locus with two alleles, A and a. Suppose that a *finite* population of N individuals consists of N_{11} individuals of genotype AA, $2N_{12}$ of genotype Aa, and N_{22} of genotype aa, with $N_{11} + 2N_{12} + N_{22} = N$. The total number of genes is $2N$; the number of A genes is $2N_{11} + 2N_{12}$. Thus the gene A appears in the proportion

$$p = \frac{2N_{11} + 2N_{12}}{2N} = \frac{N_{11}}{N} + \frac{N_{12}}{N}. \tag{4.1a}$$

Similarly, the gene a appears in the proportion

$$q = \frac{2N_{22} + 2N_{12}}{2N} = \frac{N_{22}}{N} + \frac{N_{12}}{N}. \tag{4.1b}$$

Note that $p + q = 1$.

The values p and q are the *relative frequencies* of genes A and a, respectively, but it is customary to call them briefly *gene frequencies*.

Gene frequencies can be evaluated from (relative) genotype frequencies. Putting $N_{11}/N = H_{11}$, $N_{12}/N = H_{12}$, and $N_{22}/N = H_{22}$, with $H_{11} + 2H_{12} + H_{22} = 1$, we obtain the gene frequencies in the form

$$p = H_{11} + H_{12}, \quad q = H_{22} + H_{12}, \quad \text{with } p + q = 1. \tag{4.2}$$

In this chapter, we want to derive some general probability laws for randomly mating populations. Since we have defined probabilities as limits of relative frequencies in an infinite series of independent repetitions (see section 2.2.3), we shall have in mind *infinite* (or hypothetical) populations for which these laws will hold exactly. But they will also be approximately valid if populations are very large. In our further discussion, by "population" we will understand an infinite population, so that the word "infinite" will be omitted unless ambiguity arises. Thus, if h_{11}, $2h_{12}$, h_{22}, with $h_{11} + 2h_{12} + h_{22} = 1$, represent the probability distribution of genotypes AA, Aa, aa, respectively, the gene frequencies are calculated from the formulae

$$p = h_{11} + h_{12}, \quad q = h_{22} + h_{12}. \tag{4.3}$$

Now p and q are the *probabilities* of occurrence of genes A and a, respectively, in the population, and since $p + q = 1$, they represent the *gene distribution* with respect to locus A. The gene distribution can be written in the form $pA + qa$, which is sometimes called a *gene array*.

We can easily extend our definition of gene frequencies to a locus with multiple alleles. Let $A_1, A_2, \ldots, A_s$ be the s alleles at locus A. Let

$$\Pr\{A_iA_i\} = h_{ii} \quad \text{and} \quad \Pr\{A_iA_j\} = 2h_{ij}, \tag{4.4}$$

for $i, j = 1, 2, \ldots, s$, $(i \neq j)$, with $\sum\sum h_{ij} = 1$, represent the genotype distribution in a certain population. The frequency, p_i, of gene A_i can be evaluated from the frequencies of genotypes involving A_i. Thus

$$\Pr\{A_i\} = p_i = \tfrac{1}{2}\left(2h_{ii} + 2\sum_{j \neq i}^{s} h_{ij}\right) = \sum_{j=1}^{s} h_{ij}, \tag{4.5}$$

where $i = 1, 2, \ldots, s$, with $\sum p_i = 1$.

Example 4.1. Suppose that $s = 3$ and that genotype distribution in a population is as follows:

Genotype:	A_1A_1	A_2A_2	A_3A_3	A_1A_2	A_1A_3	A_2A_3
Relative	h_{11}	h_{22}	h_{33}	$2h_{12}$	$2h_{13}$	$2h_{23}$
frequency:	0.15	0	0.20	0.25	0.35	0.05.

The gene frequencies are

$$\Pr\{A_1\} = p_1 = h_{11} + h_{12} + h_{13} = \frac{2 \times 0.15 + 0.25 + 0.35}{2} = 0.45;$$

$$\Pr\{A_2\} = p_2 = h_{22} + h_{12} + h_{23} = \frac{0.25 + 0.05}{2} = 0.15;$$

$$\Pr\{A_3\} = p_3 = h_{33} + h_{13} + h_{23} = \frac{2 \times 0.20 + 0.35 + 0.05}{2} = 0.40.$$

Suppose that there are r loci, the tth locus with s_t alleles. The frequency of the ith gene at the tth locus can be evaluated from the genotype distribution by multiplying the relative frequency of a genotype (called also simply the *genotype frequency*) by 2, 1, or 0, depending on whether the genotype includes two, one, or none of the genes in question, adding these values, and dividing the sum by 2. We notice that the sum of gene frequencies at each locus is equal to 1.

Example 4.2. Let us consider two loci, each with two alleles, and genotype frequencies as below:

$AABB$	$AABb$	$AAbb$	$AaBB$	$AaBb$	$Aabb$	$aaBB$	$aaBb$	$aabb$
h_1	h_2	h_3	h_4	h_5	h_6	h_7	h_8	h_9,

with $\sum h_k = 1$. We have

$$\Pr\{A\} = \tfrac{1}{2}(2h_1 + 2h_2 + 2h_3 + h_4 + h_5 + h_6) = p_A,$$

$$\Pr\{a\} = \tfrac{1}{2}(h_4 + h_5 + h_6 + 2h_7 + 2h_8 + 2h_9) = p_a,$$

$$\Pr\{B\} = \tfrac{1}{2}(2h_1 + 2h_4 + 2h_7 + h_2 + h_5 + h_8) = p_B,$$

$$\Pr\{b\} = \tfrac{1}{2}(h_2 + h_5 + h_8 + 2h_3 + 2h_6 + 2h_9) = p_b.$$

We also have $p_A + p_a = 1$, and $p_B + p_b = 1$.

4.2. RANDOM MATING

It is generally understood that a bisexual population mates at random if any individual of one sex has the same chance to mate with any individual of the opposite sex, within a given population.

We can regard a population as a collection of genotypes with a certain probability distribution, and define random mating in terms of these probabilities.

Let $\Gamma_1{}^{\female}, \Gamma_2{}^{\female}, \ldots, \Gamma_f{}^{\female}$ be the f distinct female genotypes (in general, with respect to several loci), with the genotype distribution

$$\Pr\{\Gamma_1{}^{\female}\}, \Pr\{\Gamma_2{}^{\female}\}, \ldots, \Pr\{\Gamma_f{}^{\female}\}, \left(\sum_{k=1}^{f}\Pr\{\Gamma_k{}^{\female}\} = 1\right),$$

and similarly, $\Gamma_1{}^{\male}, \Gamma_2{}^{\male}, \ldots, \Gamma_m{}^{\male}$ be the genotypes of males with the distribution

$$\Pr\{\Gamma_1{}^{\male}\}, \Pr\{\Gamma_2{}^{\male}\}, \ldots, \Pr\{\Gamma_m{}^{\male}\}, \left(\sum_{l=1}^{m}\Pr\{\Gamma_l{}^{\male}\} = 1\right),$$

in a certain population. In the general case $f \neq m$.

DEFINITION 4.1. *Random mating takes place when the probability of crosses between any two genotypes, $\Gamma_k{}^{\female} \times \Gamma_l{}^{\male}$, is equal to the probability that a female and a male selected at random will have genotypes $\Gamma_k{}^{\female}$ and $\Gamma_l{}^{\male}$, respectively, that is*

$$\Pr\{\Gamma_k{}^{\female} \times \Gamma_l{}^{\male}\} = \Pr\{\Gamma_k{}^{\female}\}\Pr\{\Gamma_l{}^{\male}\}, \tag{4.6}$$

where $k = 1, 2, \ldots, f; l = 1, 2, \ldots, m$.

The joint distribution of pairs $(\Gamma_k\female \times \Gamma_i\male)$ is obtained as the product of the marginal distributions for females and males (compare also Problem 2.4).

We now express Definition 4.1 in another, sometimes more convenient, form using concepts of gametes instead of genotypes.

Let $\gamma_1\female, \ldots, \gamma_{v_f}\female$ be the maternal gametic output with the gamete distribution $\Pr\{\gamma_1\female\}, \ldots, \Pr\{\gamma_{v_f}\female\}$, and similarly, $\gamma_1\male, \ldots, \gamma_{vm}\male$ be the paternal gametic output with the distribution $\Pr\{\gamma_1\male\}, \ldots, \Pr\{\gamma_{vm}\male\}$, for the next generation. Generally $v_f \neq v_m$. Let $\gamma_{ij} = \gamma_i\female\gamma_j\male$ be a zygote obtained by union of gametes $\gamma_i\female$ and $\gamma_j\male$.

DEFINITION 4.1a. *Random mating is a system in which the probability of occurrence of a zygote* $\Gamma_{ij} = \gamma_i\female\gamma_j\male$ *is equal to the probability of selecting at random the gametes* $\gamma_i\female$ *and* $\gamma_j\male$ *from the gamete pool of the population, that is*

$$\Pr\{\Gamma_{ij}\} = \Pr\{\gamma_i\female\gamma_j\male\} = \Pr\{\gamma_i\female\}\Pr\{\gamma_j\male\}, \tag{4.7}$$

where $i = 1, 2, \ldots, v_f, j = 1, 2, \ldots, v_m$.

Both definitions are equivalent in the sense that the genotype distribution derived for the next generation is the same whether we use (4.6) or (4.7) as the definition of random mating. (For proof, see Kempthorne (1957), Chapter 2.)

4.2.1. Panmictic Populations

Definition 4.1 (or 4.1a) is a very broad definition of random mating. What in several practical situations is understood by random mating requires four additional assumptions:

> (i) The fertilities and the survival abilities of each genotype and its gametes are the same for each individual in the population, and they remain the same through generations. In other words, the reproductive abilities (called fitnesses) of different genotypes and matings are the same and are not influenced by other forces such as selection, mutation, or migration. (This implies that, on the average, all mating types produce the same number of offspring.)   (4.8)
> (ii) The ratio of females to maels is $1:1$.
> (iii) The generations are nonoverlapping.
> (iv) The maternal and paternal contributions to the genotype (with respect to autosomal loci) are the same (i.e., $\gamma_i\female\gamma_j\male = \gamma_j\female\gamma_i\male$).

If assumptions (4.8) hold, the term *panmixia* for random mating has been suggested. Populations mating at random and satisfying these conditions are called *panmictic* populations. However, these terms are not very commonly used, and in this book we shall continue to use "random mating" for the special situation when assumptions (4.8) hold, at least approximately. In Chapter 10, we will consider randomly mating populations for which assumption (i) does not hold.

Some difficulties are encountered in the interpretation of random mating in human populations.

(*a*) Random mating, by definition, assigns to each individual the same chance of being the partner of a given mate. Are, then, consanguineous marriages included in a random mating? Theoretically, yes, they are. But since we consider "infinite" populations, the chance that two related individuals will mate is practically almost zero. Thus, for large populations, it is usually understood that random mating is equivalent to matings of *unrelated* partners (compare section 9.9). On the other hand, if a population is of small size, random mating will give rise to some inbreeding (see section 9.12).

(*b*) It is hard to believe that individuals mate purely by chance. A partner is usually selected on the basis of several intellectual or physical criteria. Since it is impossible to eliminate these factors, it may happen that theoretical results based on the assumption of random mating are not always in agreement with practical observations.

(*c*) To make use of formula (4.7) in evaluating the genetic structures of the next generations under random mating, we usually assume that the generations are *nonoverlapping*, that is, we consider discrete models with a "jump" of one generation. This can be true for some plant populations but not for human ones. It appears, however, that in several problems lack of this assumption has only a minor effect on the general result, so it is often neglected.

In our book, we shall discuss only models for discrete generations; continuous models with overlapping generations require more advanced mathematical theory.

▼4.3. MATRIX THEORY OF RANDOM MATING

It will be convenient to introduce some definitions and notations which can express, in an abbreviated form, the gamete, genotype, and phenotype distributions and also some other distributions which arise in the theory of random mating in large populations. Matrix algebra is the appropriate tool for this purpose. The reader who is not familiar with this subject is referred to Appendix I.

Suppose that there are v different kinds of gametes in question, which we denote by $\gamma_1, \gamma_2, \ldots, \gamma_v$. This can be written in the form of a $1 \times v$ row vector called, in this case, a *gamete vector*, that is,

$$\gamma' = (\gamma_1, \gamma_2, \ldots, \gamma_v). \qquad (4.9)$$

The symbol γ_k for gamete has here a very general meaning. It will denote any combination of genes with respect to several loci in diploid or polyploid organisms. The vector γ represents the *gametic output* in the population. We first restrict ourselves to genes located in *autosomes*.

Let Γ_{ij} denote the genotype obtained by union of female gamete γ_i and male gamete γ_j, that is,

$$\Gamma_{ij} = \gamma_i \gamma_j.$$

(Note that we now omit the symbols ♀ and ♂.) Taking into account the maternal and paternal gametic contributions to the genotype of offspring, we can present the sample space for the genotypes, Γ_{ij}, of the offspring in the form of a $v \times v$ table Γ, that is,

$$\Gamma = \begin{bmatrix} \Gamma_{11} & \Gamma_{12} & \cdots & \Gamma_{1v} \\ \Gamma_{21} & \Gamma_{22} & \cdots & \Gamma_{2v} \\ \cdots\cdots\cdots\cdots\cdots \\ \Gamma_{v1} & \Gamma_{v2} & \cdots & \Gamma_{vv} \end{bmatrix} = (\Gamma_{ij}). \qquad (4.10)$$

We notice that (4.10) is a $v \times v$ square matrix resulting from multiplication of maternal and paternal gametic outputs, that is,

$$\Gamma = \gamma\gamma'. \qquad (4.10a)$$

We will call the matrix Γ the *genotype matrix*.

Some of the genotypes may not be distinguishable from each other as far as their genetic composition is concerned. The problem of determining the actual number of distinct genotypes was discussed in Chapter 3. The sample space given by (4.10) is a convenient form of presentation and, as was pointed out in Chapter 2, it also is fairly general since it takes into account the maternal and paternal gametic contributions.

For convenience and without loss of generality we now assume that the gamete distributions for females and males are the same. Let the $1 \times v$ vector

$$g' = (g_1, g_2, \ldots, g_v), \qquad (4.11)$$

with $\Sigma g_i = 1$, represent the probabilities (called briefly frequencies) of the gametic output in the population $\mathbf{\Gamma}$. Let g_{ij} denote the probability of occurrence (called briefly frequency) of the genotype Γ_{ij} in $\mathbf{\Gamma}$. Assuming that $\mathbf{\Gamma}$ results from random mating, we have

$$g_{ij} = g_i g_j, \quad i, j = 1, 2, \ldots, v. \tag{4.12}$$

Thus under random mating the $v \times v$ matrix

$$\mathbf{\Gamma} = \mathbf{gg'} = \begin{bmatrix} g_1^2 & g_1 g_2 & \cdots & g_1 g_v \\ g_2 g_1 & g_2^2 & \cdots & g_2 g_v \\ \cdots\cdots\cdots\cdots\cdots \\ g_v g_1 & g_v g_2 & \cdots & g_v^2 \end{bmatrix} = (g_{ij}) = (g_i g_j), \tag{4.13}$$

with $\Sigma\Sigma g_i g_j = 1$, represents the distribution of the genotypes given in (4.10). We notice that the matrices $\mathbf{\Gamma}$ and $\mathbf{G}$, jointly, represent the probability space for a randomly mating population with gametic output $\mathbf{\gamma}$ and a corresponding gametic probability vector $\mathbf{g}$. The expression

$$\mathbf{g'\gamma} = \sum_{i=1}^{v} g_i \gamma_i \tag{4.14}$$

is called an *array of the gametic output* for the population $\mathbf{\Gamma}$, and the quadratic form

$$Q(\mathbf{\gamma}) = \mathbf{\gamma'G\gamma} = \sum_{i=1}^{v} \sum_{j=1}^{v} g_{ij} \gamma_i \gamma_j = \sum_{i=1}^{v} \sum_{j=1}^{v} g_{ij} \Gamma_{ij} \tag{4.15}$$

is called the *genotype array* of the population $\mathbf{\Gamma}$. In particular, when the population $\mathbf{\Gamma}$ is the result of random mating, its genotype array can be expressed as the square of the gametic output array, that is,

$$Q(\mathbf{\gamma}) = (\mathbf{g\gamma})'(\mathbf{g\gamma}) = \sum_{i=1}^{v} \sum_{j=1}^{v} g_i g_j \gamma_i \gamma_j = \left(\sum_{i=1}^{v} g_i \gamma_i \right)^2. \tag{4.15a}$$

Gamete and genotype arrays are different ways of representing gamete and genotype probability spaces, respectively.

If we want to take into account only *distinct* genotypes, the frequencies of indistinguishable genotypes can be combined and the genotype array rearranged appropriately. For instance, the genotype array with respect to two loci each with two alleles, and with frequencies as given in Example 4.2, can be written as $h_1 AABB + h_2 AABb + \cdots + h_9 aabb$. It includes only nine instead of 16 terms.

The genotype Γ_{ij} can produce a certain number of gametes among which some are different from both parental germ cells, γ_i and γ_j. For example, from the union of gametes AB and ab we obtain a genotype $ABab$. Here $\gamma_i = AB, \gamma_j = ab, \Gamma_{ij} = ABab$. The genotype $ABab$, besides gametes AB, ab, can also produce gametes Ab (γ_k, say) and aB (γ_l, say), which are different from both maternal and paternal gametes. In this example, the conditional probabilities of producing any of these gametes given that the genotype is *ABab* and loci are independent, are the same, equal to $\frac{1}{4}$. This will not be true when loci are linked.

Generally, let

$$\Pr\{\gamma_k \mid \Gamma_{ij}\} = c_{k(ij)} \tag{4.16}$$

be the conditional probability that the genotype $\Gamma_{ij} = \gamma_i \gamma_j$ produces the gamete γ_k. We shall call $c_{k(ij)}$ the *segregation probability for the gamete* γ_k in the genotype Γ_{ij}. The $v \times v$ matrix

$$\mathbf{C}_k = \begin{bmatrix} c_{k(11)} & c_{k(12)} & \cdots & c_{k(1v)} \\ c_{k(21)} & c_{k(22)} & \cdots & c_{k(2v)} \\ \cdots\cdots\cdots\cdots\cdots\cdots \\ c_{k(v1)} & c_{k(v2)} & \cdots & c_{k(vv)} \end{bmatrix} = (c_{k(ij)}) \tag{4.17}$$

represents the segregation probabilities for the gamete γ_k in the whole population. In the general case it is possible that $c_{k(ij)} \neq c_{k(ji)}$, but in most situations we have $c_{k(ij)} = c_{k(ji)}$, so that the matrix $\mathbf{C}_k$ is symmetrical.

The $v \times v^2$ matrix

$$\mathbf{C} = (\mathbf{C}_1, \mathbf{C}_2, \ldots, \mathbf{C}_v) \tag{4.18}$$

represents the segregation probabilities for all gametes in the whole population (the *segregation matrix for gametes*).

We can consider the probabilities $c_{k(ij)}$ from another point of view. The $1 \times v$ vector

$$\mathbf{c}'_{(ij)} = (c_{1(ij)}, c_{2(ij)}, \ldots, c_{v(ij)}) \tag{4.19}$$

represents the probabilities of all gametes which can be produced by the genotype Γ_{ij} (note that some of the $c_{k(ij)}$'s might be zero). It is clear that

$$\sum_{k=1}^{v} c_{k(ij)} = 1. \tag{4.20}$$

We will call vector (4.19) the (conditional) *segregation distribution for the gametes of a given genotype* Γ_{ij}.

The (unconditional) probability of occurrence of the gamete of type γ_k produced by the genotypes Γ_{ij}, in the whole population, is

$$\Pr\{\gamma_k, \Gamma_{ij}\} = \Pr\{\gamma_k \mid \Gamma_{ij}\}\Pr\{\Gamma_{ij}\} = c_{k(ij)}g_{ij} = d_{k(ij)}, \tag{4.21}$$

with $g_{ij} = g_i g_j$. The $v \times v$ matrix

$$\mathbf{D}_k = \begin{bmatrix} d_{k(11)} & d_{k(12)} & \cdots & d_{k(1v)} \\ d_{k(21)} & d_{k(22)} & \cdots & d_{k(2v)} \\ \cdots\cdots\cdots\cdots\cdots\cdots \\ d_{k(v1)} & d_{k(v2)} & \cdots & d_{k(vv)} \end{bmatrix} = \begin{bmatrix} c_{k(11)}g_{11} & \cdots & c_{k(1v)}g_{1v} \\ c_{k(21)}g_{21} & \cdots & c_{k(2v)}g_{2v} \\ \cdots\cdots\cdots\cdots\cdots\cdots \\ c_{k(v1)}g_{v1} & \cdots & c_{k(vv)}g_{vv} \end{bmatrix}, \tag{4.22}$$

for $k = 1, 2, \ldots, v$, is the probability matrix of occurrence of the gamete γ_k in the whole population $\mathbf{\Gamma}$.

Expression (4.22) can be written in abbreviated form as the term-by-term product [see Appendix I, formula (I.26)] of the matrices $\mathbf{C}_k$ and $\mathbf{G}$, that is,

$$\mathbf{D}_k = \mathbf{C}_k \boxdot \mathbf{G}, \tag{4.23}$$

for $k = 1, 2, \ldots, v$.

▼4.4. GAMETE AND GENOTYPE DISTRIBUTIONS IN THE nTH GENERATION UNDER RANDOM MATING

Let $\mathbf{\Gamma} = (\Gamma_{ij})$ be the *initial* population with the gametic output $\mathbf{\gamma}' = (\gamma_1, \gamma_2, \ldots, \gamma_v)$ and with gamete frequencies $\mathbf{g}' = (g_1, g_2, \ldots, g_v)$, respectively. Suppose that the segregation probabilities $\Pr\{\gamma_k \mid \Gamma_{ij}\}$ remain *constant* through all generations so that the segregation matrix $\mathbf{C}$, defined in (4.18), consists of the same elements in each generation. Under random mating, the genotype distribution will be given by the $v \times v$ matrix $\mathbf{G}$, defined in (4.13). The sum of the elements of the matrix $\mathbf{D}_k$ represents the probability $g_k^{(1)}$, say, of the occurrence of the gamete γ_k in the population $\mathbf{\Gamma}$, that is,

$$g_k^{(1)} = \sum_{i=1}^{v} \sum_{j=1}^{v} d_{k(ij)} = \sum_{i=1}^{v} \sum_{j=1}^{v} c_{k(ij)} g_{ij}. \tag{4.24}$$

Since $g_{ij} = g_i g_j$, formula (4.24) can be written as a quadratic form:

$$g_k^{(1)} = \mathbf{g}' \mathbf{C}_k \mathbf{g}, \qquad k = 1, 2, \ldots, v. \tag{4.24a}$$

The $1 \times v$ vector

$$\mathbf{g}^{(1)} = (g_1^{(1)}, g_2^{(1)}, \ldots, g_v^{(1)}) = (\mathbf{g}' \mathbf{C}_1 \mathbf{g}, \mathbf{g}' \mathbf{C}_2 \mathbf{g}, \ldots, \mathbf{g}' \mathbf{C}_v \mathbf{g}) \tag{4.25}$$

represents the distribution of the *gametic output* for the *next* (first) generation. Thus, under the random mating the genotype distribution in the first generation will be

$$\mathbf{G}^{(1)} = \mathbf{g}^{(1)\prime} \mathbf{g}^{(1)}. \tag{4.26}$$

The element $g_{ij}^{(1)}$ of the matrix $\mathbf{G}^{(1)}$ is now

$$g_{ij}^{(1)} = g_i^{(1)} g_j^{(1)} = (\mathbf{g}' \mathbf{C}_i \mathbf{g})(\mathbf{g}' \mathbf{C}_j \mathbf{g}) = \mathbf{g}' \mathbf{C}_i (\mathbf{g} \mathbf{g}') \mathbf{C}_j \mathbf{g} = \mathbf{g}' \mathbf{C}_i \mathbf{G} \mathbf{C}_j \mathbf{g}. \tag{4.27}$$

It is easy to show that the probability of the occurrence of the gamete γ_k in the $(n - 1)$th generation, that is, being a component of the gametic output *for* the nth generation, $g_k^{(n)}$, is

$$g_k^{(n)} = \mathbf{g}^{(n-1)\prime} \mathbf{C}_k \mathbf{g}^{(n-1)}, \tag{4.28}$$

$k = 1, 2, \ldots, v$, and the genotype distribution in the nth generation, $\mathbf{G}^{(n)}$, is

$$\mathbf{G}^{(n)} = \mathbf{g}^{(n)\prime} \mathbf{g}^{(n)}. \tag{4.29}$$

By analogy with formula (4.27), we find that the $g_{ij}^{(n)}$ element of the matrix $\mathbf{G}^{(n)}$ is given by the recurrence formula

$$g_{ij}^{(n)} = g_i^{(n)} g_j^{(n)} = \mathbf{g}^{(n-1)\prime} \mathbf{C}_i \mathbf{G}^{(n-1)} \mathbf{C}_j \mathbf{g}^{(n-1)}. \qquad (4.30)$$

It represents the probability of the occurrence of the genotype Γ_{ij} (i.e., the result of the union of gametes γ_i and γ_j) in the nth generation.

4.5. APPLICATION OF MATRIX ALGEBRA IN THE EVALUATION OF THE GENOTYPE DISTRIBUTION IN THE nTH GENERATION FOR TWO LINKED LOCI

4.5.1. Two Loci, Each with Two Alleles and Recombination Fraction λ

There are four gametes, $AB = \gamma_1$, $Ab = \gamma_2$, $aB = \gamma_3$, and $ab = \gamma_4$. Let $\mathbf{g}' = (g_1, g_2, g_3, g_4)$ be the gametic output in the initial population. The matrix of genotypes, Γ, and the matrix of their distribution, $\mathbf{G}$, after random mating are shown in the form of Tables 1a and 4.1b, respectively.

TABLE 4.1a Genotype matrix

$\Gamma =$	♀ \ ♂	$AB(\gamma_1)$	$Ab(\gamma_2)$	$aB(\gamma_3)$	$ab(\gamma_4)$
	$AB(\gamma_1)$	$AABB$	$AABb$	$AaBB$	AB/ab
	$Ab(\gamma_2)$	$AABb$	$AAbb$	Ab/aB	$Aabb$
	$aB(\gamma_3)$	$AaBB$	Ab/aB	$aaBB$	$aaBb$
	$ab(\gamma_4)$	AB/ab	$Aabb$	$aaBb$	$aabb$

TABLE 4.1b Genotype distribution

$\mathbf{G} =$	♀ \ ♂	g_1	g_2	g_3	g_4
	g_1	g_1^2	$g_1 g_2$	$g_1 g_3$	$g_1 g_4$
	g_2	$g_2 g_1$	g_2^2	$g_2 g_3$	$g_2 g_4$
	g_3	$g_3 g_1$	$g_3 g_2$	g_3^2	$g_3 g_4$
	g_4	$g_4 g_1$	$g_4 g_2$	$g_4 g_3$	g_4^2

We recall that the genotype AB/ab (in coupling) produces each of the gametes AB, ab with probabilities $\frac{1}{2}(1 - \lambda)$, and Ab, aB with probabilities $\frac{1}{2}\lambda$, where λ is the recombination fraction. For genotype aB/ab (in repulsion) the situation is reversed. The segregation matrices, $\mathbf{C}_k$, for each gamete γ_k ($k = 1, \ldots, 4$), are as follows:

For $AB(\gamma_1)$

$$\mathbf{C}_1 = \begin{bmatrix} 1 & \frac{1}{2} & \frac{1}{2} & \frac{1}{2}(1 - \lambda) \\ \frac{1}{2} & 0 & \frac{1}{2}\lambda & 0 \\ \frac{1}{2} & \frac{1}{2}\lambda & 0 & 0 \\ \frac{1}{2}(1 - \lambda) & 0 & 0 & 0 \end{bmatrix};$$

For $Ab(\gamma_2)$

$$\mathbf{C}_2 = \begin{bmatrix} 0 & \frac{1}{2} & 0 & \frac{1}{2}\lambda \\ \frac{1}{2} & 1 & \frac{1}{2}(1 - \lambda) & \frac{1}{2} \\ 0 & \frac{1}{2}(1 - \lambda) & 0 & 0 \\ \frac{1}{2}\lambda & \frac{1}{2} & 0 & 0 \end{bmatrix};$$

For $aB(\gamma_3)$

$$\mathbf{C}_3 = \begin{bmatrix} 0 & 0 & \frac{1}{2} & \frac{1}{2}\lambda \\ 0 & 0 & \frac{1}{2}(1 - \lambda) & 0 \\ \frac{1}{2} & \frac{1}{2}(1 - \lambda) & 1 & \frac{1}{2} \\ \frac{1}{2}\lambda & 0 & \frac{1}{2} & 0 \end{bmatrix};$$

For $ab(\gamma_4)$

$$\mathbf{C}_4 = \begin{bmatrix} 0 & 0 & 0 & \frac{1}{2}(1 - \lambda) \\ 0 & 0 & \frac{1}{2}\lambda & \frac{1}{2} \\ 0 & \frac{1}{2}\lambda & 0 & \frac{1}{2} \\ \frac{1}{2}(1 - \lambda) & \frac{1}{2} & \frac{1}{2} & 1 \end{bmatrix}.$$

(4.31)

We now find the gametic output for the *next* (first) generation, $\boldsymbol{\Gamma}^{(1)}$, say, when the initial population $\boldsymbol{\Gamma}$ is mating at random. Let us calculate, for instance, $g_1^{(1)}$. Using formula (4.24a), we obtain

$$g_1^{(1)} = \mathbf{g}'\mathbf{C}_1\mathbf{g}$$

$$= [g_1g_1 + \tfrac{1}{2}g_1g_2 + \tfrac{1}{2}g_1g_3 + \tfrac{1}{2}(1 - \lambda)g_1g_4]$$
$$+ [\tfrac{1}{2}g_2g_1 + \tfrac{1}{2}\lambda g_2g_3] + [\tfrac{1}{2}g_3g_1 + \tfrac{1}{2}g_3g_2] + \tfrac{1}{2}(1 - \lambda)g_4g_1$$
$$= g_1(g_1 + g_2 + g_3 + g_4) - \lambda(g_1g_4 - g_2g_3)$$
$$= g_1 - \lambda(g_1g_4 - g_2g_3).$$

(4.32)

Let us denote

$$\Delta_0 = g_1 g_4 - g_2 g_3 = \begin{vmatrix} g_1 & g_2 \\ g_3 & g_4 \end{vmatrix}. \tag{4.33}$$

Thus (4.32) can be written as

$$g_1^{(1)} = g_1 - \lambda \Delta_0. \tag{4.34}$$

In a similar way we calculate $g_2^{(1)}$, $g_3^{(1)}$, $g_4^{(1)}$, obtaining the results

$$g_1^{(1)} = g_1 - \lambda \Delta_0, \quad g_2^{(1)} = g_1 + \lambda \Delta_0, \quad g_3^{(1)} = g_3 + \lambda \Delta_0, \quad g_4^{(1)} = g_4 - \lambda \Delta_0. \tag{4.35}$$

Suppose that the population $\mathbf{\Gamma}^{(1)}$ is subject to random mating. Applying the same procedure again, we find $g_1^{(2)}$ [by analogy to (4.34)]:

$$g_1^{(2)} = g_1^{(1)} - \lambda \Delta_1, \tag{4.36}$$

where

$$\begin{aligned}
\Delta_1 &= g_1^{(1)} g_4^{(1)} - g_2^{(1)} g_3^{(1)} \\
&= (g_1 - \lambda \Delta_0)(g_4 - \lambda \Delta_0) - (g_2 + \lambda \Delta_0)(g_3 + \lambda \Delta_0) \\
&= g_1 g_4 - g_2 g_3 - \lambda \Delta_0 (g_1 + g_2 + g_3 + g_4) \\
&= \Delta_0 - \lambda \Delta_0 \\
&= (1 - \lambda)\Delta_0. \tag{4.37}
\end{aligned}$$

Repeating the same procedure n times, we can show that

$$\Delta_n = (1 - \lambda) \Delta_{n-1} = (1 - \lambda)^n \Delta_0. \tag{4.38}$$

Now

$$\begin{aligned}
g_1^{(n)} &= g_1^{(n-1)} - \lambda \Delta_{n-1} \\
&= g_1 - \lambda(\Delta_0 + \Delta_1 + \cdots + \Delta_{n-1}) \\
&= g_1 - \lambda \Delta_0 [1 + (1 - \lambda) + (1 - \lambda)^2 + \cdots + (1 - \lambda)^{n-1}]. \tag{4.39}
\end{aligned}$$

Putting $(1 - \lambda) = t$ and applying formula (II.23) to the geometric series in square brackets in (4.38), we obtain

$$g_1^{(n)} = g_1 - \lambda \Delta_0 \frac{1 - (1 - \lambda)^n}{1 - (1 - \lambda)} = g_1 - \Delta_0 [1 - (1 - \lambda)^n]. \tag{4.39a}$$

In a similar way we find $g_2^{(n)}$, $g_3^{(n)}$, $g_4^{(n)}$, so that finally we have

$$\begin{aligned}
g_1^{(n)} &= g_1 - \Delta_0 [1 - (1 - \lambda)^n]; \quad g_2^{(n)} = g_2 + \Delta_0 [1 - (1 - \lambda)^n]; \\
g_3^{(n)} &= g_3 + \Delta_0 [1 - (1 - \lambda)^n]; \quad g_4^{(n)} = g_4 - \Delta_0 [1 - (1 - \lambda)^n]. \tag{4.40}
\end{aligned}$$

Frequencies (4.40) represent the distribution of the gametic output *for the*

nth generation. Hence they are, in fact, the gamete frequencies *in* the $(n - 1)$th generation. In particular, when $\lambda = \frac{1}{2}$, linkage is not distinguished, so we assume that loci are independent. In this case we have

$$\Delta_n = (\tfrac{1}{2})^n \Delta_0. \tag{4.41}$$

Therefore, for two *independent* loci we have

$$g_1^{(n)} = g_1 - \Delta_0[1 - (\tfrac{1}{2})^n]; \quad g_2^{(n)} = g_2 + \Delta_0[1 - (\tfrac{1}{2})^n];$$
$$g_3^{(n)} = g_3 + \Delta_0[1 - (\tfrac{1}{2})^n]; \quad g_4^{(n)} = g_4 - \Delta_0[1 - (\tfrac{1}{2})^n]. \tag{4.42}$$

▼4.5.2. Two Loci, Each with Multiple Alleles

Let $A_1, A_2, \ldots, A_{s_1}$ be the s_1 alleles at locus A, and $B_1, B_2 \ldots, B_{s_2}$ be the s_2 alleles at locus B linked with A. There will be $s_1 s_2$ different gametes, and the population $\mathbf{\Gamma}$ can be presented in the form of a $s_1 s_2 \times s_1 s_2$ square matrix, as in (4.10).

Let $A_i B_k / A_j B_l$ denote a genotype for which genes $A_i B_k$ are located in one chromosome and genes $A_j B_l$ in its homologue. Assuming $i \neq j$, $k \neq l$, the segregation probabilities for the genotype $A_i B_k / A_j B_l$ are as follows:

$$
\begin{array}{cccc}
A_i B_k & A_i B_l & A_j B_k & A_j B_l \\
\tfrac{1}{2}(1 - \lambda) & \tfrac{1}{2}\lambda & \tfrac{1}{2}\lambda & \tfrac{1}{2}(1 - \lambda)
\end{array}. \tag{4.43}
$$

Let $g(A_t B_u)$ and $g^{(n)}(A_t B_u)$ denote the frequencies of gamete $A_t B_u$ ($t = 1, \ldots, s_1$; $u = 1, \ldots, s_2$) in the gametic outputs for the initial and nth ($n = 1, 2, \ldots$) generations, respectively. Of course,

$$\sum_{t=1}^{s_1} \sum_{u=1}^{s_2} g(A_t B_u) = \sum_{t=1}^{s_1} \sum_{u=1}^{s_2} g^{(n)}(A_t B_u) = 1.$$

Applying the method presented in section 4.4, we find

$$g^{(1)}(A_i B_k) = g(A_i B_k) - \lambda \Delta_0(i, k), \tag{4.44}$$

where

$$\Delta_0(i, k) = g(A_i B_k) - \left[\sum_{u=1}^{s_2} g(A_i B_u) \right]\left[\sum_{t=1}^{s_1} g(A_t B_k) \right]. \tag{4.45}$$

We notice that now $\Delta_0(i, k)$ is not the same for all gametes but depends on which alleles from A and B loci are in question.

Repeating the procedure, we can show that for the nth generation

$$\Delta_n(i, k) = (1 - \lambda)^n \Delta_0(i, k), \tag{4.46}$$

and

$$g^{(n)}(A_i B_k) = g(A_i B_k) - \Delta_0(i, k)[1 - (1 - \lambda)^n]. \tag{4.47}$$

[For details of the derivation of the above formulae see Kempthorne (1957), section 2.8. Compare also Problem 4.6, given at the end of this chapter.]

4.6. GENE FREQUENCIES IN THE nTH GENERATION

Let us return to the situation of two linked loci, each with two alleles, considered in section 4.5.1. We recall that a population is "stationary," that is, there is no migration, selection, or mutation. We will calculate the gene frequencies in the initial and the nth generations.

The frequencies of the gene A in the initial and the nth generations are

$$p_A = g_1 + g_2; \quad p_A^{(n)} = g_1^{(n)} + g_2^{(n)} = (g_1 - \Delta_n) + (g_2 + \Delta_n) = g_1 + g_2 \tag{4.48}$$

respectively. We have, then,

$$p_A = p_A^{(n)}. \tag{4.49}$$

Relations similar to that in (4.49) hold for the remaining genes. These results can be generalized to r loci, each with multiple alleles. We can summarize these results as follows.

For a "stationary" population satisfying the assumptions given in (4.8) *and subject to random mating, the gene pool remains constant through all generations.*

We see in Chapter 10, however, that this is not true when selection or mutation operates on a population.

Example 4.3. Let us consider two characters in crosspollinated plants. Suppose that the first character is controlled by a single locus, A, say, with two alleles, A and a, where A is dominant and a is recessive. Thus we distinguish two phenotypic forms, $A-$ and aa. Similarly, the second character is controlled by a locus B, with alleles B (dominant) and b (recessive) and phenotypic forms, $B-$ and bb. The characters are inherited independently of each other.

Suppose that seeds obtained from four phenotypically distinguished groups were mixed together to grow a large population of plants and, after random mating, their offspring were obtained in the following proportions:

Phenotype	Proportion	
$A-B-$	$f_{11} = 0.8125$	
$A-bb$	$f_{12} = 0.0275$	(i)
$aaB-$	$f_{21} = 0.0975$	
$aabb$	$f_{22} = 0.0625.$	

We shall call the population of plants given in (i) an initial population Γ. We wish to find the gametic and the gene frequencies for Γ. Comparing genotypes and their frequencies, after random matings, in Tables 4.1a and 4.1b, respectively, we notice that $\Pr\{aabb\} = f_{22} = g_4{}^2$; $\Pr\{A\text{-}bb\} = f_{12} = g_2{}^2 + 2g_2g_4$; $\Pr\{aaB\text{-}\} = f_{21} = g_3{}^2 + 2g_3g_4$. Using the data given in (i) we have $g_4{}^2 = 0.0625$; $g_2{}^2 + 2g_2g_4 = 0.0275$; $g_3{}^2 + 2g_3g_4 = 0.975$. Hence $\Pr\{ab\} = g_4 = 0.25$; $\Pr\{Ab\} = g_2 = 0.05$; $\Pr\{aB\} = g_3 = 0.15$; and $\Pr\{AB\} = g_1 = 1 - g_2 - g_3 - g_4 = 0.55$. The gene frequencies are $\Pr\{A\} = p_A = g_1 + g_2 = 0.60$; $\Pr\{a\} = p_a = g_3 + g_4 = 1 - p_A = 0.40$; $\Pr\{B\} = p_B = g_1 + g_3 = 0.70$; $\Pr\{b\} = p_B = g_2 + g_4 = 1 - p_B = 0.30$.

Let us find now the phenotype distribution in the *third generation*.

We calculate first the gamete frequencies for the third generation. Since the loci are independent we use formula (4.42). First we obtain $\Delta_0 = g_1g_4 - g_2g_3 = 0.1300$; $\Delta_3 = [1 - (\tfrac{1}{2})^3]\Delta_0 = 0.113750$. Hence, $g_1^{(3)} = 0.55 - 0.113750 = 0.436250$; $g_2^{(3)} = 0.05 + 0.113750 = 0.163750$; $g_3^{(3)} = 0.15 + 0.113750 = 0.263750$; $g_4^{(3)} = 0.25 - 0.113750 = 0.136250$. The frequencies of phenotypes, in the third generation, are: $\Pr\{aabb\} = [g_4^{(3)}]^2 = 0.018564$; $\Pr\{A\text{-}bb\} = [g_2^{(3)}]^2 + 2g_2^{(3)}g_3^{(3)} = 0.071436$; $\Pr\{aaB\text{-}\} = [g_3^{(3)}]^2 + 2g_3^{(3)}g_4^{(3)} = 0.141436$; $\Pr\{A\text{-}B\text{-}\} = 1 - \Pr\{A\text{-}bb\} - \Pr\{aaB\text{-}\} - \Pr\{aabb\} = 0.768564$.

4.7. THEOREM ON GENETIC EQUILIBRIUM IN A RANDOMLY MATING POPULATION

In the preceding section, we have found that the gene pool in a randomly mating population remains constant through generations but the gamete and genotype distributions do not always have this property. We now want to find what additional conditions should be satisfied to ensure that the gamete and genotype distributions also remain constant through generations, or, as it is customary to say, the population is in *equilibrium*. We now define what we mean by "equilibrium."

DEFINITION 4.2. *An arbitrary population is in genetic equilibrium with respect to r loci if the genotype distribution in the next (and the following) generations remains the same with respect to these loci. In other words, the population attains equilibrium in the nth generation if*

$$\mathbf{G}^{(n)} = \mathbf{G}^{(n+1)} = \mathbf{G}^{(n+2)} = \cdots. \tag{4.50}$$

This, of course, implies that in equilibrium the gamete frequencies also remain constant. It should be noticed that the reverse statement is not always true (see, for instance, section 9.15). However, if the population is mating *at random*, we have additionally, from (4.29), $\mathbf{G}^{(n)} = \mathbf{g}^{(n)\prime}\mathbf{g}^{(n)}$, and the condition (4.50) can, in this case, be replaced by

$$\mathbf{g}^{(n)} = \mathbf{g}^{(n+1)} = \mathbf{g}^{(n+2)} = \cdots. \tag{4.51}$$

Thus we conclude the following.

DEFINITION 4.3. *A randomly mating population is in equilibrium with respect to r loci if the gamete distribution in the next (and the following) generations remains the same with respect to these loci.*

Let us consider a few examples.

4.7.1. Hardy-Weinberg Equilibrium Law for a Single Locus

Let A be a single locus with two alleles A and a. Suppose that a population has an *arbitrary* structure

$$h_{11}AA + 2h_{12}Aa + h_{22}aa.$$

The gamete (here also the gene) frequencies are $\Pr\{A\} = g_1^{(1)} = p = h_{11} + h_{12}$, $\Pr\{a\} = g_2^{(1)} = q = h_{22} + h_{12}$. After random mating, the genotype array in the next generation, from (4.15a), is

$$(pA + qa)^2 = p^2AA + 2pqAa + q^2aa. \tag{4.52}$$

It is easy to see that the gamete frequencies are again p and q. Thus the genotype array in the second and all following generations will be the same as given by (4.52). We obtain the result that the population under random mating attains equilibrium with respect to two alleles at a single locus in the first generation.

This can easily be extended to the case of single locus with multiple alleles. Let

$$\sum_{i=1}^{s} h_{ii}A_iA_i + 2\sum\sum_{i<j} h_{ij}A_iA_j$$

represent an arbitrary structure of a population with respect to a single locus with s alleles, $A_1, A_2, \ldots, A_s$. The gamete (here also the gene) frequencies are $p_i = \sum_{j=1}^{s} h_{ij}$, $i = 1, 2, \ldots, s$. After random mating, the genotype array in the next (first) generation, from (4.15a), is

$$\left(\sum_{i=1}^{s} p_iA_i\right)^2 = \sum_{i=1}^{s} p_i^2A_iA_i + 2\sum\sum_{i<j} p_ip_jA_iA_j. \tag{4.53}$$

It is easy to see that the gamete frequencies are

$$\Pr\{A_i\} = \sum_{j=1}^{s} p_ip_j = p_i\left(\sum_{j=1}^{s} p_j\right) = p_i, \qquad i = 1, 2, \ldots, s, \tag{4.54}$$

which are again the same. Thus the genotype array in the next (and following) generation will be the same as given by (4.53). We now summarize these results.

A randomly mating population attains equilibrium with respect to a single locus, in the first generation.

This is known as the *Hardy-Weinberg equilibrium law* since it was stated independently by Hardy (1908) and Weinberg (1908). We notice that:

The genotype array of a population in equilibrium is the square of the gene array.

4.7.2. Two Linked or Independent Loci

Although genetic equilibrium with respect to each single locus separately is attained in the first generation, this does not necessarily mean that it is reached jointly with respect to all loci, and without regard to whether they are linked or independent. As we have already shown in section 4.5, the gamete frequencies, given by (4.40) for linked and by (4.42) for independent loci, depend on the number of generations, n. When n becomes large ($n \to \infty$), the term $(1 - \lambda)^n$ [or $(\frac{1}{2})^n$] tends to zero, so that the *limiting* gamete frequencies, $g_i^{(\infty)}$, ($i = 1, \ldots, 4$) for both cases of linked or independent loci, are

$$g_1^{(\infty)} = g_1 - \Delta_0, \; g_2^{(\infty)} = g_2 + \Delta_0, \; g_3^{(\infty)} = g_3 + \Delta_0, \; g_4^{(\infty)} = g_4 - \Delta_0. \quad (4.55)$$

These values are constant and represent gamete frequencies at equilibrium with respect to two loci.

We notice that if $\lambda = 0$, that is, the loci are completely linked, then $g_i^{(\infty)} = g_i^{(n)} = g_i$ ($i = 1, \ldots, 4$), so the population attains equilibrium in the first generation. On the other hand, if λ is much smaller than $\frac{1}{2}$ ($\lambda \ll \frac{1}{2}$), then $(1 - \lambda)^n \gg (\frac{1}{2})^n$ and approaching to equilibrium is slower for linked than for independent loci.

We now express the limiting gamete distribution (4.55) in terms of gene frequencies at loci A and B. Let

$$p = p_A = g_1 + g_2, \quad q = p_a = g_3 + g_4, \quad p + q = 1;$$
$$u = p_B = g_1 + g_3, \quad v = p_b = g_2 + g_4, \quad u + v = 1. \quad (4.56)$$

We calculate

$$pu = (g_1 + g_2)(g_1 + g_3) = g_1(g_1 + g_2 + g_3) + g_2 g_3$$
$$= g_1(1 - g_4) + g_2 g_3 = g_1 - (g_1 g_4 - g_2 g_3) = g_1 - \Delta_0 = g_1^{(\infty)}. \quad (4.57)$$

In a similar way we obtain

$$pv = g_2 + \Delta_0 = g_2^{(\infty)}, \quad qu = g_3 + \Delta_0 = g_3^{(\infty)}, \quad qv = g_4 - \Delta_0 = g_4.$$
(4.57a)

Hence the gamete array is

$$g_1^{(\infty)} AB + g_2^{(\infty)} Ab + g_3^{(\infty)} aB + g_4^{(\infty)} ab = (pA + qa)(uB + vb), \quad (4.58)$$

and the genotype array is

$$[g_1^{(\infty)} AB + g_2^{(\infty)} Ab + g_3^{(\infty)} aB + g_4^{(\infty)} ab]^2 = (pA + qa)^2 (uB + vb)^2. \quad (4.59)$$

It is not difficult to show that (4.59) applies to two loci, each with multiple alleles. This means that if a randomly mating population satisfying assumptions (4.8) is in genetic equilibrium with respect to two loci, the gametic array is the product of gene arrays at each locus, or equivalently, the genotype array can be expressed as the product of squares of the gene arrays at single loci. This result can be extended to r loci.

The reverse statement is also true: if the gamete array is the product of the gene arrays at single loci, then the population is in equilibrium. This can be proved by calculating the gamete arrays for the next generation, using straightforward algebra.

We summarize these results in the following theorem.

THEOREM 4.1. *Let us consider infinite, bisexual, diploid, randomly mating population satisfying assumptions (4.8). The necessary and sufficient conditions for such a population to be in genetic equilibrium with respect to r autosomal loci are that its gamete array could be expressed as the product of r gene arrays or equivalently, its genotype array could be expressed as the product of squares of r gene arrays.*

Some more advanced problems of equilibrium in polyploid populations are discussed by Elandt-Johnson (1967).

It is remarkable that the equilibrium conditions are the same whether the loci are linked or independent, but the approach to equilibrium is usually faster for independent loci. When a randomly mating population is in equilibrium, we cannot, then, detect linkage. In human populations, for instance, linkage can be detected (and estimated) from analysis of family data (see section 18.9).

Example 4.4. Suppose that it would be satisfactory to assume that the randomly mating population in Example 4.3 is in equilibrium, if the modulus of the difference between the actually calculated and the "expected" (theoretical) gamete frequencies is not more than 0.0005. We wish to know how many randomly mating generations are necessary in order to obtain this equilibrium. Our condition is, then,

$$|g_i^{(n)} - g_i^{(\infty)}| = |(\tfrac{1}{2})^n \Delta_0| < 0.0005.$$

For data in Example 4.3 we have $\Delta_0 = 0.1300$. Hence

$$(\tfrac{1}{2})^n \times 0.1300 < 0.0005$$

or

$$n > \frac{\log 0.003846}{\log 0.5000} = 8.0224 \quad \text{or} \quad n \geqslant 9.$$

Therefore, the equilibrium condition can be attained in the ninth generation.

Suppose that we consider the situation with the same $\Delta_0 = 0.1300$, but for linked loci with $\lambda = 0.1$. In this case we will have $(1 - \lambda)^n \Delta_0 < 0.0005$ or $(0.9)^n \times 0.1300 < 0.0005$. The solution is $n \geqslant 53$.

4.8. EQUILIBRIUM FOR X-LINKED GENES

We will assume that the homogametic sex, with XX chromosomes, is female, and the heterogametic sex, with chromosomes XY, is male. We will consider X-linkage, that is, we will assume that the genes under consideration are located in the X chromosomes. The transmission of heredity is as follows:

1. The *female* offspring receive their chromosomes, XX, one from the mother and one from the father; therefore, *the genotype array of the daughters is the product of the gamete arrays of the mothers and of the fathers.*

2. The *male* offspring receive the chromosome X from the mother, so *the genotype array of the sons is the gamete array of the mother.*

Let us consider a single, X-linked locus A with s alleles, $A' = (A_1, A_2, \ldots, A_s)$. Let $f_i^{(n)}$ be the "expected" frequency of the *paternal* gamete A_i for the output into the nth generation or, in other words, of the gamete A_i produced by the *fathers* of the $(n - 1)$th generation. Let $m_i^{(n)}$ be the expected frequency of *maternal* gamete A_i for the nth generation, that is, of the gamete A_i produced by the *mothers* of the $(n - 1)$th generation. In particular, for the initial population Γ (generation 0), we denote

$$f_i^{(0)} = f_i; \quad m_i^{(0)} = m_i, \quad i = 1, 2, \ldots, s.$$

By virtue of remarks 1 and 2, the genotype arrays in the initial populations of males and females are, respectively,

$$\male: \sum_{i=1}^{s} m_i A_i, \tag{4.60}$$

$$\female: \left(\sum_{i=1}^{s} f_i A_i \right) \left(\sum_{i=1}^{s} m_i A_i \right) = \sum_{i=1}^{s} \sum_{j=1}^{s} f_i m_j A_i A_j. \tag{4.61}$$

Of course, we have

$$\sum_{i=1}^{s} m_i = 1, \quad \sum_{i=1}^{s} f_i = 1, \quad \sum_{i=1}^{s} \sum_{j=1}^{s} f_i m_j = 1. \tag{4.62}$$

The frequency of the gamete A_i produced by the females of the population Γ (i.e., the maternal gamete for the first generation, $\Gamma^{(1)}$), $m_i^{(1)}$, is

$$\text{\Large\female}: m_i^{(1)} = \frac{1}{2}\left(\sum_{j=1}^{s} m_i f_j + \sum_{j=1}^{s} f_i m_j\right) = \frac{1}{2}(m_i + f_i), \qquad (4.63)$$

and the frequency of the gamete A_i produced by the males of Γ (i.e., the paternal gamete for $\Gamma^{(1)}$), $f_i^{(1)}$, is

$$\text{\Large\male}: f_i^{(1)} = m_i. \qquad (4.64)$$

Generalizing the results (using the method of mathematical induction), we obtain the recurrence formulae for the maternal and paternal gamete frequencies:

$$m_i^{(n)} = \frac{1}{2}[m_i^{(n-1)} + f_i^{(n-1)}], \qquad (4.65)$$

$$f_i^{(n)} = m_i^{(n-1)}, \qquad (4.66)$$

respectively. Of course, these formulae are valid when replacing n by $n - 1$. Thus from (4.65) we have

$$m_i^{(n-1)} = \frac{1}{2}[m_i^{(n-2)} + f_i^{(n-2)}], \qquad (4.65a)$$

$$f_i^{(n-1)} = m_i^{(n-2)}. \qquad (4.66a)$$

Substituting for $m_i^{(n-1)}$ and for $m_i^{(n-2)}$, in (4.65a), $f_i^{(n)}$ and $f_i^{(n-1)}$, respectively, we obtain

$$f_i^{(n)} = \frac{1}{2}[f_i^{(n-1)} + f_i^{(n-2)}]. \qquad (4.67)$$

Let

$$f_i^{(1)} - f_i = \Delta_0(i). \qquad (4.68)$$

Thus

$$f_i^{(2)} - f_i^{(1)} = \frac{1}{2}[f_i^{(1)} + f_i] - f_i^{(1)} = -\frac{1}{2}[f_i^{(1)} - f_i] = -\frac{1}{2}\Delta_0(i). \qquad (4.69)$$

Similarly,

$$f_i^{(3)} - f_i^{(2)} = -\frac{1}{2}[f_i^{(2)} - f_i^{(1)}] = -\frac{1}{2}(-\frac{1}{2})\Delta_0(i) = (-\frac{1}{2})^2 \Delta_0(i). \qquad (4.70)$$

Finally,

$$f_i^{(n)} - f_i^{(n-1)} = (-\frac{1}{2})^{n-1} \Delta_0(i). \qquad (4.71)$$

We can express the gamete frequency, $f_i^{(n)}$, in terms of $\Delta_0(i)$. We write

$$\begin{aligned}
f_i^{(n)} &= f_i + [f_i^{(1)} - f_i] + [f_i^{(2)} - f_i^{(1)}] + \cdots + [f_i^{(n)} - f_i^{(n-1)}] \\
&= f_i + [1 + (-\frac{1}{2}) + (-\frac{1}{2})^2 + \cdots + (-\frac{1}{2})^{n-1}] \Delta_0(i) \\
&= f_i + \frac{2}{3}[1 - (-\frac{1}{2})^n] \Delta_0(i). \qquad (4.72)
\end{aligned}$$

Taking (4.66) into account and replacing n by $n + 1$, we obtain

$$m_i^{(n)} = f_i^{(n+1)} = f_i + \frac{2}{3}[1 - (-\frac{1}{2})^{n+1}] \Delta_0(i). \qquad (4.73)$$

Letting $n \to \infty$, we obtain the limiting paternal and maternal gamete frequencies, $f_i^{(\infty)}$ and $m_i^{(\infty)}$, respectively, in the form

$$f_i^{(\infty)} = m_i^{(\infty)} = f_i + \tfrac{2}{3}\Delta_0(i). \qquad (4.74)$$

But, since $\Delta_0(i) = f_i^{(1)} - f_i = m_i - f_i$, we finally have

$$f_i^{(\infty)} = m_i^{(\infty)} = \tfrac{2}{3}m_i + \tfrac{1}{3}f_i, \qquad i = 1, 2, \ldots, s. \qquad (4.75)$$

These are the gamete frequencies in equilibrium with respect to X-linked locus. We summarize the results:

The population is in equilibrium with respect to an X-linked locus if and only if the paternal and the maternal gamete frequencies are the same.

A few remarks on the frequencies of X-linked genes will be helpful.

(i) The maternal gamete frequencies, $m_1^{(n)}, m_2^{(n)}, \ldots, m_s^{(n)}$, also represent the genotype distribution in the nth generation of males.

(ii) The maternal gamete frequency for the nth generation is the average of the maternal and the paternal frequencies for the preceding generations [formula (4.65)].

(iii) From (4.72) and (4.73) we notice that the approach to equilibrium follows a geometric series with the negative ratio $t = -\tfrac{1}{2}$, and so it is oscillatory in its character (see Fig. 4.1 in Example 4.5).

(iv) Let $p_i^{(n)}$ and p_i be the frequencies of the gene A_i in the nth generation and in the population at equilibrium respectively, in the whole population (of females and males). Then for the initial population we have

$$p_i^{(0)} = \tfrac{1}{3}(m_i + m_i + f_i) = \tfrac{2}{3}m_i + \tfrac{1}{3}f_i = p_i, \quad i = 1, 2, \ldots, s. \quad (4.76)$$

Similarly, for the first generation we have

$$p_i^{(1)} = \tfrac{2}{3}m_i^{(1)} + \tfrac{1}{3}f_i^{(1)} = \tfrac{2}{3} \cdot \tfrac{1}{2}(m_i + f_i) + \tfrac{1}{3}m_i = \tfrac{2}{3}m_i + \tfrac{1}{3}f_i = p_i \quad (4.77)$$

for $i = 1, 2, \ldots, s$.

Using mathematical induction, it follows that

$$p_i^{(n)} = p_i^{(n-1)} = \cdots = p_i, \quad i = 1, 2, \ldots, s. \qquad (4.78)$$

This means that although the female and male gene frequencies when considered separately are oscillatory, the combined gene frequencies in the whole population remain constant.

(v) For two genes, A (dominant) and a (recessive) with frequencies at equilibrium p and q respectively, the proportion of recessive males to recessive females is $q:q^2 = 1:q > 1$. However, the proportion of female *carriers* of a recessive gene a is $2pq$, and the ratio of recessive males to the female carriers is $q:2pq = 1:2p$. If $p > \tfrac{1}{2}$ this ratio is less than 1.

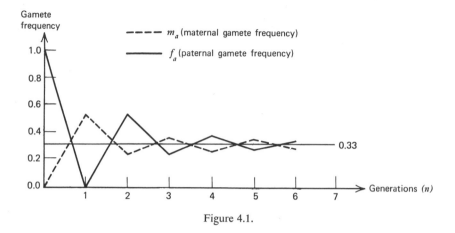

Figure 4.1.

Example 4.5.   Suppose that an initial population is obtained from matings of $♀AA \times ♂aa$ with respect to an X-linked locus, A. Then the maternal and paternal gametic outputs for the initial population are $m_A = 1$, $m_a = 0$, $f_A = 0, f_a = 1$, respectively. Thus, equilibrium will be attained when

$$f_A^{(\infty)} = m_A^{(\infty)} = \tfrac{2}{3}m_A + \tfrac{1}{3}f_A = \tfrac{2}{3} = 0.67, \quad \text{and} \quad f_A^{(\infty)} = m_a^{(\infty)} = \tfrac{1}{3} = 0.33.$$

Suppose that we observe a recessive trait controlled by the allele a. We have $\Delta_0(a) = m_a - f_a = -1$; also, from (4.73) we have $f_a^{(n)} = 1 - \tfrac{2}{3}[1 - (-\tfrac{1}{2})^n] = \tfrac{1}{3}[1 - (-\tfrac{1}{2})^{n-1}]$, and from (4.74) $m_a = \tfrac{1}{3}[1 - (-\tfrac{1}{2})^n]$. At equilibrium, $f_a^{(\infty)} = m_a^{(\infty)} = \tfrac{1}{3} = 0.33$. Figure 4.1 shows the approach to equilibrium.

SUMMARY

In this chapter, we have discussed the definition of random mating (more precisely, the definition of panmixia) and the conditions under which this term applies to the real population. A genotype distribution for a population under random mating has been presented in the matrix form and in a quadratic form (genotype array). Gamete distributions for the nth generation have been derived for two linked autosomal loci and for an X-linked locus. Necessary and sufficient conditions for genetic equilibrium have been stated in terms of gene frequencies, for these loci.

REFERENCES

Elandt-Johnson, Regina C., Equilibrium conditions in polysomic inheritance for panmictic population, *Bull. Math. Biophys.* **29** (1967), 437–449.

Falconer, D. S., *Introduction to Quantitative Genetics*, Oliver and Boyd, London, 1960.

Hardy, G H., Mendelian proportions in a mixed population, *Science* **28** (1908), 49–50.

Kempthorne, O., *An Introduction to Genetic Statistics*, John Wiley & Sons, New York, 1957.

Li, C. C., *Population Genetics*, Chicago University Press, Chicago, 1955.

Srb, A. M., Owen, R. D., and Edgar, R. C., *General Genetics*, W. H. Freeman & Co. San Francisco, 1965.

Weinberg, W., Über den Nachweis der Vererbung beim Menschen, *Jahresh. Verein vaterl. Naturk.* **64** (1908), 368–382.

PROBLEMS

4.1. Prove that Definitions 4.1 and 4.1a are equivalent.

4.2. Suppose that two populations of equal sizes, I and II, have the following distributions with respect to one autosomal locus:

$$I \quad AA \quad Aa \quad aa \qquad II \quad AA \quad Aa \quad aa$$
$$d_1 \quad 2h_1 \quad r_1 \qquad\qquad d_2 \quad 2h_2 \quad r_2,$$

where $d_i + 2h_i + r_i = 1$, $i = 1$, 2. Suppose furthermore, that these populations mate in such a way that each of the two partners is always from a different population, otherwise being selected at random.

(a) Find the genotype distribution and the gene frequencies in the first generation.

(b) If now the individuals of the first generation mate *at random*, equilibrium will be obtained in the next generation. Why? What is the genotype distribution in the population at equilibrium?

4.3. Use the following numerical data in Problem 4.2:

$$
\begin{array}{cccc ccc}
I & AA & Aa & aa & II & AA & Aa & aa \\
 & 0.20 & 0.50 & 0.30 & & 0.40 & 0.40 & 0.20.
\end{array}
$$

4.4. Let us denote the three alleles at the locus controlling the *ABO* blood group system as A, B, O. We know that alleles A and B are dominants over O, but are codominants to each other.

Suppose that in two populations, I and II, the gene frequencies are as follows:

Gene: A B O
Population I: p_1 q_1 r_1 with $p_1 + q_1 + r_1 = 1$,
Population II: p_2 q_2 r_2 with $p_2 + q_2 + r_2 = 1$.

(a) Suppose that those two populations mate in such a way that each of the two partners is from a different population, but otherwise at random. Find the genotype distribution and gene frequencies at equilibrium.

(b) Suppose that these two populations are pooled together and that the population so obtained mates at random. What are the gene frequencies and genotype distribution at equilibrium?

(c) Use the following numerical values in (a) and (b):

$$\text{I: } p_1 = 0.25, q_1 = 0.30, r_1 = 0.55;$$
$$\text{II: } p_2 = 0.27, q_2 = 0.06, r_2 = 0.67.$$

4.5. Suppose that a certain character is dominant at a single locus. Calculate the proportions of individuals possessing this character in a population in equilibrium for $p = 0.0, 0.1, \ldots, 1.0$ and represent the results graphically. How do you interpret these results?

▼**4.6.** Suppose that we consider a population with respect to two linked loci, A and B, each with three alleles, A_1, A_2, A_3 and B_1, B_2, B_3, respectively, and recombination fraction λ. Let $g(A_iB_k)$ denote the frequency of the gamete A_iB_k in the initial population Γ.

We assume that the genotype A_iB_k/A_jB_l produces gametes with the following frequencies:

$$\begin{array}{cccc} A_iB_k & A_iB_l & A_jB_k & A_jB_l \\ \tfrac{1}{2}(1-\lambda) & \tfrac{1}{2}\lambda & \tfrac{1}{2}\lambda & \tfrac{1}{2}(1-\lambda). \end{array}$$

There will be 3×3 different gametes.

(a) Represent the probability space of the genotype distribution of Γ in the form of the 9×9 matrix.

(b) Find the segregation matrix for the gamete $\gamma_1 = A_1B_1$ in Γ, that is, the matrix $\mathbf{C}_1 = \mathbf{C}(A_1B_1 \mid \Gamma)$.

(c) Using the segregation matrix for the gamete A_1B_1 and applying the theory presented in section 4.4, show that the frequency of the gamete A_1B_1 in the next generation is

$$g^{(1)}(A_1B_1) = g(A_1B_1) - \lambda\left\{g(A_1B_1) - \left[\sum_{u=1}^{3}g(A_1B_u)\right]\left[\sum_{t=1}^{3}g(A_tB_1)\right]\right\}.$$

Note that this result is consistent with formula (4.44).

(d) Using the same procedure, find the segregation matrix for the gamete A_1B_2 and its frequency in the next generation, $g^{(1)}(A_1B_2)$.

(e) Could you extend your results to the case of s_1 alleles at locus A and s_2 alleles at locus B so that the formulae given in section 4.5 can be justified?

4.7. Suppose that we have the following numerical values for the gamete frequencies in Γ defined in Problem 4.6:

Gamete (A_iB_k):	A_1B_1	A_1B_2	A_1B_3	A_2B_1	A_2B_2	A_2B_3	A_3B_1	A_3B_2	A_3B_3
$g(A_iB_k)$:	0.06	0.12	0.20	0.08	0.10	0.18	0.05	0.10	0.11,

and that $\lambda = \tfrac{1}{2}$ (i.e., the loci are independent).

(a) Calculate the gene frequencies in Γ.

(b) Calculate the gamete frequencies in $\Gamma^{(1)}$.

(c) Formula (4.47) gives the gamete frequencies for the nth generation. Letting $n \to \infty$, we find the limiting gamete distribution to be

$$g^{(\infty)}(A_iB_k) = g(A_iB_k) - \Delta_0(i, k).$$

Assuming that equilibrium is approximately reached if $|\max \Delta_0(i, k)| \leq 0.0005$, find in which generation we could expect the genotype distribution to be established.

4.8. Suppose we consider a X-linked locus A with alleles $A_1, A_2, \ldots, A_s$. Suppose that the genotype distributions for females and males in the initial population are, respectively,

$$\female: \quad \sum_{i=1}^{s} p_i^2 A_i A_i + \sum\sum_{i<j} 2p_i p_j A_i A_j, \qquad \male: \quad \sum_{i=1}^{s} p_i A_i.$$

Prove that the initial population is in equilibrium with respect to this locus.

4.9. The color of the eyes in Drosophila is X-linked. There are several alleles at this locus, where W (red) is dominant to other alleles, and these others usually show intermediate effects when they appear in heterozygote. Let us consider *three* of these alleles

$$W \text{ (red)}, \quad w^t \text{ (tinged)}, \quad \text{and} \quad w \text{ (white)}.$$

Suppose that we have a female population consisting of $0.25WW + 0.30w^t w^t + 0.45ww$, and a male population consisting of $0.40W + 0.35w^t + 0.25w$. The initial population of flies is obtained by random mating of these two parental populations given above.

(*a*) Find the paternal and maternal gametic output for the initial population.

(*b*) Find the gene frequencies and genotype distributions of females and males under equilibrium.

▼4.10. Suppose that a *tetraploid* population, Γ, with respect to one locus with two alleles and its gametic output, g, are

$$\Gamma = \begin{array}{c} \\ AA \\ Aa \\ aa \end{array} \overset{\begin{array}{ccc} AA & Aa & aa \end{array}}{\begin{bmatrix} A^4 & A^3a & A^2a^2 \\ A^3a & A^2a^2 & Aa^3 \\ A^2a^2 & Aa^3 & a^4 \end{bmatrix}}; \quad g = \begin{bmatrix} g(AA) \\ g(Aa) \\ g(aa) \end{bmatrix} = \begin{bmatrix} g_1 \\ g_2 \\ g_3 \end{bmatrix},$$

where $A^4 = AAAA$, $A^3a = AAAa$, etc. Let us assume that at the meiotic division two *chromosomes*, selected at random, pass to the gamete (this is called "random chromosomal segregation"). Thus under random chromosomal segregation the genotypes in Γ produce the following gametes:

$$A^4 \rightarrow \text{all } (AA); \qquad A^3a \rightarrow \tfrac{1}{2}(AA) + \tfrac{1}{2}(Aa);$$

$$a^4 \rightarrow \text{all } (aa); \qquad Aa^3 \rightarrow \tfrac{1}{2}(aa) + \tfrac{1}{2}(Aa);$$

$$A^2a^2 \rightarrow \tfrac{1}{6}(AA) + \tfrac{4}{6}(Aa) + \tfrac{1}{6}(aa).$$

Using the theory presented in section 4.4, solve the following problems:

(*a*) Find the gene frequencies p and q in terms of gametic output g_1, g_2, g_3.

(*b*) Show that the gametic output for the nth generation can be expressed as:

$$g_1^{(n)} = p^2 + (\tfrac{1}{3})^n \Delta_0,$$

$$g_2^{(n)} = 2pq - 2(\tfrac{1}{3})^n \Delta_0,$$

$$g_3^{(n)} = q^2 + (\tfrac{1}{3})^n \Delta_0,$$

where

$$\Delta_0 = g_1 g_3 - (\tfrac{1}{2}g_2)^2 = \begin{vmatrix} g_1 & \tfrac{1}{2}g_2 \\ \tfrac{1}{2}g_2 & g_3 \end{vmatrix}.$$

(c) Prove that, under random chromosomal segregation, the population is in equilibrium if the genotype array can be expressed as the fourth power of the gene array, that is, as $(pA + qa)^4$.

4.11. Suppose that a certain trait is controlled by two genes, A and a. Let Π_1 and Π_2 be two populations (isolates) with the following frequencies of genes A and a: $p_1 = 0.64$, $q_1 = 0.36$, in Π_1; and $p_2 = 0.00$, $q_2 = 1.00$, in Π_2, respectively. Suppose that these populations are mixed in proportions $c_1 = 0.75$ from Π_1 and $c_2 = 0.25$ from Π_2.

(a) Find the gene frequencies in the mixed population Π (sometimes called a "hybrid" population).

(b) If the "hybrid" population is subject to random mating, when will equilibrium be reached?

(c) What will be the genotype distribution in the hybrid population under equilibrium?

▼**4.12.** Let Π_1 and Π_2 be two case populations corresponding to African Negroes and Whites, respectively. Also, let $p_0 = 0.630$ and $p_0' = 0.028$ be the frequencies, in Π_1 and Π_2, respectively, of the R^0 allele in the Rhesus blood group system. Assume that the "hybrid" population Π (American Negroes) is formed by gene flow from Π_2 to Π_1 (but not vice versa) at the constant rate $\alpha = 0.036$ per generation.

Show that the gene frequency in Π for the nth generation, p_n, is

$$p_n = (1 - \alpha)^n p_0 + [1 - (1 - \alpha)^n]p_0'.$$

[*Hint:* Note that in the first generation we will have $p_1 = (1 - \alpha)p_0 + \alpha p_0'$; in the second generation: $p_2 = (1 - \alpha)p_1 + \alpha p_0' = (1 - \alpha)^2 p_0 + \alpha[1 + (1 - \alpha)]p_0'$; in the third generation: $p_3 = (1 - \alpha)p_2 + \alpha p_0' = (1 - \alpha)^3 p_0 + \alpha[1 + (1 - \alpha) + (1 - \alpha)^2]p_0'$; and so on. See also Glass and Li, *Amer. J. Hum. Genet.* **5** (1953), 1–20.]

CHAPTER 5

Random Variables and Their Distributions

5.1. DEFINITION OF A RANDOM VARIABLE

A *real variable* is a quantity which can have as its value any real number in some set. If this set is countable (i.e., can be put in one-to-one correspondence with the integers), the variable is said to be *discrete*. All the real numbers in an interval are not countable. If a variable may take any value in an interval, it is called *continuous*. Numbers of objects, people, experiments (i.e., 1, 2, . . . , n) are examples of discrete variables. A set $\{-1, -\frac{1}{2}, -\frac{1}{3}, \ldots, 0, \ldots, \frac{1}{3}, \frac{1}{2}, 1\}$ is an example of a countable set. A variable which may take any value that is in this set is a discrete variable. Measurements of length or weight are examples of continuous variables.

In classifications of materials or goods, certain attributes are selected to describe their quality. Instead of using the complete description, one may assign some numbers (e.g., class 1, 2, . . . , etc.) to express in an abbreviated form the quality of a material. In a similar way we may assign numerical values to random events, introduced in previous chapters, which are often determined in a descriptive fashion.

A real variable whose values are assigned to random events is called a *random variable;* its values depend on outcome(s) of an experiment. The synonyms *stochastic variable* and *variate* are also used. Random variables are often denoted by X, Y, Z, U, etc., to distinguish them from ordinary real variables x, y, z, u, etc. It is not essential, however, to use capital letters for random variables. As long as no confusion between random and nonrandom variables arises, either type of symbols, capital or lower-case, can be used for each of them. For convenience, we will use in this chapter capital letters for random variables and lower-case letters for nonrandom variables. We confine ourselves mostly to random variables defined in discrete sample spaces; a continuous case will be briefly discussed in section 5.3. Let us consider some examples.

Example 5.1. Let *AA*, *Aa*, *aa* be a sample space of genotypes at locus *A*, where allele *A* is dominant to *a*. Let X denote the number of dominant alleles in a genotype. Which of the three real values, $x = 0, 1, 2$, X takes when an individual is selected, depends on its genotype. We have

$$X = \begin{cases} 0 & \text{if } aa, \\ 1 & \text{if } Aa \text{ (or } aA), \\ 2 & \text{if } AA. \end{cases}$$

If instead of symbols we use for genotypes the general notation of elementary events, assigning $AA = e_1$, $Aa = e_2$, $aa = e_3$, then X is a *function* of these elementary events, $X = X(e)$, say, such that $X(e_1) = 2$, $X(e_2) = 1$, $X(e_3) = 0$.

Example 5.2. Let A, a', and a be alleles at locus A such that A is dominant over a' and a, and a' is dominant over a. The sample space of genotypes is as follows:

Genotype:	AA	Aa'	Aa	$a'a'$	$a'a$	aa
Elementary event:	e_1	e_2	e_3	e_4	e_5	e_6.

(i) Let X determine the homozygous or the heterozygous state of a genotype and be defined as

$$X = \begin{cases} 0 \text{ if a genotype is homozygous,} \\ 1 \text{ if a genotype is heterozygous.} \end{cases}$$

Again, X is a function of elementary events, $X = X(e)$, such that $X(e_1) = X(e_4) = X(e_6) = 0$ and $X(e_2) = X(e_3) = X(e_5) = 1$.

(ii) Let Y denote the phenotype as follows:

$$Y = \begin{cases} 0 \text{ if } aa, \\ 1 \text{ if } a'a' \text{ or } a'a, \\ 2 \text{ if } AA \text{ or } Aa' \text{ or } Aa. \end{cases}$$

Here Y is also a function of e, $Y = Y(e)$, but such that $Y(e_6) = 0$, $Y(e_4) = Y(e_5) = 1$, $Y(e_1) = Y(e_2) = Y(e_3) = 2$.

Y is a different function of elementary events from X, but both are defined over the same sample space. Several other functions of elementary events can be defined over the same sample space.

We now define a random variable more precisely.

DEFINITION 5.1. *A random variable, X, defined on the sample space Ω, is a real valued function of elementary events, e, on Ω, that is,*

$$X = X(e). \tag{5.1}$$

In the same sample space, Ω, one can define a function of a random variable X, $g(X)$, say. Any function $Z = g(X)$ will also be a random variable.

Example 5.3. In Example 5.1, phenotypes were defined as recessive, aa, and dominant, $A-$ (AA or Aa). Let Z be a random variable defined on Ω as

$$Z = \begin{cases} 0 \text{ if } aa \text{ (recessive)}, \\ 1 \text{ if } (A-) \text{ (dominant)}. \end{cases}$$

We can easily see that Z can be expressed as a function of the random variable X, considered in Example 5.1, as follows

$$Z = g(X) = \begin{cases} 0 \text{ if } X = 0, \\ 1 \text{ if } X > 0. \end{cases}$$

5.2. PROBABILITY DISTRIBUTION AND CUMULATIVE DISTRIBUTION FUNCTION OF A RANDOM VARIABLE

Since values of a random variable are assigned to events, the whole theory of distributions of random events, presented in Chapter 2, applies to random variables. However, when dealing with random variables, we find some new notations very convenient; therefore, we shall give a brief presentation of the theory of distributions of random variables, too. The reader is advised to study the parallelism to the results derived in Chapter 2.

DEFINITION 5.2. *If X is a random variable defined on a sample space Ω, and $\cdots x_{-1} < x_0 < x_1 \cdots$ is a finite or infinite but countable set of values which X can take, then the function $p(x)$, defined by*

$$p(x) = \begin{cases} \Pr\{X = x_i\} & \text{if } x = x_i, \\ 0 & \text{if } x_i < x < x_{i+1}, \end{cases} \tag{5.2}$$

for $i = \ldots, -1, 0, 1, \ldots$, with

$$\sum_i p(x_i) = 1, \tag{5.3}$$

is called the probability distribution of the random variable X on Ω (or, briefly, the distribution of X).

It would be more precise to write $p_X(x)$, instead of $p(x)$, to emphasize that $p_X(x)$ is not an ordinary mathematical function of a real variable x, but describes the properties of a random variable X. To simplify the notation, we omit the subscript X and write $p(x)$.

In this book, we will be concerned mostly with genetic problems in which $x_i \geq 0$ and the set of x's is finite, $x_1 < x_2 < \cdots < x_n$, say. For convenience,

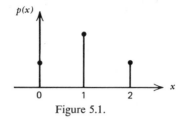

Figure 5.1.

further definitions and results will be presented for finite sets of nonnegative values of x. But there is no difficulty in generalizing the results to any arbitrary infinite and countable set of x's.

Example 5.4. Suppose that the population of genotypes considered in Example 5.1 is the F_2 progeny of two homozygous parental populations $AA \times aa$ (i.e., the offspring of $Aa \times Aa$). The distribution of the number of A genes in genotypes is identical with the distribution of genotypes, that is,

Genotype:	AA	Aa	aa
$X = x_i$:	2	1	0
$\Pr\{X = x_i\} = p(x_i)$:	$\frac{1}{4}$	$\frac{1}{2}$	$\frac{1}{4}$.

The distribution, $p(x)$, can also be represented in the plane, as in Fig. 5.1. Let us consider an event $X \leqslant x_i$. Its probability is

$$\Pr\{X \leqslant x_i\} = \Pr\{X = x_1\} + \Pr\{X = x_2\} + \cdots + \Pr\{X = x_i\}$$

$$= \sum_{t=1}^{i} \Pr\{X = x_t\} = \sum_{t=1}^{i} p(x_t). \tag{5.4}$$

Let x be a value such that $x_i \leqslant x < x_{i+1}$. We have

$$\Pr\{X \leqslant x\} = \Pr\{X \leqslant x_i\} + \Pr\{x_i < X \leqslant x\}.$$

But, since $\Pr\{x_i < X < x_{i+1}\} = 0$, then also $\Pr\{x_i < X \leqslant x\} = 0$, and

$$\Pr\{X \leqslant x\} = \sum_{t=1}^{i} p(x_t) = F(x), \quad \text{say.} \tag{5.5}$$

DEFINITION 5.3. *The function*

$$\Pr\{X \leqslant x\} = F(x) \tag{5.6}$$

is called the cumulative distribution function of the random variable X (the abbreviated form "cdf of X" is commonly used).

Here, again, a more precise notation would be $F_X(x)$ instead of $F(x)$. For all x satisfying the condition $x_i \leqslant x < x_{i+1}$,

$$F(x) = \sum_{x \leqslant x_i} \Pr\{X = x\} = \sum_{t=1}^{i} \Pr\{X = x_t\} = \sum_{t=1}^{i} p(x_t).$$

For instance, in Example 5.4, for $x = 1$ we have $F(x) = F(1) = \frac{1}{4} + \frac{1}{2} = \frac{3}{4}$;

for $x = 1.5$, $F(x) = F(1.5) = \frac{1}{4} + \frac{1}{2} = \frac{3}{4}$; for $x = 1.999$, $F(x) = F(1.999) = \frac{3}{4}$; but for $x = 2$, $F(x) = F(2) = \frac{1}{4} + \frac{1}{2} + \frac{1}{4} = 1$.

In general,

$$
\left.
\begin{aligned}
F(x) &= \sum_{t=1}^{i-1} p(x_t) && \text{for } x_{i-1} \leqslant x < x_i, \\
F(x) &= \sum_{t=1}^{i} p(x_t) = \sum_{t=1}^{i-1} p(x_t) + p(x_i) && \text{for } x_i \leqslant x < x_{i+1}.
\end{aligned}
\right\}
\tag{5.7}
$$

and

This means that $F(x)$ remains constant for $x_{i-1} \leqslant x < x_i$ and is equal to

$$\sum_{t=1}^{i-1} p(x_t)$$

but *jumps* by an amount $\Pr\{X = x_i\} = p(x_i)$ at the point $x = x_i$. Thus $F(x)$ is a *step-function*.

Example 5.5. We find the values of $F(x)$ in Example 5.4:

$$
\begin{aligned}
F(x) &= 0 && \text{for } x < 0, \\
F(x) &= \tfrac{1}{4} && \text{for } 0 \leqslant x < 1, \\
F(x) &= \tfrac{1}{4} + \tfrac{1}{2} = \tfrac{3}{4} && \text{for } 1 \leqslant x < 2, \\
F(x) &= \tfrac{3}{4} + \tfrac{1}{4} = 1 && \text{for } x \geqslant 2.
\end{aligned}
$$

Figure 5.2 represents this cumulative distribution function, $F(x)$, which has a few fundamental properties.

(i) We notice from (5.6) that, for a given value of x, $F(x)$ expresses the *probability* of the event $X \leqslant x$. Therefore, it satisfies the condition

$$0 \leqslant F(x) \leqslant 1. \tag{5.8}$$

(ii) Let $x' < x''$. By definition, we have

$$F(x'') = \Pr\{X \leqslant x''\} = \Pr\{X \leqslant x'\} + \Pr\{x' < X \leqslant x''\}.$$

Since $\Pr\{x' < X \leqslant x''\} \geqslant 0$, then

$$F(x') \leqslant F(x''). \tag{5.9}$$

This means that $F(x)$ is a *nondecreasing* function of x.

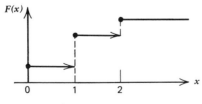

Figure 5.2.

(iii) Since we are concerned only with cases in which X is discrete, it is important always to notice whether the symbol $\leqslant$ or $<$ is used. For example, for any set of values $x_1 < x_2 < \cdots < x_n$ we have

$$\Pr\{X \leqslant x_i\} = \sum_{t=1}^{i} p(x_t) = F(x_i), \tag{5.10a}$$

but

$$\Pr\{X < x_i\} = \sum_{t=1}^{i-1} \Pr\{x_t\} = F(x_{i-1}) = \Pr\{X \leqslant x_{i-1}\}. \tag{5.10b}$$

Using (5.3) and (5.10a), we find

$$\Pr\{X > x_i\} = 1 - \Pr\{X \leqslant x_i\} = 1 - F(x_i), \tag{5.11a}$$

and from (5.3) and (5.10b), we have

$$\Pr\{X \geqslant x_i\} = 1 - \Pr\{X < x_i\} = 1 - F(x_{i-1}). \tag{5.11b}$$

Finally,

$$\Pr\{x_j < X \leqslant x_k\} = F(x_k) - F(x_j), \quad j < k \tag{5.12}$$

or, more generally,

$$\Pr\{x' < X \leqslant x''\} = F(x'') - F(x') \tag{5.13}$$

for $x' < x''$. In particular,

$$\Pr\{-\infty < x < \infty\} = F(\infty) - F(-\infty) = 1. \tag{5.14}$$

Let $g(X)$ be a strictly *monotone* (i.e., increasing or decreasing) function of X. The events $(X = x_i)$ and $[g(X) = g(x_i)]$ are, in fact, the same event expressed in terms of different random variables. Thus

$$\Pr\{X = x_i\} = \Pr\{g(X) = g(x_i)\}. \tag{5.15a}$$

Similarly, if $g(X)$ is a monotone *increasing* function of X, then

$$\Pr\{X \leqslant x\} = \Pr\{g(X) \leqslant g(x)\}. \tag{5.15b}$$

5.3. CONTINUOUS RANDOM VARIABLE. DENSITY FUNCTION

In section 2.14 we discussed briefly a continuous sample space and elementary events of the form $(x - \Delta x < X < x + \Delta x)$.

The cumulative distribution function for a *continuous* random variable is defined in the same way as for a discrete random variable, that is,

$$\Pr\{X \leqslant x\} = F(x), \quad -\infty < x < \infty.$$

If the first derivative

$$F'(x) = \frac{dF(x)}{dx} = f(x), \quad \text{say,} \tag{5.16}$$

exists, then

$$\Pr\{X \leqslant x\} = \int_{-\infty}^{x} f(t) \, dt, \qquad (5.17)$$

and

$$\Pr\{x - \Delta x < X < x + \Delta x\} = \int_{x-\Delta x}^{x+\Delta x} f(t) \, dt. \qquad (5.18)$$

The function $F'(x) = f(x)$ is called the *density function*. [A more precise notation would be $f_X(x)$ instead of $f(x)$.]

From (5.18) we immediately obtain

$$\Pr\{X = x\} = \int_{x}^{x} f(t) \, dt = 0. \qquad (5.19)$$

This means that the probability of a continuous random variable X taking any particular value x is zero.

It is important to realize that, although a random variable X is a real variable, the value which X can take is always associated with probability. Some authors define a random variable X as a quantity for which

$$\Pr\{X \leqslant x\} = F(x)$$

exists for all real values of x. A random variable X is called *discrete* if a probability distribution of the form $\Pr\{X = x\} = p(x) > 0$ exists, and is called *continuous* if the density function $F'(x) = f(x)$ exists. These definitions are quite often given in textbooks on probability and statistics.

5.4. TRUNCATED DISTRIBUTIONS

Let X be a discrete (or continuous) random variable with probability distribution (or density function) $p(x)$ and cdf $F(x)$. It is sometimes necessary to find the distribution of X over the subset of x values such that $a < x \leqslant b$. In other words, we may be interested in conditional probabilities such as

$$\Pr\{X \leqslant x \mid a < X \leqslant b\} \quad \text{and} \quad p(x \mid a < X \leqslant b).$$

Denoting the event $X \leqslant x$ by E_1 and the event $a < X \leqslant b$ by E_2, we can apply formula (2.20). We notice that from (5.13) we will have $\Pr\{a < X \leqslant b\} = F(b) - F(a)$. If x takes all possible values, we exhaust all possible events in the given sample space, so that we have (provided $\Pr\{a < X \leqslant b\} > 0$)

$$F(x \mid a < X \leqslant b) = \Pr\{X \leqslant x \mid a < X \leqslant b\} = \frac{\Pr\{X \leqslant x, a < X \leqslant b\}}{\Pr\{a < X \leqslant b\}}$$

$$= \frac{\Pr\{a < X \leqslant x\}}{\Pr\{a < X \leqslant b\}}.$$

Finally, therefore,

$$F(x \mid a < X \leqslant b) = \begin{cases} 0 & \text{if } x \leqslant a, \\ \dfrac{F(x) - F(a)}{F(b) - F(a)} & \text{if } a < x \leqslant b, \\ 1 & \text{if } x > b; \end{cases} \quad (5.20)$$

and analogously

$$p(x \mid a < X \leqslant b) = \Pr\{X = x \mid a < X \leqslant b\}$$

$$= \begin{cases} 0 & \text{if } x \leqslant a \text{ or } x > b, \\ \dfrac{p(x)}{F(b) - F(a)} & \text{if } a < x \leqslant b. \end{cases} \quad (5.21)$$

In particular,

$$F(x \mid X > x_0) = \begin{cases} 0 & \text{if } x \leqslant x_0, \\ \dfrac{F(x) - F(x_0)}{1 - F(x_0)} & \text{if } x > x_0; \end{cases} \quad (5.22)$$

and

$$p(x \mid X > x_0) = \begin{cases} 0 & \text{if } x \leqslant x_0, \\ \dfrac{p(x)}{1 - F(x_0)} & \text{if } x > x_0. \end{cases} \quad (5.23)$$

The last two formulae are especially useful in several problems of human genetics (see, for instance, Chapter 17).

The cumulative distribution functions defined in (5.20) or (5.22) are a special kind of conditional cdf, in which the condition is made with respect to *the same* variable, X, in order to reduce the sample space. They are called *truncated cumulative distribution functions of X*.

Similarly, probability distributions defined by (5.21) or (5.23) are called *truncated distributions of X*.

Example 5.6. Let s denote the number of children in a family (the family size). Let us assume that a newborn child is equally likely to be a girl or a boy. It can be shown (Chapter 8) that, for fixed s, the probability that there will be exactly x girls in such a family is

$$\Pr\{X = x\} = \binom{s}{x}(\tfrac{1}{2})^x(\tfrac{1}{2})^{s-x} = \binom{s}{x}(\tfrac{1}{2})^s, \qquad x = 0, 1, 2, \ldots, s \quad (5.24)$$

(a special case of binomial distribution).

Suppose that we are interested in the distribution of the number of girls in families, each of size s, which already have at least one girl, that is, we wish to find $\Pr\{X = x \mid X > 0\}$. From (5.23) we obtain

$$\Pr\{X = x \mid X > 0\} = \frac{\Pr\{X = x\}}{1 - \Pr\{X = 0\}} = \frac{\binom{s}{x}(\frac{1}{2})^s}{1 - (\frac{1}{2})^s}, \quad x = 1, 2, \ldots, s. \quad (5.25)$$

[The distribution given by (5.25) is a special case of the *truncated* binomial distribution.] For instance, if $s = 4$ we obtain

x:	0	1	2	3	4
$\Pr\{X = x\}$:	0.0625	0.2500	0.3750	0.2500	0.0625
$\Pr\{X = x \mid X > 0\}$:	0	0.2667	0.4000	0.2667	0.0667.

5.5. JOINT AND MARGINAL DISTRIBUTIONS IN A TWO-DIMENSIONAL SAMPLE SPACE

We shall now consider a two-dimensional sample space as defined in section 2.9. Instead of sets of compound events (E, D), we may consider a two-dimensional random variable (X, Y) and define the *joint probability distribution of X and Y*.

Suppose that, in the general case, X can take values $-\infty < x < \infty$ and Y can take values $-\infty < y < \infty$.

DEFINITION 5.4. *The joint cumulative distribution function, $F(x, y)$, of two random variables X and Y is defined as*

$$\Pr\{X \leqslant x, Y \leqslant y\} = F(x, y). \quad (5.26)$$

Analogously to the one-dimensional distribution, a more precise notation would be $F_{XY}(x, y)$ instead of $F(x, y)$.

DEFINITION 5.5. *The marginal cumulative distribution function of the random variable X, $F_1(x)$, say, is defined as*

$$\Pr\{X \leqslant x, Y < \infty\} = F_1(x) = F_X(x). \quad (5.27)$$

This is, in fact, the distribution of the variate X without regard to what the values of Y may be.

Similarly

$$\Pr\{X < \infty, Y \leqslant y\} = F_2(y) = F_Y(y) \quad (5.28)$$

is the marginal cdf of Y.

We notice that Definitions 5.4 and 5.5 apply in both continuous and discrete cases.

In particular, suppose that X takes only discrete values $x_1 < x_2 < \cdots < x_m$, and Y also only discrete values $y_1 < y_2 < \cdots < y_n$.

DEFINITION 5.6.   *The function $p(x, y)$ of ordered pairs (x, y) over Ω, defined as*

$$p_{XY}(x, y) = p(x, y) = \begin{cases} \Pr\{X = x_i, Y = y_j\} \text{ if } x = x_i, y = y_j, \\ 0 \text{ if } (x, y) \text{ is not identical with } (x_i, y_j), \\ \quad \text{for any pair } (i, j) \end{cases} \quad (5.29)$$

where $i = 1, 2, \ldots, m; j = 1, 2, \ldots, n$, and with

$$\sum_{i=1}^{m} \sum_{j=1}^{n} p(x_i, y_j) = 1,$$

is called the joint probability distribution of the discrete random variables X and Y.

Thus for any pair (x, y) such that $x_i \leqslant x < x_{i+1}$, $y_j \leqslant y < y_{j+1}$, the cdf can be evaluated from the formula

$$F(x, y) = \Pr\{X \leqslant x, Y \leqslant y\} = \sum_{x \leqslant x_i} \sum_{y \leqslant y_j} p(x, y) = \sum_{t=1}^{i} \sum_{u=1}^{j} p(x_t, y_u). \quad (5.30)$$

The definitions of joint probability distribution and joint cdf can be easily extended to k random variables.

DEFINITION 5.7.   *The probability distribution*

$$\Pr\{X = x, Y \leqslant y_n\} = \sum_{u=1}^{n} p(x, y_u) = p_1(x) = p_X(x) \quad (5.31)$$

is the marginal distribution of the random variable X, and similarly

$$\Pr\{X \leqslant x_m, Y = y\} = \sum_{t=1}^{m} p(x_t, y) = p_2(y) = p_Y(y) \quad (5.32)$$

is the marginal distribution of the random variable Y.

The reader should find it easy to see the resemblance to marginal distributions of sequences of events in Table 2.2. The probabilities $p_i.$ are now replaced by $p_1(x_i)$ $(i = 1, 2, \ldots, m)$ and $p_{.j}$ by $p_2(y_j)$ $(j = 1, 2, \ldots, n)$. We notice that

$$\sum_{i=1}^{m} p_1(x_i) = \sum_{j=1}^{n} p_2(y_j) = \sum_{i=1}^{m} \sum_{j=1}^{n} p(x_i y_j) = 1. \quad (5.33)$$

Definitions 5.6 and 5.7 can easily be generalized for infinite sets of values for x and y.

If X and Y are *continuous*, then $\Pr\{X = x, Y = y\} = 0$ for any pair of values (x, y). If the second derivative $(\partial^2 F(x, y)/\partial x \, \partial y) = (\partial^2 F(x, y)/\partial y \, \partial x) = f(x, y)$ exists almost everywhere, then $f(x, y)$ [more precisely, $f_{XY}(x, y)$] is called the *joint density function of X, Y*, and the joint cdf is

$$\Pr\{X \leqslant x, Y \leqslant y\} = \int_{-\infty}^{x} \int_{-\infty}^{y} f(t, u) \, du \, dt = F(x, y) = F_{XY}(x, y). \quad (5.34)$$

Similarly, the marginal cdf of X is

$$\Pr\{X \leqslant x, Y < \infty\} = \int_{-\infty}^{x} \left[\int_{-\infty}^{\infty} f(t, u) \, du \right] dt = \int_{-\infty}^{x} f_1(t) \, dt, \quad (5.35)$$

where

$$\int_{-\infty}^{\infty} f(t, u) \, du = f_1(t) = f_X(t) \quad (5.36)$$

is the marginal density function of X. The marginal cdf and the density of Y are defined in a similar way.

Example 5.7. Let us consider the three alleles O, A, and B in the ABO blood group system. We have assumed that A and B are dominants over O, but the heterozygote AB exhibits the properties of both antigens, so A and B are codominants (compare Example 2.2).

Suppose that the gene frequencies are as follows:

Gene:	A	B	O
Gene frequency:	$p = 0.153$	$q = 0.174$	$r = 0.673$

in a large ("infinite") population. We assume that the population is in equilibrium with respect to this locus. Let (R, D) denote a recipient-donor pair selected at random from this population. Also, let X denote the number of alleles (in both chromosomes) with respect to which a recipient and a donor may differ at this locus. Then X can take these values:

$$X = \begin{cases} 0 & \text{if } R \text{ and } D \text{ are identical,} \\ 1 & \text{if } R \text{ and } D \text{ differ with respect to one allele,} \\ 2 & \text{if } R \text{ and } D \text{ differ with respect to two alleles.} \end{cases}$$

The random variable X defines a *genetic dissimilarity* of R and D.

A donor D is suitable for (or compatible with) a recipient R if D possesses no antigen(s) which are absent in R. If D possesses at least one antigen which is absent in R, agglutination occurs and D is unsuitable for (or incompatible with) R.

Let Y be a random variable associated with incompatibility of D with R and taking values:

$$Y = \begin{cases} 0 & \text{if } D \text{ is compatible with } R, \\ 1 & \text{if } D \text{ is incompatible with } R. \end{cases}$$

TABLE 5.1a   The values (x, y) in the sample space of genotypes

Recipient \ Donor		OO p^2	OA $2pq$	AA q^2	OB $2pr$	BB r^2	AB $2qr$
OO	p^2	0, 0	1, 1	2, 1	1, 1	2, 1	2, 1
OA	$2pq$	1, 0	0, 0	1, 0	1, 1	2, 1	1, 1
AA	q^2	2, 0	1, 0	0, 0	2, 1	2, 1	1, 1
OB	$2pr$	1, 0	1, 1	2, 1	0, 0	1, 0	1, 1
BB	r^2	2, 0	2, 1	2, 1	1, 0	0, 0	1, 1
AB	$2qr$	2, 0	1, 0	1, 0	1, 0	1, 0	0, 0

Table 5.1a represents the sample space of genotypes for which the values of (X, Y) are shown.

Since we have assumed that the population is in equilibrium and (R, D) pairs are matched at random, we can evaluate the joint genotype distribution of (R, D) pairs. For instance, $\Pr\{R = AA, D = AA\} = q^2 \times q^2 = q^4$; $\Pr\{R = AB, D = OA\} = 2qr \times 2pq = 4pq^2r$; etc.

We notice that in this case it is more reasonable to use the terminology of "joint genotype distribution" of (R, D) pairs than to assign to each genotype an "artificial" random variable. But when we discuss genetic dissimilarity (X) and immunological incompatibility (Y), it is much more convenient to use the joint distribution of (X, Y).

From Table 5.1a we evaluate these probabilities. We have

$$\Pr\{X = 0, Y = 0\} = p^4 + q^4 + r^4 + (2pq)^2 + (2pr)^2 + (2qr)^2;$$

substituting $p=0.153$, $q=0.174$, $r=0.673$, we obtain $\Pr\{X=0, Y=0\}=0.307$.

The remaining probabilities are:

$$\Pr\{X = 0, Y = 1\} = 0;$$

$$\begin{aligned}
\Pr\{X = 1, Y = 0\} &= 2pq(p^2 + q^2) + q^2 \cdot 2pq + 2pr(p^2 + r^2) + r^2 \cdot 2pr \\
&\quad + 2qr(2pq + q^2 + 2pr + r^2) \\
&= 2pq(p^2 + 2q^2) + 2pr(p^2 + 2r^2) + 2qr(1 - p^2 - 2qr) \\
&= 0.370;
\end{aligned}$$

$$\begin{aligned}
\Pr\{X = 1, Y = 1\} &= p^2(2pq + 2pr) + 2pq(2pr + 2qr) + q^2 \cdot 2qr \\
&\quad + 2pr(2pq + 2qr) + r^2 \cdot 2qr \\
&= 2p^3(1 - p) + 2qr(q^2 + r^2) + 4pqr(1 + p) = 0.201;
\end{aligned}$$

$$\Pr\{X = 2, \, Y = 0\} = p^2(q^2 + r^2 + 2qr) = p^2(1 - p)^2 = 0.017;$$

$$\Pr\{X = 2, \, Y = 1\} = p^2(q^2 + r^2 + 2qr) + 2pqr^2 + q^2(2pr + r^2)$$
$$+ 2prq^2 + r^2(2pq + q^2)$$
$$= p^2(1 - p)^2 + 4pqr(1 - p) + 2q^2r^2 = 0.105.$$

The calculated probabilities are given in Table 5.1b. From this table we find that the probability of a donor D and a recipient R, selected at random, being

TABLE 5.1b Joint distribution $p(x, y)$

x \ y	0	1	$p_1(x)$
0	0.307	0.000	0.307
1	0.370	0.201	0.571
2	0.017	0.105	0.122
$p_2(y)$	0.694	0.306	1.000

genetically identical with respect to ABO locus (and so completely compatible) is

$$\Pr\{X = 0, \, Y = 0\} = 0.307.$$

This is the expected proportion of cases in which D and R are genetically identical.

The probability that D will be "immunologically compatible" with respect to this locus is

$$\Pr\{Y = 0\} = \Pr\{X = 0, \, Y = 0\} + \Pr\{X = 1, \, Y = 0\} + \Pr\{X = 2, \, Y = 0\}$$
$$= 0.694,$$

which is the value of the marginal probability $p_2(y)$ for $y = 0$.

In a similar way we can find the probabilities of other events.

5.6. CONDITIONAL DISTRIBUTIONS. INDEPENDENCE OF RANDOM VARIABLES

Finally, we define conditional distributions.

DEFINITION 5.8. *Let X and Y be two discrete random variables with joint probability distribution $p(x, y)$. Let the corresponding marginal distributions be*

$p_1(x)$ and $p_2(y)$. *The conditional distribution of X, given that Y takes the value y, is*

$$p(x \mid y) = \frac{p(x, y)}{p_2(y)} = \begin{cases} \Pr\{X = x_i \mid Y = y_j\} = \dfrac{\Pr\{X = x_i, Y = y_j\}}{\Pr\{Y = y_j\}} \\ \quad \text{if } x = x_i, y = y_j \\ 0 \quad \text{if } (x, y) \text{ is not identical with } (x_i, y_j) \end{cases} \tag{5.37}$$

for $i = 1, 2, \ldots, m; j = 1, 2, \ldots, n$.

In a similar way we define the conditional distribution of Y, given that X takes the value x:

$$p(y \mid x) = \frac{p(x, y)}{p_1(x)} = \begin{cases} \Pr\{Y = y_j \mid X = x_i\} = \dfrac{\Pr\{X = x_i, Y = y_j\}}{\Pr\{X = x_i\}} \\ \quad \text{if } x = x_i, y = y_j \\ 0 \quad \text{if } (x, y) \text{ is not identical with } (x_i, y_j) \end{cases} \tag{5.38}$$

for $i = 1, 2, \ldots, n; j = 1, 2, \ldots, n$.

DEFINITION 5.9. *The conditional cumulative distribution function $F(x \mid y_j)$, for $x_i \leqslant x < x_{i+1}$, is*

$$F(x \mid y_j) = \Pr\{X \leqslant x \mid Y = y_j\} = \sum_{t=1}^{i} \frac{\Pr\{X = x_t, Y = y_j\}}{\Pr\{Y = y_j\}}, \tag{5.39}$$

and similarly,

$$F(y \mid x_i) = \Pr\{Y \leqslant y \mid X = x_i\} = \sum_{u=1}^{j} \frac{\Pr\{X = x_i, Y = y_u\}}{\Pr\{X = x_i\}} \tag{5.40}$$

for $y_j \leqslant y < y_{j+1}$.

Suppose that

$$p(x, y) = p_1(x)p_2(y); \tag{5.41}$$

then, substituting (5.41) into (5.37) and (5.38), we obtain

$$p(x \mid y) = p_1(x), \tag{5.42}$$

$$p(y \mid x) = p_2(y), \tag{5.43}$$

respectively. This means that in this case the conditional distributions are the same as the marginal distributions of the corresponding variables or, in other words, that X and Y are *independent*.

DEFINITION 5.10. *Two random variables, X and Y, are (statistically) inde-pendent if their joint distribution, $p(x, y)$ is the product of the marginal distri-butions, $p_1(x)$ and $p_2(y)$, that is*

$$p(x, y) = p_1(x)p_2(y). \tag{5.44}$$

We notice the analogy to formula (2.49a).

The definitions and results of this section can be extended to continuous variables, using density functions $f(x, y)$, $f_1(x)$, $f_2(x)$ instead of probability functions $p(x, y)$, $p_1(x)$, $p_2(x)$, respectively, and replacing the summation sign Σ with the integration sign $\int$.

It is worthwhile to notice the difference between functional and statistical dependence (or independence).

The variables (real or random) $x_1, x_2, \ldots, x_k$ are *functionally dependent* if there exists a function ψ such that

$$\psi(x_1, x_2, \ldots, x_k) = 0. \tag{5.45}$$

Otherwise, $x_1, x_2, \ldots, x_k$ are *functionally independent*.

The (statistical) dependence or independence of random variables is solely expressed in terms of probability distributions.

Example 5.8.   Using the data from Example 5.7, Table 5.1b, we find the conditional distribution of "genetic dissimilarities" (X), *given* that a donor is compatible with a recipient ($y = 0$), that is, $p(x \mid 0) = \Pr\{X = x_i \mid Y = 0\}$, $x_i = 0, 1, 2$. We calculate

$$\Pr\{X = 0 \mid Y = 0\} = \frac{0.307}{0.694} = 0.442,$$

$$\Pr\{X = 1 \mid Y = 0\} = \frac{0.370}{0.694} = 0.533,$$

$$\Pr\{X = 2 \mid Y = 0\} = \frac{0.017}{0.694} = 0.025.$$

We note that these probabilities must add up to 1.

The variables X and Y are not independent. In fact, $p(x, y) \neq p_1(x)p_2(y)$; for instance, $0.307 \neq 0.694 \times 0.307$ and $0.370 \neq 0.694 \times 0.572$. This mathematical result is consistent with what we know about "immunological incompatibility": the joint genetical constitutions of a donor-recipient pair determine whether or not they are compatible.

5.7. DISTRIBUTION OF THE SUM OF RANDOM VARIABLES

Let X and Y be discrete random variables determined for the sets of values $x_1 < x_2 < \cdots < x_m$ and $y_1 < y_2 < \cdots < y_n$, respectively, with joint distribution $p(x, y)$. Of special interest might be a random variable Z which is the *sum* of X and Y, that is,

$$Z = X + Y. \tag{5.46}$$

In the general case Z can take mn distinct values, z_{ij}, say, such that $z_{ij} = x_i + y_j$, for $i = 1, 2, \ldots, m; j = 1, 2, \ldots, n$. In such a case we have

$$\Pr\{Z = z_{ij}\} = \Pr\{X = x_i, \, Y = y_j\} = p(x_i, y_j). \tag{5.47}$$

It happens quite often that some of the z_{ij} are identical, so that, in fact, Z takes only $s < mn$ values, $z_1, z_2, \ldots, z_s$, say. Now the probability $\Pr\{Z = z_l\}$ is the sum of probabilities $\Pr\{X = x_i, \, Y = y_j\}$ for pairs of values (x_i, y_j) satisfying the condition $x_i + y_j = z_l$. Therefore, we have

$$\Pr\{Z = z_l\} = \sum_{x_i + y_j = z_l} \Pr\{X = x_i, \, Y = y_j\} = \sum_{x_i + y_j = z_l} p(x_i, y_j). \tag{5.48}$$

If there are more than two variables, the method of evaluation of probabilities can easily be extended. For instance, if

$$Z = X_1 + X_2 + X_3,$$

we find the probabilities for $Y = X_1 + X_2$, and then for $Z = Y + X_3$.

Example 5.9.   Let us consider the example given in Problem 2.5 regarding the genetics of the color of wheat kernels. We assumed there that the color of wheat kernels is controlled by two independent loci, each with two alleles, R_1, r_1 and R_2, r_2, respectively. The intensity of red color is actually a function of genes R_i ($i = 1, 2$), regardless of whether they belong to one locus or the other.

TABLE 5.2   The color of wheat kernels as a function of R genes

Genotype	Color	X_1	X_2	Z	$p(x_{1i}, x_{2j})$	$p(z_l)$
$r_1 r_1 r_2 r_2$	White	0	0	0	$\frac{1}{16}$	$\frac{1}{16}$
$R_1 r_1 r_2 r_2$ $r_1 r_1 R_2 r_2$	Light red	1 0	0 1	1	$\frac{2}{16}$ $\frac{2}{16}$	$\frac{4}{16}$
$R_1 R_1 r_2 r_2$ $R_1 r_1 R_2 r_2$ $r_1 r_1 R_2 r_2$	Medium red	2 1 0	0 1 2	2	$\frac{1}{16}$ $\frac{4}{16}$ $\frac{1}{16}$	$\frac{6}{16}$
$R_1 R_1 R_2 r_2$ $R_1 r_1 R_2 R_2$	Medium dark red	2 1	1 2	3	$\frac{2}{16}$ $\frac{2}{16}$	$\frac{4}{16}$
$R_1 R_1 R_2 R_2$	Dark red	2	2	4	$\frac{1}{16}$	$\frac{1}{16}$

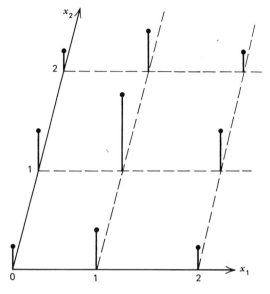

Figure 5.3.

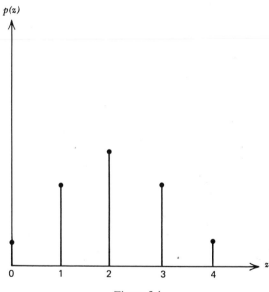

Figure 5.4.

Suppose that we are interested in the distribution of the color of wheat kernels in a large population resulting from intercrosses $R_1 r_1 R_2 r_2 \times R_1 r_1 R_2 r_2$. Let X_1 and X_2 be random variables determining the numbers of R_1 and R_2 alleles at each locus, respectively. The random variable $Z = X_1 + X_2$ determines the number of R_i genes present in the genotype, and hence the color of the wheat kernels.

Table 5.2 shows the values of X_1, X_2, and $Z = X_1 + X_2$, the joint probability distribution $\Pr\{X_1 = x_{1i}, X_2 = x_{2j}\} = p(x_{1i}, x_{2j})$, and $\Pr\{Z = z_l\} = p(z_l)$, in the F_2 generation. We notice that in this example the values of $p(z_l)$ are terms in the expansion of $(\frac{1}{2} + \frac{1}{2})^4$.

Figure 5.3 represents the probabilities $p(x_{1i}, x_{2j})$ over the two-dimensional sample space, and Fig. 5.4 the values of $p(z_l)$ over the one-dimensional sample space.

SUMMARY

A random variable as a certain numerical measure assigned to events has been introduced. The probability distribution of a random variable in one-dimensional space and the joint and marginal distributions in two-dimensional space have been defined. The reader will notice an analogy with the corresponding definitions for events, introduced in Chapter 2. Some standard distributions will be discussed in Chapter 8.

REFERENCES

Gangolli, R. A., and Ylvisaker, D., *Discrete Probability*, Harcourt, Brace and World, New York, 1966.

Johnson, N. L., and Leone, F. C., *Statistics and Experimental Design*, Vol. I, John Wiley & Sons, New York, 1964.

Lindgren, B. W., *Statistical Theory*, The Macmillan Company, New York, 1968 (second edition).

Srb, A. M., Owen, R. D., and Edgar, R. S., *General Genetics*, W. H. Freeman & Co., San Francisco, 1965.

Wilks, S. S., *Mathematical Statistics*, John Wiley & Sons, New York, 1962.

PROBLEMS

5.1. Suppose that in a certain population the gene frequencies are as follows: for the *MN* blood group system, of *M* gene, $p = 0.580$ and of *N* gene, $q = 0.420$; and for the Lutheran system, of Lu^a gene, $u = 0.036$ and of Lu^b gene, $v = 0.964$. We notice that *M* and *N* are codominant, and so are Lu^a and Lu^b, and that these systems are independent. Assume that the population is in equilibrium with respect to these loci.

(a) Find the joint genotype distribution.

Let X denote a random variable which determines the incompatibility of a donor with a recipient with respect to the MN system, namely,

$$X = \begin{cases} 0 & \text{if } D \text{ is compatible with } R, \\ 1 & \text{if } D \text{ is incompatible with } R; \end{cases}$$

and, similarly, let Y determine the incompatibility with respect to the Lutheran system.

(b) Find the distributions $p_1(x)$, $p_2(y)$, and $p(x, y)$.

(c) Find the conditional distribution $p(y \mid X = 0)$, that is, the "incompatibility distribution" with respect to the Lutheran system, *given* that the donor-recipient pair is compatible with respect to the MN system.

5.2. Find the truncated distribution of the random variable Z defined in Example 5.9, given that the white wheat kernels are eliminated.

5.3. In Example 5.9, calculate the probabilities $\Pr\{1 \leqslant Z \leqslant 2\}$, $\Pr\{0 < Z \leqslant 2\}$, $\Pr\{0 < Z < 3\}$. What events are determined by the inequalities $1 \leqslant Z \leqslant 2$, $0 < Z \leqslant 2$, and $0 < Z < 3$? Are these the same events?

CHAPTER 6

Moments of Random Variables

6.1. EXPECTED VALUES

When some properties of a population are expressed in terms of random variables, some useful characteristics can be evaluated. One of these is the population mean, also called the *expected value*. Let us consider a simple example.

Example 6.1. Consider first a finite population of N individuals in which N_1 are of genotype AA, N_2 of Aa, and N_3 of aa, with $N_1 + N_2 + N_3 = N$. The question "What is the 'average genotype' in this population?" does not make much sense, but the question "What is the average number of genes A?" appears to be reasonable. Let X denote the number of genes A in the genotype; then X takes the value 2 if AA, 1 if Aa, and 0 if aa. The average number of genes A, which we denote here by $\bar{x}$, is

$$\bar{x} = \frac{1}{N}(2 \cdot N_1 + 1 \cdot N_2 + 0 \cdot N_3) = 2 \cdot \frac{N_1}{N} + 1 \cdot \frac{N_2}{N} + 0 \cdot \frac{N_3}{N}.$$

This is the average for the finite population of size N. When we consider a model of infinite population, the relative frequencies N_i/N ($i = 1, 2, 3$) may approach limits which are probabilities $\Pr\{X = x_i\} = p(x_i)$ (where $x_1 = 2$, $x_2 = 1$, $x_3 = 0$), and $\bar{x}$ will approach a limit which is now called the expected value of X (i.e., the expected number of genes A in an infinite population). The result can be generalized in the following definition.

DEFINITION 6.1. *Let X be a discrete random variable, taking values $x_1, x_2, \ldots, x_n$, with probabilities $\Pr\{X = x_i\} = p(x_i)$, $i = 1, 2, \ldots, n$. The expected value of X, $E(X)$, is defined as*

$$E(X) = \sum_{i=1}^{n} x_i p(x_i). \tag{6.1}$$

For a continuous variable, X, with density function $f(x)$ determined for

$-\infty < x < \infty$, *the expected value is defined as*

$$E(X) = \int_{-\infty}^{\infty} xf(x)\, dx, \tag{6.2}$$

assuming that the integral in (6.2) exists.

In this chapter, we will be dealing with discrete variables taking finite sets of values. To obtain appropriate formulae for the continuous case we replace the probability function $p(x_i)$ by the density function $f(x)$ and summation by integration.

Note that:

(i) The symbol E is used because E is the first letter in the word "Expected." It has no connection with the meaning "Event," which we have also associated with E. Some authors use script $\mathscr{E}$ for the expected value. It is also customary to use the symbol $\mu(X)$ or μ_X, "the mean of X," or, if no ambiguity arises, simply the symbol μ. $E(X)$ is, in fact, a *hypothetical mean* which may never be observed but is "expected" in an infinite population.

(ii) $E(X)$ is *not a function* of X; it is a single *numerical value*. Here E plays the role of an operator. We may sometimes use the symbol E_X to indicate that this is a property of the random variable X. It will be especially necessary to use this symbol when more than one variate is considered.

Example 6.2. Let Z be a random variable determining the number of R genes in the genotype as explained in Example 5.9. Using the probability distribution $p(z_i)$, given in the last column of Table 5.2, we evaluate

$$E(Z) = \tfrac{1}{16}(0 \times 1 + 1 \times 4 + 2 \times 6 + 3 \times 4 + 4 \times 1) = \tfrac{32}{16} = 2.$$

The expected number of R genes in the F_2 generation is equal to 2.

Using Definition 6.1, we find, for any constant value a,

$$E(a) = a, \quad E(aX) = aE(X). \tag{6.3}$$

It also follows from Definition 6.1 that, if $g(X)$ is a function of a discrete variable X taking values $x_1, \ldots, x_n$, with probabilities $p(x_1), \ldots, p(x_n)$, respectively, then

$$E[g(X)] = \sum_{i=1}^{n} g(x_i)p(x_i). \tag{6.4}$$

In particular,

$$E(bX + a) = \sum_{i=1}^{n}(bx_i + a)p(x_i) = b\sum_{i=1}^{n} x_i p(x_i) + a\sum_{i=1}^{n} p(x_i) = bE(X) + a. \tag{6.5}$$

If $b = 1$ and $a = -\mu$, we obtain

$$E(X - \mu) = 0. \tag{6.6}$$

But, in general,

$$E[g(X)] \neq g[E(X)]. \tag{6.7}$$

This means that, in general, the expected value of a function $g(X)$ is *not equal* to the value of this function evaluated with X replaced by its expected value.

Example 6.3. From (6.1) we have

$$E(X) = \sum_{i=1}^{n} x_i p(x_i).$$

Let $g(X) = X^2$. From (6.4) we obtain

$$E(X^2) = \sum_{i=1}^{n} x_i^2 \, p(x_i).$$

It is easy to see that $E(X^2) \neq [E(X)]^2$. [In fact, $E(X^2) \geqslant [E(X)]^2$; compare formula (6.31).]

A useful comment may be made here. From (5.38) we have

$$p(x, y) = p(y \mid x) p_1(x). \tag{6.8}$$

We may consider $p(y \mid x)$ as a function of x, $g(x)$, say, for all possible values $x_1, x_2, \ldots$.

Taking the sums of both sides of (6.8) over all values of x, we obtain

$$p_2(y) = \sum_x p(x, y) = \sum_x [p(y \mid x)] p_1(x) = E_X[p(y \mid X)]. \tag{6.9}$$

In other words, the marginal probability $\Pr\{Y = y\} = p_2(y)$ is an *average* of the conditional probabilities $p(y \mid x)$, evaluated over all values of x.

For example, the (conditional) distributions of genotypes of offspring are different for various parental genotypes. On the average, however, the unconditional (marginal) distribution of genotypes of offspring in a large population is the same as if the offspring of all parents were pooled together and their genotype distribution evaluated regardless of the parental genotype patterns. (See sections 7.5 and 7.6.)

Let X and Y be two discrete random variables taking values $x_1, x_2, \ldots, x_m$ and $y_1, y_2, \ldots, y_n$, respectively, with joint distribution $p(x, y)$. We have

$$E(X) = \sum_{i=1}^{m} \sum_{j=1}^{n} x_i p(x_i, y_j) = \sum_{i=1}^{m} x_i \left[\sum_{j=1}^{n} p(x_i, y_j) \right] = \sum_{i=1}^{m} x_i p_1(x_i), \tag{6.10}$$

and, similarly,

$$E(Y) = \sum_{i=1}^{m} \sum_{j=1}^{n} y_j p(x_i, y_j) = \sum_{j=1}^{n} y_j p_2(y_j). \tag{6.11}$$

In other words, in order to calculate the expected value of a single variable X (or Y), it is only necessary to know the marginal distribution of this variable.

In section 5.6 we defined the conditional distribution $p(x \mid y)$; we shall now define the conditional expected value $E(X \mid Y) = E_{X \mid Y}$.

DEFINITION 6.2. *The conditional expected value of X given Y, is defined as*

$$E(X \mid Y) = E_{X \mid Y} = \sum_{i=1}^{m} x_i p(x_i \mid Y). \tag{6.12}$$

We now find the relationship between $E(X)$ and $E(X \mid Y)$:

$$E(X) = E_X = \sum_{i=1}^{m} \sum_{j=1}^{n} x_i p(x_i, y_j) = \sum_{j=1}^{n} \sum_{i=1}^{m} x_i p(x_i \mid y_j) p_2(y_j),$$

$$= \sum_{j=1}^{n} E(X \mid Y = y_j) p_2(y_j) = E_Y [E_{X \mid Y}]. \tag{6.13}$$

This means that, to evaluate the expected value with respect to X, we may calculate the conditional expected value for each given value of Y and then take the expectation of all these conditional expected values.

In a similar way we obtain

$$E(Y) = E_Y = E_X [E_{Y \mid X}]. \tag{6.14}$$

If $g(X, Y)$ is a function of two discrete random variables taking values $x_1, \ldots, x_m$ and $y_1, \ldots, y_n$, respectively, with joint distribution $p(x, y)$, then the expected value of $g(X, Y)$ is

$$E[g(X, Y)] = \sum_{i=1}^{m} \sum_{j=1}^{n} g(x_i, y_j) p(x_i, y_j). \tag{6.15}$$

Of particular interest is the case in which $g(X, Y)$ is a *sum* of two random variables, that is,

$$g(X, Y) = Z = X + Y.$$

Here we have

$$E(X + Y) = \sum_{i=1}^{m} \sum_{j=1}^{n} (x_i + y_i) p(x_i, y_j) = \sum_{i=1}^{m} \sum_{j=1}^{n} x_i p(x_i, y_j) + \sum_{i=1}^{m} \sum_{j=1}^{n} y_j p(x_i, y_j)$$

$$= \sum_{i=1}^{m} x_i p_1(x_i) + \sum_{j=1}^{m} y_j p_2(y_j) = E(X) + E(Y). \tag{6.16}$$

The result can be generalized to k random variables as follows.

The expected value of the sum of k random variables, $X_1, X_2, \ldots, X_k$, is equal to the sum of the k individual expected values, that is,

$$E(X_1 + X_2 + \cdots + X_k) = E(X_1) + E(X_2) + \cdots + E(X_k). \tag{6.17}$$

More generally, the expected value of a *linear* function of k random variables is the same linear function of the k individual expected values, that is,

$$E(a_1X_1 + a_2X_2 + \cdots + a_kX_k) = a_1E(X_1) + a_1E(X_2) + \cdots + a_kE(X_k).$$
(6.18)

Example 6.4. For the random variable Z defined in Example 5.9 we obtained $E(Z) = 2$ (see Example 6.2). But we also have

$$E(X_1) = E(X_2) = 0 \times \tfrac{1}{4} + 1 \times \tfrac{2}{4} + 2 \times \tfrac{1}{4} = 1.$$

Since $Z = X_1 + X_2$, then $E(Z) = E(X_1) + E(X_2) = 2$.

Hence we could have found $E(Z)$ without knowing the distribution of Z, using only the distributions of X_1 and X_2.

Property (6.18) applies to k random variables regardless of their dependence or independence. Another useful property applies to two *independent* random variables.

Let us find the expected value of the product $U = XY$ with probability distribution $p(x, y) = p_1(x)p_2(y)$. We have

$$E(XY) = \sum_{i=1}^{m} \sum_{j=1}^{n} x_i y_j p(x_i)p(y_j) = \sum_{i=1}^{m} x_i p_1(x_i) \sum_{j=1}^{n} y_j p_2(y_j) = E(X)E(Y). \quad (6.19)$$

This property can be extended to k variables.

The expected value of the product of k independent random variables with the joint distribution $p(x_1, x_2, \ldots, x_k) = p_1(x_1)p_2(x_2) \cdots p_k(x_k)$ *is the product of the individual expected values, that is,*

$$E(X_1X_2 \cdots X_k) = E(X_1)E(X_2) \cdots E(X_k).$$
(6.20)

6.2. DEFINITIONS OF MOMENTS

In section 6.1 we saw that the expected value of a function $g(X)$ is $E[g(X)] = \Sigma\, g(x_i)p(x_i)$. Of special interest are functions of the type $g(X) = (X - a)^r$, where a is constant.

DEFINITION 6.3. *The expected value of* $(X - a)^r$, $E[(X - a)^r]$ *(if it exists), is called the rth moment of the random variable X, or of the distribution of X, about the point a.*

If $a = 0$, $E(X^r)$ is the rth moment about zero, which is simply called the *rth moment* and is denoted as $\mu_r'(X)$ or, if no ambiguity arises, as μ_r'.

We notice that the first moment, $\mu_1'(X) = \mu_1' = E(X) = \mu_X = \mu$, is just the expected value (or mean) of X. As has been mentioned before, it is often denoted by μ (instead of μ_1').

DEFINITION 6.4.   *The rth moment about the mean*

$$\mu_r(X) = \mu_r = E\{[X - E(X)]^r\} = E[(X - \mu)^r] \qquad (6.21)$$

is called the rth central moment.

If X and Y are two discrete random variables with joint distribution $p(x, y)$ then the rth moment of X is

$$\mu_r'(X) = E(X^r) = \sum_{i=1}^{m} \sum_{j=1}^{n} x_i^r p(x_i, y_j) = \sum_{i=1}^{m} x_i^r p_1(x_i), \qquad (6.22)$$

and the rth central moment is

$$\mu_r(X) = E[(X - \mu_X)^r] = \sum_{i=1}^{m} (x_i - \mu_X)^r p_1(x_i). \qquad (6.23)$$

Similarly the rth moment and the rth central moment of Y are

$$\mu_r'(Y) = E(Y^r) = \sum_{j=1}^{n} y_j^r p_2(y_j) \qquad (6.24)$$

and

$$\mu_r(Y) = E[(Y - \mu_Y)^r] = \sum_{j=1}^{n} (y_j - \mu_Y)^r p_2(y_j), \qquad (6.25)$$

respectively.

But we can also consider joint moments of (X, Y).

DEFINITION 6.5.   *The $r_1 r_2$th mixed moment of (X, Y) is*

$$\mu_{r_1 r_2}'(X, Y) = \mu_{r_1 r_2}' = E(X^{r_1} Y^{r_2}) = \sum_{i=1}^{m} \sum_{j=1}^{n} x_i^{r_1} y_j^{r_2} p(x_i, y_j), \qquad (6.26)$$

and the $r_1 r_2$th central mixed moment is

$$\mu_{r_1 r_2}(X, Y) = \mu_{r_1 r_2} = E[(X - \mu_X)^{r_1}(Y - \mu_Y)^{r_2}]$$

$$= \sum_{i=1}^{m} \sum_{j=1}^{n} (x_i - \mu_X)^{r_1}(y_j - \mu_Y)^{r_2} p(x_i, y_j). \qquad (6.27)$$

If $Z = g(X)$ and X has the probability distribution $p(x)$, then the rth moment of $g(X)$ is $E\{[g(X)]^r\}$, and the rth central moment of $g(X)$ is $E\{[g(X) - E(g(X))]^r\}$, etc.

These definitions can also be extended to more than two random variables, but they then become a little more complicated.

6.3. VARIANCE

In most of our problems we will be concerned with the first and second moments only.

In section 6.1 we defined the expected value (mean), which is sometimes called a location characteristic or *location parameter* of the distribution. It informs us only about the average value of a certain character (X) in a population, giving no information about the variability of X.

Example 6.5. Suppose that two random variables, X_1 and X_2, are distributed as follows:

$$
\begin{array}{cccc}
X_1: & 0 & 1 & 2 \\
X_2: & -1 & 0 & 4 \\
p(x_1) = p(x_2): & \tfrac{1}{3} & \tfrac{1}{3} & \tfrac{1}{3}.
\end{array}
$$

We find $E(X_1) = E(X_2) = 1$; these two random variables have the same expected value, but they have different *spreads, dispersions*, or *variabilities*.

The *range*, defined as the difference between the greatest and the least value of X, can sometimes be used as a measure of variability. It is useful when the set of values which X can take is finite and $p(x)$ is uniform, that is, takes the same value for each x. It is not so good when $p(x)$ takes several widely different values, and has no meaning if there are no bounds on the values which X can take.

The measure of variability usually employed in statistical analysis is the variance (or its square root, the standard deviation).

DEFINITION 6.6. *The second central moment of a random variable X is called its variance, that is,*

$$
\mathrm{Var}(X) = \sigma_X{}^2 = E[(X - \mu)^2]. \tag{6.28}
$$

$\mathrm{Var}(X)$ is a single numerical value, *not a function* of X. If no ambiguity arises, the symbol σ^2, instead of $\sigma_X{}^2$, is used for the variance.

DEFINITION 6.7. *The positive square root of the variance is called the standard deviation, that is,*

$$
\sigma_X = \sigma = \sqrt{\mathrm{Var}(X)}. \tag{6.29}
$$

We notice that the standard deviation, σ_X, is measured in the same units as X, while variance, $\sigma_X{}^2$, is measured in (units)^2. The σ_X is a *measure of dispersion* of a character (X); it is sometimes called a *scale parameter*.

The variance can be calculated in two ways:

(i) By definition, we have

$$
E[(X - \mu)^2] = \sum_i (x_i - \mu)^2 p(x_i). \tag{6.30}
$$

(ii) Formula (6.30) requires us to evaluate, for each x_i, the deviation $x_i - \mu$, and then to take the square of each deviation and add the squares. More straightforward is another formula obtained as follows. Expanding

$(X - \mu)^2$ and taking expectations of each term, we obtain

$$E[(X - \mu)^2] = E(X^2) - 2E(X)\mu + \mu^2$$
$$= E(X^2) - 2[E(X)]^2 + [E(X)]^2$$
$$= E(X^2) - [E(X)]^2 = \mu_2' - \mu^2. \tag{6.31}$$

Example 6.6.   We find the expected value and the variance of the number of alleles A in a population of genotypes AA, Aa, aa at equilibrium. We have

Genotype:	AA	Aa	aa
Number of alleles A (x):	2	1	0
$p(x)$:	p^2	$2pq$	q^2.

Hence

$$E(X) = 2 \times p^2 + 1 \times 2pq + 0 \times q^2 = 2p.$$

Using formula (6.30), we find the variance:

$$Var(X) = (2 - 2p)^2 p^2 + (1 - 2p)^2 2pq + (0 - 2p)^2 q^2$$
$$= 4p^2 q^2 + (q - p)^2 \cdot 2pq + 4p^2 q^2$$
$$= 2pq(p^2 + 2pq + q^2) = 2pq.$$

It would be quicker, in this case, to calculate $Var(X)$ from formula (6.31). We have

$$\mu_2' = E(X^2) = 4 \times p^2 + 1 \times 2pq + 0 \times q^2 = 2p(2p + q) = 2p(1 + p),$$
$$Var(X) = E(X^2) - [E(X)]^2 = 2p(1 + p) - 4p^2$$
$$= 2p - 2p^2 = 2p(1 - p) = 2pq.$$

If $g(X)$ is a certain function of a random variable X, then

$$Var[g(X)] = E\{[g(X) - E[g(X)]]^2\}. \tag{6.32}$$

In particular, if $g(X) = aX$, then

$$Var(aX) = a^2 Var(X). \tag{6.33a}$$

Note also that for $g(X) = X + a$ we have

$$Var(X + a) = Var(X). \tag{6.33b}$$

6.4. COVARIANCE

When two characters are measured, it is important to know, in addition to their means and variances, whether the characters are related or not, and what is the degree of their relationship, if any exists. A useful characteristic in this regard is *covariance*.

DEFINITION 6.8. *Let X and Y be random variables with joint probability distribution p(x, y). The expected value of the product of deviations of X and Y from their respective means,*

$$\text{Cov}(X, Y) = \sigma_{XY} = E[(X - \mu_X)(Y - \mu_Y)], \tag{6.34}$$

is called the covariance between X and Y.

This is the first mixed central moment μ_{11}, sometimes called the first *central product moment*. It can be calculated from the direct formula

$$\sigma_{XY} = \mu_{11} = \sum_i \sum_j (x_i - \mu_X)(y_j - \mu_Y)p(x_i, y_j) \tag{6.35}$$

or from

$$\sigma_{XY} = \sum_i \sum_j x_i y_j p(x_i, y_j) - \mu_X \mu_Y$$

$$= \sum_i \sum_j x_i y_j p(x_i, y_j) - \left[\sum_i x_i p_1(x_i)\right]\left[\sum_j y_j p_2(y_j)\right], \tag{6.36}$$

or, briefly,

$$\sigma_{XY} = \mu_{11} = E(XY) - E(X)E(Y). \tag{6.36a}$$

The proof of (6.36) is straightforward and therefore is left to the reader.

Variance is always nonnegative ($\sigma_X^2 \geqslant 0$). However, covariance can be negative or positive and theoretically can vary from $-\infty$ to $+\infty$.

Covariance itself is not convenient as a measure of closeness of relationship. It depends on the units in which X and Y are measured. The measure of the degree of relationship between two characters should rather be some type of index, not depending on any particular kind of units. As a measure of *linear* relationship we introduce an index called the linear correlation coefficient or, briefly, the correlation coefficient.

DEFINITION 6.9. *The (linear) correlation coefficient of two random variables, X and Y, ρ_{XY}, is defined as*

$$\rho_{XY} = \frac{\text{Cov}(X, Y)}{\sqrt{\text{Var}(X)}\sqrt{\text{Var}(Y)}} = \frac{\sigma_{XY}}{\sigma_X \sigma_Y}. \tag{6.37}$$

This is sometimes called the *product moment correlation coefficient*.

If no ambiguity arises, we use ρ instead of ρ_{XY}. We notice that

$$\rho = \rho_{XY} = \frac{\mu_{11}(X, Y)}{\sqrt{\mu_2(X)}\sqrt{\mu_2(Y)}}. \tag{6.38}$$

The correlation coefficient, ρ, is dimensionless and takes values in the range

$$-1 \leqslant \rho \leqslant 1. \tag{6.39}$$

It can be shown that, if $\rho = \pm 1$, there is, with probability 1, an exact linear relationship between X and Y:

$$Y = \alpha + \beta X, \text{ say,}$$

where, for $\rho = 1$, $\beta > 0$, and for $\rho = -1$, $\beta < 0$.

The more ρ differs from -1 or $+1$, the greater the departures from a *linear* relationship between X and Y. When $\rho = 0$, X and Y are *uncorrelated*. We notice that $\rho = 0$ is equivalent to $\sigma_{XY} = 0$.

Suppose that X and Y are independent. Then $p(x, y) = p_1(x)p_2(y)$, and

$$\sigma_{XY} = \left[\sum_i (x_i - \mu_X)p_1(x_i) \right] \left[\sum_j (y_j - \mu_Y)p_2(y_j) \right] = 0.$$

Hence, *if X and Y are independent, they are also uncorrelated.*

The reverse statement is not always true—two variables can be uncorrelated, but not necessarily independent. A simple example demonstrates this.

Example 6.7. Let the joint distribution, $p(x, y)$ be as shown in the table.

x_i \ y_j	0	1	$p_1(x_i)$
-1	0	$\frac{1}{3}$	$\frac{1}{3}$
0	$\frac{1}{3}$	0	$\frac{1}{3}$
1	0	$\frac{1}{3}$	$\frac{1}{3}$
$p_2(y_j)$	$\frac{1}{3}$	$\frac{2}{3}$	1

We notice that $p(x, y) \neq p_1(x)p_2(y)$. For instance, $\Pr\{X = -1\} = \frac{1}{3}$, $\Pr\{Y = 1\} = \frac{2}{3}$, $\Pr\{X = -1\} \times \Pr\{Y = 1\} = \frac{2}{9}$, but $\Pr\{X = -1, Y = 1\} = \frac{1}{3} \neq \frac{2}{9}$ and so on. Hence X and Y are *not independent*.

We find $\mu_X = 0$, $\mu_Y = \frac{2}{3}$, and the covariance is

$$\text{Cov}(X, Y) = (-1) \times 1 \times \tfrac{1}{3} + 1 \times 1 \times \tfrac{1}{3} = 0;$$

therefore X and Y are *uncorrelated*. This is so because independence and lack of correlation are (by definition) two different properties of a pair of random variables.

Independence is the following property of the joint distribution:

$$p(x, y) = p_1(x)p_2(y).$$

Lack of correlation, on the other hand, is a property of the central product moment, namely,

$$\mu_{11} = 0.$$

Some applications of the results presented in sections 6.1–6.4 will be described in sections 6.6–6.9.

6.5. VARIANCE OF LINEAR FUNCTIONS OF TWO OR MORE RANDOM VARIABLES

Let

$$Z = a_1 X_1 + a_2 X_2 \tag{6.40}$$

be a *linear* function of two random variables, X_1 and X_2, with means μ_1 and μ_2* and variances σ_1^2 and σ_2^2, respectively, and covariance σ_{12}. We have

$$\mu_Z = a_1\mu_1 + a_2\mu_2.$$

We now evaluate the variance of Z:

$$\begin{aligned}
\text{Var}(Z) &= E\{[Z - E(Z)]^2\} = E\{[a_1(X_1 - \mu_1) + a_2(X_2 - \mu_2)]^2\} \\
&= a_1^2 E[(X_1 - \mu_1)^2] + a_2^2 E[(X_2 - \mu_2)^2] \\
&\quad + 2a_1 a_2 E[(X_1 - \mu_1)(X_2 - \mu_2)] \\
&= a_1^2 \,\text{Var}(X_1) + a_2^2 \,\text{Var}(X_2) + 2a_1 a_2 \,\text{Cov}(X_1, X_2) \\
&= a_1^2 \sigma_1^2 + a_2^2 \sigma_2^2 + 2a_1 a_2 \sigma_{12}.
\end{aligned} \tag{6.41}$$

We can generalize this result to k random variables, $X_1, X_2, \ldots, X_k$. Let μ_i and σ_i^2 ($i = 1, 2, \ldots, k$) denote the mean and the variance of the random variable X_i, and σ_{ij} ($i \neq j$) the covariance of X_i and X_j, and put

$$Z = a_1 X_1 + a_2 X_2 + \cdots + a_k X_k. \tag{6.42}$$

We showed in (6.18) that

$$E(Z) = \mu_Z = a_1\mu_1 + a_2\mu_2 + \cdots + a_k\mu_k. \tag{6.43}$$

It is easy to show (the proof is left to the reader) that

$$\text{Var}(Z) = \sigma_Z^2 = \sum_{i=1}^{k} a_i^2 \sigma_i^2 + 2 \sum\sum_{i<j} a_i a_j \sigma_{ij}. \tag{6.44}$$

If the k variables are *pairwise independent* (so that $\sigma_{ij} = 0$), then

$$\text{Var}(Z) = a_1^2 \sigma_1^2 + a_2^2 \sigma_2^2 + \cdots + a_k^2 \sigma_k^2. \tag{6.45}$$

---

* Note that here the symbols μ_1 and μ_2 are used instead of μ_{X_1} and μ_{X_2}. They denote means of random variables X_1 and X_2, respectively, and should not be confused with the notation for central moments. The same applies to formula (6.43).

In particular, for $a_1 = a_2 = \cdots = a_k = 1$, we obtain

$$Z = X_1 + X_2 + \cdots + X_k$$

and

$$\text{Var}(Z) = \sigma_1^2 + \sigma_2^2 + \cdots + \sigma_k^2, \tag{6.46}$$

that is, *the variance of the sum of k random variables which are pairwise independent is equal to the sum of the individual variances.*

In particular, when $k = 2$, and when X_1 and X_2 are independent, we obtain from (6.45)

$$\text{Var}(X_1 + X_2) = \sigma_1^2 + \sigma_2^2, \tag{6.47}$$

and for $a_1 = 1$, $a_2 = -1$,

$$\text{Var}(X_1 - X_2) = \sigma_1^2 + \sigma_2^2, \tag{6.48}$$

which is the same as $\text{Var}(X_1 + X_2)$.

Also putting $a_1 = a_2 = \cdots = a_n = (1/n)$, we obtain the variance of the arithmetic mean, $\bar{X} = (1/n)\sum X_i$, that is,

$$\text{Var}(\bar{X}) = \frac{1}{n^2} \sum_{i=1}^{n} \sigma_i^2.$$

In particular, when $\sigma_1^2 = \sigma_2^2 = \cdots = \sigma_n^2 = \sigma^2$, we have

$$\text{Var}(\bar{X}) = \sigma_{\bar{X}}^2 = \frac{\sigma^2}{n}. \tag{6.49}$$

6.6. PHENOTYPIC AND GENOTYPIC VALUES OF METRICAL CHARACTERS

So far the term "genotype" has had some conceptual meaning and has been expressed in terms of gene combinations in the diploid cell. Of course, we are interested primarily in the gene function or gene product rather than in the "gene" itself.

In plant and animal breeding valuable characters, among others, are yield and weight. These depend on the genetic constitution of an individual and also on the environmental conditions and are called *quantitative* characters.

It has been assumed that some quantitative characters are the results of additive contributions of genes at several independent loci and also of environment. The skin color of several races, for example, is due to the additive effects of genes at different loci, but dark color can be observed also on "White" individuals exposed to the sun.

Another modern theory is inclined to assume that only a few (or one) "major" genes, together with "minor" genes and environment, are responsible for quantitative traits. There is also some evidence that, for some traits, the effect of gene action is multiplicative rather than additive (Srb et al. (1965), Chapter 14).

It is not the object of this book to discuss this topic. It is useful, however, to introduce a few simple concepts from this field.

The value (measurement) of a quantitative trait which can be observed is called a *phenotypic value*. Let us denote it by Y. If there is no interaction between genetic and environmental factors, the total observed value can be constructed additively from two parts: (1) the contribution of genetic factors, the effect of which is called the *genotypic value*, G, say, and (2) the contribution of environmental factors, whose effect we denote by ε, say. We then have

$$Y = G + \varepsilon.$$

For a given population, both G and ε play the role of random variables, since G depends on the genetic constitution of the individual in which the character is observed, and ε on the environment in which the individual lives.

The *mean* phenotypic value, μ_Y, is

$$\mu_Y = \mu_G + \mu_\varepsilon, \tag{6.50}$$

and the variance of the phenotypic value, $\sigma_Y{}^2$, (provided $\sigma_{G\varepsilon} = 0$) is

$$\sigma_Y{}^2 = \sigma_G{}^2 + \sigma_\varepsilon{}^2. \tag{6.51}$$

In particular, when there is practically no influence of the environment on the phenotypic value, $\varepsilon = 0$ and we have

$$\mu_Y = \mu_G, \quad \text{and} \quad \sigma_Y{}^2 = \sigma_G{}^2, \tag{6.52}$$

that is, the phenotypic value is entirely due to genetic factors controlling the character, and its variance is due to differences in the genotype constitutions.

We now restrict ourselves to this situation in which $\varepsilon = 0$, that is, $Y = G$, and will consider a character which is controlled by one locus with two alleles.

6.7. ADDITIVE GENETIC VARIANCE

Let us first consider a model in which the genotypic value is a *linear* function of the number, X, of A genes in the genotype, that is,

$$G = g_0 + \beta X \tag{6.53}$$

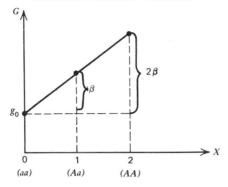

Figure 6.1.

with X taking the values $x = 0, 1, 2$, where g_0 is a certain constant and β ($\beta > 0$) is the *effect of single-gene substitution* or, briefly, the *genic effect*. Model (6.53) is an *additive* model for the genotypic value. We have

$$
\begin{aligned}
G_{AA} &= G_2 = g_0 + 2\beta, \\
G_{Aa} &= G_1 = g_0 + \beta, \\
G_{aa} &= G_0 = g_0.
\end{aligned}
\tag{6.54}
$$

Figure 6.1 represents relationship (6.53) graphically.

Suppose that the gene frequencies are p and q for A and a, respectively, with $p + q = 1$, and that the population is in equilibrium. We have

Genotype:	AA	Aa	aa
Number of A genes (x):	2	1	0
Genotypic value (G_i):	$G_{AA} = g_0 + 2\beta$	$G_{Aa} = g_0 + \beta$	$G_{aa} = g_0$
Probability:	p^2	$2pq$	q^2

The mean genotypic value, μ_G, is

$$
\mu_G = (g_0 + 2\beta)p^2 + (g_0 + \beta)2pq + g_0 \cdot q^2 = g_0 + 2p\beta.
\tag{6.55}
$$

The genotypic value has the variance

$$
\begin{aligned}
\sigma_G^2 &= (G_{AA} - \mu_G)^2 p^2 + (G_{Aa} - \mu_G)^2 2pq + (G_{aa} - \mu_G)q^2 \\
&= (2\beta q)^2 p^2 + \beta^2 (q - p)^2 \cdot 2pq + (-2\beta p)^2 q^2 = 2pq\beta^2.
\end{aligned}
\tag{6.56}
$$

In particular, for $\beta = 1$, we have

$$
\mu_G = 2p, \quad \sigma_G^2 = 2pq.
\tag{6.57}
$$

The variance, σ_G^2, calculated for the additive model (6.53) is called the *additive genetic variance*. We denote it by σ_{Ad}^2. Thus we have in this case

$$
\sigma_G^2 = \sigma_{Ad}^2.
\tag{6.58}
$$

▼ **6.8. DOMINANCE GENETIC VARIANCE**

Gene effects are not always additive. For instance, at the ABO locus of blood groups, the individuals OO form molecules with approximate molecular weight 170,000, the individuals AA form large molecules with molecular weight about 300,000, but heterozygotes AO form even larger molecules with molecular weight about 500,000 (Stern, Chapter 3). Thus, it is obvious that genic interaction occurs in heterozygote Aa and, perhaps, in homozygote AA as well.

Let us consider a *nonadditive* model of the form

$$G_{aa} = g_0, \quad G_{Aa} = g_0 + \beta + \alpha, \quad G_{AA} = g_0 + \beta, \quad \text{with } \beta > 0 \text{ and } \alpha \neq \beta.$$
(6.59)

When the model for genotypic values is nonadditive, it is customary to speak about *dominance*. The term "dominance" has here a very general meaning. It really means that, when two allelic genes combine into a zygote, they interact and their joint effect is not simply the sum of the effects of the individual alleles (compare section 3.5.3).

(i) If $\alpha = 0$, we have

$$G_{aa} = g_0, \quad G_{Aa} = g_0 + \beta, \quad G_{AA} = g_0 + \beta.$$
(6.60)

Model (6.60) determines *complete dominance*. Here, the effect of heterozygote Aa is indistinguishable from the effect of a homozygote AA. This corresponds to the originally introduced meaning of dominance for qualitative characters.

(ii) If $\alpha > \beta$, then $G_{Aa} > G_{AA}$, that is, the genotypic value of heterozygote Aa is greater than that of homozygote AA. This phenomenon is called *overdominance*.

We shall now discuss model (6.60), that is, the one for complete dominance. The genotypic mean is

$$\mu_G = (g_0 + \beta)(p^2 + 2pq) + g_0 q^2 = g_0 + (1 - q^2)\beta.$$
(6.61)

The genotypic effects are

$$G_{AA} - \mu_G = G_{Aa} - \mu_G = \beta - \beta(1 - q^2) = q^2\beta; \quad G_{aa} - \mu_G = -(1 - q^2)\beta.$$
(6.62)

The variance, σ_G^2, called the *total genetic variance*, is

$$\sigma_G^2 = (q^2\beta)^2(p^2 + 2pq) + [-(1 - q^2)\beta]^2 q^2 = q^2(1 - q^2)\beta^2.$$

$$= pq^2(1 + q)\beta^2.$$
(6.63)

We now find the correlation coefficient between the number of genes A in the genotype, X, say, and the genotypic values given in (6.60), that is, we wish to find ρ_{XG}. We have already found (Example 6.6) that $\mu_X = 2p$, $\sigma_X^2 = 2pq$. Since the joint distribution, $p(x, g)$ is the same as the distribution of G (or X) alone, we calculate the covariance:

$$
\begin{aligned}
\sigma_{XG} &= \sum (x - \mu_X)(g - \mu_G)p(x, g) \\
&= 2q \cdot q^2\beta \cdot p^2 + (q - p)q^2\beta \cdot 2pq + 2p(1 - q^2)\beta \cdot q^2 \\
&= 2pq^2\beta.
\end{aligned}
\tag{6.64}
$$

Hence, the square of the correlation coefficient is

$$
\rho_{XG}^2 = \frac{4p^2q^4\beta^2}{2pq \cdot q^2(1 - q^2)\beta^2} = \frac{2pq}{1 - q^2}.
\tag{6.65}
$$

Note that ρ_{XG} does not depend on β. In particular, for $p = q = \frac{1}{2}$, $\rho_{XG} = \sqrt{\frac{2}{3}} = \sqrt{0.6667} = 0.8165$.

The total genetic variance, σ_G^2, consists of two components: additive genetic variance, σ_{Ad}^2, and dominance genetic variance, σ_{Dom}^2, that is,

$$
\sigma_G^2 = \sigma_{Ad}^2 + \sigma_{Dom}^2.
\tag{6.66}
$$

Applying the standard method used in regression, we can split up σ_G^2 into two parts, one due to linear trend (σ_{Ad}^2) and the other due to departure from linearity (σ_{Dom}^2). Since regression is not treated in this book, we shall not give the details of evaluating σ_{Ad}^2 and σ_{Dom}^2. It turns out that the additive genetic variance is

$$
\sigma_{Ad}^2 = \rho_{XG}^2 \sigma_G^2 = \frac{2pq}{1 - q^2} q^2(1 - q^2)\beta^2 = 2pq^3\beta^2,
\tag{6.67}
$$

and the dominance variance is

$$
\sigma_{Dom}^2 = (1 - \rho_{XG}^2)\sigma_G^2 = \frac{p^2}{1 - q^2} \cdot q^2(1 - q^2)\beta^2 = p^2q^2\beta^2.
\tag{6.68}
$$

In particular, for $\beta = 1$,

$$
\sigma_{Ad}^2 = 2pq^3, \quad \sigma_{Dom}^2 = p^2q^2, \quad \sigma_G^2 = pq^2(1 + q),
\tag{6.69}
$$

and for $\beta = 1$ and $p = q = \frac{1}{2}$,

$$
\sigma_{Ad}^2 = \frac{1}{8}, \quad \sigma_{Dom}^2 = \frac{1}{16}, \quad \sigma_G^2 = \frac{3}{16}.
\tag{6.70}
$$

6.9. MIXTURES OF TWO (OR MORE) POPULATIONS

Let X and Y be two discrete characters which can be observed on each individual in a population (two discrete random variables). Suppose that X

takes only the values $x = x_1, x_2, \ldots, x_m$ and Y takes only the values $y = y_1, y_2, \ldots, y_n$.

Let us consider two populations, Π_1 and Π_2, say, in which the joint distributions of these characters are $p^{(1)}(x, y)$ and $p^{(2)}(x, y)$, respectively. In general, $p^{(1)}(x, y) \neq p^{(2)}(x, y)$, that is, the same characters have different distributions in the two populations.

For simplicity, let us denote the joint probabilities as

$$p^{(t)}(x_i, y_j) = p_{ij}^{(t)}, \tag{6.71}$$

where the superscript t denotes the population $t = 1, 2$.

Similarly, for marginal probabilities we introduce the notation

$$p_1^{(t)}(x_i) = p_{i.}^{(t)}; \quad p_2^{(t)}(y_j) = p_{.j}^{(t)}, \, t = 1, 2. \tag{6.72}$$

(Compare also section 2.11.)

Let $\mu_X^{(t)}, \mu_Y^{(t)}, \sigma_X^{(t)}, \sigma_Y^{(t)}$, and $\sigma_{XY}^{(t)}, t = 1, 2$, denote the corresponding means, variances, and covariance in each of two populations.

Suppose that a new population, Π, is a *mixture* of Π_1 and Π_2 such that

$$\Pi = c_1\Pi_1 + c_2\Pi_2, \tag{6.73}$$

with

$$c_1 + c_2 = 1. \tag{6.74}$$

The coefficients c_1, c_2 are the proportions of admixture from each population.

Let $p_{ij} = p(x_i, y_j)$ be the joint probability of (x_i, y_j) in the mixed population Π, and let $p_{i.} = p_1(x_i)$, $p_{.j} = p_2(y_j)$ be the marginal probabilities in the population Π. We notice that

$$p_{ij} = c_1 p_{ij}^{(1)} + c_2 p_{ij}^{(2)}, \quad i = 1, 2, \ldots, m; \, j = 1, 2, \ldots, n, \tag{6.75}$$

and also

$$p_{i.} = c_1 p_{i.}^{(1)} + c_2 p_{i.}^{(2)}, \quad i = 1, 2, \ldots, m, \tag{6.76a}$$

$$p_{.j} = c_1 p_{.j}^{(1)} + c_2 p_{.j}^{(2)}, \quad j = 1, 2, \ldots, n. \tag{6.76b}$$

Similarly, let μ_X, μ_Y, σ_X^2, σ_Y^2, and σ_{XY} be the corresponding means, variances, and covariance in the mixed population. We immediately find

$$\mu_X = \sum_{i=1}^m x_i p_{i.} = c_1 \sum_{i=1}^m x_i p_{i.}^{(1)} + c_2 \sum_{i=1}^m x_i p_{i.}^{(2)} = c_1 \mu_X^{(1)} + c_2 \mu_X^{(2)}. \tag{6.77a}$$

Similarly,

$$\mu_Y = c_1 \mu_Y^{(1)} + c_2 \mu_Y^{(2)}. \tag{6.77b}$$

Now we find the variance σ_X^2:

$$\sigma_X^2 = \sum_{i=1}^m (x_i - \mu_X)^2 p_{i.} = \sum_{i=1}^m (x_i - \mu_X)^2 (c_1 p_{i.}^{(1)} + c_2 p_{i.}^{(2)})$$

$$= c_1 \sum_{i=1}^m (x_i - \mu_X)^2 p_{i.}^{(1)} + c_2 \sum_{i=1}^m (x_i - \mu_X)^2 p_{i.}^{(2)}. \tag{6.78}$$

We now calculate the first term in (6.78):

$$\sum_{i=1}^{m} (x_i - \mu_X)^2 p_{i\cdot}^{(1)} = \sum_{i=1}^{m} [(x_i - \mu_X^{(1)}) + (\mu_X^{(1)} - \mu_X)]^2 p_{i\cdot}^{(1)}$$

$$= \sum_{i=1}^{m} (x_i - \mu_X^{(1)})^2 p_{i\cdot}^{(1)} + (\mu_X^{(1)} - \mu_X)^2$$

$$= \sigma_X^{(1)2} + (\mu_X^{(1)} - \mu_X)^2. \qquad (6.79)$$

The second term in (6.78) can be found in a similar way, replacing the superscript (1) by (2). Thus the variance $\sigma_X{}^2$ is

$$\sigma_X{}^2 = [c_1 \sigma_X^{(1)2} + c_2 \sigma_X^{(2)2}] + [c_1(\mu_X^{(1)} - \mu_X)^2 + c_2(\mu_X^{(2)} - \mu_X)^2], \quad (6.80a)$$

and analogously

$$\sigma_Y{}^2 = [c_1 \sigma_Y^{(1)2} + c_2 \sigma_Y^{(2)2}] + [c_1(\mu_Y^{(1)} - \mu_Y)^2 + c_2(\mu_Y^{(2)} - \mu_Y)^2]. \quad (6.80b)$$

By using similar techniques, it can be shown (the proof is left to the reader) that the covariance between X and Y, in the mixed population II, is

$$\sigma_{XY} = [c_1 \sigma_{XY}^{(1)} + c_2 \sigma_{XY}^{(2)}]$$
$$+ [c_1(\mu_X^{(1)} - \mu_X)(\mu_Y^{(1)} - \mu_Y) + c_2(\mu_X^{(2)} - \mu_X)(\mu_Y^{(2)} - \mu_Y)]. \quad (6.81)$$

The generalization to mixtures of more than two populations is straightforward.

Of particular interest is the case in which

$$\mu_X^{(1)} = \mu_X^{(2)}, \quad \mu_Y^{(1)} = \mu_Y^{(2)}$$

and

$$\sigma_X^{(1)2} = \sigma_X^{(2)2}, \quad \sigma_Y^{(1)2} = \sigma_Y^{(2)2}. \qquad (6.82)$$

In this case we will also have

$$\mu_X = \mu_X^{(1)} = \mu_X^{(2)} \quad \text{and} \quad \mu_Y = \mu_Y^{(1)} = \mu_Y^{(2)}. \qquad (6.83)$$

Substituting (6.82) and (6.83) into (6.80a), (6.80b), and (6.81), respectively, we obtain

$$\sigma_X{}^2 = \sigma_X^{(1)2} = \sigma_X^{(2)2}, \quad \sigma_Y{}^2 = \sigma_Y^{(1)2} = \sigma_Y^{(2)2}, \qquad (6.84)$$

but

$$\sigma_{XY} = c_1 \sigma_{XY}^{(1)} + c_2 \sigma_{XY}^{(2)}. \qquad (6.85)$$

We now find the correlation coefficient in the mixed population, ρ_{XY}. This is

$$\rho_{XY} = \frac{\sigma_{XY}}{\sigma_X \sigma_Y} = c_1 \frac{\sigma_{XY}^{(1)}}{\sigma_X \sigma_Y} + c_2 \frac{\sigma_{XY}^{(2)}}{\sigma_X \sigma_Y} = c_1 \rho_{XY}^{(1)} + c_2 \rho_{XY}^{(2)}, \quad (6.86)$$

since, in view of (6.84), $(\sigma_{XY}^{(1)}/\sigma_X \sigma_Y)$ and $(\sigma_{XY}^{(2)}/\sigma_X \sigma_Y)$ represent the correlation coefficients in Π_1 and Π_2, respectively.

The result obtained in (6.86) has several useful applications (see, for instance, sections 7.8.2 and 7.11).

SUMMARY

The characteristics of distributions of random variables, such as expected value, variance, covariance, or, more generally, moments, have been defined and interpreted. The correlation coefficient, as an index of the degree of linear relationship between two variables, has been introduced. Some applications to the partition of genetic variance into two components, additive genetic variance and dominance genetic variance, have been briefly presented. Mixtures of two populations and their moments have been considered.

REFERENCES

Gangolli, R. A., and Ylvisaker, D., *Discrete Probability*, Harcourt, Brace & World, New York, 1966.

Johnson, N. L., and Leone, F. C., *Statistics and Experimental Design*, Vol. I, John Wiley & Sons, New York, 1964.

Kempthorne, O., *An Introduction to Genetic Statistics*, John Wiley & Sons, New York, 1957.

Li, C. C., *Population Genetics*, The University of Chicago Press, Chicago, 1955.

Lindgren, B. W., *Statistical Theory*, The Macmillan Company, New York, 1968 (second edition).

Srb, A. M., Owen, R. D., and Edgar, R. S., *General Genetics*, W. H. Freeman & Co. San Francisco, 1965.

Stern, C., *Principles of Human Genetics*, W. H. Freeman & Co., San Francisco, 1960.

Wilks, S. S., *Mathematical Statistics*, John Wiley & Sons, New York, 1962.

PROBLEMS

6.1. Table 2.4 shows the distribution of genotypes with respect to two *linked* loci, in the form of a 3×3 matrix. Let X determine the number of alleles A at locus A, and Y the number of alleles B at locus B. Thus the genotype distribution in Table 2.4 represents also the joint distribution of (X, Y), $p(x, y)$, say.

(*a*) Calculate $E(X)$, $E(Y)$, $\text{Var}(X)$, $\text{Var}(Y)$, and $\text{Cov}(X, Y)$.

(*b*) Calculate the correlation coefficient ρ_{XY} between X and Y. What is the interpretation of this result?

(*c*) Let

$$Z = \begin{cases} 0 & \text{if } aa \\ 1 & \text{if } AA \text{ or } Aa \end{cases} \quad \text{and} \quad U = \begin{cases} 0 & \text{if } bb \\ 1 & \text{if } BB \text{ or } Bb \end{cases}$$

determine the phenotypes with respect to locus A and locus B, respectively. Find the joint phenotype distribution, $h(z, u)$, say, and calculate the correlation coefficient ρ_{ZU}. Compare it with that obtained in (*b*).

6.2. Let us consider the following population in equilibrium:

$$\begin{array}{llll} \text{Genotype:} & AA & Aa & aa \\ X: & 2 & 1 & 0 \\ p(x): & p^2 & 2pq & q^2. \end{array}$$

The random variable X denotes here the number of A genes in a genotype. Suppose that the population mates at random.

(a) For each given mating type find the (conditional) distribution of X, $E(X)$, and $\text{Var}(X)$ in the offspring.

(b) Show that the total genotypic variance in the whole population of the offspring is $\sigma_G{}^2 = 2pq$ (why?).

(c) Show that the *mean* genetic variance *within* families is $pq = \frac{1}{2}\sigma_G{}^2$.

[*Hint:* Suppose that the mating is $AA \times Aa$. Then $\text{Pr}\{AA \times Aa\} = 4p^3q$. The offspring of $AA \times Aa$ has the following distribution:

$$\left. \begin{array}{llll} \text{Offspring:} & AA & Aa & aa \\ X: & 2 & 1 & 0 \\ p(x): & \frac{1}{2} & \frac{1}{2} & 0 \end{array} \right\} \text{given the mating } AA \times Aa,$$

$E(X \mid AA \times Aa) = \frac{3}{2}$, $\text{Var}(X \mid AA \times Aa) = \frac{1}{4}$.

The contribution of this "family pattern" to total genetic variance is

$$\text{Var}(X \mid AA \times Aa) \cdot \text{Pr}\{AA \times Aa\} = \frac{1}{4} \cdot 4p^3q = p^3q.$$

The mean genetic variance is the sum of family variances, each weighted by the mating frequency.]

(d) Show that the variance *between* family means is equal to $pq = \frac{1}{2}\sigma_G{}^2$. Thus the total genetic variance consists of two equal parts, one due to the average variance within families and the other due to the variance between families, each equal to $\frac{1}{2}\sigma_G{}^2$.

6.3. Show that, if X and Y are two independent variables with joint distribution $(p_{x, y}) = p_1(x)p_2(y)$, the variance of the product, XY, is

$$\text{Var}(XY) = \mu_X{}^2\mu_Y{}^2\left(\frac{\sigma_X{}^2}{\mu_X{}^2} + \frac{\sigma_Y{}^2}{\mu_Y{}^2} + \frac{\sigma_X{}^2\sigma_Y{}^2}{\mu_X{}^2\mu_Y{}^2}\right).$$

CHAPTER 7

Genotype Distributions for Relatives in Randomly Mating Populations

7.1. INTRODUCTION

In Chapter 4, we have considered nonoverlapping, randomly mating populations and have derived some conditions for equilibrium. We have assumed that random mating implies, in practice, matings of unrelated pairs. However, in some populations certain types of matings between related pairs (e.g., between cousins or uncle and niece) may occur. There are also some other situations when it is useful to deal with related pairs. For instance, in studies of common traits (e.g., blood groups), one may find it convenient to use samples of pairs parent-child or sib-sib in order to estimate gene frequencies. In matching donor-recipient pairs for tissue or organ transplantation, a relative of the recipient is often selected as a donor since there is a greater probability of a transplant being successful. In laboratory animals brother-sister matings are often used to obtain homogeneous strains of mice or rats.

In these and similar problems it is very useful to know joint genotype and phenotype distributions of such related pairs and determine the "degree of relationship" which we will call the ancestral correlation.

This chapter is devoted to derivation of these distributions and their first and second moments. The considerable amount of algebraic detail required may be found rather tedious, especially as only a few examples of applications are given in this chapter. However, considering the present chapter as introductory to the problems treated in Chapters 9–11, 17 and 19 should put matters into clearer perspective. Without a thorough understanding of the fundamental details given in this chapter, proper appreciation of these later chapters would be difficult. The techniques developed here are extensions and applications of the theory of conditional and joint distributions of events and random variables presented in Chapters 2 and 5, respectively, and the evaluation of appropriate correlation coefficients between relatives, using the results of Chapter 6.

We first introduce so-called *a posteriori* probability and prove Bayes theorem, which is useful in several genetic problems.

7.2. BAYES THEOREM. *A POSTERIORI* PROBABILITY

Let $E_1, E_2, \ldots, E_r$ be a sequence of mutually exclusive events with probabilities $\Pr\{E_1\}, \Pr\{E_2\}, \ldots, \Pr\{E_r\}$, respectively.

Let D be an event which occurs, with $\Pr\{D\} > 0$, when and only when one of the events $E_1, E_2, \ldots, E_r$ occurs. The events $E_1, \ldots, E_r$ determine several different conditions or *causes* under which D may occur. The probabilities $\Pr\{E_1\}, \ldots, \Pr\{E_r\}$ are called *a priori* probabilities of the occurrence of these events, without regard to the event D. They can be theoretically deduced from the model represented by the sample space on which these events are defined. They are also sometimes called *hypotheses* to which the occurrence of D is related.

Let $\Pr\{D \mid E_i\}$, $i = 1, 2, \ldots, r$, be the conditional probability that the event D occurs, given that the event E_i has been observed. We assume that the probabilities $\Pr\{E_i\}$ and $\Pr\{D \mid E_i\}$, $i = 1, 2, \ldots, r$, are known.

We wish to find the probability of the event E_i, given that the event D has been observed, that is, $\Pr\{E_i \mid D\}$. This is called the *a posteriori* probability of E_i; it is calculated *after* D has been observed.

The formula from which $\Pr\{E_i \mid D\}$ can be calculated,

$$\Pr\{E_i \mid D\} = \frac{\Pr\{E_iD\}}{\sum_{j=1}^{r}\Pr\{E_jD\}} = \frac{\Pr\{D \mid E_i\}\Pr\{E_i\}}{\sum_{j=1}^{r}\Pr\{D \mid E_j\}\Pr\{E_j\}}, \tag{7.1}$$

is known as *Bayes theorem*. We now prove (7.1).

Proof. By definition

$$\Pr\{E_i \mid D\} = \frac{\Pr\{E_iD\}}{\Pr\{D\}}. \tag{7.2}$$

But $\Pr\{E_iD\} = \Pr\{D \mid E_i\} \cdot \Pr\{E_i\}$, where $\Pr\{D \mid E_i\}$ and $\Pr\{E_i\}$, $i = 1, 2, \ldots, r$, are known.

Since D occurs only in the presence of one of the events $E_1, E_2, \ldots, E_r$, then

$$\Pr\{D\} = \Pr\{E_1D\} + \Pr\{E_2D\} + \cdots + \Pr\{E_rD\}$$
$$= \Pr\{D \mid E_1\}\Pr\{E_1\} + \Pr\{D \mid E_2\}\Pr\{E_2\} + \cdots + \Pr\{D \mid E_r\}\Pr\{E_r\}$$
$$= \sum_{j=1}^{r}\Pr\{D \mid E_j\}\Pr\{E_j\}. \tag{7.3}$$

Substituting (7.3) into (7.2), we obtain (7.1). We now give an application of Bayes theorem.

7.2.1. Parental Genotypes for a Given Genotype of Offspring

Let us consider a parental population at equilibrium with respect to a single locus with genotype array $p^2AA + 2pqAa + q^2aa$, where A and a are dominant and recessive genes, respectively. Suppose that an offspring with genotype aa is observed and we wish to predict its parental genotypes. Suppose that we ask this question: what is the *a posteriori* probability that, given an offspring with genotype aa, his parents are $Aa \times Aa$?

The application of Bayes theorem is here straightforward. The observed genotype aa of an offspring corresponds to our event D. The possible matings which can produce an aa offspring are as follows:

  (i) $Aa \times Aa$ (both parents are dominants);
  (ii) $Aa \times aa$ (one parent is dominant, one recessive);
  (iii) $aa \times aa$ (both parents are recessives).

These are three events, E_1, E_2, E_3, associated with the event D. Their *a priori* probabilities are

$$\Pr\{Aa \times Aa\} = 4p^2q^2, \quad \Pr\{Aa \times aa\} = 4pq^3, \quad \Pr\{aa \times aa\} = q^4. \quad (7.4)$$

The corresponding conditional probabilities are

$$\Pr\{aa \,|\, Aa \times Aa\} = \tfrac{1}{4}, \quad \Pr\{aa \,|\, Aa \times aa\} = \tfrac{1}{2}, \quad \Pr\{aa \,|\, aa \times aa\} = 1. \quad (7.5)$$

Thus the probability that an aa offspring occurs is, from (7.3),

$$\Pr\{aa\} = \tfrac{1}{4} \times 4p^2q^2 + \tfrac{1}{2} \times 4pq^3 + 1 \times q^4 = q^2(p^2 + 2pq + q^2) = q^2. \quad (7.6)$$

The *a posteriori* probability that the parental genotypes of an aa offspring are $Aa \times Aa$ is, from (7.1),

$$\Pr\{Aa \times Aa \,|\, aa\} = \frac{\tfrac{1}{4} \times 4p^2q^2}{q^2} = p^2. \quad (7.7)$$

In a similar way we find the *a posteriori* probabilities for the remaining two possible parental matings, that is,

$$\Pr\{Aa \times aa \,|\, aa\} = \frac{\tfrac{1}{2} \times 4pq^3}{q^2} = 2pq, \quad (7.8)$$

and

$$\Pr\{aa \times aa \,|\, aa\} = \frac{q^4}{q^2} = q^2. \quad (7.9)$$

Other applications of Bayes theorem will appear in many parts of this book. A paper by Murphy and Mutalik (1969), who give several examples of the application of Bayes theorem in genetic counseling, especially for X-linked rare diseases, should also be mentioned.

▼ 7.3. GENERALIZED FORMS OF BAYES THEOREM

(i) Let $H_1, H_2, \ldots, H_r$ be a series of *hypotheses* or *categories*, and $T_1, T_2, \ldots, T_k$ a series of *attributes*. We assume that $H_1, H_2, \ldots, H_r$ are mutually exclusive events with known probabilities $\Pr\{H_1\}, \Pr\{H_2\}, \ldots, \Pr\{H_r\}$, and that $T_1, T_2, \ldots, T_k$ are not necessarily exclusive, and also not necessarily independent, events. Suppose that each of the events $H_1, \ldots, H_r$ can be accompanied by all the attributes $(T_1, T_2, \ldots, T_k)$ and that the conditional probabilities $\Pr\{T_1 T_2 \cdots T_k \mid H_i\}$, $i = 1, 2, \ldots,$ are known.

For example, $H_1, \ldots, H_r$ can denote different (exclusive) diseases, and $T_1, T_2, \ldots, T_k$ a set of certain symptoms observed when any of these diseases occurs. For a given population, $\Pr\{H_1\}, \ldots, \Pr\{H_r\}$ could correspond to their prevalence (or incidence) rates; $\Pr\{T_1 T_2 \cdots T_k \mid H_i\}$ might be found from long-run experience.

Suppose that we have observed an ill individual, with the symptoms $T_1, T_2, \ldots, T_k$. What is the probability that this diagnosis: he has the disease H_i, is correct?

This can be evaluated from the extended Bayes formula

$$\Pr\{H_i \mid T_1 T_2 \cdots T_k\} = \frac{\Pr\{T_1 T_2 \cdots T_k \mid H_i\}\Pr\{H_i\}}{\sum_{l=1}^{r} \Pr\{T_1 T_2 \cdots T_k \mid H_l\}\Pr\{H_l\}}. \tag{7.10}$$

(ii) More generally, the events $H_1, H_2, \ldots, H_r$ (in our example, the diseases) might not be exclusive events. In such cases we should also know the joint probabilities of any two, any three, $\ldots$, etc., events, that is, $\Pr\{H_i H_j\}, \Pr\{H_i H_j H_l\}, \ldots,$ and the corresponding conditional probabilities, $\Pr\{T_1 T_2 \cdots T_k \mid H_i H_j\}, \Pr\{T_1 T_2 \cdots T_k \mid H_i H_j H_l\}, \ldots$.

In particular, when there are only *two* (nonexclusive) hypotheses, the total probability $\Pr\{T_1 T_2 \cdots T_k\}$ will be calculated from the formula

$$\Pr\{T_1 T_2 \cdots T_k\} = \Pr\{T_1 \cdots T_k \mid H_1\}\Pr\{H_1\} + \Pr\{T_1 \cdots T_k \mid H_2\}\Pr\{H_2\}$$
$$- \Pr\{T_1 \cdots T_k \mid H_1 H_2\}\Pr\{H_1 H_2\}, \tag{7.11}$$

and the generalized form of Bayes formula for *two* (nonexclusive) events takes the form

$$\Pr\{H_i \mid T_1T_2 \cdots T_k\} = \frac{\Pr\{T_1T_2 \cdots T_k \mid H_i\} \Pr\{H_i\}}{\Pr\{T_1T_2 \cdots T_k\}} \quad i = 1, 2, \quad (7.12)$$

with $\Pr\{T_1T_2 \cdots T_k\}$ given by the right-hand side of (7.11).

This can be extended to the situation in which more than two hypotheses are under consideration, by using general rules for the union of nonexclusive events. [For $r = 3$, the formula analogous to (7.11) can be obtained from the rule given by (2.14).]

We now shall discuss some of the more important and common joint genotype distributions among relatives and the degree of association among these relatives.

7.4. MATING TYPES AND THEIR OFFSPRING. SINGLE AUTOSOMAL LOCUS WITH MULTIPLE ALLELES

Let us consider a population which is in equilibrium with respect to one locus with s alleles, $A_1, A_2, \ldots, A_s$. Let the corresponding gene frequencies be $p_1, p_2, \ldots, p_s$, with $\Sigma p_i = 1$.

We represent the population and its genotype distribution in the form of a

TABLE 7.1   Mating types

Type	Mating	Description of mating type
1	$A_iA_i \times A_iA_i$	Both parents the same homozygotes
2	$A_iA_i \times A_jA_j, i \neq j$	Both parents different homozygotes
3	$A_iA_i \times A_iA_j$	One parent a homozygote, the other a heterozygote, with one allele in common
4	$A_iA_i \times A_jA_k, i \neq j \neq k$	One parent a homozygote, the other a heterozygote, with no allele in common
5	$A_iA_j \times A_iA_j$	Both parents the same heterozygotes
6	$A_iA_j \times A_iA_k$	Both parents heterozygotes, with one allele in common
7	$A_iA_j \times A_kA_l, i \neq j \neq k \neq l$	Both parents heterozygotes, with no allele in common

genotype array (which seems to be convenient in this situation), that is,

$$\sum_{i=1}^{s} p_i^2 A_i A_i + 2 \sum_{i<j} \sum p_i p_j A_i A_j = \sum_{i=1}^{s} \sum_{j=1}^{s} p_i p_j A_i A_j. \tag{7.13}$$

In such a population there are $\frac{1}{2}s(s+1)$ different genotypes: s homozygotes of type $A_i A_i$ and $\frac{1}{2}s(s-1)$ heterozygotes of type $A_i A_j$. We notice that

$$\Pr\{A_i A_i\} = p_i^2, \qquad i = 1, 2, \ldots, s,$$
$$\Pr\{A_i A_j\} = 2p_i p_j, \qquad i, j = 1, 2, \ldots, s, i \neq j. \tag{7.14}$$

We shall distinguish *seven types* of parental genotype matings, as in Table 7.1. The corresponding probabilities for each type, when the population is mating at random are as follows:

(1) $\Pr\{A_i A_i \times A_i A_i\} = p_i^4$;

(5) $\Pr\{A_i A_j \times A_i A_j\} = 2p_i p_j \times 2p_i p_j = 4p_i^2 p_j^2, \qquad i \neq j$; $\tag{7.15}$

but the probabilities

(2) $\Pr\{A_i A_i \times A_j A_j\} = \Pr\{♀A_i A_i \times ♂A_j A_j\} + \Pr\{♀A_j A_j \times ♂A_i A_i\}$,
$\qquad\qquad = p_i^2 p_j^2 + p_i^2 p_j^2 = 2 \times p_i^2 p_j^2, \qquad i \neq j$;

and similarly,

(3) $\Pr\{A_i A_i \times A_i A_j\} = 2 \times 2p_i^3 p_j, \qquad i \neq j$;

(4) $\Pr\{A_i A_i \times A_j A_k\} = 2 \times 2p_i^2 p_j p_k, \qquad i \neq k$;

(6) $\Pr\{A_i A_j \times A_i A_k\} = 2 \times 4p_i^2 p_j p_k, \qquad i \neq k$; $\tag{7.16}$

(7) $\Pr\{A_i A_j \times A_k A_l\} = 2 \times 4p_i p_j p_k p_l, \qquad i \neq j \neq k \neq l$.

We now write down the seven kinds of offspring genotypes, in order to obtain the seven types of joint genotype combinations for parent-child

TABLE 7.2 Conditional probabilities of offspring
for a given parental mating type

Parental mating type	Offspring (conditional probabilities)						
	$A_i A_i$	$A_j A_j$	$A_i A_j$	$A_i A_k$	$A_j A_k$	$A_i A_l$	$A_j A_l$
1. $A_i A_i \times A_i A_i$	1	0	0	0	0	0	0
2. $A_i A_i \times A_j A_j$	0	0	1	0	0	0	0
3. $A_i A_i \times A_i A_j$	$\frac{1}{2}$	0	$\frac{1}{2}$	0	0	0	0
4. $A_i A_i \times A_j A_k$	0	0	$\frac{1}{2}$	$\frac{1}{2}$	0	0	0
5. $A_i A_j \times A_i A_j$	$\frac{1}{4}$	$\frac{1}{4}$	$\frac{1}{2}$	0	0	0	0
6. $A_i A_j \times A_i A_k$	$\frac{1}{4}$	0	$\frac{1}{4}$	$\frac{1}{4}$	$\frac{1}{4}$	0	0
7. $A_i A_j \times A_k A_l$	0	0	0	$\frac{1}{4}$	$\frac{1}{4}$	$\frac{1}{4}$	$\frac{1}{4}$

(section 7.5) or full-sib pairs (section 7.11). These offspring genotypes are A_iA_i, A_jA_j, A_iA_j, A_iA_k, A_jA_k, A_iA_l, A_jA_l, $i \neq j \neq k \neq l$.

For each kind of offspring genotype we find the conditional probability of its occurrence with a given parental mating type. For example,

$$\Pr\{A_iA_i \mid A_iA_i \times A_iA_i\} = 1, \quad \Pr\{A_iA_i \mid A_iA_i \times A_iA_j\} = \tfrac{1}{2},$$
$$\Pr\{A_iA_i \mid A_iA_j \times A_iA_j\} = \tfrac{1}{4}, \text{ etc.}$$

These conditional probabilities are given in Table 7.2.

7.5. JOINT PARENT-CHILD DISTRIBUTION

We wish to find general formulae for calculating the probabilities of the joint occurrence of child-parent genotype combinations according to the types given in Table 7.1. A few general remarks will be useful.

1. Since the population is in equilibrium, the joint probabilities of a certain genotype combination are the same for parent-child as for child-parent. For example,

$$\Pr\{\text{Child} = A_iA_i, \text{Parent} = A_iA_j\} = \Pr\{\text{Child} = A_iA_j, \text{Parent} = A_iA_i\}.$$

2. Since child and parent have *always* one gene in common, some of the types given in Table 7.1 never occur. The only types of child-parent (or parent-child) combinations which can occur are the following:

$$
\begin{array}{ccl}
\text{Type} & \text{Child-Parent} & \\
1 & A_iA_i,\ A_iA_i & \\
3 & A_iA_i,\ A_iA_j & \\
5 & A_iA_j,\ A_iA_j & \\
6 & A_iA_j,\ A_iA_k. &
\end{array}
\right\} \quad (7.17)
$$

3. The joint genotype distribution of child and a randomly chosen parent is the same as the distribution of child-mother or child-father. The following example gives the justification for this statement.

Let us consider, for instance, the parental mating $A_iA_i \times A_iA_j$. In a population under equilibrium, the probability of this mating, without distinguishing which is the mother's and which is the father's genotype, is $\Pr\{A_iA_i \times A_iA_j\} = 4p_i^3p_j$, and the probability of selecting the A_iA_i parent is $\tfrac{1}{2}$. The conditional probability that a child will be A_iA_i is $\Pr\{A_iA_i \mid A_iA_i \times A_iA_j\} = \tfrac{1}{2}$. Hence the joint probability of selecting a child A_iA_i, a parent (mother or father) A_iA_i, and the parental mating $A_iA_i \times A_iA_j$ is

$$\Pr\{A_iA_i \text{ and } (\underline{A_iA_i^*} \times A_iA_j)\} = \tfrac{1}{2} \cdot \tfrac{1}{2} \cdot 4p_i^3 p_j = p_i^3 p_j. \qquad (7.18)$$

Suppose that the mother is A_iA_i and the father is A_iA_j. Then $\Pr\{\female A_iA_i \times \male A_iA_j\} = 2p_i^3 p_j$. Now

$$\Pr\{A_iA_i \text{ and } (\female A_iA_i \times \male A_iA_j)\} = \tfrac{1}{2} \cdot 2p_i^3 p_j = p_i^3 p_j, \qquad (7.19)$$

which is the same as (7.18). The same formula is valid for mating $\male A_iA_i \times \female A_iA_j$.

We now derive formulae from which the joint probabilities of child-parent genotypes, for each type given in (7.17), can be calculated.

Type 1. $\Pr\{A_iA_i, A_iA_i\}$

The child A_iA_i and the parent A_iA_i can occur jointly when the parental matings are $A_iA_i \times A_iA_i$ or $A_iA_i \times A_iA_j, j = 1, 2, \ldots, s, j \neq i$. We have

$$\Pr\{A_iA_i \text{ and } A_iA_i\} = \Pr\{A_iA_i \text{ and } (\underline{A_iA_i} \times A_iA_i)\} + \sum_{j \neq i}^{s} \Pr\{A_iA_i \text{ and } (\underline{A_iA_i} \times A_iA_j)\}$$

$$= \Pr\{A_iA_i \mid A_iA_i \times A_iA_i\}\Pr\{A_iA_i \times A_iA_i\}$$

$$+ \tfrac{1}{2}\sum_{j \neq i}^{s} \Pr\{A_iA_i \mid A_iA_i \times A_iA_j\}\Pr\{A_iA_i \times A_iA_j\}$$

$$= 1 \cdot p_i^4 + \tfrac{1}{2}\sum_{j \neq i}^{s} \tfrac{1}{2} \cdot 4p_i^3 p_j = p_i^4 + p_i^3 \sum_{j \neq i}^{s} p_j = p_i^3[p_i + (1 - p_i)] = p_i^3.$$

$$(7.20)$$

Type 3. $\Pr\{A_iA_i, A_iA_j\}$

$$\Pr\{A_iA_i \text{ and } A_iA_j\} = \Pr\{A_iA_i \text{ and } (\underline{A_iA_j} \times A_iA_i)\} + \Pr\{A_iA_i \text{ and } (\underline{A_iA_j} \times A_iA_j)\}$$

$$+ \sum_{k \neq j \neq i}^{s} \Pr\{A_iA_i \text{ and } (\underline{A_iA_j} \times A_iA_k)\}$$

$$= \tfrac{1}{2} \cdot \tfrac{1}{2} \cdot 4p_i^3 p_j + \tfrac{1}{4} \cdot 4p_i^2 p_j^2 + \tfrac{1}{4}\sum_{k \neq j \neq i}^{s} \tfrac{1}{2} \cdot 8p_i^2 p_j p_k = p_i^3 p_j + p_i^2 p_j^2 + p_i^2 p_j \sum_{k \neq j \neq i}^{s} p_k$$

$$= p_i^2 p_j[p_i + p_j + (1 - p_i - p_j)] = p_i^2 p_j, \qquad i, j = 1, 2, \ldots, s, i \neq j.$$

$$(7.21)$$

* The underlined genotype is that of the parent; more precisely, the left-hand side of (7.18) should be:

$\Pr\{\text{Child} = A_iA_i \text{ and Parent} = A_iA_i \text{ and mating } (A_iA_i \times A_iA_j)\}$
$\quad = \tfrac{1}{2}\Pr\{\text{Child} = A_iA_i \text{ and } (A_iA_i \times A_iA_j)\} = \tfrac{1}{2}\Pr\{A_iA_i \mid A_iA_i \times A_iA_j\}\Pr\{A_iA_i \times A_iA_j\}$

In view of Remark 1 we also obtain

$$\Pr\{A_iA_j \text{ and } A_iA_i\} = p_i^2 p_j. \tag{7.22}$$

Type 5. $\Pr\{A_iA_j, A_iA_j\}$

$\Pr\{A_iA_j \text{ and } A_iA_j\}$

$$= \Pr\{A_iA_j \text{ and } (\underline{A_iA_j} \times A_iA_i)\} + \Pr\{A_iA_j \text{ and } (\underline{A_iA_j} \times A_jA_j)\}$$

$$+ \Pr\{A_iA_i \text{ and } (\underline{A_iA_j} \times A_iA_j)\} + \sum_{k \neq j \neq i} \Pr\{A_iA_j \text{ and } (\underline{A_iA_j} \times A_iA_k)\}$$

$$+ \sum_{k \neq j \neq i} \{A_iA_j \text{ and } (\underline{A_iA_j} \times A_jA_k)\}$$

$$= p_i^3 p_j + p_i p_j^3 + 2p_i^2 p_j^2 + p_i^2 p_j \sum_{k \neq j \neq i} p_k + p_i p_j^2 \sum_{k \neq j \neq i} p_k$$

$$= p_i p_j [p_i^2 + p_j^2 + 2p_i p_j + p_i(1 - p_i - p_j) + p_j(1 - p_i - p_j)]$$

$$= p_i p_j (p_i + p_j), \quad i, j = 1, 2, \ldots, s, \, i \neq j. \tag{7.23}$$

Type 6. $\Pr\{A_iA_j, A_iA_k\}$

$\Pr\{A_iA_j \text{ and } A_iA_k\}$

$$= \Pr\{A_iA_j \text{ and } (\underline{A_iA_k} \times A_iA_j)\} + \Pr\{A_iA_j \text{ and } (\underline{A_iA_k} \times A_jA_j)\}$$

$$+ \Pr\{A_iA_j \text{ and } (\underline{A_iA_k} \times A_jA_k)\} + \sum_{\substack{l \neq k \\ \neq j \neq i}} \Pr\{A_iA_j \text{ and } (\underline{A_iA_k} \times A_jA_l)\}$$

$$= p_i^2 p_j p_k + p_i p_j^2 p_k + p_i p_j p_k^2 + p_i p_j p_k \sum_{\substack{l \neq k \\ \neq j \neq i}} p_l$$

$$= p_i p_j p_k [p_i + p_j + p_k + (1 - p_i - p_j - p_k)] = p_i p_j p_k. \tag{7.24}$$

Summarizing, the formulae for joint probabilities of the four types of child-parent (or parent-child) combinations are as follows:

Type	Joint probabilities	
1	$\Pr\{A_iA_i \text{ and } A_iA_i\} = p_i^3,$	
3	$\Pr\{A_iA_i \text{ and } A_iA_j\} = \Pr\{A_iA_j \text{ and } A_iA_i\} = p_i^2 p_j,$	$i \neq j,$
5	$\Pr\{A_iA_j \text{ and } A_iA_j\} = p_i p_j(p_i + p_j),$	$i \neq j,$
6	$\Pr\{A_iA_j \text{ and } A_iA_k\} = \Pr\{A_iA_k \text{ and } A_iA_j\} = p_i p_j p_k,$	
		$i \neq j \neq k.$

$$\tag{7.25}$$

All types will appear when $s \geqslant 4$. Table 7.3 represents the joint probability space of parent-child genotypes for $s = 4$. It can be considered as a 10×10 [in general, a $\frac{1}{2}s(s + 1) \times \frac{1}{2}s(s + 1)$] matrix. We denote it by **M**. The

TABLE 7.3   Joint genotype distribution for parent-child; $s = 4$. (Matrix $\mathbf{M}$)

Parent \ Child	A_1A_1	A_2A_2	A_3A_3	A_4A_4	A_1A_2	A_1A_3	A_1A_4	A_2A_3	A_2A_4	A_3A_4	Total
A_1A_1	p_1^3	0	0	0	$p_1^2p_2$	$p_1^2p_3$	$p_1^2p_4$	0	0	0	p_1^2
A_2A_2	0	p_2^3	0	0	$p_2^2p_1$	0	0	$p_2^2p_3$	$p_2^2p_4$	0	p_2^2
A_3A_3	0	0	p_3^3	0	0	$p_3^2p_1$	0	$p_3^2p_2$	0	$p_3^2p_4$	p_3^2
A_4A_4	0	0	0	p_4^3	0	0	$p_4^2p_1$	0	$p_4^2p_2$	$p_4^2p_3$	p_4^2
A_1A_2	$p_1^2p_2$	$p_2^2p_1$	0	0	$p_1p_2(p_1+p_2)$	$p_1p_2p_3$	$p_1p_2p_4$	$p_1p_2p_3$	$p_1p_2p_4$	0	$2p_1p_2$
A_1A_3	$p_1^2p_3$	0	$p_3^2p_1$	0	$p_1p_2p_3$	$p_1p_3(p_1+p_3)$	$p_1p_3p_4$	$p_1p_2p_3$	0	$p_1p_3p_4$	$2p_1p_3$
A_1A_4	$p_1^2p_4$	0	0	$p_4^2p_1$	$p_1p_2p_4$	$p_1p_3p_4$	$p_1p_4(p_1+p_4)$	0	$p_1p_2p_4$	$p_1p_3p_4$	$2p_1p_4$
A_2A_3	0	$p_2^2p_3$	$p_3^2p_2$	0	$p_1p_2p_3$	$p_1p_2p_3$	0	$p_2p_3(p_2+p_3)$	$p_2p_3p_4$	$p_2p_3p_4$	$2p_2p_3$
A_2A_4	0	$p_2^2p_4$	0	$p_4^2p_2$	$p_1p_2p_4$	0	$p_1p_2p_4$	$p_2p_3p_4$	$p_2p_4(p_2+p_4)$	$p_2p_3p_4$	$2p_2p_4$
A_3A_4	0	0	$p_3^2p_4$	$p_4^2p_3$	0	$p_1p_3p_4$	$p_1p_3p_4$	$p_2p_3p_4$	$p_2p_3p_4$	$p_3p_4(p_3+p_4)$	$2p_3p_4$
Total	p_1^2	p_2^2	p_3^2	p_4^2	$2p_1p_2$	$2p_1p_3$	$2p_1p_4$	$2p_2p_3$	$2p_2p_4$	$2p_3p_4$	1

matrix $\mathbf{M}$ has the following properties:

1. $\mathbf{M}$ is symmetrical, so that
$$\mathbf{M} = \mathbf{M}'. \tag{7.26}$$

2. It can be partitioned into four submatrices. In the $s \times s$ submatrix, $\mathbf{M}_{11}$, both parent and child are homozygotes. In the $s \times \frac{1}{2}s(s-1)$ submatrix, $\mathbf{M}_{12}$, the parental genotypes are homozygotes and the genotypes of the children are heterozygotes. The situation is reversed for the $\frac{1}{2}s(s-1) \times s$ submatrix, $\mathbf{M}_{21}$; we notice also that
$$\mathbf{M}_{21} = \mathbf{M}_{12}'. \tag{7.27}$$
Finally, there is the $\frac{1}{2}s(s-1) \times \frac{1}{2}s(s-1)$ submatrix, $\mathbf{M}_{22}$, where both parent and child are heterozygotes.

The partition of matrix $\mathbf{M}$ is marked in Table 7.3 and can be given schematically as follows:
$$\mathbf{M} = \begin{bmatrix} \mathbf{M}_{11} & \mathbf{M}_{12} \\ \mathbf{M}_{21} & \mathbf{M}_{22} \end{bmatrix}. \tag{7.28}$$

3. The marginal distributions are the genotype distribution of the parents (the last column in Table 7.3) and the genotype distribution of the children (the last row). As would be expected from the equilibrium law, they are both the same.

If the allelic properties (dominance, codominance, etc.) are known, it is easy to find the joint phenotype distribution from Table 7.3. (See Problem 7.1a.)

In particular, if there are only two alleles, A and a, with gene frequencies p and q, respectively, Table 7.3 takes the form of Table 7.4. We notice that Table 7.4 is not sufficiently general to cover all *types* of joint parent-child genotypes given in Table 7.3.

If there is a dominance (A over a), the joint phenotype distribution for parent-child is as shown in Table 7.5a or 7.5b. It is easy to see that Table 7.5b can be obtained from Table 7.5a by using the fact that $p + q = 1$.

TABLE 7.4 Joint genotype distribution for parent-child, $s = 2$

Parent (x) \ Child (y)	AA (2)	Aa (1)	aa (0)	Total
AA (2)	p^3	p^2q	0	p^2
Aa (1)	p^2q	pq	pq^2	$2pq$
aa (0)	0	pq^2	q^3	q^2
Total	p^2	$2pq$	q^2	1

TABLE 7.5*a*   Joint parent-child phenotype distribution

Child (*y*) Parent (*x*)	$A-$ (1)	aa (0)	Total
$A-$ (1) aa (0)	$p(1 + pq)$ pq^2	pq^2 q^3	$p^2 + 2pq$ q^2
Total	$p^2 + 2pq$	q^2	1

TABLE 7.5*b*   (in terms of *q*)

Child (*y*) Parent (*x*)	$A-$ (1)	aa (0)	Total
$A-$ (1) aa (0)	$(1 - q)(1 + q - q^2)$ $q^2(1 - q)$	$q^2(1 - q)$ q^3	$1 - q^2$ q^2
Total	$1 - q^2$	q^2	1

7.6.   CONDITIONAL DISTRIBUTIONS OF CHILD GENOTYPES WHEN THE PARENT'S GENOTYPE IS GIVEN, AND VICE VERSA

From Table 7.3 it is easy to find the conditional probability that, for example, the child is of genotype A_iA_j, *given* that the parent is A_iA_k:

$$\Pr\{A_iA_j \mid A_iA_k\} = \frac{\Pr\{A_iA_j \text{ and } A_iA_k\}}{\Pr\{A_iA_k\}}. \tag{7.29}$$

In particular, we find

$$\left.\begin{aligned}
\Pr\{A_iA_j \mid A_iA_i\} &= \frac{p_i^2 p_j}{p_i^2} = p_j, \\
\Pr\{A_iA_j \mid A_iA_j\} &= \frac{p_i p_j(p_i + p_j)}{2p_i p_j} = \tfrac{1}{2}(p_i + p_j), \\
\Pr\{A_iA_j \mid A_iA_k\} &= \frac{p_i p_j p_k}{2p_i p_k} = \tfrac{1}{2}p_j, \\
\Pr\{A_iA_j \mid A_kA_l\} &= 0.
\end{aligned}\right\} \tag{7.30}$$

TABLE 7.6 Conditional genotype distributions of offspring when a genotype of a parent is given, Pr{Child | Parent}.
(Matrix **T**)

Child / (given) Parent	A_1A_1	A_2A_2	A_3A_3	A_4A_4	A_1A_2	A_1A_3	A_1A_4	A_2A_3	A_2A_4	A_3A_4	Total
A_1A_1	p_1	0	0	0	p_2	p_3	p_4	0	0	0	1
A_2A_2	0	p_2	0	0	p_1	0	0	p_3	p_4	0	1
A_3A_3	0	0	p_3	0	0	p_1	0	p_2	0	p_4	1
A_4A_4	0	0	0	p_4	0	0	p_1	0	p_2	p_3	1
A_1A_2	$\frac{1}{2}p_1$	$\frac{1}{2}p_2$	0	0	$\frac{1}{2}(p_1+p_2)$	$\frac{1}{2}p_3$	$\frac{1}{2}p_4$	$\frac{1}{2}p_3$	$\frac{1}{2}p_4$	0	1
A_1A_3	$\frac{1}{2}p_1$	0	$\frac{1}{2}p_3$	0	$\frac{1}{2}p_2$	$\frac{1}{2}(p_1+p_3)$	$\frac{1}{2}p_4$	$\frac{1}{2}p_2$	0	$\frac{1}{2}p_4$	1
A_1A_4	$\frac{1}{2}p_1$	0	0	$\frac{1}{2}p_4$	$\frac{1}{2}p_2$	$\frac{1}{2}p_3$	$\frac{1}{2}(p_1+p_4)$	0	$\frac{1}{2}p_2$	$\frac{1}{2}p_3$	1
A_2A_3	0	$\frac{1}{2}p_2$	$\frac{1}{2}p_3$	0	$\frac{1}{2}p_1$	$\frac{1}{2}p_1$	0	$\frac{1}{2}(p_2+p_3)$	$\frac{1}{2}p_4$	$\frac{1}{2}p_4$	1
A_2A_4	0	$\frac{1}{2}p_2$	0	$\frac{1}{2}p_4$	$\frac{1}{2}p_1$	0	$\frac{1}{2}p_1$	$\frac{1}{2}p_3$	$\frac{1}{2}(p_2+p_4)$	$\frac{1}{2}p_3$	1
A_3A_4	0	0	$\frac{1}{2}p_3$	$\frac{1}{2}p_4$	0	$\frac{1}{2}p_1$	$\frac{1}{2}p_1$	$\frac{1}{2}p_2$	$\frac{1}{2}p_2$	$\frac{1}{2}(p_3+p_4)$	1

The conditional distributions of genotypes of offspring for a given genotype of a parent are represented in Table 7.6. We notice that this table can be obtained from Table 7.3 by dividing each element in a row by the last element of this row.

Table 7.6 is a $\frac{1}{2}s(s+1) \times \frac{1}{2}s(s+1)$ matrix; it is sometimes called the *transition matrix* T [Li and Sacks (1954)].

1. Each row of the matrix T represents the conditional distribution of the genotypes of offspring for a given genotype of a parent.

2. The matrix T is not symmetrical, so that

$$T \neq T'. \tag{7.31}$$

3. It can be also partitioned into four submatrices:

$$T = \begin{bmatrix} T_{11} & T_{12} \\ T_{21} & T_{22} \end{bmatrix}, \tag{7.32}$$

where

$$T_{21} \neq T'_{12}. \tag{7.33}$$

4. The conditional distributions of parental genotypes for a given genotype of offspring are just T'. The matrix T can be obtained from the matrix M by dividing each element in a column by the last element of this column.

When $s = 2$, the transition matrix T takes the form shown in Table 7.7.

TABLE 7.7   Conditional genotype distribution
Pr{Child | Parent}; $s = 2$

Parent (given) \ Child	AA	Aa	aa	Total
AA	p	q	0	1
Aa	$\frac{1}{2}p$	$\frac{1}{2}$	$\frac{1}{2}p$	1
aa	0	p	q	1

Let us now introduce a square $\frac{1}{2}s(s+1) \times \frac{1}{2}s(s+1)$ matrix R with identical columns, each representing a genotype distribution for a population in equilibrium. For our example with $s = 4$, the matrix R will be the 10×10 matrix:

$$R = \begin{bmatrix} p_1^2 & \cdots & p_1^2 & \cdots & p_1^2 \\ \cdots\cdots\cdots\cdots\cdots\cdots\cdots\cdots \\ p_4^2 & \cdots & p_4^2 & \cdots & p_4^2 \\ 2p_1p_2 & \cdots & 2p_1p_2 & \cdots & 2p_1p_2 \\ \cdots\cdots\cdots\cdots\cdots\cdots\cdots\cdots \\ 2p_3p_4 & \cdots & 2p_3p_4 & \cdots & 2p_3p_4 \end{bmatrix}. \tag{7.34}$$

Its transpose, $\mathbf{R}'$, is

$$\mathbf{R}' = \begin{bmatrix} p_1^2 & \cdots & p_4^2 & 2p_1p_2 & 2p_1p_3 & \cdots & 2p_3p_4 \\ p_1^2 & \cdots & p_4^2 & 2p_1p_2 & 2p_1p_3 & \cdots & 2p_3p_4 \\ \cdots\cdots\cdots\cdots\cdots\cdots\cdots\cdots\cdots\cdots\cdots \\ p_1^2 & \cdots & p_4^2 & 2p_1p_2 & 2p_1p_3 & \cdots & 2p_3p_4 \end{bmatrix}. \qquad (7.35)$$

We notice that the joint genotype distribution represented by matrix $\mathbf{M}$ is the term-by-term product of $\mathbf{T}$ by $\mathbf{R}$, that is,

$$\mathbf{M} = \mathbf{T} \ \square \ \mathbf{R}.$$

(For definition of the term-by-term product of two matrices see I.26.)

7.7. JOINT GRANDPARENT-GRANDCHILD GENOTYPE DISTRIBUTION

We now wish to find the joint genotype distribution of grandparent and grandchild combinations. For simplicity, we introduce the following abbreviations: GP = grandparent; Pt = parent; GC = grandchild.

Let us first find the conditional probability that $GC = A_1A_1$, given that $GP = A_1A_1$, that is, $\Pr\{GC = A_1A_1 \mid GP = A_1A_1\}$. If the grandparent is A_1A_1, the probabilities of the possible genotypes of a parent are the elements of the *first row* of the matrix $\mathbf{T}$ as given in Table 7.6. For *each* genotype of a parent the probabilities that a child should be A_1A_1 are the elements of the *first column* of $\mathbf{T}$. Thus, the total conditional probability that $GC = A_1A_1$, given $GP = A_1A_1$, is the *sum* of the products of the corresponding elements of the *first row by the first column* of the matrix $\mathbf{T}$. Mathematically this is expressed as

$\Pr\{GC = A_1A_1 \mid GP = A_1A_1\}$

$$= \sum_{i \leqslant j}^{s} \sum^{s} \Pr\{Pt = A_iA_j \mid GP = A_1A_1\} \Pr\{GC = A_1A_1 \mid Pt = A_iA_j\}$$

$$= \tfrac{1}{2}p_1 + \tfrac{1}{2}p_1^2. \qquad (7.36)$$

Similarly,

$\Pr\{GC = A_1A_1 \mid GP = A_2A_2\}$

$$= \sum_{i \leqslant j}^{s} \sum^{s} \Pr\{Pt = A_iA_j \mid GP = A_2A_2\} \Pr\{GC = A_1A_1 \mid Pt = A_iA_j\} = \tfrac{1}{2}p_1^2,$$

$$(7.37)$$

which is obtained by multiplication of the *second row by the first column* in the matrix **T**.

Following the same procedure for each GC-GP combination of genotypes, we obtain the complete matrix of *conditional* probabilities $\Pr\{GC \mid GP\}$, which we will denote by $\mathbf{T}_2$ (transition matrix for the *second* generation).

We notice that this matrix can be expressed as

$$\mathbf{T}_2 = \mathbf{T}^2 = \tfrac{1}{2}\mathbf{T} + \tfrac{1}{2}\mathbf{R}', \tag{7.38}$$

where $\mathbf{R}'$ is defined in (7.35). This formula was first obtained by Li and Sacks (1954), who used a slightly different notation.

The joint grandparent-grandchild genotype distribution is

$$\mathbf{M}_2 = \mathbf{T}_2 \boxdot \mathbf{R} = \tfrac{1}{2}\mathbf{T} \boxdot \mathbf{R} + \tfrac{1}{2}\mathbf{R}' \boxdot \mathbf{R} = \tfrac{1}{2}\mathbf{M} + \tfrac{1}{2}\mathbf{R}' \boxdot \mathbf{R}. \tag{7.39}$$

We now investigate the interpretation of the term-by-term product $\mathbf{R}' \boxdot \mathbf{R}$. Let us, for convenience, denote the genotype vector for a population in equilibrium with respect to one locus, by $\mathbf{h}$, that is,

$$\mathbf{h}' = (p_1{}^2, p_2{}^2, \ldots, p_s{}^2, 2p_1p_2, 2p_1p_3, \ldots, 2p_{s-1}p_s). \tag{7.40}$$

We notice that

$$\mathbf{R}' \boxdot \mathbf{R} = \mathbf{R} \boxdot \mathbf{R}' = \mathbf{hh}' = \mathbf{H}. \tag{7.41}$$

The $\tfrac{1}{2}s(s + 1) \times \tfrac{1}{2}s(s + 1)$ matrix $\mathbf{H}$ represents the joint distribution of independent pairs in a population at equilibrium. Substituting (7.41) into (7.39), we obtain

$$\mathbf{M}_2 = \tfrac{1}{2}\mathbf{M} + \tfrac{1}{2}\mathbf{H}. \tag{7.42}$$

It can be shown [see Li and Sacks (1954)] that the "transition" matrix for the nth generation, $\mathbf{T}_n$, can be obtained from the formula

$$\mathbf{T}_n = \mathbf{T}^n = (\tfrac{1}{2})^{n-1}\mathbf{T} + [1 - (\tfrac{1}{2})^{n-1}]\mathbf{R}'; \tag{7.43}$$

hence

$$\mathbf{M}_n = (\tfrac{1}{2})^{n-1}\mathbf{M} + [1 - (\tfrac{1}{2})^{n-1}]\mathbf{H}, \tag{7.44}$$

where

$$\mathbf{T}_1 = \mathbf{T} \quad \text{and} \quad \mathbf{M}_1 = \mathbf{M}.$$

Let us consider the case of one locus with two alleles, A and a, with gene frequencies p and q, respectively.

The matrix $\mathbf{M}$ is given in Table 7.4. The genotype vector is $\mathbf{h}' = (p^2, 2pq, q^2)$. The matrices $\mathbf{R}$ and $\mathbf{R}'$ are

$$\mathbf{R} = \begin{bmatrix} p^2 & p^2 & p^2 \\ 2pq & 2pq & 2pq \\ q^2 & q^2 & q^2 \end{bmatrix}, \quad \mathbf{R}' = \begin{bmatrix} p^2 & 2pq & q^2 \\ p^2 & 2pq & q^2 \\ p^2 & 2pq & q^2 \end{bmatrix}, \tag{7.45}$$

and

$$\mathbf{R} \ \square \ \mathbf{R'} = \mathbf{hh'} = \mathbf{H} = \begin{bmatrix} p^4 & 2p^3q & p^2q^2 \\ 2p^3q & 4p^2q^2 & 2pq^3 \\ p^2q^2 & 2pq^3 & q^4 \end{bmatrix}. \qquad (7.46)$$

Using formula (7.42), we find the joint genotype distribution for pairs (GP, PC) as in Table 7.8.

TABLE 7.8   Joint genotype distribution for grandparent-grandchild, $s = 2$

GP \ GC	AA	Aa	aa	Total
AA	$\frac{1}{2}p^3(1+p)$	$\frac{1}{2}p^2q(1+2p)$	$\frac{1}{2}p^2q^2$	p^2
Aa	$\frac{1}{2}p^2q(1+2p)$	$\frac{1}{2}pq(1+4pq)$	$\frac{1}{2}pq^2(1+2q)$	$2pq$
aa	$\frac{1}{2}p^2q^2$	$\frac{1}{2}pq^2(1+2q)$	$\frac{1}{2}q^3(1+q)$	q^2
Total	p^2	$2pq$	q^2	1

The (GP, GC) phenotype distribution (in terms of q) is given in Table 7.9 (compare Problem 7.3b).

TABLE 7.9   Joint phenotype distribution for grandparent-grandchild; $s = 2$

GP \ GC	$A-$	aa	Total
$A-$	$1 - 2q^2 + \frac{1}{2}q^3 + \frac{1}{2}q^4$	$\frac{1}{2}q^2(1-q)(2+q)$	$1 - q^2$
aa	$\frac{1}{2}q^2(1-q)(2+q)$	$\frac{1}{2}q^3(1+q)$	q^2
Total	$1 - q^2$	q^2	1

7.8.   ANCESTRAL CORRELATION

From the preceding discussion it is clear that the genotype of the child depends on the genotypes of his parents or even his grandparents. We therefore say that the genotypes of offspring and their ancestors are *correlated*. In order to calculate a correlation coefficient, which is defined for random variables, we must assign to the genotypes some numerical values which will correspond to the values of a random variable. The usual way in which this is done is to use the genotypic values, even if the traits corresponding to the genotypes are of qualitative character (see section 6.6). Of course, the value of the correlation coefficient depends on what genotypic values are assigned to the genotypes.

We shall consider here two cases of one locus with two alleles, A and a:

(i) when the genic effects are *additive*;

(ii) when there is complete *dominance* of A over a.

We shall discuss first correlation between parent and child.

7.8.1. Correlation between Parent and Child

(i) Let us consider a joint genotype distribution for parent-child as given in Table 7.4. "Totals" represent the (marginal) parental and offspring distributions, respectively, under equilibrium. Let X and Y denote the genotypic values of parental and offspring genotypes, respectively. On the assumption that gene effects are *additive*, and assigning the value 0 to the gene a and 1 to the gene A, X and Y take values 0, 1, 2 as shown in Table 7.4. (Note that these values are equal to the numbers of gene A present in a genotype.) Using the results of Example 6.6, we obtain

$$\mu_X = \mu_Y = 2p; \quad \sigma_X{}^2 = \sigma_Y{}^2 = 2pq. \tag{7.47}$$

From formula (6.36) we calculate the covariance between X and Y. We have

$$\mathrm{Cov}(X, Y) = \sigma_{XY} = 2 \cdot 2 \cdot p^3 + 2 \cdot 1 \cdot p^2 q + 1 \cdot 2 \cdot p^2 q + 1 \cdot 1 \cdot pq - 2p \cdot 2p$$

$$= 4p^2(p + q) + pq - 4p^2 = pq. \tag{7.48}$$

The correlation coefficient between parent (X) and child (Y) genotypic values which we denote by ρ_M, and which we will call the *genotypic correlation coefficient*, is from (6.37)

$$\rho_{XY} = \rho_M = \frac{pq}{\sqrt{2pq}\,\sqrt{2pq}} = \frac{1}{2}. \tag{7.49}$$

Note that ρ_M does not depend on the gene frequencies.

(ii) Assuming complete *dominance* of gene A over a and assigning the values 1 and 0 to phenotypes $A-$ and aa, respectively, we can evaluate the *phenotypic* correlation coefficient, ρ_M^*, say. The sample space given in Table 7.5a or 7.5b exhibits the appropriate parent-child phenotype distribution. In this case we obtain

$$\mu_X = \mu_Y = 1 - q^2; \quad \sigma_X{}^2 = \sigma_Y{}^2 = q^2(1 - q^2); \quad \sigma_{XY} = q^3(1 - q). \tag{7.50}$$

Hence,

$$\rho_M^* = \frac{q^3(1-q)}{q^2(1-q^2)} = \frac{q}{1+q}. \tag{7.51}$$

Now the correlation coefficient is no longer independent of the gene frequency. It is very nearly 0 if q is small, and it is equal to $\frac{1}{2}$ when $q = 1$.

When there are *more than two alleles*, the correlation coefficient depends on the values assigned to particular genotypes. However, if the genic effects are *additive*, it turns out that $\rho_M = \frac{1}{2}$; that is, in this case is independent of the number of alleles.

7.8.2.   Correlation between Grandparent and Grandchild

To obtain the correlation coefficient between genotypes of grandparent and grandchild, ρ_{M_2}, say, we use the results of section 6.9. First we notice that the joint GP-GC genotype distribution, M_2, can be presented in the following form:

$$M_2 = \tfrac{1}{2}M + \tfrac{1}{2}H. \tag{7.42}$$

This means that the population represented by M_2 is a "mixture" of the population represented by parent-child pairs (matrix M) and random pairs (matrix H). It resembles (6.73) with $c_1 = c_2 = \frac{1}{2}$. The distribution M_2 also satisfies conditions (6.82). Hence ρ_{M_2} can be calculated from (6.86), that is, we will have

$$\rho_{M_2} = \tfrac{1}{2}\rho_M + \tfrac{1}{2}\rho_H = \tfrac{1}{2}\rho_M, \tag{7.52}$$

since ρ_H is the genotypic correlation coefficient between two randomly mating populations, so that we have $\rho_H = 0$.

(i) In particular, when the gene effects are *additive*, we obtain

$$\rho_{M_2} = \tfrac{1}{2} \cdot \tfrac{1}{2} = (\tfrac{1}{2})^2. \tag{7.53}$$

This can be extended to the case in which the correlation between an ancestor and its nth generation is considered. We will have

$$\rho_{M_n} = (\tfrac{1}{2})^{n-1}\rho_M = (\tfrac{1}{2})^n. \tag{7.54}$$

Formulae (7.53) and (7.54) hold whatever the number, s, of alleles at a single locus.

(ii) For *phenotypic* ancestral correlation when there is complete dominance of A over a, we obtain

$$\rho_{M_2}^* = \tfrac{1}{2}\rho_M^* = \tfrac{1}{2}\frac{q}{1+q}, \tag{7.55}$$

and

$$\rho_{M_n}^* = (\tfrac{1}{2})^{n-1}\rho_M^* = (\tfrac{1}{2})^{n-1}\frac{q}{1+q} \ . \tag{7.56}$$

7.9. JOINT PARENT-CHILD DISTRIBUTION AND THEIR CORRELATION FOR X-LINKED GENES

We assume that the heterogametic sex is male and will consider a X-linked locus with two alleles, A and a. We will have four kinds of parent-child pairs: (1) mother-daughter, (2) mother-son, (3) father-daughter, and (4) father-son. There will be 3×3 combinations of genotypes for mother-daughter, 3×2 for mother-son, 2×3 for father-daughter, and 2×2 for father-son. Let us denote the matrices $\mathbf{T}$, $\mathbf{R}$, and $\mathbf{M}$, discussed in preceding sections, by $\mathbf{T}_{mn}$, $\mathbf{R}_{mn}$, and $\mathbf{M}_{mn}$, $m = 1, 2, 3$ and $n = 1, 2$, respectively, where the subscript m refers to the parent and n to the child. The general relation

$$\mathbf{M}_{mn} = \mathbf{T}_{mn} \ \boxdot \ \mathbf{R}_{mn} \tag{7.57}$$

is valid for each of the four parent-child combinations.

Our task will be to evaluate these matrices and the corresponding genotypic and phenotypic correlation coefficients, ρ_{mn} and ρ_{mn}^*, respectively. Although the algebra is straightforward, we find it too lengthy to give all the details of derivations. We present only the results, but the reader should be able to derive them by using arguments similar to those of section 7.6.

1. Mother-Daughter Joint Genotype Distribution. It is easy to see that $\mathbf{T}_{33}$ is the same as in Table 7.3, $\mathbf{R}_{33}$ is the same as in (7.45), and $\mathbf{M}_{33} = \mathbf{T}_{33} \ \boxdot \ \mathbf{R}_{33}$ is the same as in Table 7.4.

2. Mother-Son Joint Genotype Distribution

$$\mathbf{T}_{32} = \begin{array}{c} \\ AA \\ Aa \\ aa \end{array}\begin{array}{cc} A & a \\ \begin{bmatrix} 1 & 0 \\ \tfrac{1}{2} & \tfrac{1}{2} \\ 0 & 1 \end{bmatrix} \end{array}, \quad \mathbf{R}_{32} = \begin{bmatrix} p^2 & p^2 \\ 2pq & 2pq \\ q^2 & q^2 \end{bmatrix}, \quad \mathbf{M}_{32} = \begin{bmatrix} p^2 & 0 \\ pq & pq \\ 0 & q^2 \end{bmatrix}. \tag{7.58}$$

3. Father-Daughter Genotype Distribution

$$\mathbf{T}_{23} = \begin{array}{c} \\ A \\ a \end{array}\begin{array}{ccc} AA & Aa & aa \\ \begin{bmatrix} p & q & 0 \\ 0 & p & q \end{bmatrix} \end{array}, \quad \mathbf{R}_{23} = \begin{bmatrix} p & p & p \\ q & q & q \end{bmatrix}, \quad \mathbf{M}_{23} = \begin{bmatrix} p^2 & pq & 0 \\ 0 & pq & q^2 \end{bmatrix}.$$

$$\tag{7.59}$$

We notice that

$$\mathbf{M}_{23} = \mathbf{M}'_{32} \tag{7.60}$$

but

$$\mathbf{T}_{23} \neq \mathbf{T}'_{32} \quad \text{and} \quad \mathbf{R}'_{23} \neq \mathbf{R}_{32}. \tag{7.61}$$

4. Father-Son Genotype Distribution

$$
\begin{array}{cc}
 & A \quad a \\
\end{array}
$$

$$
\mathbf{T}_{22} = \begin{array}{c} A \\ a \end{array}\begin{bmatrix} p & q \\ p & q \end{bmatrix}, \quad
\mathbf{R}_{22} = \begin{bmatrix} p & p \\ q & q \end{bmatrix}, \quad
\mathbf{M}_{22} = \begin{bmatrix} p^2 & pq \\ pq & q^2 \end{bmatrix}. \tag{7.62}
$$

Using matrices $\mathbf{M}_{mn}$ and assigning the values 1 and 0 to genes A and a, respectively, we can easily calculate the genotypic and phenotypic correlation coefficients.

The genotypic correlation coefficients, ρ_{mn}, are

$$\rho_{33} = \tfrac{1}{2}, \quad \rho_{32} = \rho_{23} = \frac{1}{\sqrt{2}} = \tfrac{1}{2}\sqrt{2}, \quad \rho_{22} = 0. \tag{7.63}$$

If there is complete dominance of A over a, it can be shown that the phenotypic correlation coefficients, ρ^*_{mn}, are

$$\rho^*_{33} = \frac{q}{1+q}, \quad \rho^*_{32} = \rho^*_{23} = \sqrt{\frac{q}{1+q}}, \quad \rho^*_{22} = 0. \tag{7.64}$$

(See also Problem 7.8.)

7.10. JOINT GENOTYPE DISTRIBUTION OF GRANDPARENT AND GRANDCHILD WHEN THE GENES ARE X-LINKED

Since the grandchild may receive a gene from its grandparent through its mother or through its father, both possibilities must be taken into account in the derivation of appropriate formulae. The algebra is even more tedious here than in section 7.9, so we restrict ourselves to presenting the results in Table 7.10. Here $\mathbf{T}_{mn(2)}$ is the total transition matrix for the second generation (i.e., for grandparent-grandchild), while $[\mathbf{T}_{mn(2)}(\text{mother})]$ and $[\mathbf{T}_{mn(2)}(\text{father})]$ are the corresponding components due to the mother and the father, respectively. We have

$$\mathbf{T}_{mn(2)} = \tfrac{1}{2}[\mathbf{T}_{mn(2)}(\text{mother}) + \mathbf{T}_{mn(2)}(\text{father})], \tag{7.65}$$

and

$$\mathbf{M}_{mn} = \mathbf{T}_{mn} \ \square \ \mathbf{R}_{mn}. \tag{7.66}$$

TABLE 7.10 Conditional (given the genotype of grandparent) and joint genotype distributions of GP-GC combinations for X-linked genes

	Grandmother-granddaughter	Grandmother-grandson	Grandfather-granddaughter	Grandfather-grandson
$\mathbf{T}_{mn(2)}$(mother)	$\mathbf{T}_{33}\mathbf{T}_{33} = \frac{1}{2}\mathbf{T}_{33} + \frac{1}{2}\mathbf{R}'_{33}$	$\mathbf{T}_{33}\mathbf{T}_{32} = \frac{1}{2}\mathbf{T}_{32} + \frac{1}{2}\mathbf{R}'_{32}$	$\mathbf{T}_{23}\mathbf{T}_{32} = \frac{1}{2}\mathbf{T}_{23} + \frac{1}{2}\mathbf{R}'_{23}$	$\mathbf{T}_{23}\mathbf{T}_{32} = \frac{1}{2}\mathbf{I}_{22} + \frac{1}{2}\mathbf{R}'_{22}$
$\mathbf{T}_{mn(2)}$(father)	$\mathbf{T}_{32}\mathbf{T}_{23} = \mathbf{T}_{33}$	$\mathbf{T}_{32}\mathbf{T}_{22} = \mathbf{R}'_{32}$	$\mathbf{T}_{22}\mathbf{T}_{23} = \mathbf{R}'_{23}$	$\mathbf{T}_{22}\mathbf{T}_{22} = \mathbf{R}'_{22}$
$\mathbf{T}_{mn(2)}$	$\frac{3}{4}\mathbf{T}_{33} + \frac{1}{4}\mathbf{R}'_{33}$	$\frac{1}{4}\mathbf{T}_{32} + \frac{3}{4}\mathbf{R}'_{32}$	$\frac{1}{4}\mathbf{T}_{23} + \frac{3}{4}\mathbf{R}'_{23}$	$\frac{1}{4}\mathbf{I}_{22} + \frac{3}{4}\mathbf{R}'_{22}$
$\mathbf{H}_{mn}$	$\mathbf{R}_{33} \boxdot \mathbf{R}'_{33}$	$\mathbf{R}_{32} \boxdot \mathbf{R}'_{32}$	$\mathbf{R}_{23} \boxdot \mathbf{R}'_{23}$	$\mathbf{R}_{22} \boxdot \mathbf{R}'_{22}$
$\mathbf{M}_{mn(2)}$	$\frac{3}{4}\mathbf{M}_{33} + \frac{1}{4}\mathbf{H}_{33}$	$\frac{1}{4}\mathbf{M}_{32} + \frac{3}{4}\mathbf{H}_{32}$	$\frac{1}{4}\mathbf{M}_{23} + \frac{3}{4}\mathbf{H}_{23}$	$\frac{1}{4}(\mathbf{I}_{22} \boxdot \mathbf{R}_{22}) + \frac{3}{4}\mathbf{H}_2$
$\rho_{mn(2)}$	$\frac{3}{8}$	$\frac{1}{4}\sqrt{2}$	$\frac{1}{8}\sqrt{2}$	$\frac{1}{4}$
$\rho^*_{mn(2)}$	$\frac{3}{4}\frac{q}{1+q}$	$\frac{1}{4}\sqrt{\frac{q}{1+q}}$	$\frac{1}{4}\sqrt{\frac{q}{1+q}}$	$\frac{1}{4}$

The matrix

$$\mathbf{I}_{22} = \begin{bmatrix} 1 & 0 \\ 0 & 1 \end{bmatrix} \tag{7.67}$$

is the 2 × 2 identity matrix.

The matrices $\mathbf{M}_{mn(2)}$ represent the joint genotype distributions for all four kinds of grandparent-grandchild combinations, and $\rho_{mn(2)}$ and $\rho^*_{mn(2)}$ the corresponding genotypic and phenotypic (assuming complete dominance of gene A over a) correlation coefficients.

The reader interested in further details about the derivation of transition matrices is referred to a paper by Li and Sacks (1954).

7.11. JOINT FULL-SIB DISTRIBUTION AND THEIR GENOTYPIC CORRELATION COEFFICIENT

We again consider a randomly mating population in equilibrium with respect to a single locus with s alleles, $A_1, A_2, \ldots, A_s$, and their frequencies $p_1, p_2, \ldots, p_s$, respectively, with $\sum p_i = 1$.

There are seven types of parental matings (Table 7.1), and seven types of full-sib pairs can occur. Of course, for a given mating type of parents only certain types of sib pairs are possible. Table 7.2 will be useful in finding these types and in evaluating their conditional probabilities.

Let us take, for example, mating type 3, that is, $A_iA_i \times A_iA_j$. In row 3 of Table 7.2 the only nonzero probabilities of offspring are $\frac{1}{2}$ for A_iA_i and $\frac{1}{2}$ for A_iA_j. Therefore, only the following types of sib pairs: A_iA_i, A_iA_i (type 1), A_iA_i, A_iA_j (type 3), and A_iA_j, A_iA_j (type 5), can occur with $A_iA_i \times A_iA_j$ parents. Since the genotypes of the first and the second child are independent events for a given parental mating type, we have:

$$\Pr\{(A_iA_i, A_iA_i) \mid A_iA_i \times A_iA_j\} = \tfrac{1}{2} \cdot \tfrac{1}{2} = \tfrac{1}{4},$$

and

$$\Pr\{(A_iA_i, A_iA_i) \text{ and } A_iA_i \times A_iA_j\} = \tfrac{1}{4} \cdot 4p_i^3 \sum_{j \neq i}^{s} p_j = p_i^3(1 - p_i) = p_i^3 - p_i^4. \tag{7.68}$$

The probability of the *ordered* pair A_iA_i, A_iA_j (i.e., the first child A_iA_i and the second A_iA_j) and the mating $A_iA_i \times A_iA_j$ is

$$\Pr\{(A_iA_i, A_iA_j) \text{ and } A_iA_i \times A_iA_j\} = \Pr\{(A_iA_j, A_iA_i) \text{ and } A_iA_i \times A_iA_j\}$$
$$= \tfrac{1}{2} \cdot \tfrac{1}{2} \cdot 4p_i^3 p_j = p_i^3 p_j, \qquad i \neq j, \tag{7.69}$$

and similarly,

$$\Pr\{(A_iA_j, A_iA_j) \text{ and } A_iA_i \times A_iA_j\} = p_i^3 p_j, \qquad i \neq j. \tag{7.70}$$

On the other hand, the pairs of parallel cells with nonzero probabilities along the columns indicate within which parental mating a particular sib-pair type can occur. For example, type 2 (A_iA_i, A_jA_j) can occur only within matings $A_iA_j \times A_iA_j$ (type 5). But sib-pair type 1 (A_iA_i, A_iA_i) occurs within four matings: $A_iA_i \times A_iA_i$ (type 1), $A_iA_i \times A_iA_j$ (type 3), $A_iA_j \times A_iA_j$ (type 5), and $A_iA_j \times A_iA_k$ (type 6).

Special attention should also be paid to mating types 3, 4, 6, and 7. For example, sib-pair type 5 (A_iA_j, A_iA_j) can occur with mating type 3 ($A_iA_i \times A_iA_j$) and, moreover, with mating type 7, since type 7 can be written as $A_iA_k \times A_jA_l$. This is so because the order of the subscripts is immaterial; the only condition they must satisfy is $i \neq j \neq k \neq l$.

Evaluation of the probabilities for each (ordered) sib-pair type needs rather careful algebraic manipulations. Here we present only the derivation for the most complicated type, type 5, that is, A_iA_j, A_iA_j (Table 7.11).

TABLE 7.11   Probabilities of occurrence of sib-pair type 5 (A_iA_j, A_iA_j) with different mating types

Type	Parental Mating	Sib-sib $\Pr\{(A_iA_j, A_iA_j)$ and mating type$\}$
2	$A_iA_i \times A_jA_j$	$1 \cdot 2p_i^2 p_j^2$
3	$A_iA_i \times A_iA_j$	$\frac{1}{4} \cdot 4p_i^3 p_j$
	$A_jA_j \times A_iA_j$	$\frac{1}{4} \cdot 4p_i p_j^3$
4	$A_iA_i \times A_jA_k$	$\frac{1}{4} \cdot 4p_i^2 p_j \sum\limits_{k \neq i \neq j} p_k$
	$A_jA_j \times A_iA_k$	$\frac{1}{4} \cdot 4p_i p_j^2 \sum\limits_{k \neq i \neq j} p_k$
5	$A_iA_j \times A_iA_j$	$\frac{1}{4} \cdot 4p_i^2 p_j^2$
6	$A_iA_j \times A_iA_k$	$\frac{1}{16} \cdot 8p_i^2 p_j \sum\limits_{k \neq i \neq j} p_k$
	$A_iA_j \times A_jA_k$	$\frac{1}{16} \cdot 8p_i p_j^2 \sum\limits_{k \neq i \neq j} p_k$
	$A_iA_k \times A_jA_k$	$\frac{1}{16} \cdot 8p_i p_j \sum\limits_{k \neq i \neq j} p_k^2$
7	$A_iA_k \times A_jA_l$	$\frac{1}{16} \cdot 8p_i p_j \sum\limits_{l \neq k \neq j \neq i} \sum p_k p_l$

Adding all these probabilities and combining the appropriate terms, we obtain

$$\Pr\{A_iA_j, A_iA_j\}$$

$$= (2p_i^2p_j^2 + p_i^2p_j^2) + (p_i^3p_j + p_ip_j^3) + (\tfrac{3}{2}p_i^2p_j + \tfrac{3}{2}p_ip_j^2)\left(\sum_{k\neq j\neq i} p_k\right)$$

$$+ \tfrac{1}{2}p_ip_j\left(\sum_{k\neq j\neq i} p_k^2 + \sum_{l\neq k\neq j\neq i}\sum p_kp_l\right)$$

$$= \tfrac{1}{2}p_ip_j\{6p_ip_j + 2(p_i^2 + p_j^2) + 3(p_i + p_j)[1 - (p_i + p_j)]$$

$$+ [1 - (p_i + p_j)]^2\}$$

$$= \tfrac{1}{2}p_ip_j[2p_ip_j + 2(p_i + p_j)^2 + 3(p_i + p_j)$$

$$- 3(p_i + p_j)^2 + 1 - 2(p_i + p_j) + (p_i + p_j)^2]$$

$$= \tfrac{1}{2}p_ip_j[1 + (p_i + p_j) + 2p_ip_j]. \tag{7.71}$$

The derivations of the appropriate formulae for the remaining ordered sib-pair types are left to the reader as exercises in algebra. They are as follows:

1. $\Pr\{A_iA_i, A_iA_i\} = \tfrac{1}{4}p_i^2(1 + p_i)^2,$
2. $\Pr\{A_iA_i, A_jA_j\} = \tfrac{1}{4}p_i^2p_j^2,$
3. $\Pr\{A_iA_i, A_iA_j\} = \tfrac{1}{2}p_i^2p_j(1 + p_i),$
4. $\Pr\{A_iA_i, A_jA_k\} = \tfrac{1}{2}p_i^2p_jp_k,$
5. $\Pr\{A_iA_j, A_iA_j\} = \tfrac{1}{2}p_ip_j[1 + (p_i + p_j) + 2p_ip_j],$
6. $\Pr\{A_iA_j, A_iA_k\} = \tfrac{1}{2}p_ip_jp_k(1 + 2p_i),$
7. $\Pr\{A_iA_j, A_kA_l\} = p_ip_jp_kp_l.$

$$\tag{7.72}$$

Note that the order of birth is immaterial, so that, for example,

$$\Pr\{A_iA_i, A_iA_j\} = \Pr\{A_iA_j, A_iA_i\}$$

and also

$$\Pr\{A_iA_j, A_jA_k\} = \Pr\{A_jA_k, A_iA_j\}, \text{ etc.}$$

Using formulae (7.72), we can construct the joint probability space in the form of a matrix, similar to the matrix **M**, which we will now call matrix **B**. The transition matrix for sibs, called now the matrix **S**, can be obtained by arguments analogous to those given in section 7.6 for transition matrix **T**.

Tables 7.12 and 7.13 represent matrices **B** and **S** for number of alleles $s = 2$ (alleles A and a with gene frequencies p and q, $p + q = 1$, respectively).

We can also obtain the joint distribution, **B**, of sib-sib pairs in a different way.

TABLE 7.12   Joint genotype distribution for full-sib pairs
in a random mating population; $s = 2$. (Matrix **B**)

II Sibling / I Sibling	AA	Aa	aa	Total
AA	$\frac{1}{4}p^2(1 + p)^2$	$\frac{1}{2}p^2q(1 + p)$	$\frac{1}{4}p^2q^2$	p^2
Aa	$\frac{1}{2}p^2q(1 + p)$	$pq(1 + pq)$	$\frac{1}{2}pq^2(1 + q)$	$2pq$
aa	$\frac{1}{4}p^2q^2$	$\frac{1}{2}pq^2(1 + q)$	$\frac{1}{4}q^2(1 + q)^2$	q^2
Total	p^2	$2pq$	q^2	1

TABLE 7.13   Conditional genotype distribution of one sibling, given
the genotype of another. (Matrix **S**)

II Sibling / I Sibling (given)	AA	Aa	aa	Total
AA	$\frac{1}{4}(1 + p)^2$	$\frac{1}{2}q(1 + p)$	$\frac{1}{4}q^2$	1
Aa	$\frac{1}{4}p(1 + p)$	$\frac{1}{2}(1 + pq)$	$\frac{1}{4}q(1 + q)$	1
aa	$\frac{1}{4}p^2$	$\frac{1}{2}p(1 + q)$	$\frac{1}{4}(1 + q)^2$	1

Let us introduce a diagonal matrix **D** whose elements are the genotype
frequencies of the population in equilibrium, that is,

$$\mathbf{D} = \begin{bmatrix} p_1^2 & 0 & 0 & \cdots & 0 & \cdots & 0 \\ 0 & p_2^2 & 0 & \cdots & 0 & & \\ \cdots & \cdots & \cdots & \cdots & \cdots & \cdots & \cdots \\ 0 & 0 & 0 & \cdots & 2p_1p_2 & \cdots & 0 \\ \cdots & \cdots & \cdots & \cdots & \cdots & \cdots & \cdots \\ 0 & 0 & 0 & \cdots & 0 & \cdots & 2p_{s-1}p_s \end{bmatrix}. \tag{7.73}$$

We notice that

$$\mathbf{D} = \mathbf{I} \square \mathbf{R}, \tag{7.74}$$

where **I** is the identity matrix defined in Appendix I. The matrix **S** ("Sibs")
represents conditional probabilities of genotypes of a second member of

sib-pair, given the genotype of a first member of sib-pair. Li and Sacks (1954) have shown that $\mathbf{S}$ can be expressed in terms of matrices $\mathbf{I}$, $\mathbf{T}$, and $\mathbf{R}'$, namely,

$$\mathbf{S} = \tfrac{1}{4}\mathbf{I} + \tfrac{1}{2}\mathbf{T} + \tfrac{1}{4}\mathbf{R}'. \tag{7.75}$$

Thus, the matrix $\mathbf{B}$, representing the joint probability of sib-sib pairs, is

$$\mathbf{B} = \mathbf{S} \boxdot \mathbf{R}. \tag{7.76}$$

Substituting (7.75) into (7.76) and multiplying term by term each component in the right-hand side of (7.76) by $\mathbf{R}$, we obtain

$$\mathbf{B} = \tfrac{1}{4}\mathbf{D} + \tfrac{1}{2}\mathbf{M} + \tfrac{1}{4}\mathbf{H}. \tag{7.77}$$

Let E be an event which can be defined over different sample spaces. Let the probabilities of E defined over $\mathbf{B}$, $\mathbf{D}$, $\mathbf{M}$, and $\mathbf{H}$ be $\Pr\{E \mid \mathbf{B}\}$, $\Pr\{E \mid \mathbf{D}\}$, $\Pr\{E \mid \mathbf{M}\}$, $\Pr\{E \mid \mathbf{H}\}$, respectively. It is easy to see that

$$\Pr\{E \mid \mathbf{B}\} = \tfrac{1}{4}\Pr\{E \mid \mathbf{D}\} + \tfrac{1}{2}\Pr\{E \mid \mathbf{M}\} + \tfrac{1}{4}\Pr\{E \mid \mathbf{H}\} \tag{7.78}$$

also holds.

Using formula (6.86), we find the correlation coefficient, $\rho_{\mathbf{B}}$, that is

$$\rho_{\mathbf{B}} = \tfrac{1}{4}\rho_{\mathbf{D}} + \tfrac{1}{2}\rho_{\mathbf{M}} + \tfrac{1}{4}\rho_{\mathbf{H}}. \tag{7.79}$$

It is easy to show that $\rho_{\mathbf{D}} = 1$. We have already found that $\rho_{\mathbf{H}} = 0$, and when gene effects are additive, $\rho_{\mathbf{M}} = \tfrac{1}{2}$. Thus

$$\rho_{\mathbf{B}} = \tfrac{1}{4} + \tfrac{1}{2} \cdot \tfrac{1}{2} = \tfrac{1}{2}, \tag{7.80}$$

that is, the genotypic correlation coefficient between full-sibs is the same as between parents and children (assuming additive gene effects).

For $s = 2$, and when there is complete dominance of A over a, we easily obtain the phenotypic correlation coefficient

$$\rho_{\mathbf{B}}^* = \frac{1 + 3q}{4(1 + q)} \tag{7.81}$$

(see Problem 7.13).

For X-linked genes methods similar to those in section 7.9 could be used. Here we can consider the joint distributions of pairs: sister-sister, sister-brother, brother-sister, and brother-brother.

The joint genotype distributions for other relatives, such as uncle-nephew, first cousins, and second cousins, can be evaluated simply by using the three basic matrices: $\mathbf{D}$, $\mathbf{M}$, $\mathbf{H}$, or $\mathbf{I}$, $\mathbf{T}$, $\mathbf{R}$. The methods were originally suggested by Li and Sacks (1954). The distributions were expressed in terms of $\mathbf{I}$, $\mathbf{T}$, $\mathbf{R}'$ matrices (the authors used the symbol $\mathbf{O}$ for our $\mathbf{R}'$ matrix), and the method was called the ITO method.

The reader is referred to this original paper, in which it was shown that the joint genotype distributions for half-sibs, first cousins, and uncle-nephew are the same as for grandparent-grandchild, and that the transition matrix for second cousins is

$$\mathbf{S}^2 = \tfrac{1}{16}\mathbf{I} + \tfrac{6}{16}\mathbf{T} + \tfrac{9}{16}\mathbf{R}'. \tag{7.82}$$

7.12. INTERPRETATION OF JOINT DISTRIBUTIONS OF RELATIVES IN LARGE POPULATIONS

In previous sections we saw that a certain parent-child genotype combination occurs in different families with different probabilities. For instance, the combination $\text{Pt} = A_iA_i$, $\text{C} = A_iA_i$ occurs in mating $A_iA_i \times A_iA_i$ with probability 1, but in mating $A_iA_i \times A_iA_j$ only with probability $\tfrac{1}{2}$. As we shall see in Chapter 8, the probability of occurrence of a child having a particular genotype, in a family with known parental genotypes, also depends on the *size* of the family.

In large populations, however, if we pool all possible outcomes from families of different sizes and calculate the *expected* proportion (probability) in which a particular parent-child combination occurs, we find that this is exactly the joint probability obtained from the matrix $\mathbf{M}$. The same situation exists with full-sib pairs. In a given family of size $n \geqslant 2$ we have $\binom{n}{2} = \dfrac{n(n-1)}{2}$ possible pairs. A certain combination of sib-pair genotypes has a definite probability of occurrence in a family of a given size and known genotypes of the parents. These probabilities are different for different families. But if we pool the results together and find the *expected proportion*, we find that this is the element of the matrix $\mathbf{B}$ corresponding to the sib-sib pair.

The situation will change considerably if we deal only with small samples. In such cases, the number of families of given size with given genotypes of the parents plays an important role in calculating the appropriate probabilities. Problems of this kind will be discussed in Chapters 11, 17, and 18.

7.13. SNYDER'S RATIOS

Let D and d be dominant and recessive genes, respectively. There are some (often deleterious) traits in human populations which are supposed to be simple recessive. If the trait is rare, it is usually necessary to search for it using methods known as "family study," in order to estimate its gene frequency. However, if the trait is not very rare, the expected proportion of recessive offspring can be calculated if the parental mating type is known.

Let us consider a population in equilibrium with respect to locus D, that is, $p^2DD + 2pqDd + q^2dd$. We shall distinguish three types of phenotypic matings:

(i) Dominant × Dominant. The expected proportion (probability) of these matings is:

$$\Pr\{\text{Dom} \times \text{Dom}\} = \Pr\{DD \times DD\} + \Pr\{DD \times Dd\} + \Pr\{Dd \times Dd\}$$
$$= p^4 + 2 \times p^2 \cdot 2pq + 4p^2q^2 = p^2(p^2 + 4pq + 4q^2)$$
$$= p^2(p + 2q)^2$$
$$= p^2(1 + q)^2 = (1 - q)^2(1 + q)^2 = (1 - q^2)^2. \qquad (7.83)$$

The expected proportion of recessive offspring and Dom × Dom parents is

$$\Pr\{\text{Offspring Rec and Dom} \times \text{Dom}\} = \tfrac{1}{4}\Pr\{Dd \times Dd\} = p^2q^2. \qquad (7.84)$$

Hence the expected proportion, denoted by S_{11}, of recessive offspring, given that the parents are Dom × Dom, is

$$S_{11} = \Pr\{\text{Rec} \mid \text{Dom} \times \text{Dom}\} = \frac{\Pr\{\text{Rec and Dom} \times \text{Dom}\}}{\Pr\{\text{Dom} \times \text{Dom}\}}$$

$$= \frac{p^2q^2}{p^2(1 + q)^2} = \frac{q^2}{(1 + q)^2}. \qquad (7.85)$$

(ii) Dominant × Recessive. We have

$$\Pr\{\text{Dom} \times \text{Rec}\} = \Pr\{DD \times dd\} + \Pr\{Dd \times dd\}$$
$$= 2 \times p^2q^2 + 2 \times 2pq \cdot q^2 = 2pq^2(p + 2q)$$
$$= 2pq^2(1 + q) = 2q^2(1 - q^2). \qquad (7.86)$$

Then

$$\Pr\{\text{Offspring Rec and Dom} \times \text{Rec}\} = \tfrac{1}{2}\Pr\{Dd \times dd\} = \tfrac{1}{2} \cdot 4pq^3 = 2pq^3, \qquad (7.87)$$

and the expected proportion, denoted by S_{10}, of recessive offspring, given that the parents are Dom × Rec, is

$$S_{10} = \Pr\{\text{Rec} \mid \text{Dom} \times \text{Rec}\} = \frac{\Pr\{\text{Rec and Dom} \times \text{Rec}\}}{\Pr\{\text{Dom} \times \text{Rec}\}}$$

$$= \frac{2pq^3}{2pq^2(1 + q)} = \frac{q}{1 + q}. \qquad (7.88)$$

(iii) Recessive × Recessive. In this case it is easy to see that

$$S_{00} = \Pr\{\text{Rec} \mid \text{Rec} \times \text{Rec}\} = 1. \qquad (7.89)$$

The probabilities S_{10} and S_{11} are known as *Snyder's ratios*. The subscripts indicate the number of dominant parents. We notice that

$$S_{11} = S_{10}^2. \qquad (7.90)$$

If the gene frequency, q, is unknown, Snyder's ratios, calculated from appropriate data obtained from two generations, can be used to estimate q. (See sections 12.7 and 12.8.)

SUMMARY

Definitions of conditional and joint probabilities of the events introduced in Chapter 1 and concepts of genotypic values, considered as random variables, their means, variances, and covariances, discussed in Chapter 6, have found full applications in this chapter. We have obtained the joint genotype distributions and genotypic and phenotypic correlations between parent-child, grandparent-grandchild, and full-sib pairs for an autosomal and sex-linked locus. Bayes theorem on *a posteriori* probability has been proved and used on these occasions. Also the **DMH** method (or, equivalently, the **ITR** method) has been introduced in order to show how the joint distributions among such relatives as grandparent-grandchild, full sibs, and others can be calculated by simple arithmetic from the basic joint distribution for parent-child (matrix **M**). The results obtained here apply only to large populations of families of different sizes. They relate to theoretical or *expected proportions* (probabilities) and should be carefully distinguished from the values actually obtained in family studies.

REFERENCES

Kempthorne, O., *An Introduction to Genetic Statistics*, John Wiley & Sons, New York, 1957.

Li, C. C., and Sacks, L., The derivation of joint distribution and correlation between relatives by the use of stochastic matrices, *Biometrics* **10** (1954), 347–360.

Li, C. C., *Population Genetics*, The University of Chicago Press, Chicago, 1955.

Murphy, E. A., and Mutalik, G. S., The application of Bayesian methods in genetic counseling, *Hum. Heredity* **19** (1969), 126–151.

Snyder, L. H., Studies in human inheritance, X. A table to determine the proportion of recessives to be expected in various matings involving a unit character, *Genetics* **19** (1934), 1–17.

Wright, S., The biometric relation between parent and offspring, *Genetics* **6** (1921), 111–123.

PROBLEMS

7.1. Suppose that A is a dominant and a a recessive allele at a single locus with gene frequencies p and q, respectively, and that the population in question is in equilibrium with respect to this locus.

(a) Prove that the joint parent-child phenotype distribution can be expressed in terms of q as in Table 7.5b.

(b) Express the full-sib genotype distribution given in Table 7.12 in terms of q.

7.2. Let us consider three alleles, A, B, and O, in the ABO blood group locus, having gene frequencies p, q, r, with $p + q + r = 1$. It is assumed that A and B are codominants, but each is dominant over O. As we pointed out in Chapter 4, there will be four phenotypes to which the four blood groups are assigned:

Genotype:	AA, AO	BB, BO	AB	OO
Phenotype:	$A-$	$B-$	AB	OO
Blood group:	A	B	AB	O.

It happens that, if blood is transfused from one individual (donor) to another (recipient), the transfusion will be successful only if the donor is of the same group as the recipient or is of group O.

(a) Find the joint phenotype distribution of recipient-donor, when the recipient is a child and the donor one of the parents, in the whole population of child-parent pairs.

(b) Find the joint phenotype distribution of recipient-donor for full-sib pairs. [*Hint:* Construct first the genotype and then the phenotype distribution using formulae (7.72), or use formula (7.78).]

(c) Let us define the random variable X over the sample space for recipient-donor such that

$$X = \begin{cases} 0 & \text{if the donor is suitable for the recipient}; \\ 1 & \text{if the donor is unsuitable for the recipient}. \end{cases}$$

Find the distribution functions of X in cases (a) and (b).

(d) What is the probability $\Pr\{X = 0\}$ in both cases, (a) and (b)? Compare the results with those obtained for a random population of recipient-donor pairs, considered in Example 5.7.

(e) Calculate numerical results in each case for $p = 0.153, q = 0.174, r = 0.673$.

7.3. (a) Express the joint probability distribution of grandparent-grandchild genotypes, given in Table 7.8, in terms of q.

(b) Prove that the phenotype distribution of pairs (GP, GC) is that in Table 7.9.

7.4. Using formula (7.44), find the joint genotype distribution of great-grand-parent and great-grandchild when there are two alleles at a locus.

7.5. Using Table 7.9 of the joint phenotype distribution of grandparents and grand-children, prove that their phenotypic correlation coefficient, $\rho_{M_2}^*$, is as given in (7.55), that is,

$$\rho_{M_2}^* = \tfrac{1}{2} \frac{q}{1 + q}.$$

7.6. Represent graphically (on the same graph) the coefficients ρ_M^*, $\rho_{M_2}^*$, $\rho_{M_3}^*$, and $\rho_{M_4}^*$ as functions of q. How do you interpret the results?

7.7. Prove that for an X-linked locus with two alleles the joint genotype distribution is the same for mother-son as for father-daughter, that is, $\mathbf{M}_{32} = \mathbf{M}'_{23}$.

7.8. Prove that for X-linked genes with complete dominance the correlation coefficient between the phenotypes of mother and son (or father and daughter) is $\sqrt{q/(1 + q)}$, where q is the frequency of the recessive gene.

7.9. Prove that the joint probabilities of full-sib pairs of type 6 (A_iA_j, A_iA_k) are given by the formula

$$\Pr\{A_iA_j, A_iA_k\} = \tfrac{1}{2}p_ip_jp_k(1 + 2p_i).$$

[*Hint:* Prepare a table similar to Table 7.11, from which the probability for type 5, (A_iA_j, A_iA_j), has been derived.]

7.10. (*a*) Find the formulae for the joint probabilities for the seven sib-pair types when all gene frequencies are equal, that is, $p_i = 1/s$, $i = 1, 2, \ldots, s$.

(*b*) Find the matrix $\mathbf{B}$ for $s = 4$ in this case.

7.11. (*a*) Show that in the case $s = 2$ the joint genotype distributions of (i) half-sibs, (ii) first cousins, (ii) uncle-nephew (or uncle-niece, aunt-nephew, aunt-niece) are the same as those for grandparent-grandchild.

(*b*) Could you show that these results can be extended to any $s > 2$?

7.12. Using the results of Table 7.12, show that the phenotype distribution of sib-sib pairs is as follows:

Sib I \ Sib II	Dom ($A-$)	Rec (aa)
Dom ($A-$)	$\tfrac{1}{4}(1 - q)(4 + 4q - 3q^2 - q^3)$	$\tfrac{1}{4}q^2(1 - q)(3 + q)$
Rec (aa)	$\tfrac{1}{4}q^2(1 - q)(3 + q)$	$\tfrac{1}{4}q^2(1 + q)^2$

7.13. (*a*) Show that the phenotypic correlation for sib-sib pairs, ρ_B^*, is

$$\rho_B^* = \frac{1 + 3q}{4(1 + q)}.$$

(*b*) Derive ρ_B^*, using the table given in Problem 7.13.

(*c*) Use formula (6.86).

7.14. Find the joint genotype distribution at a single locus with two alleles, A_1, A_2, for triplets: (*a*) parent-child-child and (*b*) sib-sib-sib, in a large randomly mating population.

(*Hint:* Denoting by p_i the frequency of the gene A_i, $i = 1, 2$, prove that there are the following types of triplets with corresponding probabilities)

Genotype types of triplets			Probabilities	
			Parent-child-child	Sib-sib-sib
1. A_iA_i	A_iA_i	A_iA_i	$\frac{1}{2}p_i^3(1 + p_i)$	$\frac{1}{16}p_i^2(1 + 3p_i)^2$
2. A_iA_i	A_iA_i	A_jA_j	0	$\frac{1}{16}p_i^2p_j^2$
3. A_iA_i	A_iA_i	A_iA_j	$\frac{1}{2}p_i^3p_j$	$\frac{1}{8}p_i^2p_j(1 + 3p_i)$
4. A_iA_i	A_iA_j	A_jA	0	$\frac{1}{8}p_i^2p_j^2$
5. A_iA_i	A_iA_j	A_iA_j	$\frac{1}{2}p_i^2p_j(1 + p_j)$	$\frac{1}{4}p_i^2p_j(1 + p_i)$
6. A_iA_j	A_iA_i	A_iA_i	$\frac{1}{4}p_i^2p_j(1 + p_i)$	As type 3
7. A_iA_j	A_iA_i	A_jA_j	$\frac{1}{4}p_i^2p_j^2$	As type 4
8. A_iA_j	A_iA_i	A_iA_j	$\frac{1}{2}p_i^2p_j$	As type 5
9. A_iA_j	A_iA_j	A_iA_j	$\frac{1}{2}p_ip_j$	$\frac{1}{2}p_ip_j(1 + 3p_ip_j)$

7.15. Find the joint genotype distributions for (a) sister-sister, (b) sister-brother, and (c) brother-brother, for an X-linked locus with alleles A and a.

7.16. Calculate genotypic and phenotypic correlation coefficients for Problem 7.15, assuming additive effects of genes (i.e., assign the values 1 and 0 to genes A and a, respectively).

CHAPTER 8

Some Standard Distributions and
Their Properties

8.1. INTRODUCTION

In Chapters 5 and 6, we introduced general concepts and definitions of random variables, their probability distributions, and moments. There is a number of distributions which can be derived under fairly general assumptions and which can be applied to several real situations. They serve as *models*. The observed distributions, called also *empirical* distributions, never fit the models exactly but often approach them closely.

As was mentioned in Chapter 5, we distinguish distributions of *continuous* and *discrete* random variables. In this book we are concerned mostly with discrete distributions. However, we will introduce a few standard continuous distributions since some discrete distributions can be approximated by certain continuous distributions.

The derivation of these distributions sometimes requires rather advanced mathematics. We have tried to reduce the mathematics of this chapter, confining ourselves mostly to giving results in the form of definitions or statements. The derivations of probability or density functions for standard distributions and proofs of important theorems can be found in books on mathematical statistics. Some of these books are listed in the references at the end of this chapter.

Although the rigorous mathematical procedures are reduced, the reader may still find this chapter highly theoretical. We have deliberately included all this theoretical material in one chapter. Applications will be spread widely over the rest of the book, and it will be helpful to return to Chapter 8 and review the basic mathematical knowledge needed for a particular genetic application.

8.2. THE NORMAL DISTRIBUTION

The *normal* (or *Gaussian*) distribution is the most common and useful in applied statistics. It is defined as follows.

DEFINITION 8.1. *A random variable X is normally distributed if its density function, f(x), is*

$$f(x) = \frac{1}{\sigma\sqrt{2\pi}} \, e^{-(x-\mu)^2/2\sigma^2}, \qquad -\infty < x < \infty, \tag{8.1}$$

where $\pi = 3.14159 \cdots$ *and* $e = 2.71828 \cdots$ *are two irrational numbers.*

The constants μ and σ^2 are called the *parameters* of the distribution, and it can be shown that μ is the expected value (mean) and σ^2 is the variance of the normal variable X, that is,

$$E(X) = \mu, \quad \text{Var}(X) = \sigma^2. \tag{8.2}$$

The function $f(x)$ is represented in Fig. 8.1. It is a bell-shaped curve, symmetrical about the vertical line $x = \mu$. At the point $x = \mu$, $f(x)$ attains its maximum. The curve approaches the x-axis as x tends to $\pm \infty$. The cumulative distribution function (cdf), is

$$\Pr\{X \leqslant x\} = \int_{-\infty}^{x} f(t)\, dt = F(x). \tag{8.3}$$

It is represented by the shaded area in Fig. 8.1. But $F(x)$ as a function of x can be also shown as in Fig. 8.2. The probability $\Pr\{X \leqslant x\} = F(x)$ can be, then, represented by the ordinate $F(x)$ as in Fig. 8.2. Of course, $F(x)$ satisfies condition (5.14), that is,

$$\Pr\{-\infty < X < \infty\} = \int_{-\infty}^{\infty} f(t)\, dt = F(\infty) - F(-\infty) = 1. \tag{8.4}$$

The integral in (8.4) represents the total area under the curve $f(x)$.

We introduce the symbol $\frown$ for "is distributed as" and the symbol $\stackrel{\cdot}{\frown}$ for "is approximately distributed as." The sentence "The random variable X is normally distributed with mean μ and variance σ^2" will sometimes be

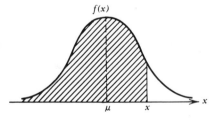

Figure 8.1

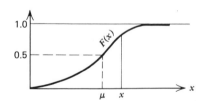

Figure 8.2

written in the abbreviated form "$X \frown N(\mu, \sigma^2)$," where N stands for "normal."

In practice, many continuous characters whose frequencies follow, at least approximately, a bell-shaped distribution are assumed to be normally distributed (e.g., height, weight, and intelligence). On the other hand, there can be bell-shaped distributions which are not necessarily normal.

8.3. THE STANDARD NORMAL DISTRIBUTION

The distribution given in (8.1) represents, in fact, a *family* of distributions, depending on the values μ and σ. It would not be feasible to calculate the probability $\Pr\{X \leqslant x\}$ for each individual case. Let

$$Z = \frac{X - \mu}{\sigma}, \quad z = \frac{x - \mu}{\sigma}, \tag{8.5a}$$

or

$$X = \mu + Z\sigma, \quad x = \mu + z\sigma, \tag{8.5b}$$

where $X \frown N(\mu, \sigma^2)$. The random variable Z is called a *standard normal variable*. It can be shown that

$$E(Z) = 0, \quad \text{Var}(Z) = 1. \tag{8.6}$$

Let $\varphi(z)$ and $\Phi(z)$ denote the density and the cdf of Z, respectively. We have

$$\varphi(z) = \frac{1}{\sqrt{2\pi}} e^{-z^2/2}, \quad -\infty < z < \infty \tag{8.7}$$

and

$$\Pr\{Z \leqslant z\} = \int_{-\infty}^{z} \varphi(t) \, dt = \Phi(z).^* \tag{8.8}$$

Using property (5.15b), we have

$$\Phi(z) = \Pr\{Z \leqslant z\} = \Pr\left\{\frac{X - \mu}{\sigma} \leqslant z\right\}$$

$$= \Pr\{X \leqslant \mu + z\sigma\} = \Pr\{X \leqslant x\} = F(x). \tag{8.9}$$

8.4. TABLE OF PERCENTAGE POINTS FOR NORMAL DISTRIBUTION

In the calculation of probabilities (8.8) we can use Table I. Let z_γ be the value of z such that the probability $\Pr\{Z \leqslant z_\gamma\}$ is equal to γ, that is,

$$\Pr\{Z \leqslant z_\gamma\} = \int_{\infty}^{z_\gamma} f(t) \, dt = \Phi(z_\gamma) = \gamma. \tag{8.10}$$

* Note that, for continuous variables, $\Pr\{Z \leqslant z\} = \Pr\{Z < z\}$.

TABLE I.   [a]The cumulative standard normal distribution function $\Pr\{Z \leqslant z_\gamma\} = \gamma$

z	0.00	0.01	0.02	0.03	0.04	0.05	0.06	0.07	0.08	0.09
0.0	0.5000	0.5040	0.5080	0.5120	0.5160	0.5199	0.5239	0.5279	0.5319	0.5359
0.1	0.5398	0.5438	0.5478	0.5517	0.5557	0.5596	0.5636	0.5675	0.5714	0.5753
0.2	0.5793	0.5832	0.5871	0.5910	0.5948	0.5987	0.6026	0.6064	0.6103	0.6141
0.3	0.6179	0.6217	0.6255	0.6293	0.6331	0.6368	0.6406	0.6443	0.6480	0.6517
0.4	0.6554	0.6591	0.6628	0.6664	0.6700	0.6736	0.6772	0.6808	0.6844	0.6879
0.5	0.6915	0.6950	0.6985	0.7019	0.7054	0.7088	0.7123	0.7157	0.7190	0.7224
0.6	0.7257	0.7291	0.7324	0.7357	0.7389	0.7422	0.7454	0.7486	0.7517	0.7549
0.7	0.7580	0.7611	0.7642	0.7673	0.7703	0.7734	0.7764	0.7794	0.7823	0.7852
0.8	0.7881	0.7910	0.7939	0.7967	0.7995	0.8023	0.8051	0.8078	0.8106	0.8133
0.9	0.8159	0.8186	0.8212	0.8238	0.8264	0.8289	0.8315	0.8340	0.8365	0.8389
1.0	0.8413	0.8438	0.8461	0.8485	0.8508	0.8531	0.8554	0.8577	0.8599	0.8621
1.1	0.8643	0.8665	0.8686	0.8708	0.8729	0.8749	0.8770	0.8790	0.8810	0.8830
1.2	0.8849	0.8869	0.8888	0.8907	0.8925	0.8944	0.8962	0.8980	0.8997	0.90147
1.3	0.90320	0.90490	0.90658	0.90824	0.90988	0.91149	0.91309	0.91466	0.91621	0.91774
1.4	0.91924	0.92073	0.92220	0.92364	0.92507	0.92647	0.92785	0.92922	0.93056	0.93189
1.5	0.93319	0.93448	0.93574	0.93699	0.93822	0.93943	0.94062	0.94179	0.94295	0.94408
1.6	0.94520	0.94630	0.94738	0.94845	0.94950	0.95053	0.95154	0.95254	0.95352	0.95449
1.7	0.95543	0.95637	0.95728	0.95818	0.95907	0.95994	0.96080	0.96164	0.96246	0.96327
1.8	0.96407	0.96485	0.96562	0.96638	0.96712	0.96784	0.96856	0.96926	0.96995	0.97062
1.9	0.97128	0.97193	0.97257	0.97320	0.97381	0.97441	0.97500	0.97558	0.97615	0.97670
2.0	0.97725	0.97778	0.97831	0.97882	0.97932	0.97982	0.98030	0.98077	0.98124	0.98169
2.1	0.98214	0.98257	0.98300	0.98341	0.98382	0.98422	0.98461	0.98500	0.98537	0.98574
2.2	0.98610	0.98645	0.98679	0.98713	0.98745	0.98778	0.98809	0.98840	0.98870	0.98899
2.3	0.98928	0.98956	0.98983	$0.9^{2}0097$	$0.9^{2}0358$	$0.9^{2}0613$	$0.9^{2}0863$	$0.9^{2}1106$	$0.9^{2}1344$	$0.9^{2}1576$
2.4	$0.9^{2}1802$	$0.9^{2}2024$	$0.9^{2}2240$	$0.9^{2}2451$	$0.9^{2}2656$	$0.9^{2}2857$	$0.9^{2}3053$	$0.9^{2}3244$	$0.9^{2}3431$	$0.9^{2}3613$

158

TABLE 1. (continued)

z	0.00	0.01	0.02	0.03	0.04	0.05	0.06	0.07	0.08	0.09
2.5	$0.9^2 3790$	$0.9^2 3963$	$0.9^2 4132$	$0.9^2 4297$	$0.9^2 4457$	$0.9^2 4614$	$0.9^2 4766$	$0.9^2 4915$	$0.9^2 5060$	$0.9^2 5201$
2.6	$0.9^2 5339$	$0.9^2 5473$	$0.9^2 5604$	$0.9^2 5731$	$0.9^2 5855$	$0.9^2 5975$	$0.9^2 6093$	$0.9^2 6207$	$0.9^2 6319$	$0.9^2 6427$
2.7	$0.9^2 6533$	$0.9^2 6636$	$0.9^2 6736$	$0.9^2 6833$	$0.9^2 6928$	$0.9^2 7020$	$0.9^2 7110$	$0.9^2 7197$	$0.9^2 7282$	$0.9^2 7365$
2.8	$0.9^2 7445$	$0.9^2 7523$	$0.9^2 7599$	$0.9^2 7673$	$0.9^2 7744$	$0.9^2 7814$	$0.9^2 7882$	$0.9^2 7948$	$0.9^2 8012$	$0.9^2 8074$
2.9	$0.9^2 8134$	$0.9^2 8193$	$0.9^2 8250$	$0.9^2 8305$	$0.9^2 8359$	$0.9^2 8411$	$0.9^2 8462$	$0.9^2 8511$	$0.9^2 8559$	$0.9^2 8605$
3.0	$0.9^2 8650$	$0.9^2 8694$	$0.9^2 8736$	$0.9^2 8777$	$0.9^2 8817$	$0.9^2 8856$	$0.9^2 8893$	$0.9^2 8930$	$0.9^2 8965$	$0.9^2 8999$
3.1	$0.9^3 0324$	$0.9^3 0646$	$0.9^3 0957$	$0.9^3 1260$	$0.9^3 1553$	$0.9^3 1836$	$0.9^3 2112$	$0.9^3 2378$	$0.9^3 2636$	$0.9^3 2886$
3.2	$0.9^3 3129$	$0.9^3 3363$	$0.9^3 3590$	$0.9^3 3810$	$0.9^3 4024$	$0.9^3 4230$	$0.9^3 4429$	$0.9^3 4623$	$0.9^3 4810$	$0.9^3 4991$
3.3	$0.9^3 5166$	$0.9^3 5335$	$0.9^3 5499$	$0.9^3 5658$	$0.9^3 5811$	$0.9^3 5959$	$0.9^3 6103$	$0.9^3 6242$	$0.9^3 6376$	$0.9^3 6505$
3.4	$0.9^3 6631$	$0.9^3 6752$	$0.9^3 6869$	$0.9^3 6982$	$0.9^3 7091$	$0.9^3 7197$	$0.9^3 7299$	$0.9^3 7398$	$0.9^3 7493$	$0.9^3 7585$
3.5	$0.9^3 7674$	$0.9^3 7759$	$0.9^3 7842$	$0.9^3 7922$	$0.9^3 7999$	$0.9^3 8074$	$0.9^3 8146$	$0.9^3 8215$	$0.9^3 8282$	$0.9^3 8347$
3.6	$0.9^3 8409$	$0.9^3 8469$	$0.9^3 8527$	$0.9^3 8583$	$0.9^3 8637$	$0.9^3 8689$	$0.9^3 8739$	$0.9^3 8787$	$0.9^3 8834$	$0.9^3 8879$
3.7	$0.9^3 8922$	$0.9^3 8964$	$0.9^4 0039$	$0.9^4 0426$	$0.9^4 0799$	$0.9^4 1158$	$0.9^4 1504$	$0.9^4 1838$	$0.9^4 2159$	$0.9^4 2468$
3.8	$0.9^4 2765$	$0.9^4 3052$	$0.9^4 3327$	$0.9^4 3593$	$0.9^4 3848$	$0.9^4 4094$	$0.9^4 4331$	$0.9^4 4558$	$0.9^4 4777$	$0.9^4 4988$
3.9	$0.9^4 5190$	$0.9^4 5385$	$0.9^4 5573$	$0.9^4 5753$	$0.9^4 5926$	$0.9^4 6092$	$0.9^4 6253$	$0.9^4 6406$	$0.9^4 6554$	$0.9^4 6696$
4.0	$0.9^4 6833$	$0.9^4 6964$	$0.9^4 7090$	$0.9^4 7211$	$0.9^4 7327$	$0.9^4 7439$	$0.9^4 7546$	$0.9^4 7649$	$0.9^4 7748$	$0.9^4 7843$
4.1	$0.9^4 7934$	$0.9^4 8022$	$0.9^4 8106$	$0.9^4 8186$	$0.9^4 8263$	$0.9^4 8338$	$0.9^4 8409$	$0.9^4 8477$	$0.9^4 8542$	$0.9^4 8605$
4.2	$0.9^4 8665$	$0.9^4 8723$	$0.9^4 8778$	$0.9^4 8832$	$0.9^4 8882$	$0.9^4 8931$	$0.9^4 8978$	$0.9^5 0226$	$0.9^5 0655$	$0.9^5 1066$
4.3	$0.9^5 1460$	$0.9^5 1837$	$0.9^5 2199$	$0.9^5 2545$	$0.9^5 2876$	$0.9^5 3193$	$0.9^5 3497$	$0.9^5 3788$	$0.9^5 4066$	$0.9^5 4332$
4.4	$0.9^5 4587$	$0.9^5 4831$	$0.9^5 5065$	$0.9^5 5288$	$0.9^5 5502$	$0.9^5 5706$	$0.9^5 5902$	$0.9^5 6089$	$0.9^5 6268$	$0.9^5 6439$
4.5	$0.9^5 6602$	$0.9^5 6759$	$0.9^5 6908$	$0.9^5 7051$	$0.9^5 7187$	$0.9^5 7318$	$0.9^5 7442$	$0.9^5 7561$	$0.9^5 7675$	$0.9^5 7784$
4.6	$0.9^5 7888$	$0.9^5 7987$	$0.9^5 8081$	$0.9^5 8172$	$0.9^5 8258$	$0.9^5 8340$	$0.9^5 8419$	$0.9^5 8494$	$0.9^5 8566$	$0.9^5 8634$
4.7	$0.9^5 8699$	$0.9^5 8761$	$0.9^5 8821$	$0.9^5 8877$	$0.9^5 8931$	$0.9^5 8983$	$0.9^6 0320$	$0.9^6 0789$	$0.9^6 1235$	$0.9^6 1661$
4.8	$0.9^6 2067$	$0.9^6 2453$	$0.9^6 2822$	$0.9^6 3173$	$0.9^6 3508$	$0.9^6 3827$	$0.9^6 4131$	$0.9^6 4420$	$0.9^6 4696$	$0.9^6 4958$
4.9	$0.9^6 5208$	$0.9^6 5446$	$0.9^6 5673$	$0.9^6 5889$	$0.9^6 6094$	$0.9^6 6289$	$0.9^6 6475$	$0.9^6 6652$	$0.9^6 6821$	$0.9^6 6981$

Example: $\Phi(3.57) = 0.9^3 8215 = 0.9998215$.

a Abridged from Table II of *Statistical Tables and Formulas* by A. Hald, John Wiley & Sons, New York, 1952.

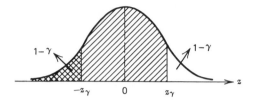

Figure 8.3

Table I gives only probabilities (8.10) for $z_\gamma > 0$, that is for $\gamma > 0.50$. For instance, $\Pr\{Z \leqslant 1.16\} = 0.8770$ is obtained directly from Table I. However, Table I can also be used for $z_\gamma < 0$. Since the normal curve is symmetrical about the vertical line $\mu = 0$, the double-shaded (left) and the blank (right) tails in Fig. 8.3 are equal. The area of each tail is

$$\Pr\{Z \leqslant -z_\gamma\} = \Pr\{Z > z_\gamma\} = 1 - \Pr\{Z \leqslant z_\gamma\} = 1 - \Phi(z_\gamma) = 1 - \gamma.$$

$$(8.11)$$

Thus, for instance, $\Pr\{Z \leqslant -1.16\} = 1 - 0.8770 = 0.1230$; also $\Pr\{Z > 1.16\} = 0.1230$.

Let z_{γ_1} and z_{γ_2} be two numbers such that $z_{\gamma_1} < z_{\gamma_2}$ (and so $\gamma_1 < \gamma_2$). We have

$$\Pr\{z_{\gamma_1} < Z \leqslant z_{\gamma_2}\} = \Pr\{Z \leqslant z_{\gamma_2}\} - \Pr\{Z \leqslant z_{\gamma_1}\}$$

$$= \Phi(z_{\gamma_2}) - \Phi(z_{\gamma_1}) = \gamma_2 - \gamma_1. \qquad (8.12)$$

(See Fig. 8.4.) Thus, for instance, $\Pr\{0.21 \leqslant Z \leqslant 1.33\} = \Pr\{Z \leqslant 1.33\} - \Pr\{Z \leqslant 0.21\} = 0.9082 - 0.5832 = 0.3250$. Of special use (in the theory of interval estimation and testing hypotheses) are notations where $\gamma = 1 - \frac{1}{2}\alpha$,

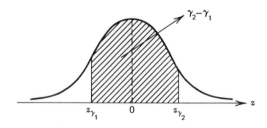

Figure 8.4

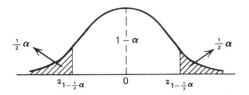

Figure 8.5

with $\frac{1}{2}\alpha < \frac{1}{2}$. From Fig. 8.5 we can see that

$$\Pr\{Z \leqslant z_{\frac{1}{2}\alpha}\} = \Pr\{Z \leqslant -z_{1-\frac{1}{2}\alpha}\} = \Pr\{Z > z_{1-\frac{1}{2}\alpha}\} = \tfrac{1}{2}\alpha, \qquad (8.13a)$$

$$\Pr\{|Z| \leqslant z_{1-\frac{1}{2}\alpha}\} = \Pr\{-z_{1-\frac{1}{2}\alpha} \leqslant Z \leqslant z_{1-\frac{1}{2}\alpha}\} = 1 - \alpha, \qquad (8.13b)$$

$$\Pr\{|Z| > z_{1-\frac{1}{2}\alpha}\} = \Pr\{Z < -z_{1-\frac{1}{2}\alpha}\} + \Pr\{Z > z_{1-\frac{1}{2}\alpha}\} = \alpha. \qquad (8.13c)$$

Example 8.1. We wish to find the probability that the variable X is between the limits $\mu - \sigma$ and $\mu + \sigma$. We have

$$\Pr\{\mu - \sigma < X < \mu + \sigma\} = \Pr\left\{-1 < \frac{X - \mu}{\sigma} < 1\right\}$$

$$= \Pr\{-1 < Z < 1\} = 1 - \alpha.$$

Here $z_{1-\frac{1}{2}\alpha} = 1.0$. From Table I, we find $1 - \tfrac{1}{2}\alpha = 0.8413$. Hence $1 - \alpha = 0.6826$. In a similar way we find

$$\Pr\{\mu - 2\sigma < X < \mu + 2\sigma\} = \Pr\{-2 < Z < 2\} = 0.9545,$$
$$\Pr\{\mu - 3\sigma < X < \mu + 3\sigma\} = \Pr\{-2 < Z < 3\} = 0.9973.$$

Although X can take values from $-\infty$ to $+\infty$, in practice almost all (precisely, 99.73%) possible values are included in an interval $(\mu - 3\sigma, \mu + 3\sigma)$.

Example 8.2. It has been found that the average birth weight of Chinese infants born in Singapore in 1950–53 was approximately $\mu \doteq 6.34$ lb and the standard deviation $\sigma \doteq 1.03$ lb [Mills (1959)]. Assuming that the weight is normally distributed, we wish to find:

(a) The proportion of babies whose weight is between 5.0 and 7.0 lb. We calculate

$$\frac{5.00 - 6.34}{1.03} = -1.30, \quad \text{and} \quad \frac{7.00 - 6.34}{1.03} = 0.64.$$

Then

$$\Pr\{5.0 < X < 7.0\} = \Pr\{-1.30 < Z < 0.64\} = \Pr\{Z < 0.64\}$$
$$- [1 - \Pr\{Z < 1.30\}] = 0.7389 - 0.0968 = 0.6421.$$

(b) The lower bound of weights of the 20% heaviest babies. We have $\Pr\{Z > z_{0.80}\} = 0.20$. From Table I we find $z_{0.80} = 0.84$. Thus

$$x_{0.80} = \mu + z_{0.80}\sigma = 6.34 + 0.84 \times 1.03 = 6.34 + 0.87 = 7.21 \text{ lb.}$$

We now give (without proof) a useful property of the sum of k independent normal variables. [For proof, see Johnson and Leone (1964), Chapter 5.]

THEOREM 8.1. *Let $X_1, X_2, \ldots, X_k$ be k independent normal variables such that $X_i \frown N(\mu_i, \sigma_i^2)$, $i = 1, 2, \ldots, k$. The linear function*

$$Y = a_1 X_1 + a_2 X_2 + \cdots + a_k X_k \tag{8.14}$$

is also normally distributed with mean

$$\mu_Y = a_1 \mu_1 + a_2 \mu_2 + \cdots + a_k \mu_k \tag{8.15}$$

and variance

$$\sigma_Y^2 = a_1^2 \sigma_1^2 + a_2^2 \sigma_2^2 + \cdots + a_k^2 \sigma_k^2. \tag{8.16}$$

In particular, when each X_i has the same distribution, $N(\mu, \sigma^2)$, and $a_1 = a_2 = \cdots = a_k = 1$, then the sum $Y = X_1 + X_2 + \cdots + X_k$ is normally distributed with mean $\mu_Y = k\mu$ and variance $\sigma_Y^2 = k\sigma^2$. Applications of this theorem will appear in later sections.

8.5. SOME DISTRIBUTIONS RELATED TO THE NORMAL DISTRIBUTION

We now introduce three standard distributions, called χ^2-, t-, and F-distributions, which are related to the normal distribution and play an important role in applied statistics. Among these the χ^2-distribution is widely used in genetic problems.

8.5.1. The Chi-Square Distribution

Let $Z_1, Z_2, \ldots, Z_\nu$ be ν independent normal variables, each distributed $N(0, 1)$. The sum of their squares, denoted by χ^2, that is,

$$\chi^2 = Z_1^2 + Z_2^2 + \cdots + Z_\nu^2, \tag{8.17}$$

has the density $h(\chi^2; \nu)$, say, with ($\chi^2 > 0$). [The explicit form of $h(\chi^2; \nu)$ can be found in books on statistics.] Here ν is the only parameter of this distribution. It is called the number of *degrees of freedom* (briefly, d.f.) and indicates the number of independent variables in (8.17). It can be shown that

$$E(\chi^2) = \nu, \quad \text{Var}(\chi^2) = 2\nu, \tag{8.18}$$

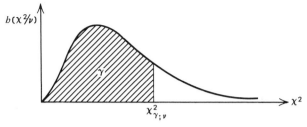

Figure 8.6

which means that the mean and the variance of a random variable χ^2 depend only on the number of degrees of freedom. The χ^2-distribution is positively skewed (Fig. 8.6).

Let $\chi^2_{\gamma;\nu}$ be the value of χ^2, with ν d.f., such that

$$\Pr\{\chi^2 \leqslant \chi^2_{\gamma;\nu}\} = \int_0^{\chi^2_{\gamma;\nu}} h(u, \nu)\, du = \gamma. \tag{8.19}$$

Function (8.19) (regarded as a function of $\chi^2_{\gamma;\nu}$) is the cdf of χ^2. It is represented by the shaded area in Fig. 8.6.

Table II gives the values of $\chi^2_{\gamma;\nu}$ for different ν and different percentage points of γ.

In particular, if a random variable X is distributed as $N(\mu, \sigma^2)$, then

$$Z^2 = \frac{(X - \mu)^2}{\sigma^2} \tag{8.20}$$

is distributed as χ^2 with 1 d.f.

Let $\chi^2(\nu_1), \chi^2(\nu_2), \ldots, \chi^2(\nu_m)$ be m *independent* χ^2-variates, with $\nu_1, \nu_2, \ldots, \nu_m$ d.f., respectively. It follows from (8.17) that the sum

$$\chi^2(\nu) = \chi^2(\nu_1) + \chi^2(\nu_2) + \cdots + \chi^2(\nu_m) \tag{8.21}$$

is also distributed as χ^2, with $\nu = \nu_1 + \nu_2 + \cdots + \nu_m$ d.f.

Applications of this distribution will be shown in later chapters, especially Chapters 12 and 13.

8.5.2. "Student's" t-distribution

Let Z and U be two independent random variables such that Z is distributed normally as $N(0, 1)$ and U is distributed as χ^2 with ν d.f. The ratio

$$t = \frac{Z}{\sqrt{U/\nu}}, \qquad -\infty < t < \infty \tag{8.22}$$

has the t-distribution with ν d.f. This distribution depends only on the

TABLE II.[a] Percentage points of the χ^2 distribution. Values of $\chi^2_{\gamma;\nu}$ such that $Pr\{\chi^2 \leqslant \chi^2_{\gamma;\nu}\} = \gamma$

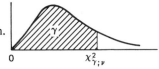

ν \ γ	0.005	0.010	0.025	0.05	0.10	0.20	0.30	0.40	0.50
1	0.^{4}4	0.^{3}16	0.^{3}98	0.^{2}39	0.0158	0.0642	0.1485	0.2750	0.4549
2	0.0100	0.0201	0.0506	0.1026	0.2107	0.4463	0.7134	1.022	1.386
3	0.0717	0.1148	0.2158	0.3518	0.5844	1.005	1.424	1.869	2.366
4	0.2070	0.2971	0.4844	0.7107	1.064	1.649	2.195	2.753	3.357
5	0.4117	0.5543	0.8312	1.145	1.610	2.343	3.000	3.656	4.351
6	0.6757	0.8721	1.237	1.635	2.204	3.070	3.828	4.570	5.348
7	0.9893	1.239	1.690	2.167	2.833	3.822	4.671	5.493	6.346
8	1.344	1.646	2.180	2.733	3.490	4.594	5.527	6.423	7.344
9	1.735	2.088	2.700	3.325	4.168	5.380	6.393	7.357	8.343
10	2.156	2.588	3.247	3.940	4.865	6.179	7.267	8.295	9.342
11	2.603	3.053	3.816	4.575	5.578	6.989	8.148	9.237	10.34
12	3.074	3.571	4.404	5.226	6.304	7.807	9.034	10.18	11.34
13	3.565	4.107	5.009	5.892	7.041	8.634	9.926	11.13	12.34
14	4.075	4.660	5.629	6.571	7.790	9.467	10.82	12.08	13.34
15	4.601	5.229	6.262	7.261	8.547	10.31	11.72	13.03	14.34
16	5.142	5.812	6.908	7.962	9.312	11.15	12.62	13.98	15.34
17	5.697	6.408	7.564	8.672	10.09	12.00	13.53	14.94	16.34
18	6.265	7.015	8.231	9.390	10.86	12.86	14.44	15.89	17.34
19	6.844	7.633	8.907	10.12	11.65	13.72	15.35	16.85	18.34
20	7.434	8.260	9.591	10.85	12.44	14.58	16.27	17.80	19.34
21	8.034	8.897	10.28	11.59	13.24	15.44	17.18	18.76	20.34
22	8.643	9.542	10.98	12.34	14.04	16.31	18.10	19.73	21.34
23	9.260	10.20	11.69	13.09	14.85	17.19	19.02	20.69	22.34
24	9.866	10.86	12.40	13.85	15.66	18.06	19.94	21.65	23.34
25	10.52	11.52	13.12	14.61	16.47	18.94	20.87	22.62	24.34
26	11.16	12.20	13.84	15.38	17.29	19.82	21.79	23.58	25.34
27	11.81	12.88	14.57	16.15	18.11	20.70	22.72	24.54	26.34
28	12.46	13.56	15.31	16.93	18.94	21.59	23.65	25.51	27.34
29	13.12	14.26	16.05	17.71	19.77	22.48	24.58	26.48	28.34
30	13.79	14.95	16.79	18.49	20.60	23.36	25.51	27.44	29.34
40	20.71	22.16	24.43	26.51	29.05	32.35	34.87	37.13	39.34
50	27.99	29.71	32.36	34.76	37.69	41.45	44.31	46.86	49.33
60	35.53	37.48	40.48	43.19	46.46	50.64	53.81	56.62	59.33
70	43.28	45.44	48.76	51.74	55.33	59.90	63.35	66.40	69.33
80	51.17	53.54	57.15	60.39	64.28	69.21	72.92	76.19	79.33
90	59.20	61.75	65.65	69.13	73.29	78.56	82.51	85.99	89.33
100	67.33	70.06	74.22	77.93	82.36	87.95	92.13	95.81	99.33

[a] Taken with permission from H. L. Harter, *Biometrika* **51** (1964), 231–239.

TABLE II.—(continued)

γ $\backslash$ ν	0.60	0.70	0.80	0.90	0.95	0.975	0.990	0.995	0.999
1	0.7083	1.074	1.642	2.706	3.841	5.024	6.635	7.879	10.83
2	1.833	2.408	3.219	4.605	5.991	7.378	9.210	10.60	13.82
3	2.946	3.665	4.642	6.251	7.815	9.348	11.34	12.84	16.27
4	4.045	4.878	5.989	7.779	9.488	11.14	13.28	14.86	18.47
5	5.132	6.064	7.289	9.236	11.07	12.83	15.09	16.75	20.52
6	6.211	7.231	8.558	10.64	12.59	14.45	16.81	18.55	22.46
7	7.283	8.383	9.803	12.02	14.07	16.01	18.48	20.28	24.32
8	8.350	9.524	11.03	13.36	15.51	17.53	20.09	21.96	26.12
9	9.414	10.66	12.24	14.68	16.92	19.02	21.67	23.59	27.88
10	10.47	11.78	13.44	15.99	18.31	20.48	23.21	25.19	29.59
11	11.53	12.90	14.63	17.28	19.68	21.92	24.72	26.76	31.26
12	12.58	14.01	15.81	18.55	21.03	23.34	26.22	28.30	32.91
13	13.64	15.12	16.98	19.81	22.36	24.74	27.69	29.82	34.53
14	14.69	16.22	18.15	21.06	23.68	26.12	29.14	31.32	36.12
15	15.73	17.32	19.31	22.31	25.00	27.49	30.58	32.80	37.70
16	16.78	18.42	20.46	23.54	26.30	28.85	32.00	34.27	39.25
17	17.82	19.51	21.62	24.77	27.59	30.19	33.41	35.72	40.79
18	18.87	20.60	22.76	25.99	28.87	31.53	34.81	37.16	42.31
19	19.91	21.69	23.90	27.20	30.14	32.85	36.19	38.58	43.82
20	20.95	22.77	25.04	28.41	31.41	34.17	37.57	40.00	45.32
21	21.99	23.86	26.17	29.62	32.77	35.48	38.93	41.40	46.80
22	23.03	24.94	27.30	30.81	33.92	36.78	40.29	42.80	48.27
23	24.07	26.02	28.43	32.01	35.17	38.08	41.64	44.18	49.73
24	25.11	27.10	29.55	33.20	36.42	39.36	42.98	45.56	51.18
25	26.14	28.17	30.68	34.38	37.65	40.65	44.31	46.93	52.62
26	27.18	29.25	31.80	35.56	38.89	41.92	45.64	48.29	54.05
27	28.21	30.32	32.91	36.74	40.11	43.19	46.96	49.64	55.48
28	29.25	31.39	34.03	37.92	41.34	44.46	48.28	50.99	56.89
29	30.28	32.46	35.14	39.09	42.56	45.72	49.59	52.34	58.30
30	31.32	33.53	36.25	40.26	43.77	46.98	50.89	53.67	59.70
40	41.62	44.16	47.27	51.81	55.76	59.34	63.69	66.77	73.40
50	51.89	54.72	58.16	63.17	67.50	71.42	76.15	79.49	86.66
60	62.13	65.23	68.97	74.40	79.08	83.30	88.38	91.95	99.61
70	72.36	75.69	79.71	85.53	90.53	95.02	100.4	104.2	112.3
80	82.57	86.12	90.41	96.58	101.9	106.6	112.3	116.3	124.8
90	92.76	96.52	101.1	107.6	113.1	118.1	124.1	128.3	137.2
100	102.9	106.9	111.7	118.5	124.3	129.6	135.8	140.2	149.4

parameter ν and is symmetrical about the vertical line $t = 0$. Its density, $g(t; \nu)$ say, was originally derived by W. S. Gosset, who used the pseudonym "Student." Tables of values $t_{\gamma;\nu}$, such that $\Pr\{t \leqslant t_{\gamma;\nu}\} = \gamma$, for different γ and ν can be found in books on statistics. For large ν (usually > 30), $t_{\gamma;\nu} \doteq z_\gamma$ (actually $t_{\gamma;\nu} > z_\gamma$).

8.5.3. The F-Distribution

Let χ_1^2 and χ_2^2 be two independent random variables distributed as χ^2 with ν_1 and ν_2 d.f., respectively. The variable

$$F = \frac{\chi_1^2/\nu_1}{\chi_2^2/\nu_2}, \qquad 0 < F < \infty \tag{8.23}$$

has an F-distribution with ν_1 and ν_2 d.f. It depends on two parameters, ν_1 and ν_2. It is positively skewed, similarly to χ^2. The formula for the density function and tables of the values $F_{\gamma;\nu_1;\nu_2}$ such that $\Pr\{F \leqslant F_{\gamma;\nu_1;\nu_2}\} = \gamma$ can be found in books on statistics.

8.6. BERNOULLI TRIALS

There are some situations in which only two outcomes of an experiment are of interest: an event C occurs or does not occur. If, for example, we are interested in the white eye color of Drosophila, then event C corresponds to white color; any other eye color is $\bar{C}$ (non-C).

When an experiment is performed and event C occurs, it is customary to call the experiment a "success"; when C does not occur, it is termed a "failure." The meaning of these words is rather formalistic and does not always correspond to the usual sense. If, for instance, event C means a certain disease, and $\bar{C}$ the absence of this disease, the presence of the disease is called a "success" even though it is not really a success to be ill.

We define a random variable U as follows:

$$U = \begin{cases} 1 & \text{if } C \text{ occurs,} \\ 0 & \text{if } \bar{C} \text{ occurs (i.e., } C \text{ does not occur).} \end{cases} \tag{8.24}$$

Let P be the probability of occurrence of event C in a single trial, that is, $\Pr\{C\} = P$. Then $\Pr\{\bar{C}\} = 1 - P$. Hence

$$\Pr\{U = u\} = P^u(1 - P)^{1-u}, \qquad u = 0, 1. \tag{8.25}$$

The expected value of U is

$$E(U) = 1 \cdot P + 0 \cdot (1 - P) = P, \tag{8.26}$$

and the variance

$$\operatorname{Var}(U) = E(U^2) - [E(U)]^2 = P - P^2 = P(1 - P). \qquad (8.27)$$

When an experiment is repeated over and over again, the sequence of such experiments is called *Bernoulli trials*. We will be interested in the distributions of certain random variables defined over the sample space formed by Bernoulli trials.

8.7. THE BINOMIAL DISTRIBUTION

Suppose that n *independent* Bernoulli trials, each with the same probability, P, of success, have been performed. The basic sample space, Ω, for this finite sequence is n-dimensional, each dimension including only two points: 0 and 1, so that there are 2^n points in Ω. We denote

$$U_i = \begin{cases} 1 & \text{if } C \text{ occurs in the } i\text{th trial,} \\ 0 & \text{if } \bar{C} \text{ occurs in the } i\text{th trial,} \end{cases}$$

where $i = 1, 2, \ldots, n$. The U_i's are mutually independent.

Let

$$X = U_1 + U_2 + \cdots + U_n. \qquad (8.28)$$

Then X *is the number of* "*successes*" *in n independent Bernoulli trials*.

In n independent trials, the probability of a *given sequence* of r successes and $n - r$ failures is

$$P^r(1 - P)^{n-r}.$$

There are $\binom{n}{r}$ possible different orderings for which $X = r$; hence the probability that event C occurs r times and does not occur $n - r$ times in n independent trials is

$$\Pr\{X = r\} = \binom{n}{r} P^r(1 - P)^{n-r}, \qquad r = 0, 1, \ldots, n. \qquad (8.29)$$

It should be noted that (8.29) is the $(r + 1)$th term in the expansion of the binomial $[P + (1 - P)]^n$. The distribution is called the *binomial distribution* and is sometimes denoted as $b(r; n, P)$. The constants n and P are the *parameters* of the distribution. The cumulative distribution function is

$$B(r) = \Pr\{X \leqslant r\} = \sum_{i=0}^{r} \binom{n}{i} P^i(1 - P)^{n-i}, \qquad r = 0, 1, \ldots, n. \qquad (8.30)$$

Of course,

$$\sum_{i=0}^{n} \binom{n}{i} P^i(1 - P)^{n-i} = [P + (1 - P)]^n = 1. \qquad (8.31)$$

There exist several tables for calculating either $\Pr\{X = r\}$ or cdf $\Pr\{X \leqslant r\}$; for example, National Bureau of Standards, *Tables of the Binomial Probability Distribution*, Government Printing Office, 1949.

The expected value of X is

$$E(X) = E(U_1) + E(U_2) + \cdots E(U_n) = P + P + \cdots + P = nP, \quad (8.32)$$

and the variance of X is (since the U_i's are mutually independent)

$$\text{Var}(X) = \text{Var}(U_1) + \text{Var}(U_2) + \cdots + \text{Var}(U_n)$$
$$= P(1 - P) + P(1 - P) + \cdots + P(1 - P) = nP(1 - P). \quad (8.33)$$

Let $X_1, X_2, \ldots, X_k$ be k independent binomial random variables, each distributed binomially with the same parameter P but different values of n, that is, $X_i \frown b(r_i; P, n_i)$, $i = 1, 2, \ldots, k$. The sum of these k independent binomial variables,

$$Y = X_1 + X_2 + \cdots + X_k, \quad (8.34)$$

is also binomially distributed with the parameters P and $n = n_1 + n_2 + \cdots + n_k$, that is,

$$Y \frown b(y; P, n). \quad (8.35)$$

This is evident from the definition of the binomial distribution.

Example 8.3. There are in human beings, animals, and plants several characters for which the heterozygote is distinguishable from both homozygotes. For example, thalassemia is a blood disease associated with irregular shape of the blood cells (sickle-cell anemia). It is a simple recessive trait. Although more than two alleles are present at this locus, we will restrict ourselves to a population in which there are only two alleles, C (normal) and c (sickle-cell). Usually homozygotes cc die in early life (thalassemia major), whereas heterozygotes are only slightly affected (thalassemia minor).

Suppose that two parents are heterozygotes $Cc \times Cc$ and the number of offspring in the family (the family size) is $s = 4$.

(a) What is the probability that in such a family there will be no child cc (i.e., with thalassemia major)?

The probability that any child born in this family will be cc is $P = \frac{1}{4}$. The births of four children follow the model of independent Bernoulli trials, so we have

$$\Pr\{X = 0\} = \binom{4}{0} \left(\frac{1}{4}\right)^0 \left(\frac{3}{4}\right)^4 = \left(\frac{3}{4}\right)^4 = \frac{81}{256} = 0.3164.$$

(b) What is the probability that exactly one child will be cc?

$$\Pr\{X = 1\} = \binom{4}{1} \left(\frac{1}{4}\right) \left(\frac{3}{4}\right)^3 = \left(\frac{3}{4}\right)^3 = \frac{27}{64} = 0.4219.$$

(c) What is the probability that *at most* one child will be cc?

$$\Pr\{X \leqslant 1\} = \Pr\{X = 0\} + \Pr\{X = 1\} = 0.3164 + 0.4219 = 0.7383.$$

(d) What is the probability that *at least* one child will be cc?

$$\Pr\{X \geqslant 1\} = 1 - \Pr\{X < 1\} = 1 - \Pr\{X = 0\} = 0.6836.$$

(e) Suppose that $N = 10$ families with parental genotypes $Cc \times Cc$, each having $s = 4$ children, are selected. What is the probability that none of these families will have cc children?

Let X_1 be the number of families in which no cc child has been born. Now, the probability of no cc child in a $Cc \times Cc$ family is $P_1 = 0.3164$. Thus, the probability that all ten families will have no cc child, is

$$\Pr\{X_1 = 10\} = \binom{10}{10} P_1^{10}(1 - P_1)^0 = (0.3164)^{10} = 0.00001.$$

In general, the probability that there will be k families with no cc children born, is

$$\Pr\{X_1 = k\} = \binom{N}{k}(0.3164)^k(0.6836)^{N-k}, \qquad k = 0, 1, \ldots, 10.$$

8.8. PROPORTIONS OF SEGREGATING AND NONSEGREGATING FAMILIES FOR A SIMPLE RECESSIVE TRAIT

The study of the distributions of certain qualitative characters in families is one of several applications of the binomial distribution.

Let dd determine a simple *recessive*, abnormal trait, and suppose that the genotypes DD and Dd are indistinguishable. Suppose further that the population under consideration is in equilibrium and that the frequencies of D and d are p and q, respectively. Two kinds of parental phenotypic matings are observed: Dom $\times$ Rec and Dom $\times$ Dom, and families of size s are examined with respect to children of the dd genotype. Let X denote the number of dd (called "affected") children in a family of size s. Table 8.1 shows the probabilities that there will be r affected children in a family of size s, $\Pr\{X = r\}$, for different mating types.

By *nonsegregating* families we will mean the families in which no affected (i.e., no recessive dd) children are observed; other families will be called *segregating*. Let us introduce a new random variable U defined as

$$U = \begin{cases} 0 & \text{if a family is nonsegregating,} \\ 1 & \text{if a family is segregating.} \end{cases} \qquad (8.36)$$

TABLE 8.1 Probabilities of r affected children in a family of size s for different mating types

Mating type	Probability of mating	Conditional probability of offspring					Probability of affected children	Expected proportion of families of size s with exactly r affected children
		Genotypes			Phenotypes			
		DD	Dd	dd	$D-$	dd		
Dom × Rec								
$DD \times dd$	$2p^2q^2$	0	1	0	1	0	0	0
$Dd \times dd$	$4pq^3$	0	$\frac{1}{2}$	$\frac{1}{2}$	$\frac{1}{2}$	$\frac{1}{2}$	$\binom{s}{r}(\frac{1}{2})^s$	$4pq^3\binom{s}{r}(\frac{1}{2})^s$
Total	$2q^2(1 - q^2)$							
Dom × Dom								
$DD \times DD$	p^4	1	0	0	1	0	0	0
$DD \times Dd$	$4p^3q$	$\frac{1}{2}$	$\frac{1}{2}$	0	1	0	0	0
$Dd \times Dd$	$4p^2q^2$	$\frac{1}{4}$	$\frac{1}{2}$	$\frac{1}{4}$	$\frac{3}{4}$	$\frac{1}{4}$	$\binom{s}{r}(\frac{1}{4})^r(\frac{3}{4})^{s-r}$	$4p^2q^2\binom{s}{r}(\frac{1}{4})^r(\frac{3}{4})^{s-r}$
Total	$(1 - q^2)^2$							

Mating Type Dom × Rec. As we can see in Table 8.1, for this mating type nonsegregating families are $DD \times dd$ (with conditional probability 1) and also include those among $Dd \times dd$ matings which do not have affected children ($r = 0$). We find $\Pr\{X = 0 \mid Dd \times dd\} = (\frac{1}{2})^s$. Therefore, the total probability of nonsegregating families of size s, given mating type Dom × Rec, $P_{10}(U = 0) = P_{10}(0)$, say, is

$$
\begin{aligned}
P_{10}(0) &= \frac{2p^2q^2 + 4pq^3(\frac{1}{2})^s}{2q^2(1 - q^2)} = \frac{2pq^2[p + 2q(\frac{1}{2})^s + (2q - 2q)]}{2q^2(1 - q)(1 + q)} \\
&= \frac{(1 - q)\{(1 + q) - 2q[1 - (\frac{1}{2})^s]\}}{(1 - q)(1 + q)} \\
&= 1 - 2\left(\frac{q}{1 + q}\right)[1 - (\frac{1}{2})^s].
\end{aligned}
\tag{8.37}
$$

We recall from section 7.13 that $q/(1 + q) = S_{10}$ is Snyder's ratio, or the expected proportion of affected children *in a population* of Dom × Rec. Thus, (8.37) takes the form

$$
P_{10}(0) = 1 - 2S_{10}[1 - (\frac{1}{2})^s].
\tag{8.37a}
$$

Probability (8.37) or (8.37a) is, in fact, the expected proportion of nonsegregating families of size s in a population of given mating type, Dom × Rec. Of course, the expected proportion of segregating families (i.e., with at least one affected child), $P_{10}(1)$, is

$$
P_{10}(1) = 1 - P_{10}(0) = 2S_{10}[1 - (\frac{1}{2})^s].
\tag{8.38}
$$

We notice that the distribution of affected children in a segregating family of size s and given mating type Dom × Rec, $\Pr\{X = r \mid X > 0\}$, is

$$
\Pr\{X = r \mid X > 0\} = \frac{\binom{s}{r}(\frac{1}{2})^r(\frac{1}{2})^{s-r}}{1 - (\frac{1}{2})^s} = \frac{(\frac{1}{2})^s}{1 - (\frac{1}{2})^s}\binom{s}{r}, \quad r = 1, 2, \ldots, s.
\tag{8.39}
$$

This is a binomial distribution *truncated* at the point $r = 0$ (compare section 5.4).

Suppose that N families, each of size s and parental mating type Dom × Rec, are observed. Let X_1 denote the number of nonsegregating families among these N families. Then the distribution of the number of nonsegregating families, X_1, is binomial with parameters N and $P_{10}(0)$, that is,

$$
\Pr\{X_1 = k\} = \binom{N}{k}[P_{10}(0)]^k[P_{10}(1)]^{N-k}, \quad k = 0, 1, \ldots, N.
\tag{8.40}
$$

Mating Type Dom × Dom. Similar probabilities can be obtained for mating type Dom × Dom. If $P_{11}(0)$ and $P_{11}(1)$ denote, respectively, the expected proportions of nonsegregating and segregating families, each of size s, in a population of Dom $\times$ Dom, then, from Table 8.1, we find

$$P_{11}(0) = \frac{p^4 + 4p^3q + 4p^2q^2(\frac{3}{4})^s}{(1 - q^2)^2} = \frac{p^2[p^2 + 4pq + 4q^2 - 4q^2 + 4q^2(\frac{3}{4})^s]}{p^2(1 + q)^2}$$

$$= \frac{(1 + q)^2 - 4q^2[1 - (\frac{3}{4})^s]}{(1 + q)^2} = 1 - 4\frac{q^2}{(1 + q)^2}[1 - (\frac{3}{4})^s]. \qquad (8.41)$$

Again, recalling section 7.13, we notice that

$$\frac{q^2}{(1 + q)^2} = S_{11} = S_{10}{}^2$$

is Snyder's ratio, or the expected proportion of affected children in a population of Dom $\times$ Dom matings. Thus

$$P_{11}(0) = 1 - 4S_{11}[1 - (\tfrac{3}{4})^s] = 1 - (2S_{10})^2[1 - (\tfrac{3}{4})^s]. \qquad (8.41a)$$

The expected proportion of segregating families in a population of Dom $\times$ Dom is

$$P_{11}(1) = 1 - P_{11}(0) = 4S_{11}[1 - (\tfrac{3}{4})^s] = (2S_{10})^2[1 - (\tfrac{3}{4})^s]. \qquad (8.42)$$

Let X be the number of affected children in a segregating family of size s of the mating type Dom $\times$ Dom. The distribution of X is

$$\Pr\{X = r \mid X > 0\} = \binom{s}{r}\frac{(\frac{1}{4})^r(\frac{3}{4})^{s-r}}{1 - (\frac{3}{4})^s}, \qquad r = 1, 2, \ldots, s. \qquad (8.43)$$

Example 8.4. Inability to taste the substance phenylthiocarbamide (PTC) or related compounds is assumed to be a single recessive trait. If we denote the gene for nontaste of PTC by t, and its allelomorph by T, then tt individuals are "nontasters" (PTC$-$) and TT or Tt are "tasters" (PTC$+$). Using Snyder's data (1932) it has been found that for American Whites the frequency of gene t is (approximately) $q = 0.545$. The trait under consideration is tt (i.e., the affected individuals are nontasters).

(a) Suppose that the parental mating is taster $\times$ nontaster. What is the probability that their first child will be a taster?

This probability is the same as the probability of selecting a nonsegregating family, from the whole population of matings taster $\times$ nontaster, each family having one child. From formula (8.37), for $s = 1$, we obtain.

$$P_{10}(0) = 1 - 2 \times \frac{0.545}{1.545}(1 - \tfrac{1}{2}) = 1 - 0.3528 = 0.6472.$$

(b) What is the probability that the second child will be a taster if the first child was a taster?

Let E_1 and E_2 denote the events that the first and the second children are tasters respectively. Thus we have

$$\Pr\{E_2 \mid E_1\} = \frac{\Pr\{E_1 E_2\}}{\Pr\{E_1\}} = \frac{P_{10}(0 \text{ for } s = 2)}{P_{10}(0 \text{ for } s = 1)} = \frac{1 - 2\dfrac{q}{1+q} \cdot \dfrac{3}{4}}{1 - 2\dfrac{q}{1+q} \cdot \dfrac{1}{2}} = \frac{2 - q}{2}$$

$$= 1 - \tfrac{1}{2}q = 1 - 0.2725 = 0.7275.$$

(c) What is the probability that the second child will be a taster if the first was a nontaster?

In this case, we know that the mating was $Tt \times tt$, so the probability that the second child will be a taster is $\tfrac{1}{2}$.

(d) What is the expected proportion of nonsegregating families (i.e., with tasters only) in the population of one-child families and of matings taster $\times$ taster?

We apply formula (8.41) for $s = 1$, and obtain

$$P_{11}(0) = 1 - 4 \times (0.3528)^2 \times \tfrac{1}{4} = 1 - 0.1245 = 0.8755.$$

(e) What is the probability that among $N = 5$ families, each with one child, selected from parental type taster $\times$ taster, there will be exactly $k = 3$ families in which the child is a taster? Using formula analogous to that in (8.40), we find

$$\Pr\{X_1 = 3\} = \binom{5}{3}(0.8755)^3(0.1245)^2 = 0.1040.$$

(f) What is the probability that the child in each of the five families will be a taster?

$$\Pr\{X_1 = 5\} = (0.8755)^5 = 0.5144.$$

8.9. DISTRIBUTION OF PROPORTIONS

We are sometimes interested in the distribution of the proportion of successes rather than the number of successes in n independent trials. Let $\bar{X} = (X/n)$ and $\bar{x} = (x/n)$. Thus, the events $\bar{X} = \bar{x}$ and $X = x$ are the same events and

$$\Pr\{\bar{X} = \bar{x}\} = \Pr\{X = x\} = \binom{n}{n\bar{x}} P^{n\bar{x}}(1 - P)^{n(1-\bar{x})}, \qquad (8.44)$$

where $\bar{x} = 0, (1/n), (2/n), \dots, 1$.

It is easy to show that

$$E(\overline{X}) = E(X/n) = P, \quad \mathrm{Var}(\overline{X}) = \mathrm{Var}(X/n) = \frac{P(1-P)}{n}. \quad (8.45)$$

8.10. THE GEOMETRIC AND NEGATIVE BINOMIAL DISTRIBUTIONS

We consider again a sequence of independent Bernoulli trials in which the probability of a certain event C occurring in a single trial is $\mathrm{Pr}\{C\} = P$. Let Y_1 now denote the number of failures *before the first success*, when the trials are continued as long as necessary to obtain a success. Now the number of trials is a random variable, equal to the number of failures plus 1, that is, $Y_1 + 1$. The probability that exactly y_1 failures will occur before the first success is

$$\mathrm{Pr}\{Y_1 = y_1\} = (1-P)\cdot(1-P)\cdots(1-P)P = P(1-P)^{y_1}, \quad (8.46)$$

for $y_1 = 0, 1, \ldots$.
(Note that y_1 is a nonnegative integer.) The probabilities defined by (8.46) constitute the *geometric distribution*. These probabilities are terms of the geometric series [see formula (II.21)]

$$P + P(1-P) + P(1-P)^2 + \cdots = P\cdot\frac{1}{P} = 1.$$

Let X be a random variable which represents the number of failures *before the kth success*. Then

$$X = Y_1 + Y_2 + \cdots + Y_k, \quad (8.47)$$

where Y_j is the number of failures between the $(j-1)$th and the jth success. The number of independent trials *before* the kth success is $X + k - 1$. In $x + k - 1$ independent trials the probability of a *given sequence* of x failures and $k - 1$ successes is

$$P^{k-1}(1-P)^x.$$

There are $\binom{x+k-1}{k-1}$ different such orderings, so that the total probability of x failures and $k - 1$ successes in the first $x + k - 1$ trials is

$$\binom{x+k-1}{k-1}P^{k-1}(1-P)^x,$$

and the probability that in $x + k$ trials exactly x failures will occur before the kth success is

$$\mathrm{Pr}\{X = x\} = \binom{x+k-1}{k-1}P^{k-1}(1-P)^x P = \binom{x+k-1}{k-1}P^k(1-P)^x$$

$$(8.48)$$

for $x = 0, 1, 2, \ldots$.

The number of independent trials required to obtain the kth success is a random variable

$$Z = X + k.$$

Of course, the events $Z = z = x + k$ and $X = x$ are the same events, so we easily find that

$$\Pr\{Z = z\} = \binom{z-1}{k-1} P^k (1 - P)^{z-k}, \qquad (8.49)$$

for $z = k, k + 1, \ldots$, and $\Pr\{X = x\}$, given by (8.48), are the same probabilities.

Functions (8.48) or (8.49) represent the *negative binomial distribution or binomial waiting time distribution*. They give the probability that one must *wait* through z independent trials in order to obtain k successes.

It can be shown that

$$E(X) = \frac{k(1-P)}{P} ; \quad \text{Var}(X) = \frac{k(1-P)}{P^2} . \qquad (8.50)$$

Thus,

$$E(Z) = E(X) + k = \frac{k}{P} ; \quad \text{Var}(Z) = \text{Var}(X) = \frac{k(1-P)}{P^2} . \qquad (8.51)$$

We also notice that

$$\Pr\{Z \leqslant z\} = \sum_{i=k}^{z} \binom{i-1}{k-1} P^k (1 - P)^{i-k}. \qquad (8.52)$$

Example 8.5. Assume that each child born is equally likely to be a boy or a girl.

(a) What average size of a family should be expected if the parents wish to have exactly two boys?

We have $P = \frac{1}{2}$, $k = 2$, so $E(Z) = 2 : \frac{1}{2} = 4$.

(b) What is the probability that a couple who want *exactly* two boys will have to wait until they have four children?

We have $z = 4$, $k = 2$, $P = \frac{1}{2}$. Thus,

$$\Pr\{Z = 4\} = \binom{3}{1}\left(\frac{1}{2}\right)^4 = \frac{3}{16} = 0.1875$$

(less than 20%).

(c) But the probability that they will need to wait until they have *at least* four children is

$$\Pr\{Z \geqslant 4\} = 1 - \Pr\{Z < 4\} = 1 - \Pr\{Z = 2\} - \Pr\{Z = 3\}$$

$$= 1 - 0.2500 - 0.2500 = 0.5000,$$

that is, 50%.

(*d*) Moreover, the probability that they will need to wait only until they have *at most* four children is

$$\Pr\{Z \leqslant 4\} = \Pr\{Z = 2\} + \Pr\{Z = 3\} + \Pr\{Z = 4\} = 0.6875,$$

that is, about 70%.

8.11. THE MULTINOMIAL DISTRIBUTION

Another very useful distribution in genetics, besides the binomial one, is the multinomial distribution. This can be considered a generalization of the binomial distribution.

Let $C_1, C_2, \ldots, C_k$ be an exhaustive set of k mutually exclusive events (or classes, or categories) and $P_1, P_2, \ldots, P_k$, with $\sum P_i = 1$, be the corresponding probabilities associated with occurrence of these events in a single trial.

Let X_i be a random variable corresponding to the number of occurrences of the event C_i, and r_i the observed number of outcomes of the event C_i ($i = 1, 2, \ldots, k$), in n independent trials.

We must have

$$\sum_{i=1}^{k} X_i = n, \quad \sum_{i=1}^{k} r_i = n, \quad \text{and} \quad \sum_{i=1}^{k} P_i = 1. \tag{8.53}$$

The joint probability function

$$\Pr\{X_1 = r_1, X_2 = r_2, \ldots, X_k = r_k\} = \frac{n!}{r_1! r_2! \cdots r_k!} P_1^{r_1} P_2^{r_2} \cdots P_k^{r_k} \tag{8.54}$$

or briefly,

$$\Pr\{r_1, r_2, \ldots, r_k\} = n! \prod_{i=1}^{k} \frac{P_i^{r_i}}{r_i!} \tag{8.54a}$$

constitutes the *multinomial distribution*. We notice that (8.54) is a term in the expansion of $(P_1 + P_2 + \cdots + P_k)^n$ [compare formula (3.12)].

The results of the n trials can always be divided into the two classes: r_i outcomes of the event C_i and $(n - r_i)$ outcomes in the remaining events "non-C_i" (or $\bar{C}_i$). Thus the random variable X_i is binomially distributed,

$$X_i \frown b(r_i; n, P_i), \qquad i = 1, 2, \ldots, k.$$

This implies that

$$E(X_i) = nP_i \quad \text{and} \quad \text{Var}(X_i) = nP_i(1 - P_i), \qquad i = 1, 2, \ldots, n. \tag{8.55}$$

If we are interested in the proportion, X_i/n, rather than in the total number of outcomes, we have

$$E\left(\frac{X_i}{n}\right) = P_i, \quad \text{Var}\left(\frac{X_i}{n}\right) = \frac{P_i(1 - P_i)}{n}, \qquad i = 1, 2, \ldots, n. \quad (8.56)$$

We may divide our k classes into two different groups: $(C_i + C_j)$ and the remaining classes excluding C_i and C_j. Thus $(X_i + X_j)$ is binomially distributed, that is,

$$(X_i + X_j) \frown b[(r_i + r_j); n, (P_i + P_j)], \qquad (8.57)$$

and, therefore,

$$E(X_i + X_j) = P_i + P_j; \quad \text{Var}(X_i + X_j) = n(P_i + P_j)[1 - (P_i + P_j)]. \quad (8.58)$$

The variables $X_1, X_2, \ldots, X_k$ are subject to the condition $\sum X_i = n$; this means that they are *functionally dependent*. Functional dependence of resulting outcomes suggests that the random variables $X_1, X_2, \ldots, X_k$ are correlated. Let us find the covariance and correlation coefficient of X_i and X_j.

For any two random variables, X_i and X_j, we have in general [see formula (6.41)],

$$\text{Var}(X_i + X_j) = \text{Var}(X_i) + \text{Var}(X_j) + 2\text{Cov}(X_i, X_j).$$

Hence,

$$2\text{Cov}(X_i, X_j) = \text{Var}(X_i + X_j) - \text{Var}(X_i) - \text{Var}(X_j). \quad (8.59)$$

In particular, substituting from (8.58) and (8.55) into the right-hand side of (8.59), we obtain

$$\begin{aligned}
2\text{Cov}(X_i, X_j) &= n(P_i + P_j)[1 - (P_i + P_j)] - nP_i(1 - P_i) - nP_j(1 - P_j) \\
&= n[P_i(1 - P_i) - P_iP_j + P_j(1 - P_j) - P_jP_i - P_i(1 - P_i) \\
&\quad - P_j(1 - P_j)] = -2nP_iP_j
\end{aligned}$$

or

$$\text{Cov}(X_iX_j) = -nP_iP_j. \qquad (8.60)$$

For proportions

$$\text{Cov}\left(\frac{X_i}{n}, \frac{X_j}{n}\right) = -\frac{P_iP_j}{n}. \qquad (8.60a)$$

The correlation coefficient, ρ, is

$$\rho = \frac{\text{Cov}(X_i, X_j)}{\sqrt{\text{Var}(X_i)}\sqrt{\text{Var}(X_j)}} = \frac{-P_iP_j}{\sqrt{P_i(1 - P_i)}\sqrt{P_j(1 - P_j)}} = -\sqrt{\frac{P_iP_j}{(1 - P_i)(1 - P_j)}}. \qquad (8.61)$$

We notice that ρ does not depend on n. It is negative, since the more outcomes fall into class C_i, the fewer are available to fall into class C_j, in view of the restriction that the total number of trials is equal to n.

From what has been said above, we can see that the multinomial distribution can be regarded as a *joint* distribution of k *dependent binomial* variables.

There will be several applications of multinomial distributions in later chapters.

8.12. THE POISSON DISTRIBUTION

It sometimes occurs that the parameter P in the binomial distribution is very small but n is large enough so that the mean nP is appreciable. Let us consider a sequence of binomial distributions such that $n \to \infty$ and $P \to 0$ while $nP = \lambda$ remains constant. It can be shown that

$$\lim_{\substack{n \to \infty \\ nP=\lambda}} \binom{n}{x} P^x(1-P)^{n-x} = \frac{\lambda^x}{x!} e^{-\lambda}. \tag{8.62}$$

The probabilities defined by

$$p(x) = \frac{\lambda^x}{x!} e^{-\lambda}, \qquad x = 0, 1, 2, \ldots \tag{8.63}$$

constitute the *Poisson distribution*. Thus, *the Poisson distribution can be regarded as the limiting case of a sequence of binomial distributions* in which, as $n \to \infty$, the mean $nP = \lambda$ remains constant. It can be used in the approximate evaluation of binomial probabilities with large n and small P.

There are several tables of the probability function $p(x)$ and the cumulative distribution function

$$\Pr\{X < x\} = F(x-1) = \sum_{i=1}^{x-1} \frac{\lambda^i}{i!} e^{-\lambda} \quad \text{or} \quad P(X \geqslant x) = 1 - F(x-1) \tag{8.64}$$

for different λ [see, for instance, Johnson and Leone (1964)].

It can also be shown that for the Poisson distribution

$$E(X) = \lambda \quad \text{and} \quad \text{Var}(X) = \lambda, \tag{8.65}$$

that is, the mean and the variance are each equal to the same value, λ, which is called the parameter of the Poisson distribution.

Example 8.6. It is usually assumed that albinism is a rare recessive trait. It has been estimated (Stern, (1960), p. 103), that in various European countries about 1 person in 20,000 is albino. Suppose that we observe 10,000 persons. What is the probability that at least 2 will be albinos?

The distribution of the number of albinos, X, say, follows the binomial distribution with $P = (1/20,000) = 5 \times 10^{-5}$ and $n = 10,000$. We notice that P is very small and n rather large, so we use the Poisson approximation with $\lambda = 5 \times 10^{-5} \cdot 10,000 = 0.5$. We have

$$P\{X \geqslant 2\} \doteq \sum_{i=2}^{\infty} \frac{(0.5)^i}{i!}\, e^{-0.5} = 1 - [e^{-0.5} + 0.5e^{-0.5}] = 0.0902.$$

There is only a small chance (about 9%) that in even such a large sample there will be 2 or more albinos.

8.13. POPULATION AND SAMPLE. SAMPLING FROM INFINITE AND FINITE POPULATIONS

In biological sciences the term *population* usually means a rather large group or collection of individuals of the same kind. This is a *biological population*. We may also consider inanimate objects such as tins of preserved food or electric bulbs or automobiles. These are examples of *populations of objects*. These populations are always *finite*, although quite often very large.

Suppose that one tosses a coin. Either of two outcomes, head or tail, can be observed. One can imagine an infinite series of independent tosses—this will correspond to an *infinite or hypothetical population of outcomes*. Sometimes the term *universe* is used, instead of infinite population.

We will call a part of the population to which the theory of random variables and their distributions can be applied a *random sample*. This is a very broad definition, and it is necessary to give more detail. Using a sample we wish to make some statements about the population. These problems will be discussed in Chapters 11, 12, and 13. Here we only outline a simple method of obtaining a sample.

Let X be a certain character observed in an infinite or finite population. Thus X can be represented by a random variable X having a certain distribution, $p(x)$, in a given population. Suppose, for example, that we consider a population of offspring of matings $Aa \times Aa$. Let X be a random variable such that $X = 0$ if aa, $X = 1$ if Aa, and $X = 2$ if AA. The genotype distributions, $p(x)$, in infinite (hypothetical) and finite (of size N) populations are as follows:

Genotype:	AA	Aa	aa	
$X = x$:	2	1	0	
$p(x)$ (infinite):	$\frac{1}{4}$	$\frac{1}{2}$	$\frac{1}{4}$	(8.66)
$p(x)$ (finite):	N_1/N	N_2/N	N_3/N,	

where $N_1 + N_2 + N_3 = N$ are (absolute) frequencies of genotypes.

Suppose that we wish to obtain a set of n observations on X or, in other words, to draw a sample of size n. This can be done in several ways; one of these can be defined as follows.

DEFINITION 8.2.   *If a set of n observations on a random variable X is obtained in such a manner that at each stage of sampling the observation is made independently of the outcomes of the preceding observations, the sampling technique is called simple random sampling and a set of n values so obtained represents a simple random sample of size n.*

We should distinguish between sampling from infinite and finite populations.

8.13.1. Sampling from Infinite Populations

In this case, the distribution $p(x)$ is the same at each stage of the sampling process. Each observation represents a value of a random variable having the distribution $p(x)$. The following definition of a simple random sample is sometimes used.

DEFINITION 8.3.   *A set of values obtained on n independent and identically distributed random variables represents a simple random sample of size n.*

8.13.2. Sampling from Finite Populations

In this case, we are interested in selection from a population of units such as objects or individuals. The essential condition for simple random sampling technique is that at each drawing all the units have the same probability of being selected in an independent manner.

We may consider two kinds of simple selection or drawing the units from a finite population.

1. *Sampling with replacement* takes place when the selected unit is returned to the population before the next drawing is made. Thus before the next drawing, the distribution of X is exactly the same as before the first drawing. The procedure can be repeated infinitely, and, in fact, Definition 8.3 applies here exactly.

2. *Sampling without replacement* takes place when the unit, once selected, is not returned to the population. Thus at the second drawing there are only $N - 1$ objects, but each has a probability $1/(N - 1)$ of being selected. The distribution of X depends on which object has been selected at the first drawing. On the other hand, the probability of selecting any sample of size n is $1 \Big/ \binom{N}{n}$. The sampling procedure is again an independent process and according to Definition 8.2, the sample so obtained is called a simple random sample. A few remarks are useful:

1. Simple random sampling is often called briefly *random sampling*, and we will sometimes do so.

2. How can a simple random sample be obtained? It appears, from experience, that pointing "at random" or other "haphazard" methods does not give samples selected in an independent manner (see Kendall and Stuart (1961), Vol. I, Chapter 9). A simple random sample should be obtained by using tables of random numbers whenever possible. Such tables and explanations of how to use them can be found in many introductory books on statistics.

3. When we perform experiments on plants or laboratory animals, we have in mind an infinite (or "model") population, and the results of an experiment represent a simple random sample as discussed in section 8.13.1. On the other hand, natural populations (e.g., human populations) are usually finite, and in practice the sampling is without replacement. If the population is large, however, sampling with or without replacement does not change the probabilities substantially. Definition 8.3 is then used for practical purposes.

4. It is very important for research workers always to be aware of what sampling procedure they are using in collecting data. This point will be discussed further in sections 12.5–12.9 (sampling from unrelated and related pairs) and in sections 17.3 and 17.4 (methods of ascertainment). Here we give only an example to demonstrate how different sampling procedures affect the evaluation of probability.

Example 8.7. Suppose that in a certain large population the frequency for the nontasting PTC recessive gene t is $q = 0.545$, and let us assume that the population is in equilibrium with respect to the PTC-taste locus.

Suppose that we select $s = 4$ individuals and want to know the chance that in such a sample there will be at least one nontaster. (i.e., a recessive tt.) Let r denote the number of nontasters in the sample of size $s = 4$. Note that r is now a random variable taking the values 0, 1, 2, 3, 4.

We might select the four individuals by using different methods (i.e., obtain the sample from different experimental designs) and calculate the probabilities $\Pr\{r \geqslant 1\}$ for each experimental design. Let us consider the following experiments.

Experiment I. A *simple random sample* of size $s = 4$ is drawn from the whole population. In this case, the proportion of recessives is $P = q^2$, and $\Pr\{r \geqslant 1\}$ is the appropriate binomial probability with $P = q^2$:

$$\Pr\{r \geqslant 1 \mid \text{Exp. I}\} = 1 - (1 - q^2)^s.$$

Substituting $s = 4$ and $q = 0.545$, we obtain

$$\Pr\{r \geqslant 1 \mid \text{Exp. I}\} = 1 - (1 - 0.545^2)^4 = 1 - 0.2442 = 0.7558.$$

Experiment II. A *family* of size $s = 4$ is selected from the whole population. We note that in this case the families which are able to produce nontasters in offspring must be from these matings: $tt \times tt$ or $Tt \times tt$ or $Tt \times Tt$. Then, using the results from Table 8.1, we find

$$\Pr\{r \geqslant 1 \mid \text{Exp. II}\} = q^2 \times 1 + 4pq^3[1 - (\tfrac{1}{2})^s] + 4p^2q^2[1 - (\tfrac{1}{4})^s]$$
$$= q^2\{1 + 4pq[1 - (\tfrac{1}{2})^s] + 4p^2[1 - (\tfrac{1}{4})^s]\}.$$

Again, substituting $s = 4$ and $q = 0.545$, we obtain $\Pr\{r \geqslant 1 \mid \text{Exp. II}\} = 0.8182$.

Experiment III. Suppose that some additional information on the *phenotypes* of the parents is available, for example, that both parents are tasters (i.e., Dom × Dom), and we select a *family* with $s = 4$ children. The only segregating families of this class are $Tt \times Tt$. The probability of more than one dd child in such a family is $1 - (\tfrac{1}{4})^s$. The probability of selecting such a family among all Dom × Dom families is $4p^2q^2/(1 - q^2)^2$, so that the total probability of selecting at least one nontaster in Experiment III is

$$\Pr\{r \geqslant 1 \mid \text{Exp. III}\} = \frac{[1 - (\tfrac{1}{4})^s]4p^2q^2}{(1 - q^2)^2} = 4S_{11}[1 - (\tfrac{1}{4})^s],$$

and for our numerical values of q and s we obtain $\Pr\{r \geqslant 1 \mid \text{Exp. III}\} = 0.4958$.

Experiment IV. Suppose that the offspring of all parents in the population who are both tasters are pooled together and a simple random sample of size s is drawn from this part of the generation of offspring. Now, the proportion of nontasters among the offspring of Dom × Dom is Snyder's ratio, $S_{11} = q^2/(1 + q)^2$. This is a binomial parameter in s independent trials. Then

$$\Pr\{r \geqslant 1 \mid \text{Exp. IV}\} = 1 - \left[1 - \frac{q^2}{(1 + q)^2}\right]^s = 1 - \left[\frac{1 + 2q}{(1 + q)^2}\right]^s,$$

and for our values of s and q we obtain

$$\Pr\{r \geqslant 1 \mid \text{Exp. IV}\} = 0.4123.$$

8.14. THE DISTRIBUTION OF THE SAMPLE

Let $x_1, x_2, \ldots, x_n$ be the n values *observed* in a sample. We can regard them as the values of a random variable X. Assigning to each value a weight equal to $(1/n)$, we obtain the distribution of the observed sample or the empirical distribution of X.

DEFINITION 8.4. *If f_x is the number of sample values which are less than or equal to x, then the cumulative relative frequencies of the sample, $F^*(x)$,* *defined as*

$$F^*(x) = \frac{f_x}{n},\tag{8.68}$$

is called the empirical distribution or the sample distribution of the random variable X. $F^*(x)$ is a step function with jumps of height $(1/n)$ at each observed x_i.

For such empirical distributions we can calculate various *characteristics of the sample.*

DEFINITION 8.5.   *The rth moment of the sample (or the rth sample moment) is*

$$m_r' = \frac{1}{n}\sum_{i=1}^{n} x_i^r, \qquad r = 1, 2, \ldots.\tag{8.69}$$

It is customary to denote the characteristics of the population by Greek letters, and those of the sample by corresponding Roman letters. An exception is the sample mean, which is usually denoted by $\bar{x}$ instead of m_1', that is,

$$m_1' = \bar{x} = \frac{1}{n}\sum_{i=1}^{n} x_i.$$

These considerations are valid whether X is continuous or discrete. If, of course, X is discrete and takes only k possible values, $x_1', x_2', \ldots, x_k'$, say, then the "empirical probabilities" (i.e., the observed relative frequencies) are n_j/n $(j = 1, 2, \ldots, k)$, where n_j is the observed number of values x_j'.

Thus the rth sample moment can be calculated from the formula

$$m_r' = \sum_{j=1}^{k} x_j'^r \frac{n_j}{n} = \frac{1}{n}\sum_{j=1}^{k} n_j x_j'^r.\tag{8.70}$$

The central rth sample moment is defined as

$$m_r = \frac{1}{n}\sum_{i=1}^{n} (x_i - \bar{x})^r, \qquad r = 1, 2, \ldots,\tag{8.71}$$

or, if X is discrete,

$$m_r = \frac{1}{n}\sum_{j=1}^{k} n_j(x_j' - \bar{x})^r.\tag{8.72}$$

Example 8.8.   Suppose that, as a result of crossings of $Aa \times Aa$, 200 individuals are obtained with the following frequencies of genotypes:

x_j':	AA (2)	Aa (1)	aa (0)
n_j:	55	98	47
n_j/n:	0.2750	0.4900	0.2350 .

The last line represents the empirical distribution of the genotypes for this particular sample. From (8.70) and (8.72) we obtain $m_1' = \bar{x} = 1.03$ and $m_2 = 0.5084$, respectively.

Let $(x_1, y_1), (x_2, y_2), \ldots, (x_n, y_n)$ be a two-dimensional sample. This represents n pairs of observations of two random variables (X, Y), each pair observed on the same individual. The *central first product moment of the sample* is

$$m_{11} = \frac{1}{n} \sum_{i=1}^{n} (x_i - \bar{x})(y_i - \bar{y}). \tag{8.73}$$

8.15. SAMPLING DISTRIBUTIONS

Let $x_1, x_2, \ldots, x_n$ be a random sample. Before the sample was observed, this set represents n independent and identically distributed random variables. From now on we shall use, on several occasions when it is convenient, small letters to denote random variables as well as observed values of actually obtained samples. We are now sufficiently acquainted with concepts of random variables and their observed values that we shall be able to recognize what role they play on any particular occasion.

DEFINITION 8.6.   *Any function, $g(x_1, x_2, \ldots, x_n)$ of a random sample is called a statistic.*

A statistic is, of course, also a random variable and therefore has all the properties of random variables. For instance, sample mean, $\bar{x} = n^{-1} \sum x_i$, is a statistic. Sample moments, defined in preceding sections, are also statistics.

DEFINITION 8.7.   *The distribution of a statistic is called its sampling distribution.*

For instance, we can imagine all possible samples of size n from a certain population. The sample mean, $\bar{x}$, can take values in different intervals with different probabilities. These define the sampling distribution of $\bar{x}$. We emphasize the difference between the *distribution of the sample* and the *sampling distribution*. The former is the empirical distribution or the distribution of actually observed values in a particular sample, defined in (8.68), whereas the latter is a probability distribution of a statistic which itself is a random variable.

8.16. SOME EXACT SAMPLING DISTRIBUTIONS FROM NORMAL POPULATION

In section 8.5 we discussed some distributions of functions of independent normal variables. We can now see that these distributions can be used as sampling distributions when the sample is drawn from a normal population.

8.16.1. Distribution of the Sample Mean

Let

$$\bar{x} = \frac{1}{n}(x_1 + x_2 + \cdots + x_n) \tag{8.74}$$

be a sample mean, where $x_i \frown N(\mu, \sigma^2)$, $i = 1, 2, \ldots, n$. Applying Theorem 8.1 (see section 8.4), we find that the sum $(x_1 + x_2 + \cdots + x_n)$ is also normally distributed, with mean $n\mu$ and variance $n\sigma^2$. Thus,

$$E(\bar{x}) = \frac{1}{n} n\mu = \mu \quad \text{and} \quad \text{Var}(\bar{x}) = \frac{1}{n^2} n\sigma^2 = \frac{\sigma^2}{n}.$$

We have obtained the following theorem.

THEOREM 8.2. *If a random variable x is normally distributed with mean μ and variance σ^2, then the sample mean, $\bar{x}$, is also normally distributed with the same mean μ and variance σ^2/n.*

In particular, the variable $z = (\bar{x} - \mu)/(\sigma/\sqrt{n})$ is distributed as $N(0, 1)$. Thus, the probability $\Pr\{\bar{x} \leqslant \bar{x}_y\}$ or $\Pr\{\bar{x}_{y_1} < \bar{x} < \bar{x}_{y_2}\}$, etc., can be evaluated by using tables of the standard normal distribution.

8.16.2. Distribution of the Sample Variance

The statistic

$$s^2 = \frac{1}{n-1} \sum_{i=1}^{n} (x_i - \bar{x})^2 \tag{8.75}$$

is called the sample *variance*. It is not precisely the second central sample moment, defined in (8.71), since it is divided by $n - 1$ instead of by n. The reason will be explained in Chapter 11.

The number $n - 1$ is called the *degrees of freedom* (d.f.). This term will be used in connection with other functions (statistics), and we shall give a rule for calculating the d.f. In (8.75) the n variables, x_i, $i = 1, 2, \ldots, n$, are not functionally independent, since they satisfy one relationship, $\bar{x} = n^{-1} \sum x_i$; only $n - 1$ of them are functionally independent. In most situations, the number of *degrees of freedom*, ν, is equal to the number of variables reduced by the number of constraints among these variables. It can be shown [for a simple proof see, for instance, Nemenyi (1969)] that the statistic

$$\chi^2 = \frac{(n-1)s^2}{\sigma^2} = \frac{\sum\limits_{i=1}^{n}(x_i - \bar{x})^2}{\sigma^2} \tag{8.76}$$

is distributed as χ^2 with $v = n - 1$ d.f.

8.16.3. Distribution of the Ratio $(\bar{x} - \mu)/(s/\sqrt{n})$

We notice that $z = (\bar{x} - \mu)/(\sigma/\sqrt{n})$ is distributed as $N(0, 1)$ and s/σ is distributed as $\chi/\sqrt{n-1}$. It can be proved that $\bar{x}$ and s^2 are independent [see, for instance, Cramér (1946), Chapter 29]. Thus, using (8.22), we find that the statistic

$$t = \frac{\bar{x} - \mu}{s/\sqrt{n}} \tag{8.77}$$

has the t-distribution with $v = n - 1$ d.f.

8.16.4. Distribution of the Ratio of Two Independent Sample Variances

Let s_1^2 and s_2^2 be two independent sample variances obtained from two random samples of sizes n_1 and n_2, respectively, both drawn from normal populations with the same variance, σ^2. Applying (8.23), we find that the ratio

$$F = \frac{s_1^2}{s_2^2} \tag{8.78}$$

has the F-distribution with $v_1 = n_1 - 1$ and $v_2 = n_2 - 1$ d.f.; respectively.

8.17. LARGE-SAMPLE THEORY. THE CENTRAL LIMIT THEOREM

As we shall see in Part II, several statistics are used in estimating and testing hypotheses about population parameters. To construct a test, knowledge of the distribution of a statistic used as the test criterion is required. It is not always easy to find the exact distribution of a statistic, but very often some approximate distribution can be used, especially when the sample is sufficiently large. Approximation by normal distribution is the one commonly used. A theorem that gives conditions under which the normal approximation can be used is known as the *central limit theorem*.

There are quite a number of forms of the central limit theorem, but we give (without proof) the most common one.

CENTRAL LIMIT THEOREM. *Let* $X_1, X_2, \ldots, X_n$ *be n random variables which are mutually independent and arbitrarily but identically distributed with mean μ and finite, nonzero variance σ^2. Let*

$$Y_n = X_1 + X_2 + \cdots + X_n \tag{8.79}$$

be the sum of these variables, and

$$\bar{X}_n = \frac{1}{n} Y_n = \frac{1}{n} \sum_{i=1}^{n} X_i \tag{8.80}$$

be their mean. It can be proved that, as $n \to \infty$, the standard random variable

$$Z_n = \frac{Y_n - n\mu}{\sigma\sqrt{n}} = \frac{\bar{X}_n - \mu}{\sigma/\sqrt{n}} \tag{8.81}$$

is approximately normally distributed with mean 0 and variance 1.
[Cramér (1946), Chapter 17].

The central limit theorem has very useful practical aspects: if $\bar{X}$ is a sample mean, we can evaluate the probabilities $\Pr\{a < \bar{X} \leqslant b\}$ approximately by using tables of standard normal distribution, whatever the distribution of X is, provided σ is known. We have

$$\Pr\{a < \bar{X} \leqslant b\} = \Pr\{a - \mu < \bar{X} - \mu \leqslant b - \mu\}$$

$$= \Pr\left\{ \frac{a - \mu}{\sigma/\sqrt{n}} < \frac{\bar{X} - \mu}{\sigma/\sqrt{n}} \leqslant \frac{b - \mu}{\sigma/\sqrt{n}} \right\}$$

$$\doteq \Phi\left(\frac{b - \mu}{\sigma/\sqrt{n}} \right) - \Phi\left(\frac{a - \mu}{\sigma/\sqrt{n}} \right). \tag{8.82}$$

We notice that, if the sample is from the normal population, (8.82) holds exactly (compare Theorem 8.2).

We shall see in Chapter 11 that the distributions of several other statistics can be approximated by the normal distributions, assuming that n is sufficiently large. A rather vague point in the application of the central limit theorem is what is meant by "sufficiently large." In some cases a fairly good approximation is obtained when $n = 10$, whereas for others $n = 100$ is not sufficient. We shall demonstrate in several examples how this theory can be used, and it is hoped that the reader will get some feeling as to when he can apply the central limit theorem.

8.18.   NORMAL APPROXIMATION TO THE BINOMIAL DISTRIBUTION

We have already seen that if, in a binomial distribution, $n \to \infty$ and $P \to 0$, so that $\lambda = nP$ remains constant, the binomial distribution converges to the Poisson distribution. The binomial distribution can also be approximated by the normal.

Let X be a binomial variable defined as $X = \sum U_i$, where the U_i's are independent and identically distributed variates with $E(U_i) = P$ and $\mathrm{Var}(U_i) = P(1 - P)$, $i = 1, 2, \ldots, n$. We have $E(X) = \mu_X = nP$ and $\mathrm{Var}(X) = \sigma_X^2 = nP(1 - P)$. Thus, X satisfies the conditions for the central limit theorem, and for large n the variable

$$Z = \frac{X - nP}{\sqrt{nP(1 - P)}} = \frac{(X/n) - P}{\sqrt{P(1 - P)/n}} \tag{8.83}$$

is approximately distributed as $N(0, 1)$. Therefore the percentage points of the binomial cumulative distribution function can be evaluated by using tables of standard normal cdf. We will have

$$\Pr\{X \leqslant r\} \doteq \Pr\{Z \leqslant z\} = \Pr\left\{Z \leqslant \frac{r - nP}{\sqrt{nP(1 - P)}}\right\}$$

$$= \Phi\left\{\frac{r - nP}{\sqrt{nP(1 - P)}}\right\} = \Phi\left\{\frac{(r/n) - P}{\sqrt{P(1 - P)/n}}\right\}. \tag{8.84}$$

Noting that a discrete (binomial) distribution is approximated by a continuous (normal) distribution, it has been found [for detail, see Brownlee (1960), Chapter 4] that a better approximation is obtained by calculating

$$\Pr\{X \leqslant r\} \doteq \Phi\left[\frac{(r + \frac{1}{2}) - nP}{\sqrt{nP(1 - P)}}\right] = \Phi\left[\frac{(r + \frac{1}{2})/n - P}{\sqrt{P(1 - P)/n}}\right], \tag{8.85a}$$

$$\Pr\{X < r\} \doteq \Phi\left[\frac{(r - \frac{1}{2}) - nP}{\sqrt{nP(1 - P)}}\right] = \Phi\left[\frac{(r - \frac{1}{2})/n - P}{\sqrt{P(1 - P)/n}}\right], \tag{8.85b}$$

$$\Pr\{X \geqslant r\} = 1 - \Pr\{X < r\}, \tag{8.85c}$$

$$\Pr\{r_1 \leqslant X \leqslant r_2\} \doteq \Phi\left[\frac{(r_2 + \frac{1}{2}) - nP}{\sqrt{nP(1 - P)}}\right] - \Phi\left[\frac{(r_1 - \frac{1}{2}) - nP}{\sqrt{nP(1 - P)}}\right], \tag{8.85d}$$

$$\Pr\{r_1 < X \leqslant r_2\} \doteq \Phi\left[\frac{(r_2 + \frac{1}{2}) - nP}{\sqrt{nP(1 - P)}}\right] - \Phi\left[\frac{(r_1 + \frac{1}{2}) - nP}{\sqrt{nP(1 - P)}}\right], \tag{8.85e}$$

etc. The term $\pm 1/(2n)$ is called a "correction for continuity."

Example 8.9. We now use the normal approximation in evaluating $\Pr\{X \geqslant 2\}$ in Example 8.6. We have $P = 0.00005$, $n = 10,000$, $nP = 0.5$ and $nP(1 - P) = 0.499975$, so $\sqrt{nP(1 - P)} \doteq 0.7071$. We calculate $z = [(r - \frac{1}{2}) - nP]/\sqrt{nP(1 - P)} = [(2 - \frac{1}{2}) - 0.5]/0.7071 = 1.41$. Then

$$\Pr\{X \geqslant 2\} = 1 - \Pr\{X < 2\} \doteq 1 - \Phi(1.41) = 1 - 0.9207 = 0.0793.$$

If we had not used the continuity correction, we would have found $z = (2 - 0.5)/0.7071 = 2.12$, and $\Pr\{X \geqslant 2\} = 0.0170$. We recall that the probability obtained by the Poisson approximation is 0.0902, so the continuity correction improves the approximation.

Normal approximation can be recommended even for n as small as 15, provided that P is close to $\frac{1}{2}$. When P is very small and n is large, the Poisson approximation is more appropriate. For most practical purposes n may be assumed to be "sufficiently large" to use a normal approximation if $nP(1 - P)$ is at least 3.

8.19. THE APPROXIMATE χ^2-DISTRIBUTION FOR A FUNCTION OF MULTINOMIAL VARIABLES

Let $r_1, r_2, \ldots, r_k$, with $\sum r_l = n$, be multinomial variables. (Note that now r's are used for random variables as well as for observed class frequencies.) Let $P_1, P_2, \ldots, P_k$, with $\sum P_l = 1$, be the corresponding multinomial parameters. It has been shown by K. Pearson that the statistic

$$X^2 = \sum_{l=1}^{k} \frac{(r_l - nP_l)^2}{nP_l} \tag{8.86}$$

is approximately distributed as χ^2 with $k - 1$ d.f. (Note that the variables $r_1, r_2, \ldots, r_k$ are subject to one constraint, $\sum r_l = n$.) The statistic in (8.86) is well known as "Pearson's chi-square," and its applications will be widely discussed in Chapter 13.

In particular, when $k = 2$, we have a binomial distribution. If P is a binomial parameter, and r is the number of successes and $n - r$ the number of failures, then (8.86) takes the form

$$X^2 = \frac{(r - nP)^2}{nP} + \frac{[(n - r) - n(1 - P)]^2}{n(1 - P)}, \tag{8.87}$$

and the statistic X^2 is approximately distributed as χ^2 with 1 d.f. (Note that now r represents a binomial variable, denoted in previous sections by X, as well as the observed number of successes.)

On the other hand, using the approximately normal variable, defined in

(8.83) (with X now replaced by r), we find

$$Z^2 = \frac{(r - nP)^2}{nP(1 - P)} \tag{8.88}$$

is approximately distributed as χ^2 with 1 d.f. It can be shown that (8.87) and (8.88) are identical (see Problem 8.1).

8.20. APPROXIMATE MEAN AND VARIANCE OF A FUNCTION OF RANDOM VARIABLES

Let X be a random variable (discrete or continuous) with the values of $E(X) = \mu_X$ and $\text{Var}(X) = \sigma_X^2 > 0$, both finite and known.

It is convenient, in some situations, to deal with a certain function of X, $Y = g(X)$, say, rather than with X itself. On the other hand, the function $g(X)$ may be quite a complicated function of X, and the evaluation of $E[g(X)]$ and $\text{Var}[g(X)]$ by straightforward methods such as those discussed in Chapter 6 may then be rather tedious. Approximate formulae obtained by a method called *statistical differentials* are sometimes useful.

Let $Y = g(X)$ be a strictly monotonic function of X, with first derivative*

$$\frac{dg}{dX}\bigg|_{X=\mu_X} = g'(\mu_X) \neq 0.$$

The function $g(X)$ can be expanded by Taylor series about $X = \mu_X$ (Appendix II) as

$$g(X) = g(\mu_X) + (X - \mu_X)\frac{dg}{dX}\bigg|_{X=\mu_X} + \cdots. \tag{8.89}$$

It may happen that the terms including higher-order derivatives are negligible. This is quite often the case when X itself is a certain function of a random sample, and the sample size, n, is large (see, for instance, Example 8.10). In such circumstances we can write (8.89) as:

$$g(X) \doteq g(\mu_X) + (X - \mu_X)\frac{dg}{dX}\bigg|_{X=\mu_X}. \tag{8.90}$$

Taking expected values of both sides of (8.90), we obtain

$$E[g(X)] \doteq g(\mu_X). \tag{8.91}$$

---

* In the following expresion $\dfrac{dg}{dX}\bigg|_{X=\mu_X}$ means that the first derivative of $g(X)$ with respect to X has been evaluated at the point $X = \mu_X$.

Subtracting (8.91) from (8.90), we obtain

$$g(X) - E[g(X)] = (X - \mu_X) \left. \frac{dg}{dX} \right|_{X=\mu_X}. \tag{8.92}$$

Squaring both sides of (8.92) and again taking expected values, we obtain an approximate formula for the variance of the function of $g(X)$ as

$$\text{Var}[g(X)] \doteq \left(\frac{dg}{dX} \right)^2 \bigg|_{X=\mu_X} \sigma_X^{\,2}. \tag{8.93}$$

Example 8.10. Let p and q be the gene frequencies of A and a genes, respectively, in a very large ("infinite") population under equilibrium. Suppose that a simple random sample of size n has been selected and the number of *recessives* counted. Let r denote the number of recessives. (Here r will be treated both as a random variable from a random sample and as a certain value in an observed sample.) Let $y = r/n$ denote the proportion of recessives in a random sample.

The expected proportion of recessives under equilibrium is

$$E(y) = E(r/n) = q^2,$$

and the variance of this proportion is

$$\text{Var}(y) = \frac{q^2(1 - q^2)}{n}$$

(note that r/n is a binomial proportion).

Let $\hat{q}$ denote the frequency of a recessive gene a in the sample; $\hat{q}$ is a random variable. We have $y = \hat{q}^2$, or

$$\hat{q} = \sqrt{y} = g(y).$$

We now wish to find the expected value, $E(\hat{q})$, and the variance, $\text{Var}(\hat{q})$. For $n \to \infty$, we will have, approximately,

$$E(\hat{q}) \doteq q; \quad \text{Var}(\hat{q}) \doteq \left(\frac{d\hat{q}}{dy} \right)^2 \bigg|_{y=q^2} \text{Var}(y) \doteq \left(\frac{1}{2\sqrt{y}} \right)^2 \bigg|_{y=q^2} \text{Var}(y)$$

$$= \frac{1}{4q^2} \frac{q^2(1 - q^2)}{n} = \frac{1 - q^2}{4n}.$$

The approximate formulae for mean and variance given in (8.91) and (8.93), respectively, can be extended to monotonic functions of k variables, $g(X_1, X_2, \ldots, X_k)$, say. In this case, using Taylor's expansion and assuming that at least the first derivatives be in the neighborhood of $(\mu_1, \mu_2, \ldots, \mu_k)$ and that the sum of terms with higher than first-order partial derivatives is negligible, we obtain

$$E[g(X_1, X_2, \ldots, X_k)] \doteq g(\mu_1, \mu_2, \ldots, \mu_k) \tag{8.94}$$

and

$$\text{Var}[g(X_1, X_2, \ldots, X_k)] \doteq \sum_{i=1}^{k} \left(\frac{\partial g}{\partial X_i}\right)^2 \bigg|_{\{X\}=\{\mu\}} \sigma_i^2 + 2 \sum_{i<j} \sum \left(\frac{\partial g}{\partial X_i} \frac{\partial g}{\partial X_j}\right) \bigg|_{\{X\}=\{\mu\}} \sigma_{ij},$$

(8.95)

where $E(X_i) = \mu_i$, $\text{Var}(X_i) = \sigma_i^2$, $\text{Cov}(X_i, X_j) = \sigma_{ij}$, with $i, j = 1, 2, \ldots, k$ ($i \neq j$), and $\{X\} = \{\mu\}$ means that in the set of variables $(X_1, X_2, \ldots, X_k)$ the variable X_i takes the value μ_i, with $i = 1, 2, \ldots, k$. We shall sometimes use these approximate formulae in problems of estimating genetic parameters and testing genetic hypotheses when rather complicated functions of random samples are employed.

SUMMARY

Several common continuous and discrete distributions, very useful in applied statistics, have been introduced and their statistical properties discussed. Among these, the normal, chi-square, binomial, and multinomial distributions are of the greatest importance for mathematical geneticists. The concepts of random samples and certain functions of random samples, called statistics, have been defined. The distributions of statistics, called sampling distributions, have been introduced. The central limit theorem which states a very important result—that simple sample means are approximately normally distributed as the sample size, n, becomes large—has been stated. Some useful approximate formulae for calculating expected values and variances, using the method of "statistical differentials," have been given. This chapter is concerned mostly with results obtained in the theory of probability and statistics; applications will be demonstrated in later chapters in which problems of estimating genetic parameters and fitting genetic models to the data are discussed.

REFERENCES

Brownlee, K. A., *Statistical Theory and Methodology in Science and Engineering*, John Wiley & Sons, New York, 1960.

Cramér, H., *Mathematical Methods of Statistics*, Princeton University Press, Princeton, N.J., 1946.

Cotterman, C. W., and Snyder, L. H., Tests of simple Mendelian inheritance in randomly collected data of one and two generations, *J. Amer. Statist. Assoc.* **34** (1939), 513–523.

Johnson, N. L., and Leone, F. C., *Statistics and Experimental Designs*, Vol. I, John Wiley & Sons, New York, 1964.

Kendall, M. G., and Stuart, A., *The Advanced Theory of Statistics*, Vol. I, Hafner Publishing Co. New York, 1961.

Lindgren, B. W., *Statistical Theory*, The Macmillan Company, New York, 1968.

Mills, J., Distribution of birth weights of Chinese and Indian infants born in Singapore: birth weight as an index of maturity, *Amer. J. Hum. Genet.* **23** (1959), 164–174.

Morton, N. E., Models and evidence in human population genetics. In: *Genetics Today* Vol. 3 (Proceedings of the XI International Congress on Genetics, The Hague, The Netherlands, 1963), ed. by S. J. Geerts, Pergamon Press, Oxford, pp. 935–951.

Nemenyi, P., Variances: an elementary proof and a nearly distribution-free test, *Amer. Statistician* **23**, No. 5 (1969), 35–37.

Snyder, L. H., Studies in human inheritance, IX. The inheritance of taste deficiency, *Ohio J. Sci.* **32** (1932), 12–24.

Stern, C., *Human Genetics*, W. H. Freeman & Co., San Francisco, 1960.

PROBLEMS

8.1. Show that

$$\frac{(n_1 - n\theta_1)^2}{n\theta_1\theta_2} = \frac{(n_1 - n\theta_1)^2}{n\theta_1} + \frac{(n_2 - n\theta_2)^2}{n\theta_2},$$

where $\theta_1 + \theta_2 = 1$ and $n_1 + n_2 = n$.

8.2. Suppose that, in a finite population of size N, N_1 individuals possess a property C and the remaining $N_2 = N - N_1$ individuals do not possess property C. Suppose also that a random sample of size n is selected from this population and that the number, X, of individuals with property C is observed.

(*a*) Show that the probabilities that exactly r individuals with property C will be observed in the sample of size n are:

(i) For sampling with replacement:

$$\Pr\{X = r\} = \binom{n}{r}\left(\frac{N_1}{N}\right)^r \left(\frac{N - N_1}{N}\right)^{n-r}, \qquad r = 0, 1, \ldots, n$$

(the term of binomial distribution) with

$$E(X) = n\frac{N_1}{N} \quad \text{and} \quad \text{Var}(X) = n\frac{N_1(N - N_1)}{N^2}.$$

(ii) For sampling without replacement:

$$\Pr\{X = r\} = \frac{\binom{N_1}{r}\binom{N - N_1}{n - r}}{\binom{N}{n}}.$$

This last probability is a term in a so-called *hypergeometric distribution*.

(*b*) Show that for hypergeometric distribution

$$E(X) = n\frac{N_1}{N} \quad \text{and} \quad \text{Var}(X) = n\frac{N_1(N - N_1)}{N^2} \cdot \frac{N - n}{N - 1}.$$

8.3. Suppose that a population of size $N = 500$ consists of 300 dominants and 200 recessives. Using the results of Problem 8.2, calculate, for sampling with and without replacement and sample size $n = 10$, the following probabilities:

(a) Exactly 2 individuals will be recessives.

(b) At least 2 individuals will be recessives.

(c) The number of recessives, X, will be between 1 and 3 inclusive ($1 \leqslant X \leqslant 3$).

(d) Calculate the expected numbers of recessives and the variances for both sampling methods.

8.4. Let us consider a single locus with two codominant alleles, A_1, A_2, and families of the parental type $A_1A_1 \times A_1A_2$. The probability that a child born into such a family will be A_1A_1 is $\frac{1}{2}$. The probability that 2 children born into such a family will both be A_1A_1 is $\frac{1}{2} \times \frac{1}{2} = \frac{1}{4}$, since these events are independent.

Let $F_{r;s}$ denote a "family pattern" such that, in a family with s children, r children are of type A_1A_1. Thus, $F_{0;4}$ means that in a family with 4 children there are no A_1A_1 children; $F_{2;4}$, that there are 2 A_1A_1 children, and so on. Let X be the number of A_1A_1 children in a family of size s. Thus the expected proportion of families of pattern $F_{2;4}$ in a large population of families, each with 4 children is

$$\Pr\{F_{2;4}\} = \Pr\{X = 2\} = \binom{4}{2}(\tfrac{1}{2})^4 = 6 \times (\tfrac{1}{2})^4.$$

Suppose that we select at random a pair of children. There are $\binom{4}{2} = 6$ possible pairs, but among them only 1 pair is (A_1A_1, A_1A_1). Thus, the probability that we select this pair is

$$\Pr\{(A_1A_1, A_1A_1) \,|\, F_{2;4}\} = \tfrac{1}{6}.$$

The probability of selecting a family of pattern $F_{2;4}$ from families with 4 children, *and* the sib pair (A_1A_1, A_1A_1), is

$$\Pr\{(A_1A_1, A_1A_1) \text{ and } F_{2;4}\} = \Pr\{(A_1A_1, A_1A_1) \,|\, F_{2;4}\} \times \Pr(F_{2;4})$$

$$= \tfrac{1}{6} \times 6 \times (\tfrac{1}{2})^4 = (\tfrac{1}{2})^4.$$

(a) Show that

$$\Pr\{(A_1A_1, A_1A_1) \text{ and } F_{3;4}\} = 2 \times (\tfrac{1}{2})^4,$$

$$\Pr\{(A_1A_1, A_1A_1) \text{ and } F_{4;4}\} = (\tfrac{1}{2})^4.$$

(b) Show that the average probability of selecting a pair (A_1A_1, A_1A_1) is then

$$(\tfrac{1}{2})^4 + 2(\tfrac{1}{2})^4 + (\tfrac{1}{2})^4 = \tfrac{1}{4},$$

which is the same as for the whole population.

8.5. Show that the property described in Problem 8.4b does not depend on s. (*Hint:* Prove it by using methods similar to those described in Problem 8.4, but with s unspecified.)

8.6. It has been suggested that left-handedness is a single recessive trait [Ramaley, *Amer. Naturalist* **47** (1913), 730]. Let R denote the dominant gene (for right-handedness) and r the recessive gene (for left-handedness). Let q be the frequency of gene r.

Suppose that in a certain population the prevalence of left-handed people is 8% (i.e., $q^2 = 0.08$), and that a sample of 305 married couples (parents) have been selected.

(a) Find the expected numbers of families in which both parents are right-handed, one right- and the other left-handed, both left-handed.

(b) Suppose we observe 20 families such that (i) both parents are right-handed, and (ii) each family has 5 children among whom at least one is left-handed (i.e., segregating families Dom × Dom, each of size s). Find the expected numbers of families in which exactly 1, 2, 3, 4, 5 children are left-handed.

▼**8.8.** Suppose a volume V of fluid contains a large number, N, of small organisms (e.g., bacteria) uniformly distributed in the fluid. Suppose also that we examine a drop of volume D under the microscope. We assume that V is very large in comparison to D.

Show that the number of organisms, X, in this drop will have the Poisson distribution with $\lambda = N(D/V)$. [*Hint:* Assuming "random" distribution of organisms in the fluid, we can take $P_N = D/V$. Thus $X \frown b(x; N, P_N)$. Since P_N is small, and we assume $P_N N = \lambda$ is constant, we can apply the result given in (8.62).]

8.9. Let $Z = X/Y$ be a ratio of two random variables with means μ_X, μ_Y, variances σ_X^2, σ_Y^2, respectively, and covariance σ_{XY}. Using approximate formulae (8.94) and (8.95), show that

$$E(Z) = \mu_Z \doteq \frac{\mu_X}{\mu_Y}$$

and

$$\mathrm{Var}(Z) = \sigma_Z^2 \doteq \frac{\mu_X^2}{\mu_Y^2}\left(\frac{\sigma_Y^2}{\mu_X^2} + \frac{\sigma_Y^2}{\mu_Y^2} - \frac{2\sigma_{XY}}{\mu_X\mu_Y}\right).$$

8.7. Show that in Example 8.4 the *a posteriori* probability of matings $TT \times tt$ and $Tt \times tt$, given that a child observed in a family taster × nontaster was a taster, are p and q, respectively.

CHAPTER 9

Inbreeding and Nonrandom Mating

9.1. INBREEDING. IDENTITIES OF GENES

9.1.1. Concepts of Inbreeding

"Inbreeding" is generally used to indicate a certain form of mating which leads to an increase of homozygosity and a decrease of heterozygosity in a population. Commonly the term is used for matings between consanguineous partners. An offspring of two related parents has some probability of carrying genes which have their origin in a gene present in the common ancestor of both parents. Such individual will be *inbred*. This kind of mating is a certain form of *nonrandom* mating and in an "infinite" population will result in deviations from Hardy-Weinberg equilibrium proportions.

However, if a population is finite, inbreeding can take place even if the population mates *at random* (compare section 9.12). Similarly, if a population is divided into heterogeneous (with different gene frequencies) sub-populations, with random mating within each group, we observe a decrease of heterozygosity in comparison to the whole randomly mating population (section 9.13). [For discussion, see also Allen (1965).]

Inbreeding has been developed, in large part, to improve several characters in plants and animals. We do not attempt to present here all the theoretical and practical results which have been achieved in this field. We confine ourselves only to the basic definitions and results, which can be applied mostly to qualitative traits.

9.1.2. Identity of Genes by Descent and in State

Let us first consider two homologous chromosomes, 1 and 2, say, and a single locus A on these chromosomes. Suppose that we observe a situation such as that presented in Fig. 9.1(a).* We denote the genes at locus A, in individual Z, by the symbol $\bigcirc$ in chromosome 1, and by the symbol $\otimes$ in its

* A large circle denotes a female, and a large square a male.

196

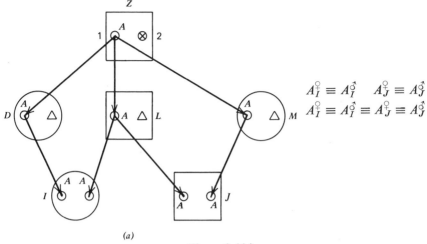

$$A_I^{\,\circ} \equiv A_I^{\,\circ} \qquad A_J^{\,\circ} \equiv A_J^{\,\circ}$$
$$A_I^{\,\circ} \equiv A_I^{\,\circ} \equiv A_J^{\,\circ} \equiv A_J^{\,\circ}$$

(a)

Figure 9.1(a)

homologue. Let, for example, A be the gene assigned the symbol ○ in Z. Suppose that Z is a sire which has been mated to several dams and that some of its offspring (D, L, M, say) received gene A from Z. (Symbol △ denotes any gene from another parent, not traced in the pedigree.) Individual I, an offspring of D and L, received both genes which are exact copies of gene A derived from chromosome 1 of Z, which is called an *ancestor* of I. Individual I received one gene A through its mother, D—we will call this gene the *maternal gene A* and denote it by $A^{\,\circ}$; another gene A came through I's father, L—this is the *paternal gene A*, and we will denote it by $A^{\,\circ}$. Similarly, J, an offspring of L and M, also has both genes derived from the same chromosome of the ancestor Z. Here Z is a *common ancestor* to I and J.

DEFINITION 9.1.   *Genes which are copies of the same gene present in the ancestor or, in other words, are derived from the same location on a single chromosome of an ancestor, are called identical by descent.*

We will use the symbol $\equiv$ to denote identity by descent. Thus, the genes A in individual I are identical by descent, and we write $A_I^{\,\circ} \equiv A_I^{\,\circ}$. Similarly, $A_J^{\,\circ} \equiv A_J^{\,\circ}$. It should be emphasized that identity by descent is defined in terms of *the same ancestral chromosome.**

Figure 9.1(b) represents a situation in which the ancestor Z possesses in

---

* Malécot (1966) uses the term "locus" to determine the location of a gene in a single chromosome. Thus, in a cell of an individual two locations lying exactly side by side in two homologous chromosomes are called by Malécot two "homologous loci." If there are two individuals, they have, in Malécot's terminology, $2 \times 2 = 4$ "homologous loci." In his terminology, genes are identical by descent if they are derived from the same "locus," (i.e., location in a single chromosome) of an ancestor.

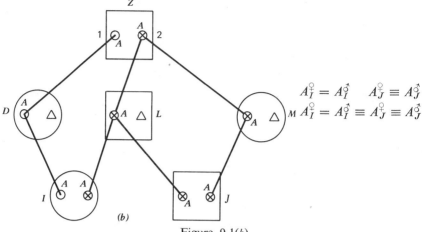

$$A_I^{\female} = A_I^{\male} \qquad A_J^{\female} \equiv A_J^{\male}$$
$$_M\, A_I^{\female} = A_I^{\male} \equiv A_J^{\female} \equiv A_J^{\male}$$

Figure 9.1(b)

both chromosomes the same gene A. Individual J received both genes A from the same ancestral chromosome, 2, so these genes are identical by descent, that is, $A_J^{\female} \equiv A_J^{\male}$.

On the other hand, the genes A present in I, even though they were derived from one ancestor and are alike in their function, since both are A, are not copies of the ancestral gene A located in the same chromosome. Hence they are not identical by descent.

In Fig. 9.1(c), individual I received both genes A from the same chromosome

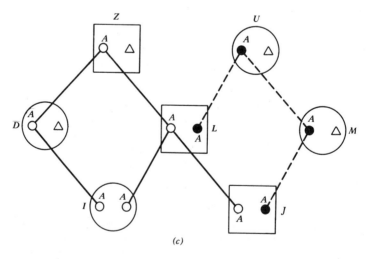

Figure 9.1(c)

$$A_I^{\female} \equiv A_I^{\male} \qquad A_J^{\female} = A_J^{\male}$$
$$A_I^{\female} \equiv A_I^{\male} \equiv A_J^{\male} = A_J^{\female}$$

of the ancestor Z, so these are identical by descent. On the other hand, individual J also possesses two genes A, but each was derived from a different ancestor (the gene A from Z is denoted by $\bigcirc$, and that from U by $\bullet$).

DEFINITION 9.2.   *Genes which are the same in function but are not derived from the same ancestral chromosome are called alike (or identical) in state.*

We will use the symbol $=$ to denote identity in state. Thus, in Fig. 9.1(*b*) $A_I^{\female} = A_I^{\male}$, and similarly in Fig. 9.1(*c*) $A_J^{\female} = A_J^{\male}$—these genes are identical in state.

Let us now consider *two* individuals and their genetic patterns, which depend on their origin. They can be unrelated or related because they may not have, or may have, some ancestor(s) in common. Even if the two individuals have all four genes the same, they can be in a different *situation of identity*.

Figure 9.1(*a*) represents the situation in which all four genes are identical by descent, that is, $A_I^{\female} \equiv A_I^{\male} \equiv A_J^{\female} \equiv A_J^{\male}$. In Fig. 9.1(*c*) the two genes in I and the paternal gene in J are identical by descent, but the maternal gene in J is identical in state with the previous ones, that is, $A_I^{\female} \equiv A_I^{\male} \equiv A_J^{\male} = A_J^{\female}$.

Figure 9.1(*d*) represents the situation. $A_I^{\female} \equiv A_J^{\male}$ $A_I^{\male} \equiv A_J^{\female}$. We also have $A_L^{\male} \equiv A_I^{\female}$ $A_L^{\female} \equiv A_I^{\male}$, etc.

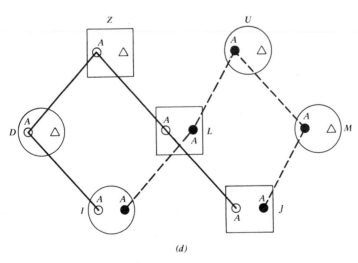

(*d*)

Figure 9.1(*d*)

$$A_I^{\female} = A_I^{\male} \qquad A_J^{\female} = A_J^{\male}$$
$$A_I^{\female} \equiv A_J^{\male} \qquad A_I^{\male} \equiv A_J^{\female}$$
$$A_L^{\male} \equiv A_I^{\female} \qquad A_L^{\female} \equiv A_I^{\male}$$
$$A_L^{\male} \equiv A_L^{\female} \qquad A_L^{\female} \equiv A_J^{\female}$$

As has been shown by Gillois (1966), fifteen different identity situations for two individuals, I and J, can be recognized. The probability of realization of a given situation of identity is called a *coefficient of identity*. It depends on the relationship of I and J and on the number of stages in reproduction from common ancestor(s). For example, in Fig. 9.1(c) there are two stages from Z to I or to J, but only one stage from Z to L.

9.2. COEFFICIENT OF INBREEDING OF AN INDIVIDUAL. MALÉCOT'S DEFINITION

When a certain form of inbreeding is carried on through several generations, it is useful to introduce a certain measure (or index) which will inform us about the "degree" of inbreeding to be expected in a given individual. This measure will be called a *coefficient of inbreeding of an individual*. Before we introduce a definition, we make the additional assumption that mutation and selection during the reproductive stages are very rare and can be neglected.

A few authors have defined this index, using different approaches. We discuss here only two of these definitions, due to Malécot (1948) and Wright (1922), respectively. As we shall see later, they lead, in certain situations, to the same results. We give first the definition due to Malécot, since it seems to be easy to understand and has a general meaning and convenient form in terms of probability.

DEFINITION 9.3. (MALÉCOT'S DEFINITION). *The coefficient of inbreeding of an individual I, F_I, say, is the probability that the two genes present in this individual at a given locus are identical by descent.*

Since we confine ourselves to a single autosomal locus, the words "single autosomal" will be omitted unless ambiguity arises.

Let $\alpha\beta$ denote a genotype of I, where α stands for any gene present on one chromosome and β for its allelomorph present on the homologous chromosome, at a given locus.

Definition 9.3 can then be expressed in the mathematical form

$$F_I = \Pr\{\alpha \equiv \beta\}. \tag{9.1}$$

9.3. CONSANGUINITY OF TWO INDIVIDUALS. MALÉCOT'S COEFFICIENT OF PARENTAGE

Let us now consider two individuals, I and J, say. Let $\alpha\beta$ be a genotype of I, where α and β stand for any gene in each of two homologous chromosomes in I, and let $\gamma\delta$ be a genotype of J, where γ and δ stand for any gene in each of two homologous chromosomes in J, at the same locus A.

DEFINITION 9.4 (MALÉCOT'S COEFFICIENT OF PARENTAGE). *The probability that a gene selected at random from I is identical by descent with a gene selected at random from J is called the coefficient of parentage between I and J. If we denote this by φ_{IJ}, we shall have*

$$\varphi_{IJ} = \tfrac{1}{4}[\Pr\{\alpha \equiv \gamma\} + \Pr\{\alpha \equiv \delta\} + \Pr\{\beta \equiv \gamma\} + \Pr\{\beta \equiv \delta\}]. \quad (9.2)$$

The coefficient φ_{IJ} measures the degree of relationship between I and J. In fact, an index $\rho_{IJ} = 2\varphi_{IJ}$ (although defined in a different way) was first introduced by Wright (1922) and is called the *coefficient of relationship* (see also section 9.7). Other names for φ_{IJ}, the *coefficient of kinship* or the *coefficient of consanguinity*, are also used.

9.4. RELATIONS BETWEEN F AND φ

We now derive a few useful relations between F and φ which represent basic formulae in developing the theory of inbreeding.

9.4.1. Coefficient φ_{II}

Suppose that I has the coefficient of inbreeding $F_I = \Pr\{\alpha \equiv \beta\}$. The coefficient of parentage of I with itself is by definition

$$\begin{aligned}\varphi_{II} &= \tfrac{1}{4}[\Pr\{\alpha \equiv \alpha\} + \Pr\{\alpha \equiv \beta\} + \Pr\{\beta \equiv \alpha\} + \Pr\{\beta \equiv \beta\}] \\ &= \tfrac{1}{4}(1 + F_I + F_I + 1) = \tfrac{1}{2}(1 + F_I).\end{aligned} \quad (9.3)$$

In particular, when I is not inbred, that is, when $F_I = 0$, we have $\varphi_{II} = \tfrac{1}{2}$.

These results can be interpreted as follows: if an individual I is inbred, with the coefficient of inbreeding F_I, the probability that it will transmit a copy of an ancestral gene is $(1 + F_I)/2$ and is larger than when I is not inbred. In the latter case, we obtain the known result that the probability of transmitting any of two homologous genes is only $\tfrac{1}{2}$.

9.4.2. Relation between φ_{LM} and $F_{(L\times M)}$

Let L and M be father and mother, respectively, of an individual K. It will sometimes be convenient to denote an offspring of L and M by $(L \times M)$ [i.e., $K \equiv (L \times M)$].

Suppose that L has a genotype $\alpha\beta$ and M has a genotype $\gamma\delta$, at the same locus. The coefficient of parentage between L and M, φ_{LM}, by definition, is given by the right-hand side of formula (9.2). An offspring of L and M can have a genotype $\alpha\gamma$ or $\alpha\delta$ or $\beta\gamma$ or $\beta\delta$. Thus the coefficient of inbreeding of an offspring K is

$$F_K = \tfrac{1}{4}[\Pr\{\alpha \equiv \gamma\} + \Pr\{\alpha \equiv \delta\} + \Pr\{\beta \equiv \gamma\} + \Pr\{\beta \equiv \delta\}]. \qquad (9.4)$$

But (9.4) is exactly the same as (9.2), so we obtain the relation

$$F_K = F_{(L \times M)} = \varphi_{LM}, \qquad (9.5)$$

which means:

The inbreeding coefficient of an offspring of mating $L \times M$, $F_{(L \times M)}$, is equal to the coefficient of parentage, φ_{LM}, of its parents.

From formulae (9.3) and (9.5) we immediately obtain the coefficient of parentage of an offspring K with itself, expressed in terms of the coefficient of parentage of its parents, L and M, that is,

$$\varphi_{KK} = \tfrac{1}{2}(1 + F_K) = \tfrac{1}{2}(1 + \varphi_{LM}). \qquad (9.6)$$

Example 9.1. Find the coefficient of parentage of a brother, B, and a sister, S, whose parents are unrelated and not inbred, and the coefficient of inbreeding of each of their offspring, K, at a single autosomal locus.

Let $\alpha\beta$ be the genotype of the father, and $\gamma\delta$ be the genotype of the mother. The probability that a son and a daughter both received the gene α from their father is $(\tfrac{1}{2})^2 = \tfrac{1}{4}$. The same is true for the gene β. Similarly, the probability that both received the gene γ from their mother is $\tfrac{1}{4}$, and the same holds for the gene δ. Hence,

$$\varphi_{BS} = \tfrac{1}{4}(\tfrac{1}{4} + \tfrac{1}{4} + \tfrac{1}{4} + \tfrac{1}{4}) = \tfrac{1}{4}.$$

Thus, from (9.5) we have

$$F_K = F_{(B \times S)} = \varphi_{BS} = \tfrac{1}{4}.$$

From formula (9.6) we find also that the coefficient of parentage of K with itself, φ_{KK}, is

$$\varphi_{KK} = \tfrac{1}{2}(1 + F_K) = \tfrac{1}{2}(1 + \tfrac{1}{4}) = \tfrac{5}{8}.$$

9.4.3. Coefficient of Parentage of an Offspring K of Mating $L \times M$ with an Individual T Which is Related to L or M or to Both L and M

Let L and M be the father and the mother of an individual $K \equiv (L \times M)$, and T be another individual whose coefficient of parentage with L is φ_{TL},

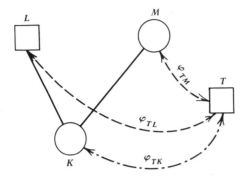

Figure 9.2

and with M is φ_{TM}. We want to find the coefficient of parentage between the offspring K of $L \times M$ mating with T, that is, φ_{TK} (Fig. 9.2).

The probability that the father, L, transmits to K a gene identical by descent with a gene of T is $\frac{1}{2}\varphi_{TL}$; similarly, the probability that the mother, M, transmits to K a gene identical by descent with a gene of T is $\frac{1}{2}\varphi_{TM}$. The total probability that K has a gene identical by descent with a gene of T is then

$$\varphi_{TK} = \varphi_{T,(L \times M)} = \tfrac{1}{2}(\varphi_{TL} + \varphi_{TM}). \tag{9.7}$$

We have obtained the following result:

The coefficient of parentage of an offspring of two individuals, L and M, with an individual T is equal to the average of the coefficients of parentage of T with L and of T with M.

Example 9.2. Suppose that L and T are two brothers, M is an (unrelated) wife of L, and K is a daughter of $L \times M$. (In Fig. 9.3, the "dotted" individuals are not included in calculations of φ_{TK}.) In other words, T and K are uncle and niece, respectively.

(a) We wish to find the coefficient of parentage for uncle-niece. We have (from Example 9.1) $\varphi_{TL} = \frac{1}{4}$ and, since T and M are unrelated, $\varphi_{TM} = 0$. Thus $\varphi_{TK} = \varphi_{T,(L \times M)} = \frac{1}{2}(\frac{1}{4} + 0) = \frac{1}{8}$.

(b) If uncle and niece are married, the coefficient of inbreeding of each of their offspring, according to formula (9.5), will be $\frac{1}{8}$.

(c) Suppose that L and M are brother and sister, respectively (and so T and M). Then, $\varphi_{TL} = \varphi_{TM} = \frac{1}{4}$, and hence

$$\varphi_{TK} = \varphi_{T,(L \times M)} = \tfrac{1}{2}(\tfrac{1}{4} + \tfrac{1}{4}) = \tfrac{1}{4},$$

which is the same as the coefficient of parentage for brother-sister calculated in Example 9.1.

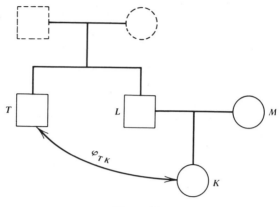

Figure 9.3

Example 9.3. Find the coefficient of inbreeding for each of the offspring of first-cousin marriages. Figure 9.4 represents one of the four possible arrangements for such offspring.

From Example 9.1 the coefficient of parentage between B and S (brother and sister) is $\varphi_{BS} = \frac{1}{4}$. Using formula (9.7), we calculate

$$\varphi_{SL} = \varphi_{S,(B \times D)} = \tfrac{1}{2}(\varphi_{SB} + \varphi_{SD}) = \tfrac{1}{2}(\tfrac{1}{4} + 0) = \tfrac{1}{8}.$$

Again using (9.7), we find

$$\varphi_{LM} = \varphi_{L,(E \times S)} = \tfrac{1}{2}(\varphi_{LE} + \varphi_{LS}) = \tfrac{1}{2}(0 + \tfrac{1}{8}) = \tfrac{1}{16}.$$

From (9.5) we have $F_K = \varphi_{LM} = \frac{1}{16}$.

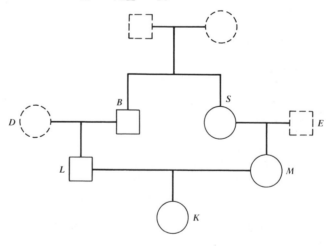

Figure 9.4

Since there are four possible arrangements for first-cousin pairs (list them!), the total coefficient of parentage is $F_K = \frac{1}{4} \times 4 \times \frac{1}{16} = \frac{1}{16}$.

From the general formula (9.7) we obtain some interesting particular results, which will be discussed in sections 9.4.4 and 9.4.5.

9.4.4. Coefficient of Parentage between Parent and Offspring

Let L (or M) be the same individual as T. Then substituting L for T in (9.7) and applying (9.6) to φ_{LL}, we obtain

$$\varphi_{LK} = \varphi_{L,(L \times M)} = \tfrac{1}{2}(\varphi_{LL} + \varphi_{LM}) = \tfrac{1}{4}(1 + F_L + 2\varphi_{LM}). \qquad (9.8)$$

Similarly, we have

$$\varphi_{MK} = \varphi_{M,(L \times M)} = \tfrac{1}{2}(\varphi_{MM} + \varphi_{LM}) = \tfrac{1}{4}(1 + F_M + 2\varphi_{LM}). \qquad (9.9)$$

In particular, when $F_L = 0$, we have

$$\varphi_{LK} = \varphi_{L,(L \times M)} = \tfrac{1}{4}(1 + 2\varphi_{LM}), \qquad (9.10)$$

and, similarly, for $F_M = 0$

$$\varphi_{MK} = \varphi_{M,(L \times M)} = \tfrac{1}{4}(1 + 2\varphi_{LM}). \qquad (9.11)$$

If, additionally, the parents, L and M, are unrelated, that is, if $\varphi_{LM} = 0$, we obtain

$$\varphi_{L,(L \times M)} = \varphi_{M,(L \times M)} = \tfrac{1}{4}. \qquad (9.12)$$

9.4.5. Coefficient of Parentage between Two Offspring

Let K_1 and K_2, say, be two offspring of parents L and M. From (9.8) and (9.9) we have $\varphi_{K_1 L} = \tfrac{1}{2}(\varphi_{LL} + \varphi_{LM})$ and $\varphi_{K_1 M} = \tfrac{1}{2}(\varphi_{MM} + \varphi_{LM})$, respectively.

Now K_1 takes the role of T and the other offspring, K_2, takes the role of K in Fig. 9.2. Thus, from formula (9.7) we have

$$\begin{aligned}
\varphi_{K_1 K_2} = \varphi_{(L \times M),(L \times M)} &= \tfrac{1}{2}(\varphi_{K_1 L} + \varphi_{K_1 M}) \\
&= \tfrac{1}{4}(\varphi_{LL} + \varphi_{MM} + 2\varphi_{LM}) \\
&= \tfrac{1}{4}[\tfrac{1}{2}(1 + F_L) + \tfrac{1}{2}(1 + F_M) + 2\varphi_{LM}]. \qquad (9.13)
\end{aligned}$$

Again, if the parents are not inbred, so that $F_L = F_M = 0$, we have

$$\varphi_{K_1 K_2} = \varphi_{(L \times M),(L \times M)} = \tfrac{1}{4}(1 + 2\varphi_{LM}) = \tfrac{1}{2}(\tfrac{1}{2} + \varphi_{LM}), \qquad (9.14)$$

which is the same as (9.10) or (9.11). This means that the coefficient of parentage of two independent offspring from the same cross is the same as the coefficient of parentage of parent and offspring, provided that the parents are not inbred.

It should be pointed out that, if T and K are the same individual, formula (9.7) does not hold and φ_{KK} must be obtained from (9.6). Comparing (9.14) with (9.6), we notice that

$$\varphi_{KK} - \varphi_{K_1 K_2} = \tfrac{1}{2}(1 + \varphi_{LM}) - \tfrac{1}{2}(\tfrac{1}{2} + \varphi_{LM}) = \tfrac{1}{4}. \qquad (9.15)$$

Example 9.4. Find the coefficient of inbreeding in the third generation of brother-sister mating. [*Note:* We will call the population of (unrelated) parents generation 0, and the brother-sister pairs generation 1. Thus, the offspring produced by this mating will represent generation 2, etc. For convenience, we denote the coefficient of parentage in the generation n by φ_n, and the coefficient of inbreeding of an individual in the generation n by F_n.]

The coefficient of parentage between brother-sister (i.e., in generation 1) is $\varphi_1 = \tfrac{1}{4}$ (see Example 9.1). Applying formula (9.14), we obtain $\varphi_2 = \tfrac{1}{4}(1 + \tfrac{1}{2}) = \tfrac{3}{8}$. But from (9.5) we have $F_3 = \varphi_2 = \tfrac{3}{8}$, so that the coefficient of inbreeding in the third generation of brother-sister mating is $\tfrac{3}{8}$.

9.5. WRIGHT'S FORMULA FOR EVALUATION OF AN INBREEDING COEFFICIENT IN AN IRREGULAR PEDIGREE

By *regular mating* we will understand a system of mating in which, for each generation, the mating partners are relatives of the same type (for instance, brother-sister or parent-child matings). Any other mating will be called *irregular*.

Using basic formula (9.7) and those derived from (9.7) in the previous sections, one can calculate the coefficient of inbreeding of an individual resulting from any regular or irregular mating. Another formula has been given by Wright (1922, 1950). It is especially useful for irregular mating when the coefficient of inbreeding of an individual with given ancestral tree has to be evaluated. But, of course, it applies to regular mating systems as well.

Let us first consider a simple pedigree with one ancestor Z in common as in Fig. 9.5.

An *ancestral line* is a sequence of individuals (or "steps") between a specified individual I, say, and its ancestor Z. Thus the *paternal* (ancestral) *line X*, say, consists of individuals from the father of I (X_n, say) back to the ancestor Z. Let n denote the number of steps in the paternal line X, that is, $X_n \rightarrow X_{n-1} \rightarrow \cdots \rightarrow X_1 \rightarrow Z$.

Similarly, the *maternal* (ancestral) *line, Y*, say, consists of individuals from the mother of I ($Y_{n'}$, say) to Z. Let n' denote the number of steps in the

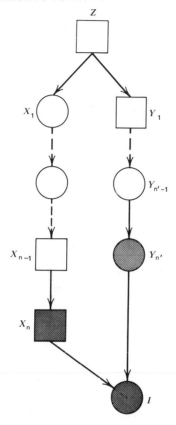

Figure 9.5

maternal line, that is, $Y_{n'} \to Y_{n'-1} \to \cdots \to Y_1 \to Z$. The numbers n and n' must satisfy the condition $n + n' \neq 0$.

We first assume that none of the individuals in the paternal and in the maternal lines were inbred previously, but the ancestor Z can have its own inbreeding coefficient, F_Z. The probability that Z transmits the same gene to both lines (X and Y) is $\frac{1}{2}(1 + F_Z)$. Since the transmission of a gene through the paternal line is independent of transmission through the maternal line, the probability that I will receive both genes identical by descent from Z is

$$(\tfrac{1}{2})^n \cdot (\tfrac{1}{2})^{n'} \cdot \tfrac{1}{2}(1 + F_Z) = (\tfrac{1}{2})^{n+n'+1}(1 + F_Z). \qquad (9.16)$$

In particular, when Z is not inbred, that is, when $F_Z = 0$, formula (9.16) becomes

$$(\tfrac{1}{2})^n \cdot (\tfrac{1}{2})^{n'} \cdot \tfrac{1}{2} = (\tfrac{1}{2})^{n+n'+1}. \qquad (9.17)$$

If individual I has more than one ancestor in common or if the individuals in paternal lines are inbred or related, formula (9.16) [or (9.17)] represents only the *contribution* of the ancestor Z to the coefficient of inbreeding of individual I. The contributions of other ancestors will be evaluated in a similar manner.

The sequence from the father through the paternal line to the common ancestor, and back through the maternal line to the mother, is called a *loop* [Kudo (1962)]. Thus, the loop in Fig. 9.5 is $X_n \to \cdots \to X_2 \to X_1 \to Z \to Y_1 \to Y_2 \to \cdots \to Y_{n'}$ or, briefly, $X_n \cdots X_2 X_1; Z; Y_1 Y_2 \cdots Y_{n'}$. The number of individuals in the loop is called the *length of the loop* and is equal to $n + n' + 1$. In a pedigree tree some common ancestors of the parents of an individual I may have their own common ancestors, which will also, of course, be common ancestors of I. This situation is called *rejoining*. The common ancestors of I which also have their own common ancestors will be inbred, and their coefficient of inbreeding can be calculated from the earlier ancestral loops.

In general, there can be r loops in a pedigree with $Z_1, Z_2, \ldots, Z_r$ common (independent or joint) ancestors. Let F_{Z_l} denote the coefficient of inbreeding of the ancestor Z_l, and $n_l + n'_l + 1$ be the length of the loop assigned to this ancestor. The total coefficient of inbreeding of I, F_I, will be the sum of contributions of all loops, that is,

$$F_I = \sum_{l=1}^{r} (\tfrac{1}{2})^{n_l + n'_l + 1}(1 + F_{Z_l}). \qquad (9.18)$$

Formula (9.18) can be used to calculate F_I for any regular or irregular mating.

Example 9.5.   In Example 9.2, using the method of coefficient of parentage, we found that the coefficient of inbreeding of an offspring of an uncle-niece marriage is $\tfrac{1}{8}$. We shall now prove this result, using formula (9.18).

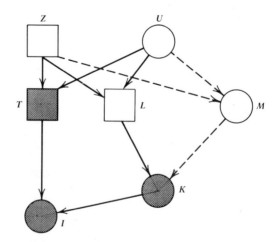

Figure 9.6

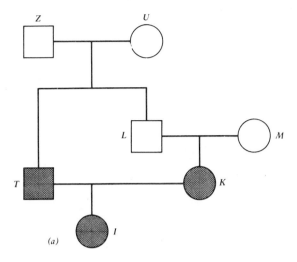

Figure 9.6(a)

Let Z and U be the common ancestors of I. The symbols for the remaining members in the pedigree in Fig. 9.6 are the same as those in Fig. 9.3.

(a) We have from Z only one loop, T; Z; LK of length $n + n' + 1 = 1 + 2 + 1 = 4$. Thus the contribution of ancestor Z to the inbreeding of I is $(\frac{1}{2})^4 = \frac{1}{16}$. The contribution from U is also $\frac{1}{16}$ (loop T; U; LK with length 4). Thus the total coefficient of inbreeding is $F_I = \frac{1}{16} + \frac{1}{16} = \frac{1}{8}$.

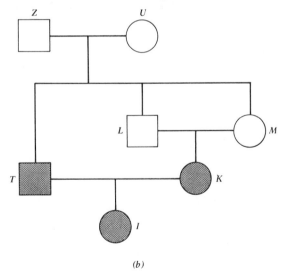

(b)

Figure 9.6(b)

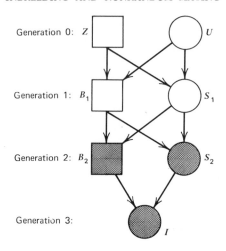

Generation 0: Z U

Generation 1: B_1 S_1

Generation 2: B_2 S_2

Generation 3: I

Figure 9.7

(*b*) But if M is also an offspring of Z and U [so the mating $(L \times M)$ is a brother-sister mating] two other loops, $T; Z; MK$ and $T; U; MK$, each of length 4, denoted by dotted lines in Fig. 9.6, are present. Thus the total coefficient of inbreeding of I is the sum of four contributing loops, that is, $(\frac{1}{2})^4 + (\frac{1}{2})^4 + (\frac{1}{2})^4 + (\frac{1}{2})^4 = \frac{1}{4}$. This is, of course, the same result as was obtained in Example 9.2(*c*).

Note: Figure 9.6 can also be represented in a more commonly used form of pedigree diagram. Figures 9.6(*a*) and 9.6(*b*) are the pedigree diagrams for the situations discussed in (*a*) and (*b*), respectively.

Example 9.6. In Example 9.4 we showed that the coefficient of inbreeding in the third generation of brother-sister mating is $\frac{3}{8}$. We will now prove this, using Wright's formula (9.18). Figure 9.7 represents the scheme for this mating and the calculations are given in the table.

l	Loop	Length of the loop, $n_l + n_l' + 1$	Contribution to F_I $(\frac{1}{2})^{n_l + n_l' + 1}$
1	$B_2 B_1; Z; S_1 S_2$	5	$\frac{1}{32}$
2	$B_2 S_1; Z; B_1 S_2$	5	$\frac{1}{32}$
3	$B_2 B_1; U; S_1 S_2$	5	$\frac{1}{32}$
4	$B_2 S_1; U; B_1 S_2$	5	$\frac{1}{32}$
5	$B_2; B_1; S_2$	3	$\frac{1}{8}$
6	$B_2; S_1; S_2$	3	$\frac{1}{8}$
	Total		$F_I = F_3 = \frac{1}{8}$

When calculating the inbreeding coefficient for the fourth generation in brother-sister mating, we must notice that the new common ancestors, B_2 and S_2, will be inbred with $F_2 = \frac{1}{4}$. When formula (9.18) is used, the calculations become rather tedious as the number of generations increases. In section 9.10 we will derive, for several regular breeding systems, recurrence formulae from which F can be obtained by a straightforward method. On the other hand, Wright's formula is often useful (although still tedious) when inbreeding is irregular and we need to trace an individual through different ancestors. Example 9.7 represents such a case.

Example 9.7.   Figure 9.8 represents a pedigree of the Shorthorn bull Comet (C), given by Wright (1922, 1950). Here L and M are father and mother, respectively. The loops from the three non-inbred ancestors, Z_1, Z_2, Z_3, are straightforward. Special attention should be paid to the loop L; L; M. Here L is the father of C but is also a common ancestor because M is his daughter. Also, L is inbred. This inbreeding coefficient, F_L, can be calculated from the loops of his parents; that is, the contribution to F_L from

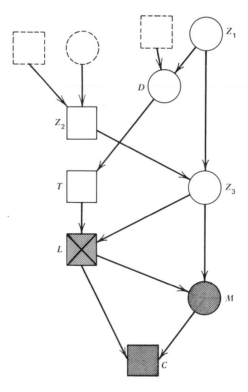

Figure 9.8. Adapted from Wright (1950).

l	Loop	$n_l + n'_l + 1$	$(\frac{1}{2})^{n_l+n'_l+1}(1 + F_{Z_l})$	
1	$LTD; Z_1; Z_3M$	6	$(\frac{1}{2})^6$	$= 0.0156$
2	$LT; Z_2; Z_3M$	5	$(\frac{1}{2})^5$	$= 0.0312$
3	$L; Z_3; M$	3	$(\frac{1}{2})^3$	$= 0.1250$
4	$L; L; M$	2	$(\frac{1}{2})^2(1 + 0.1875)$	$= 0.2969$
	Total		$F_C = 0.4687$	

$TD_4; Z_1; Z_3$ is $(\frac{1}{2})^4$, and from $T; Z_2; Z_3$ is $(\frac{1}{2})^3$, so that $F_L = (\frac{1}{2})^4 + (\frac{1}{2})^3 = 0.0625 + 0.1250 = 0.1875$. The coefficient of inbreeding of C, F_C, has been calculated from (9.16), and the calculations are presented in the table above. We have obtained $F_C = 0.4687$.

When the number of ancestors in a pedigree increases, it is often difficult to count correctly all possible loops. Therefore the calculated inbreeding coefficient may be subject to error. Another not too complicated method, based on counting individuals which occur in different paternal and maternal lines simultaneously, has been developed by Kudo (1962). Readers interested in this technique are referred to his paper.

9.6. THE X-LINKED GENES

Let us assume that the heterogametic sex is male, and consider an X-linked trait. We obtain the following rules for calculating the inbreeding coefficient of an individual I with respect to an X-linked locus.

(i) There is no inbreeding in a male (he cannot have two genes identical by descent, since only one gene is present in a male). If, then, I is a male, we have $F_I = 0$. Similarly, if Z_j is a male ancestor, $F_{Z_j} = 0$.

We now are interested in calculating an inbreeding coefficient of a *female* individual, I.

(ii) Suppose that in the loop associated with the ancestor Z_k (male or female) there are two males, one after the other. Because the male offspring receives X-linked genes from his mother, and not from his father, the contribution of Z_k to the coefficient of inbreeding of I is zero.

(iii) Let $Z_1, Z_2, \ldots, Z_r$ be r common ancestors of a female individual I, with the inbreeding coefficients $F_{Z_1}, F_{Z_2}, \ldots, F_{Z_r}$, respectively, and such that (ii) is not satisfied [note that in view of (i) some of these coefficients may be zero]. In counting a length of the loop in the ancestral chain, paternal line $\rightarrow$ ancestor $\rightarrow$ maternal line, we omit all male individuals. Let m_l be the number of female individuals in the loop of the ancestor Z_l; in other words, m_l is the length of the loop associated with the ancestor Z_l. The inbreeding coefficient of I is then calculated from the formula

$$F_I = \sum_{l=1}^{r} (\tfrac{1}{2})^{m_l}(1 + F_{Z_l}). \tag{9.19}$$

More details about these rules can be found in papers by Haldane and Moshinsky (1939) and Wright (1950). Some graphical presentations are given in Chapter 16 of the book by C. C. Li (1955).

9.7. THE STRUCTURE OF A POPULATION SUBJECT TO INBREEDING. WRIGHT'S DEFINITIONS OF THE COEFFICIENTS OF INBREEDING AND RELATIONSHIP

Let us consider a panmictic population

$$(pA + qa)^2 = p^2AA + 2pqAa + q^2aa \tag{9.20}$$

as an initial or "base" population. Suppose that this population is practicing inbreeding with an average coefficient of inbreeding F. We can find the structure of the next generation by using the following argument.

If selection and mutation are excluded, the probability that a gene A is present in an offspring is p. The probability that an offspring possesses two genes identical by descent is F, and the probability that both of them are A is Fp. Similarly, the probability that an offspring possesses two genes a identical by descent is Fq.

The probability that any two genes present in an individual of the next generation are not identical by descent is $1 - F$, and the probabilities that they will be AA, Aa, or aa are p^2, $2pq$, and q^2, respectively. Thus, the structure of the next generation will be

$$F(pAA + qaa) + (1 - F)(p^2AA + 2pqAa + q^2aa)$$
$$= (p^2 + Fpq)AA + 2(1 - F)pqAa + (q^2 + Fpq)aa. \tag{9.21}$$

The coefficient of inbreeding, F, for a population is also sometimes called the *fixation index* (hence the symbol F), while $1 - F = P$ is called the *panmictic index* [Wright (1950)].

In inbred populations, as mentioned in section 9.1, we always observe increased homozygosity and decreased heterozygosity in comparison to randomly mating populations. The coefficient $1 - F = P$ measures the proportion of heterozygotes [see (9.21)]. In the limiting case, when $F = 1$ (and so $P = 1 - F = 0$), there are no heterozygotes and the population is said to be *fixed*. In this case, the proportion of homozygotes AA is $(p^2 + pq) = p(p + q) = p$, and similarly the proportion of homozygotes aa in a fixed population is q. However, one randomly mating generation removes all accumulation of homozygosity.

TABLE 9.1   Genotype distribution in a population with inbreeding

♀ (X) \ ♂ (Y)	A (1)	a (0)	Total
A (1)	$p^2 + Fpq$	$(1 - F)pq$	p
a (0)	$(1 - F)pq$	$q^2 + Fpq$	q
Total	p	q	1

We now represent the genotype array (9.21) in the form of the joint distribution resulting from uniting maternal and paternal gametes (Table 9.1). Let X and Y be random variables assigned to maternal and paternal gamete arrays, respectively, each variable taking value 0 if a occurs and 1 if A occurs. From Table 9.1 we find $\mu_X = \mu_Y = p$, $\sigma_X = \sigma_Y = pq$, and $\sigma_{XY} = p^2 + Fpq - p^2 = Fpq$. Hence the correlation coefficient between X and Y is

$$\rho_{XY} = \frac{\sigma_{XY}}{\sigma_X \sigma_Y} = F. \qquad (9.22)$$

We obtain the following definition.

DEFINITION 9.5 (WRIGHT'S DEFINITION OF THE COEFFICIENT OF INBREEDING) *The correlation coefficient of uniting gametes is called the inbreeding coefficient of an individual in a population.*

More precisely, it should be called the expected (mean) coefficient of inbreeding. By using similar arguments, formula (9.21) can be extended to s alleles at a single locus. If p_i is the frequency of the gene A_i, and F is the inbreeding coefficient in a population, then the structure of the next generation is

$$F \sum_{i=1}^{s} p_i A_i A_i + (1 - F) \left(\sum_{i=1}^{s} p_i A_i \right)^2 = \sum_{i=1}^{s} [(1 - F)p_i^2 + Fp_i] A_i A_i$$
$$+ 2(1 - F) \sum_{i<j} \sum p_i p_j A_i A_j, \qquad (9.23)$$

or

$$= F \sum_{i=1}^{s} p_i A_i A_i + (1 - F) \sum_{i=1}^{s} \sum_{j=1}^{s} p_i p_j A_i A_j. \qquad (9.23a)$$

A few remarks may be helpful at this point.

1. Suppose that we start with an arbitrary population (generation 0, say)

$$dAA + hAa + raa, \qquad (9.24)$$

with $d + h + r = 1$. The frequencies of the A and a genes are

$$p = d + \tfrac{1}{2}h \quad \text{and} \quad q = +\tfrac{1}{2}h, \tag{9.25}$$

respectively.

By using the same arguments as for a panmictic population given by (9.20), we can prove that the genotypic array for the next generation is the same as in (9.21), where p and q are given by (9.25).

2. Since gene frequencies remain the same through all generations, then again, using the same arguments, we can show that after n generations of inbreeding the genotypic array is

$$(p^2 + F_n pq)AA + 2(1 - F_n)pqAa + (q^2 + F_n pq)aa, \tag{9.26}$$

where F_n is the coefficient of inbreeding in the nth generation ($n = 1, 2, \ldots$), and p, q are given by (9.25). However, if the initial population is given by (9.20), then (9.26) holds for $n = 0, 1, 2, \ldots$, since $F_0 = 0$. Extension to s alleles is straightforward.

3. The coefficient of inbreeding, as defined by Wright, is the same for each individual in a population only if a simple regular inbreeding system is in operation (e.g., parent × offspring or brother × sister). If proportions $c_1, c_2, \ldots, c_k$, with $\sum c_i = 1$, are practicing inbreeding with corresponding inbreeding coefficients $F_1, F_2, \ldots, F_k$, the mean (expected) inbreeding coefficient in the population is

$$\bar{F} = \sum_{i=1}^{k} c_i F_i. \tag{9.27}$$

DEFINITION 9.6 (WRIGHT'S DEFINITION OF RELATIONSHIP). *The correlation coefficient between the genotypes of individuals $I \times J$ is called the coefficient of relationship between I and J and is denoted as ρ_{IJ}.*

In fact, it is only under a regular mating system that ρ_{IJ} is the same for each pair; if the population is subject to irregular mating we can only speak about an *average* (or expected) coefficient of relationship for any two individuals in a population. Comparing ρ_{IJ} with Malécot's coefficient of parentage, evaluated under the same circumstances, it turns out that for two non inbred individuals

$$\rho_{IJ} = 2\varphi_{IJ}. \tag{9.28}$$

If, however, I and J are inbred

$$\rho_{IJ} = \frac{2\varphi_{IJ}}{\sqrt{1 + F_I}\,\sqrt{1 + F_J}}. \tag{9.28a}$$

Example 9.8. The genotype distribution for parent-offspring pairs is given in Table 7.4. The coefficient of relationship between parent-child is then $\tfrac{1}{2}$ (compare formula 7.49). To find the coefficient of inbreeding, using

TABLE 9.2   Distribution of uniting gametes in
parent-offspring mating

♀ \ ♂	A	a	Total
A	$p^2 + \frac{1}{4}pq$	$pq - \frac{1}{4}pq$	p
a	$pq - \frac{1}{4}pq$	$q^2 + \frac{1}{4}pq$	q
Total	p	q	1

Wright's approach, we notice, for instance, that the conditional probabilities that an offspring will be AA, given the matings are $AA \times AA$, $AA \times Aa$ (or $Aa \times AA$), and $Aa \times Aa$, are 1, $\frac{1}{2}$, and $\frac{1}{4}$, respectively. Multiplying these by the appropriate probabilities of matings, as given in Table 7.4, we obtain

$$\Pr\{AA\} = 1 \cdot p^3 + \tfrac{1}{2}(p^2q + p^2q) + \tfrac{1}{4}pq = p^2 + \tfrac{1}{4}pq.$$

This is the probability of uniting the gametes A and A in the mating parent $\times$ offspring. In a similar way we find the probabilities $\Pr\{Aa\}$, $\Pr\{aA\}$, and $\Pr\{aa\}$.

Table 9.2 shows all of these probabilities. They represent, of course, the distribution of genotypes in the first generation of parent-offspring (P $\times$ O) matings. We find from Table 9.2 that the inbreeding coefficient in the first generation of parent-offspring mating is $F_{(P \times O)} = \frac{1}{4}$. The coefficient of correlation between parent and offspring is, from (7.49), $\rho_{PO} = \frac{1}{2}$—this is, then, Wright's coefficient of relationship. We found [see formula (9.12)] that the coefficient of parentage of parent-offspring, $\varphi_{PO} = \frac{1}{4}$. So we have here $\rho_{PO} = 2\varphi_{PO}$.

In a similar way we can calculate the inbreeding coefficient for offspring of sib $\times$ sib matings (see Problem 9.3) and also for other cases.

9.8. GENOTYPIC MEAN AND VARIANCE IN A POPULATION WITH INBREEDING

Let us now consider a single locus with s alleles. Let G_{ij} represent the genotypic value of a certain character associated with genotype A_iA_j.

In a randomly mating population ($F = 0$), the genotype array is

$$\left(\sum_{i=1}^{s} p_i A_i \right)^2 = \sum_{i=1}^{s} \sum_{j=1}^{s} p_i p_j A_i A_j. \tag{9.29}$$

Let μ_0 and σ_0^2 be the mean and the variance, respectively, of this character in a randomly mating population, that is,

$$\mu_0 = \sum_{i=1}^{s} \sum_{j=1}^{s} G_{ij} p_i p_j \tag{9.30}$$

and

$$\sigma_0^2 = \sum_{i=1}^{s} \sum_{j=1}^{s} G_{ij}^2 p_i p_j - \left(\sum_{i=1}^{s} \sum_{j=1}^{s} G_{ij} p_i p_j \right)^2. \tag{9.31}$$

In a completely fixed population (i.e., one including only homozygotes, so that $F = 1$) obtained by a certain kind of inbreeding, the genotype array is $\sum p_i A_i A_i$.

Let μ_1 and σ_1^2 be the mean and the variance, respectively, of this character in a fixed population, that is,

$$\mu_1 = \sum_{i=1}^{s} G_{ii} p_i \tag{9.32}$$

and

$$\sigma_1^2 = \sum_{i=1}^{s} G_{ii}^2 p_i - \left(\sum_{i=1}^{s} G_{ii} p_i \right)^2. \tag{9.33}$$

In a population with the coefficient of fixation F, the genotype array is given by (9.23a). Let $\mu_{(F)}$ and $\sigma_{(F)}^2$ be the mean and the variance in this population. Thus we have

$$\mu_{(F)} = (1 - F) \sum_{i=1}^{s} \sum_{j=1}^{s} G_{ij} p_i p_j + F \sum_{i=1}^{s} G_{ii} p_i$$

$$= (1 - F)\mu_0 + F\mu_1 = \mu_0 + F(\mu_1 - \mu_0), \tag{9.34}$$

and

$$\sigma_{(F)}^2 = (1 - F) \sum_{i=1}^{s} \sum_{j=1}^{s} G_{ij}^2 p_i p_j + F \sum_{i=1}^{s} G_{ii}^2 p_i - \mu_{(F)}^2$$

$$= (1 - F)(\sigma_0^2 + \mu_0^2) + F(\sigma_1^2 + \mu_1^2) - [(1 - F)\mu_0 + F\mu_1]^2$$

$$= (1 - F)\sigma_0^2 + F\sigma_1^2 + F(1 - F)(\mu_1 - \mu_0)^2. \tag{9.35}$$

Thus if, instead of random mating, one applies inbreeding, the change in mean will be proportional to F, but the variance will also include a term with the coefficient F^2.

Of particular interest is the case in which the effects of alleles are *additive*, that is, $G_{ij} = G_i + G_j$. We then have

$$\mu_0 = \mu_1 = \mu_{(F)} = 2 \sum_{i=1}^{s} G_i p_i. \tag{9.36}$$

Note that the mean is not affected by inbreeding. The variances are

$$\sigma_0^2 = \sum_{i=1}^{s} \sum_{j=1}^{s} (G_i + G_j)^2 p_i p_j - 4 \left(\sum_{k=1}^{s} G_k p_k \right)^2$$

$$= 2 \left[\sum_{k=1}^{s} G_k^2 p_k - \left(\sum_k^s G_k p_k \right)^2 \right], \tag{9.37}$$

$$\sigma_1^2 = \sum_{k=1}^{s} (G_k + G_k)^2 p_k - 4 \left(\sum_{k=1}^{s} G_k p_k \right)^2$$

$$= 4 \left[\sum_{k=1}^{s} G_k^2 p_k - \left(\sum_{k=1}^{s} G_k p_k \right)^2 \right] = 2\sigma_0^2, \tag{9.38}$$

and

$$\sigma_{(F)}^2 = (1 - F)\sigma_0^2 + F\sigma_1^2 = (1 + F)\sigma_0^2. \tag{9.39}$$

This means that the variance increases in the proportion $(1 + F)$ and reaches $2\sigma_0^2$ under complete fixation. In other words, if a character is the result of the additive effects of genes, its average genotypic value in a population remains the same regardless of whether the population breeds at random or by a certain mating system. However, the variance increases in the ratio $(1 + F)$ for mating with inbreeding as compared to random mating.

9.9. GENE FREQUENCY OF RARE RECESSIVE TRAITS, EVALUATED FROM CONSANGUINEOUS MARRIAGES

As was explained in section 4.2, random mating in "infinite" populations is, in practice, equivalent to mating among unrelated pairs. If some proportion of consanguineous marriages is observed, the reason may be that the size of the population is small or some proportion of the population is practicing inbreeding, or both. For human populations, the only consanguineous marriages of appreciable frequency are those of first cousins.

Let us consider an "infinite" population in equilibrium with respect to a single locus with alleles A and a, having gene frequencies p and q, respectively. Suppose that aa denotes an affected individual. Let c be the proportion of first-cousin marriages, and $1 - c$ of unrelated marriages. In the next generation, the proportion of individuals aa from the first cousin marriages will be $c(\frac{1}{16}q + \frac{15}{16}q^2)$, since for first-cousin offspring we have $F = \frac{1}{16}$ (see Example 9.3). Assuming that the inbreeding due to other causes is negligible, the proportion of all recessives aa due to random and nonrandom mating is $(1 - c)q^2 + c(\frac{1}{16}q + \frac{15}{16}q^2)$. Let κ be the proportion of first-cousin marriages among the parents of affected individuals. We then have

$$\kappa = \frac{c(\frac{1}{16}q + \frac{15}{16}q^2)}{(1 - c)q^2 + c(\frac{1}{16}q + \frac{15}{16}q^2)} = \frac{c(1 + 15q)}{16q + c(1 - q)}. \tag{9.40}$$

If c is very small, (9.40) is sometimes approximated by

$$\kappa \doteq \frac{c(1 + 15q)}{16q}. \tag{9.40a}$$

Solving (9.40) with respect to q, we obtain

$$q = \frac{c(1 - \kappa)}{16\kappa - 15c - c\kappa}. \tag{9.41}$$

Formula (9.41) is sometimes used to estimate the gene frequency of rare recessive abnormal trait from the proportion of first-cousin marriages. This method requires estimation of the frequency of first-cousin marriages, c, in the population and the frequency of marriages with affected children, κ, but does not require estimation of the frequency of affected individuals in the total population. It should be noticed, however, that formula (9.41) is based on several assumptions, for example, that the "base" population is very large and is in Hardy-Weinberg equilibrium, that the trait is due entirely to genetic causes, and that no mutation or selection occurs. Thus, q estimated from (9.41) may be only a rough approximation to the true frequency (see also the discussion in sections 9.1.1 and 9.14).

9.10. REGULAR BREEDING SYSTEMS

9.10.1. Recurrence Relations

To obtain a general formula for the coefficient of inbreeding in the nth generation, we will use Malécot's method of coefficient of parentage. But first it will be useful to give the general solution for a certain recurrence relation.

The linear relation of the type

$$c_0 u_n + c_1 u_{n+1} + \cdots + c_r u_{n+r} = 0, \tag{9.42}$$

where $c_0, c_1, \ldots, c_r$ are known constants, is called in mathematics a *recurrence relation* or *difference equation*. Let us consider a so-called *auxiliary equation*:

$$c_0 + c_1 y + c_2 y^2 + \cdots + c_r y^r = 0. \tag{9.43}$$

This is a polynomial of rth order. Let $\varepsilon_1, \varepsilon_2, \ldots, \varepsilon_r$ be the roots of this polynomial. We first assume that all the ε's are different. Then

$$u_n = \lambda_1 \varepsilon_1^{\,n} + \lambda_2 \varepsilon_2^{\,n} + \cdots + \lambda_r \varepsilon_r^{\,n}, \tag{9.44}$$

where the values of $\lambda_1, \lambda_2, \ldots, \lambda_r$ are determined from the values $u_0, u_1, \ldots, u_{r-1}$, which must be given in the problem.

If the roots $\varepsilon_1, \varepsilon_2, \ldots, \varepsilon_r$ are not all different, and the root ε_i occurs k times, then the part in solution for u_n, derived from this root, will be

$$(\lambda_{i0} + \lambda_{i1}n + \lambda_{i2}n^2 + \cdots + \lambda_{i(k-1)}n^{k-1})\varepsilon_i^n. \tag{9.45}$$

We will now consider some common regular systems of inbreeding.

9.10.2. Selfing

Many plants are fertilized by self-pollination. In selfing, the paternal and maternal genotypes are, of course, the same. In the case of two alleles, A and a, only the matings $AA \times AA$, $Aa \times Aa$, and $aa \times aa$ are possible.

Let the initial population (generation 0) have inbreeding coefficient F_0. The coefficient of parentage of an individual with itself is, from (9.3), $\frac{1}{2}(1 + F_0)$. This is also the coefficient of inbreeding of the offspring, that is, the coefficient F_1, say, in the first generation, that is,

$$F_1 = \tfrac{1}{2}(1 + F_0).$$

Repeating the same argument, we find the coefficient of inbreeding in the second generation to be

$$F_2 = \tfrac{1}{2}(1 + F_1) = \tfrac{1}{2}(1 + \tfrac{1}{2} + \tfrac{1}{2}F_0) = \tfrac{1}{2} + (\tfrac{1}{2})^2 + (\tfrac{1}{2})^2 F_0.$$

Similarly, for the third generation

$$F_3 = \tfrac{1}{2}(1 + F_2) = \tfrac{1}{2} + (\tfrac{1}{2})^2 + (\tfrac{1}{2})^3 + (\tfrac{1}{2})^3 F_0,$$

and so on. Finally, for the nth generation, we obtain

$$\begin{aligned}
F_n &= \tfrac{1}{2} + (\tfrac{1}{2})^2 + (\tfrac{1}{2})^3 + \cdots + (\tfrac{1}{2})^n + (\tfrac{1}{2})^n F_0 \\
&= \tfrac{1}{2}[1 + \tfrac{1}{2} + (\tfrac{1}{2})^2 + (\tfrac{1}{2})^{n-1} + (\tfrac{1}{2})^{n-1} F_0].
\end{aligned} \tag{9.46}$$

Since

$$1 + \tfrac{1}{2} + (\tfrac{1}{2})^2 + \cdots + (\tfrac{1}{2})^{n-1} = \frac{1 - (\tfrac{1}{2})^n}{1 - \tfrac{1}{2}} = 2[1 - (\tfrac{1}{2})^n], \tag{9.47}$$

by substituting (9.47) into (9.46) we obtain

$$F_n = 1 - (\tfrac{1}{2})^n + (\tfrac{1}{2})^n F_0 = 1 - (\tfrac{1}{2})^n (1 - F_0). \tag{9.48}$$

Or, using the panmictic index, $P_n = 1 - F_n$, for the nth generation, we obtain

$$P_n = (\tfrac{1}{2})^n P_0. \tag{9.49}$$

Assuming $F_0 = 0$ (and therefore $P_0 = 1$), we obtain

$$F_n = 1 - (\tfrac{1}{2})^n \quad \text{and} \quad P_n = (\tfrac{1}{2})^n. \tag{9.50}$$

It is easy to see that P_n in (9.50) satisfies the recurrence formula

$$P_{n+1} = \tfrac{1}{2}P_n \tag{9.51}$$

for $n = 1, 2, \ldots$ and $P_0 = 1$.

Let us now consider an initial population, $d_0 AA + h_0 Aa + r_0 aa$, with $d_0 + h_0 + r_0 = 1$, and assume that it reproduces itself by self-pollination. Without loss of generality, we can assume that $F_0 = 0$ and so $P_0 = 1$. The only matings which can produce heterozygotes are $Aa \times Aa$. By utilizing this fact, it can be shown that after n generations of self-pollination the genotype array takes the form

$$(d_0 + \tfrac{1}{2}F_n h_0)AA + (1 - F_n)h_0 Aa + (r_0 + \tfrac{1}{2}F_n h_0)aa, \tag{9.52}$$

with $n = 0, 1, 2, \ldots$, and F_n given by (9.50).

The proportion of heterozygotes in the nth generation is then

$$h_n = (1 - F_n)h_0 = P_n h_0 = (\tfrac{1}{2})^n h_0, \tag{9.53}$$

and the proportion of both types of homozygotes (AA and aa) is

$$d_n + r_n = 1 - h_n = 1 - (\tfrac{1}{2})^n h_0. \tag{9.54}$$

If the initial population consists only of heterozygotes Aa so that $h_0 = 1$, (9.54) takes the form

$$d_n + r_n = 1 - (\tfrac{1}{2})^n. \tag{9.55}$$

As $n \to \infty$, $F_n \to 1$ and $h_n \to 0$, so that a self-pollinated population will reach equilibrium with the genotype array

$$(d_0 + \tfrac{1}{2}h_0)AA + (r_0 + \tfrac{1}{2}h_0)aa = pAA + qaa. \tag{9.56}$$

Suppose that we consider a population with respect to m independent loci. If h_{j0} is the proportion of heterozygotes at the jth locus in the initial population, the proportion of homozygotes with respect to m loci in the nth generation is

$$\Pr\{\text{Hom}\} = \prod_{j=1}^{m} [1 - (\tfrac{1}{2})^n h_{j0}], \tag{9.57}$$

and, in particular, when the initial population is heterozygous with respect to these m loci, we have

$$\Pr\{\text{Hom}\} = [1 - (\tfrac{1}{2})^n]^m. \tag{9.58}$$

9.10.3. Brother-Sister Mating

Let φ_n be the coefficient of parentage between two offspring in the nth generation of regular brother-sister mating. From (9.5) we can calculate the coefficient of inbreeding of an individual in the $(n + 1)$th generation, that is,

$$F_{n+1} = \varphi_n, \qquad (9.59)$$

and analogously

$$F_{n+2} = \varphi_{n+1}. \qquad (9.59a)$$

But from (9.13) we have

$$\varphi_{n+1} = \tfrac{1}{4}[\tfrac{1}{2}(1 + F_n) + \tfrac{1}{2}(1 + F_n) + 2\varphi_n]. \qquad (9.60)$$

Substituting (9.59) and (9.59a) into (9.60), we obtain

$$F_{n+2} = \tfrac{1}{4}(1 + F_n + 2F_{n+1}). \qquad (9.61)$$

Formula (9.61) can be applied directly if we wish to calculate coefficients of inbreeding for successive generations, assuming that the coefficients in two preceding generations are known. We already have $F_1 = 0$, $F_2 = \tfrac{1}{4}$. Then, from (9.61) we obtain

$$F_3 = \tfrac{1}{4}(1 + 0 + 2 \cdot \tfrac{1}{4}) = \tfrac{3}{8}, \quad F_4 = \tfrac{1}{4}(1 + \tfrac{1}{4} + 2 \cdot \tfrac{3}{8}) = \tfrac{8}{16},$$
$$F_5 = \tfrac{1}{4}(1 + \tfrac{3}{8} + 2 \cdot \tfrac{1}{8}) = \tfrac{19}{32}, \quad \text{and so on.}$$

If we want to calculate the coefficient of inbreeding for the nth generation, omitting the preceding steps, we have to find the general solution for a certain recurrence equation. Let $P_{n+r} = 1 - F_{n+r}$ $(r = 0, 1, 2)$ be the pan-mictic index. Then (9.61) can be written in the form

$$1 - F_{n+2} = \tfrac{1}{4}(4 - 1 - F_n - 2F_{n+1}) = \tfrac{1}{4}[2(1 - F_{n+1}) + (1 - F_n)]$$

or

$$P_{n+2} = \tfrac{1}{4}(2P_{n+1} + P_n) = \tfrac{1}{2}P_{n+1} + \tfrac{1}{4}P_n \qquad (9.62)$$

or

$$P_{n+2} - \tfrac{1}{2}P_{n+1} - \tfrac{1}{4}P_n = 0. \qquad (9.62a)$$

Using the theory of recurrence equations given in section 9.10.1, we have to solve the auxiliary equation

$$y^2 - \tfrac{1}{2}y - \tfrac{1}{4} = 0 \quad \text{or} \quad 4y^2 - 2y - 1 = 0. \qquad (9.63)$$

The roots of this equation are

$$\varepsilon_1 = \tfrac{1}{4}(1 + \sqrt{5}) \quad \text{and} \quad \varepsilon_2 = \tfrac{1}{4}(1 - \sqrt{5}). \qquad (9.64)$$

Thus the general solution for P_n is of the form

$$P_n = \lambda_1 \varepsilon_1{}^n + \lambda_2 \varepsilon_2{}^n = (\tfrac{1}{4})^n[(1 + \sqrt{5})^n \lambda_1 + (1 - \sqrt{5})^n \lambda_2]. \qquad (9.65)$$

The quantities λ_1 and λ_2 in (9.65) are arbitrary and can be evaluated from the initial conditions for two first populations (generation 0 and generation 1).

(i) Suppose that generation 0 represents a population of unrelated parents, and that generation 1 is a population of their offspring. Thus, we have $F_0 = 0$,

$F_1 = 0$, and so $P_0 = 1$, $P_1 = 1$. Substituting these values into (9.65), we obtain two simultaneous equations:

$$\lambda_1 + \lambda_2 = 1, \\ (1 + \sqrt{5})\lambda_1 + (1 - \sqrt{5})\lambda_2 = 4.$$

Their solutions are

$$\lambda_1 = \frac{3 + \sqrt{5}}{2\sqrt{5}}, \quad \lambda_2 = -\frac{3 - \sqrt{5}}{2\sqrt{5}}, \tag{9.66}$$

and so (9.65) takes the form

$$P_n = \left(\frac{1}{4}\right)^n \frac{1}{2\sqrt{5}} [(3 + \sqrt{5})(1 + \sqrt{5})^n - (3 - \sqrt{5})(1 - \sqrt{5})^n],$$

$$n = 0, 1, 2, \ldots . \tag{9.67}$$

(ii) Suppose that we take as generation 0 the offspring of a randomly mating population (i.e., a population of brothers and sisters), and as generation 1, the offspring of brother × sister matings. In this case $F_0 = 0$, $F_1 = \frac{1}{4}$, and so $P_0 = 1$, $P_1 = \frac{3}{4}$. Substituting these into (9.64), we have

$$\lambda_1 + \lambda_2 = 1, \\ (1 + \sqrt{5})\lambda_1 + (1 - \sqrt{5})\lambda_2 = 3,$$

with the solutions

$$\lambda_1 = \frac{2 + \sqrt{5}}{2\sqrt{5}}, \quad \lambda_2 = -\frac{2 - \sqrt{5}}{2\sqrt{5}}. \tag{9.68}$$

Formula (9.65) now takes the form

$$P'_{n'} = \left(\frac{1}{4}\right)^{n'} \frac{1}{2\sqrt{5}} [(2 + \sqrt{5})(1 + \sqrt{5})^{n'} - (2 - \sqrt{5})(1 - \sqrt{5})^{n'}], \tag{9.69}$$

where $n' = 0, 1, 2, \ldots$ denotes the renumbered generations. Since $n = n' + 1$, we have

$$P_n = P'_{n-1}, \quad n = 1, 2, \ldots . \tag{9.70}$$

In fact, we may start with any (inbred) generation k and evaluate λ_1 and λ_2 from such initial conditions. Formula (9.70) will be generalized as

$$P_n = P'_{n-k}, \quad \text{where} \quad n - k = n' \quad \text{for} \quad k = 0, 1, \ldots, n \tag{9.71}$$

The proportion of heterozygotes in the nth generation, h_n, will be calculated from the formula

$$h_n = 2(1 - F_n)pq = 2P_npq, \tag{9.72}$$

where p and q are the gene frequencies in the initial population. [Note: if generation 0 is panmictic of type (9.20), then formula (9.72) is valid for $n = 0, 1, 2, \ldots$; for an arbitrary population, (9.72) holds for $n = 1, 2, \ldots .]

Example 9.9. Let the initial population (generation 0) consist of heterozygotes Aa, so that $h_0 = 1$ and $p = q = \frac{1}{2}$. Find the proportion of heterozygotes in the second generation.

(a) To calculate P_2 we may use formula (9.67). We obtain $P_2 = \frac{3}{4}$. Applying (9.72), we obtain $h_2 = \frac{3}{8}$.

(b) We now use another approach. The offspring of random matings $Aa \times Aa$ is $\frac{1}{4}AA + \frac{1}{2}Aa + \frac{1}{4}aa$. Considering the population of offspring (brothers and sisters) as a generation 0, we calculate P_1' from (9.69). Again it turns out that $P_1' = \frac{3}{4}$, so $h_1' = \frac{3}{8}$.

Example 9.10. Find the coefficient of inbreeding in the 20th generation of brother-sister mating.

From formula (9.67) we have

$$P_{20} = \left(\frac{1}{4}\right)^{20} \frac{1}{2\sqrt{5}} [(3 + \sqrt{5})(1 + \sqrt{5})^{20} - (3 - \sqrt{5})(1 - \sqrt{5})^{20}] = 0.0169.$$

Hence $F_{20} = 1 - P_{20} = 0.9831$.

It is left to the reader to show that the same result will be obtained from formula (9.69) for $n' = 19$.

9.10.4. Parent-Offspring Mating

Here we may distinguish two kinds of mating.

(i) The first kind is shown in Fig. 9.9. The mating is between a fixed sire and his daughter, granddaughter, great-granddaughter, etc. Let $Z \times A$

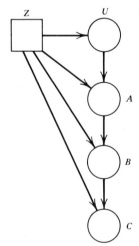

Figure 9.9

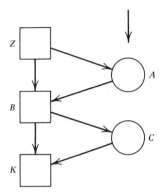

Figure 9.10

represent the nth generation. We have $\varphi_{ZU} = \varphi_{n-1}$ and hence $F_A = F_n = \varphi_{ZU}$.

$$F_{n+1} = F_B = \varphi_{ZA} = \varphi_n = \tfrac{1}{2}(\varphi_{ZZ} + \varphi_{ZU}) = \tfrac{1}{2}(\varphi_{ZZ} + F_n);$$

$$F_{n+2} = F_C = \varphi_{ZC} = \varphi_{n+1} = \tfrac{1}{2}(\varphi_{ZZ} + F_{n+1}).$$

Hence,

$$F_{n+2} - F_{n+1} = \tfrac{1}{2}F_{n+1} - \tfrac{1}{2}F_n$$

or

$$F_{n+2} - \tfrac{3}{2}F_{n+1} + \tfrac{1}{2}F_n = 0. \qquad (9.73)$$

Putting $P_{n+r} = 1 - F_{n+r}$ $(r = 0, 1, 2)$, we obtain

$$P_{n+2} - \tfrac{3}{2}P_{n+1} + \tfrac{1}{2}P_n = 0. \qquad (9.74)$$

(ii) A more common type of parent-offspring mating is that in which each individual is mated successively with its (younger) parent and with its offspring (Fig. 9.10). Again, by using Malécot's method of coefficient of parentage, it can be shown (the proof is left to the reader) that

$$P_{n+2} - \tfrac{1}{2}P_{n+1} - \tfrac{1}{4}P_n = 0,$$

which is the same as formula (9.62a) for full-sib mating.

9.11. MIXTURE OF SELFING AND RANDOM MATING IN A POPULATION

Complete self-pollination in plants is not a common phenomenon. Actually, most so-called self-pollinated plants practice random mating also.

Let us first consider a population under equilibrium with the usual array

$$p^2AA + 2pqAa + q^2aa = (pA + qa)^2.$$

Let ϕ be the proportion of individuals in this population practicing selfing, and $1 - \phi$ the proportion practicing random mating. Then in the first generation the structure of the population will be. [from (9.21) with $F_1 = \frac{1}{2}$ for selfing]

$$(1 - \phi)(p^2 AA + 2pq Aa + q^2 aa) + \phi[(p^2 + \tfrac{1}{2}pq)AA + pq Aa + (q^2 + \tfrac{1}{2}pq)aa]$$
$$= (p^2 + \tfrac{1}{2}\phi pq)AA + 2pq(1 - \tfrac{1}{2}\phi)Aa + (q^2 + \tfrac{1}{2}\phi pq)aa$$
$$= d_1 AA + h_1 Aa + r_1 aa. \tag{9.75}$$

The proportion of heterozygotes, h_1, in this first generation is

$$h_1 = 2pq(1 - \tfrac{1}{2}\phi). \tag{9.76}$$

In the second generation it is not necessary that the same plants perform selfing; therefore, in some plants, the inbreeding will be removed by random mating. Thus, in the second generation, the structure of the population will be

$$(1 - \phi)(p^2 AA + 2pq Aa + q^2 aa) + \phi[(d_1 + \tfrac{1}{4}h_1)AA$$
$$+ \tfrac{1}{2}h_1 Aa + (r_1 + \tfrac{1}{4}h_1)aa]. \tag{9.77}$$

Therefore, the proportion of heterozygotes, h_2, in the second generation is

$$h_2 = 2pq(1 - \phi) + \tfrac{1}{2}h_1\phi = 2pq(1 - \phi) + pq(1 - \tfrac{1}{2}\phi)\phi$$
$$= 2pq(1 - \phi + \tfrac{1}{2}\phi - \tfrac{1}{4}\phi^2) = 2pq[1 - \tfrac{1}{2}\phi - (\tfrac{1}{2}\phi)^2]. \tag{9.78}$$

Repeating this process, we find the proportion of heterozygotes in the nth generation:

$$h_n = 2pq[1 - \tfrac{1}{2}\phi - (\tfrac{1}{2}\phi)^2 - \cdots - (\tfrac{1}{2}\phi)^n]$$
$$= 2pq\{1 - \tfrac{1}{2}\phi[1 + \tfrac{1}{2}\phi + \tfrac{1}{2}\phi)^2 + (\tfrac{1}{2}\phi)^{n-1}]\}$$
$$= 2pq\left[1 - \tfrac{1}{2}\phi \frac{1 - (\tfrac{1}{2}\phi)^n}{1 - \tfrac{1}{2}\phi}\right]. \tag{9.79}$$

As $n \to \infty$, the population reaches equilibrium with the proportion of heterozygotes

$$h_\infty = 2pq\left(1 - \frac{\tfrac{1}{2}\phi}{1 - \tfrac{1}{2}\phi}\right) = 2pq\left(1 - \frac{\phi}{2 - \phi}\right). \tag{9.80}$$

The quantity

$$F_\infty = \frac{\phi}{2 - \phi} \tag{9.81}$$

is then the coefficient of inbreeding of the population at equilibrium, with selfing occurring in proportion ϕ.

More advanced problems, in which there are mixtures of selfing and random mating, have been discussed, with two linked loci taken into account, by Bennett and Binet (1956), and with several independent loci by Ghai (1964). A general review of advanced problems on this subject can be found in the expository paper by Karlin (1968).

9.12. INBREEDING IN RANDOMLY MATING POPULATION OF FINITE SIZE

When the population is *finite* (and rather small in size), a certain amount of inbreeding can occur even under random mating. Here we restrict ourselves to a simple situation.

Let N_1 be the number of male genes and N_2 the number of female genes in the nth generation. Let F_{n-2} and F_{n-1} be the coefficients of inbreeding in the $(n-2)$th and $(n-1)$th generations, respectively (Fig. 9.11). The probability that two genes in an individual belonging to the nth generation come from a single individual (male or female) belonging to the $(n-2)$th generation is

$$\frac{1}{4} \cdot \frac{1}{N_1} + \frac{1}{4} \cdot \frac{1}{N_2} = \frac{1}{N}. \tag{9.82}$$

Here N is called the "effective size" of the population.

The probability that these genes will also be identical by descent is

$$\frac{1}{N} \cdot \frac{1 + F_{n-2}}{2}.$$

The probability that two genes in an individual belonging to the nth generation come from different individuals of the $(n-2)$th generation is $1 - (1/N)$, and the probability that these genes will be identical by descent, that is, come from the same individual of the $(n-1)$th generation is $[1 - (1/N)]F_{n-1}$. Thus, the total probability that an individual in the nth generation will have two genes identical by descent, that is, its inbreeding

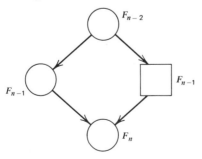

Figure 9.11

coefficient F_n, satisfies the recurrence relation

$$F_n = \frac{1}{N} \cdot \frac{1 + F_{n-2}}{2} + \left(1 - \frac{1}{N}\right) F_{n-1}. \tag{9.83}$$

9.13. INBREEDING IN A POPULATION DIVIDED INTO ISOLATES. WAHLUND'S PRINCIPLE

Human populations are often divided into different ethnic, religious, or economic groups with a tendency to intermarry within each group. When these differences are sharp enough, such groups are sometimes called *isolates*. Isolates, especially when they are not very large, represent populations with a certain degree of inbreeding.

Suppose that the total population is divided into M isolates. Let p_i and $q_i = 1 - p_i$ be the frequencies of genes A and a, respectively, and w_i be the weight (proportional to the size in the ith isolate), with $\sum w_i = 1$. We assume that these subpopulations are sufficiently large so that equilibrium law, $p_i^2 AA + 2 p_i q_i Aa + q_i^2$, for the ith subpopulation holds approximately for all i.

Let

$$\bar{p} = \sum_{i=1}^{M} w_i p_i \quad \text{and} \quad \bar{q} = 1 - \bar{p} = \sum_{i=1}^{M} w_i q_i \tag{9.84}$$

be the mean frequencies of genes A and a, respectively, in the total population.

We notice that the variances of the group gene frequencies, p_i and q_i $(i = 1, 2, \ldots, M)$, are, respectively,

$$\sigma_p^2 = \sum_{i=1}^{M} w_i (p_i - \bar{p})^2 = \sum_{i=1}^{M} w_i p_i^2 - \bar{p}^2, \tag{9.85a}$$

and

$$\sigma_q^2 = \sum_{i=1}^{M} w_i q_i^2 - \bar{q}^2 = \sum_{i=1}^{M} w_i (1 - p_i)^2 - (1 - \bar{p})^2$$

$$= 1 - 2 \sum_{i=1}^{M} w_i p_i + \sum_{i=1}^{M} w_i p_i^2 - 1 + 2\bar{p} - \bar{p}^2$$

$$= \sum_{i=1}^{M} w_i p_i^2 - \bar{p}^2 = \sigma_p^2. \tag{9.85b}$$

Thus, these variances are equal, that is,

$$\sigma_p^2 = \sigma_q^2 = \sigma^2. \tag{9.86}$$

From (9.85a), (9.85b), and (9.86) we have

$$\sum_{i=1}^{M} w_i p_i^2 = \bar{p}^2 + \sigma^2, \tag{9.87a}$$

$$\sum_{i=1}^{M} w_i q_i^2 = \bar{q}^2 + \sigma^2. \tag{9.87b}$$

Similarly,

$$\sum_{i=1}^{M} w_i p_i q_i = 2 \sum_{i=1}^{M} w_i p_i (1 - p_i) = 2 \sum_{i=1}^{M} w_i p_i - 2 \sum_{i=1}^{M} w_i p_i$$

$$= 2\bar{p} - 2\bar{p}^2 - 2\sigma^2 = 2\bar{p}\bar{q} - 2\sigma^2. \tag{9.87c}$$

Formulae (9.87a), (9.87b), and (9.87c) represent the proportions of genotypes AA, aa, and Aa, respectively, in the total population.

Hence the structure of the total population is

$$(\bar{p}^2 + \sigma^2)AA + 2(\bar{p}\bar{q} - \sigma^2)Aa + (\bar{q}^2 + \sigma^2)aa. \tag{9.88}$$

Comparing (9.88) with (9.21), we can write

$$\bar{F}\bar{p}\bar{q} = \sigma^2;$$

hence

$$\bar{F} = \frac{\sigma^2}{\bar{p}\bar{q}}. \tag{9.89}$$

Formula (9.88) means that, if the population is divided into distinct subpopulations, with random mating within each group, the decrease in the proportion of heterozygotes, as compared to that occurring under random mating of the whole population, is proportional to the variance of gene frequencies of the isolates. This is known as *Wahlund's principle*.

Formula (9.89) may be used as another definition of an expected coefficient of inbreeding in the total population. Here $\bar{F}$ is not the inbreeding coefficient used in the sense defined before, but is rather a certain index determining the decrease in heterozygosity (or the increase in homozygosity) occurring because of the isolation of the subpopulations. In other words, if a population is divided into groups mating randomly *within* each group, and the gene frequencies in each group are different, the result is the same decrease in heterozygosity would occur if the total population were to practice inbreeding with an average inbreeding coefficient $\bar{F}$ defined by (9.89).

If the isolates are small, the situation becomes more complicated and the results of section 9.12 must be applied.

9.14. GENERAL REMARKS ON DIFFERENT DEFINITIONS AND APPLICABILITY OF THE INBREEDING COEFFICIENT

Inbreeding has been broadly defined as a certain form of mating which results in an increase of homozygosity. The coefficient of inbreeding, F, has been introduced as a certain measure of the degree of homozygosity. In natural populations, two components should be distinguished: nonrandom, due to matings among relatives; and random, resulting from finite population size and subdivision of a population into isolates, where mating is random. We have discussed these aspects only separately. The size of the book does not permit going into the details of such problems as the estimation of inbreeding in human populations when both components should be taken into account.

We have presented two definitions of the coefficient of inbreeding, F_I, of an individual I. Malécot's definition is expressed in terms of probabilities and applies to a single individual; Wright's definition is based on the concept of correlation between the uniting gametes of consanguineous parents and considers an individual as a member of a population mating according to a certain system. These definitions are applied chiefly when the inbreeding is resulting from consanguineous matings. For random mating, with subdivision of a population into isolates, we have Wahlund's formula for F, given in (9.89).

The reader interested in the history of the inbreeding coefficient may note that this term was first introduced by Pearl (1913), but not in a sufficiently precise way to be adopted. Bernstein's definition (1930) has some common elements with Wright's definition and leads, in fact, to the same results as obtained by the methods of Malécot and Wright when an individual is regarded as a member of a population subject to a regular mating system with constant inbreeding. A general treatment of inbreeding, which takes into account several independent or linked loci, has been presented by Shikata (1962, 1965). From the point of view of his theory (which is beyond the scope of this book), the results obtained by the previous authors appear to be particular cases.

The coefficient of parentage, φ_{IJ}, between two individuals, I and J, which is defined as the probability that I and J possess genes identical by descent, is a useful concept introduced by Malécot. It is equivalent to one-half of the coefficient of relationship (correlation) among mating as defined by Wright, provided that the mating population is not inbred.

We have presented three methods of calculating a coefficient of inbreeding.

1. One method utilizes Malécot's coefficient of parentage between the parents, L and M, and their offspring, K. The basic formula is given in (9.5). that is, $F_K = F_{(L \times M)} = \varphi_{LM}$. In order to obtain the coefficient of parentage of two related individuals we recall the "key" formulae (9.6), (9.7), (9.10), and (9.13). Applying these formulae, one can evaluate a coefficient of inbreeding for any individual in regular or irregular breeding. As we have seen in section 9.10, this is especially useful in regular breeding systems.

2. Wright's formula (9.18) is also of general use. It becomes more complicated, however, as the number of common ancestors with their own coefficients of inbreeding increases. On the other hand, it is simple for irregular breeding with few ancestors.

3. The method utilizing Wright's correlation coefficient requires a knowledge of the joint distribution of parental mating. It gives an expected (mean) coefficient of inbreeding in a population; if the population is subjected to regular systems of inbreeding, this will also be the coefficient of inbreeding for a single member of this population. In this theory it is assumed that the initial or "base" population of the parents practicing inbreeding is in Hardy-Weinberg equilibrium. We can never be sure that this assumption applies to natural, particularly to human, populations.

In addition to these three methods, another approach based on generation matrix theory has been given by R. A. Fisher (1948), and later presented in an elegant form by O. Kempthorne (1957, Chapter 6).

9.15.   ASSORTATIVE MATING

We have defined inbreeding in large (infinite) populations as a special type of nonrandom mating which is based on the *genetic* properties of mates: the number of genes identical by descent. Another form of nonrandom mating is based on the *phenotypic* properties of mates: two individuals mate because they are alike or because they are different from each other. This type is usually defined as an *assortative mating*. When the mating is based on phenotypic resemblance—so that, for instance, dominants mate with dominants, and recessives with recessives—it is called *positive* assortative, while the mating with respect to opposing characters (e.g., dominants with recessives) is called *negative* assortative mating. In the light of these definitions selfing might be considered as positive assortative mating. Inbreeding leads to homozygosity with respect to all loci at the same rate. On the other hand, assortative mating may take place with respect to only one or a few characters, and can be random with respect to others. Positive assortative mating and inbreeding both lead ultimately to homozygosity, but a longer time is required with assortative mating. For further discussion of the similarities and differences between inbreeding and assortative mating see a short paper by Lewontin et al. (1968).

A great deal of mathematical analysis on the evaluation of proportions of homozygotes and heterozygotes with respect to qualitative traits under assortative mating has been done by Jennings (1916, 1917), Robbins (1917, 1918), and Wentworth and Remick (1916). We restrict ourselves here to some very simple cases of one locus with two alleles, A and a.

9.15.1. Positive Assortative Mating

Suppose that the initial population is $d_0 AA + h_0 Aa + r_0 aa$, with $d_0 + h_0 + r_0 = 1$. The proportion of dominants is $d_0 + h_0 = 1 - r_0$, and the proportion of recessives is r_0. Table 9.3 represents schematically mating Dom $\times$ Dom and Rec $\times$ Rec.

In the first generation of mating Dom $\times$ Dom we find the following proportion of genotypes:

$$\Pr\{AA \,|\, \text{Dom} \times \text{Dom}\} = \Pr\{AA \,|\, AA \times AA\}\Pr\{AA \times AA \,|\, \text{Dom} \times \text{Dom}\}$$
$$+ \Pr\{AA \,|\, AA \times Aa\}\Pr\{AA \times Aa \,|\, \text{Dom} \times \text{Dom}\}$$
$$+ \Pr\{AA \,|\, Aa \times Aa\}\Pr\{Aa \times Aa \,|\, \text{Dom} \times \text{Dom}\}$$

$$= 1 \cdot \frac{d_0^2}{(d_0 + h_0)^2} + \frac{1}{2} \cdot 2 \frac{d_0 h_0}{(d_0 + h_0)^2} + \frac{1}{4} \frac{h_0^2}{(d_0 + h_0)^2}$$

$$= \frac{(d_0 + \frac{1}{2}h_0)^2}{(d_0 + h_0)^2} \, . \tag{9.90}$$

Then

$$\Pr\{AA \text{ and } \text{Dom} \times \text{Dom}\} = \frac{(d_0 + \frac{1}{2}h_0)^2}{(d_0 + h_0)^2} (d_0 + h_0) = \frac{(d_0 + \frac{1}{2}h_0)^2}{d_0 + h_0} \, . \tag{9.91a}$$

In a similar way we obtain

$$\Pr\{Aa \text{ and } \text{Dom} \times \text{Dom}\} = \frac{h_0(d_0 + \frac{1}{2}h_0)}{d_0 + h_0} \, , \tag{9.91b}$$

$$\Pr\{aa \text{ and } \text{Dom} \times \text{Dom}\} = \frac{1}{4} \frac{h_0^2}{d_0 + h_0} \, . \tag{9.91c}$$

TABLE 9.3   Positive
assortative mating

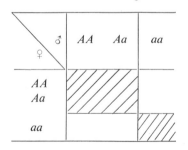

In mating Rec $\times$ Rec we obtain only recessives, that is,

$$\Pr\{aa \text{ and Rec} \times \text{Rec}\} = r_0. \tag{9.91d}$$

Then, the genotypic structure in the first generation will be $d_1 AA + h_1 Aa + r_1 aa$, where

$$d_1 = \frac{(d_0 + \frac{1}{2}h_0)^2}{d_0 + h_0}, \tag{9.92a}$$

$$h_1 = \frac{h_0(d_0 + \frac{1}{2}h_0)}{d_0 + h_0}, \tag{9.92b}$$

$$r_1 = \frac{1}{4}\frac{h_0^2}{d_0 + h_0} + r_0 = \frac{1}{4}\frac{h_0^2 + 4r_0(d_0 + h_0)}{d_0 + h_0}. \tag{9.92c}$$

Let $p_0 + \frac{1}{2}h_0 = p$ and $q_0 = r_0 + \frac{1}{2}h_0 = q$ be the frequencies of genes A and a respectively in the initial population; and let p_n and q_n be the corresponding gene frequencies in the nth generation. Thus the frequency, p_1, of the gene A in the first generation is

$$p_1 = d_1 + \frac{1}{2}h_1 = \frac{(d_0 + \frac{1}{2}h_0)^2 + \frac{1}{2}h_0(d_0 + \frac{1}{2}h_0)}{d_0 + h_0} = \frac{(d_0 + \frac{1}{2}h_0)(d_0 + h_0)}{d_0 + h_0}$$

$$= d_0 + \frac{1}{2}h_0 = p_0 = p,$$

which is the same as the frequency in the initial population. It is easy to see that this is also true for the nth generation. We then have

$$p_n = p_0 = p, \quad q_n = q_0 = q. \tag{9.93}$$

The proportion of heterozygotes, given by (9.92b), can be written as

$$h_1 = \frac{(d_0 + \frac{1}{2}h_0)h_0}{(d_0 + \frac{1}{2}h_0) + \frac{1}{2}h_0} = \frac{ph_0}{p + \frac{1}{2}h_0} = \frac{2ph_0}{2p + h_0}. \tag{9.94}$$

Repeating this procedure, we obtain the following recurrence formula for the proportion of heterozygotes, h_n, in the nth generation:

$$h_n = \frac{2ph_{n-1}}{2p + h_{n-1}}. \tag{9.95}$$

Substituting successively $h_1, h_2, \ldots, h_{n-1}$, we find

$$h_n = \frac{2ph_0}{2p + nh_0}, \quad n = 1, 2, \ldots. \tag{9.96}$$

In particular, when $h_0 = 2pq$, we obtain

$$h_n = \frac{2pq}{1 + nq}. \tag{9.97}$$

As $n \to \infty$, $h_n \to 0$, and the population will have the equilibrium structure

$$pAA + qaa. \tag{9.98}$$

9.15.2. Negative Assortative Mating

When the mating is Dom $\times$ Rec, it is called *completely negative assortative mating*. We notice that the first generation of matings $AA \times aa$ and $Aa \times aa$ yields only genotypes Aa and aa. Thus the matings in the first generation will be only of type $Aa \times aa$, which leads immediately to the equilibrium state in the next generation, that is, $\frac{1}{2}Aa + \frac{1}{2}aa$.

9.16. APPROACH TO HOMOZYGOSITY BY SUCCESSIVE BACKCROSSES

We notice that usually nonrandom matings increase homozygosity, so that finally the structure of the population under equilibrium is $pAA + qaa$ (except for the negative mating).

Sometimes one may wish to have a homogeneous population of one type of homozygotes, AA or aa, only. This can be obtained by repeating backcrosses to one or another parental genotype.

Let us consider two parental strains, Pt_1 of genotype AA and Pt_2 of genotype aa, and their offspring F_1 with genotype Aa (do not confuse F_1 with the inbreeding coefficient!).

9.16.1. Multiple backcrosses to Pt_2

In the first generation of crosses $Aa \times aa$ we obtain $\frac{1}{2}Aa + \frac{1}{2}aa$. Crossing again $(Aa \times aa) \times aa$, we obtain in the second generation $\frac{1}{4}Aa + \frac{1}{4}aa + \frac{1}{4}aa = \frac{1}{4}Aa + \frac{3}{4}aa$ or $(\frac{1}{2})^2Aa + [1 - (\frac{1}{2})^2]aa$. Repeating this procedure n times, we obtain in the nth generation

$$(\tfrac{1}{2})^nAa + [1 - (\tfrac{1}{2})^n]aa. \tag{9.99}$$

For m independent loci, the proportion of m-tuple recessive homozygotes, $aabbcc \ldots$, after n successive backcrosses to Pt_2, will be

$$\Pr\{aabbcc \cdots \mid F_1 \times \underbrace{Pt_2 \times Pt_2 \times \cdots \times Pt_2}_{n \text{ times}}\} = [1 - (\tfrac{1}{2})^n]^m.$$

$$\tag{9.100}$$

9.16.2. Multiple Backcrosses to Pt_1

The same argument applies to backcrosses $Aa \times AA$. Thus, after n successive backcrosses to Pt_1, the proportion of m-tuple homozygotes $AABBCC \cdots$ will be

$$\Pr\{AABBCC \cdots \mid F_1 \times \underbrace{Pt_1 \times Pt_1 \times \cdots \times Pt_1}_{n \text{ times}}\} = [1 - (\tfrac{1}{2})^n]^m. \quad (9.101)$$

SUMMARY

Definitions of the coefficient of inbreeding of an individual and of the coefficient of parentage (or relationship) between two individuals have been introduced. Two approaches—one probabilistic and due to Malécot, and the other based on correlation among uniting gametes and due to Wright—have been presented. Methods of evaluating coefficients of inbreeding in regular and irregular mating systems have been described. The effects of inbreeding on the mean and variance of a genotypic value in a population have been discussed. It has also been shown that random mating in isolates or in populations of finite size results in an increase of homozygosity and a decrease in heterozygosity. Other kinds of nonrandom, such as those based on phenotypic resemblance or successive backcrosses to parental strains, have been briefly presented. It has been shown that almost all kinds of nonrandom mating (except mixture of selfing and random mating and negative assortative mating) leads to homozygosity. The equilibrium structure of a population when mating is based on phenotypic resemblance is $pAA + qaa$, while in successive backcrosses to one of two kinds of homozygotes, aa or AA, the population will consist of homozygotes aa or AA, only.

REFERENCES

Allard, R. W., Jain, S. K., and Workman, P. L., Genetics of inbreeding populations. In: *Advances in Genetics* **14**, ed. by E. W. Caspari (1968), 55–132.

Allen, G., Random and non-random inbreeding, *Eugen. Quart.* **12** (1965), 181–198.

Bennett, J. H. and Binet, F. E., Association between Mendelian factors with mixed selfing and random mating, *Heredity* **10** (1956) 51–55.

Bernstein, F., Fortgesetzte Untersuchungen aus der Theorie der Blutgruppen, *Z. Abstamm. Vererbgsl.* **56** (1930), 233–273.

Crow, J. F., Breeding structure of population. Effective population number. In: *Statistics and Mathematics in Biology*, ed. by Kempthorne et al., Hafner Publishing Co., New York, 1964, Chapter 43, pp. 543–556.

Fisher, R. A., *The Theory of Inbreeding*, Oliver and Boyd, London, 1948 (second edition, 1965).

Garber, M. J., Approach to genetic equilibrium with varying percentages of selfing and cross-classification, *J. Heredity* 42 (1951), 299–300.

Ghai, G. L., The genotypic composition and variability in plant populations under mixed self-fertilization and random mating, *J. Indian Soc. Agr. Statist.* 16 (1964), 94–125.

Gillois, M., Le concept d'identité et son importance en génétique, *Ann. de Génét.* 9 (1966), 58–65.

Green, E. L., Breeding systems. In: *Biology of the Laboratory Mouse*, ed. by E. L. Green, McGraw-Hill Book Company, New York, 1966, pp. 11–22.

Haldane, J. B. S., and Moshinsky, P., Inbreeding in Mendelian populations with special reference to human cousin marriage, *Ann. Eugen.* 9 (1939), 321–340.

Jacquard, A., Liaison génétique entre individus apparentés, *Population* 23 (1968a), 94–128.

Jacquard, A., Evolution des populations d'effectif limité, *Population* 23 (1968b), 279–300.

Jennings, H. S., Production of pure homozygotic organisms from heterozygotes by self-fertilization, *Amer. Nat.* 46 (1912), 487–491.

Jennings, H. S., The numerical results of diverse systems of breeding, *Genetics* 1 (1916), 53–89; 2 (1917), 97–154.

Karlin, S., Equilibrium behavior of population genetic models with non-random mating, I. Preliminary and special mating systems, *J. Appl. Prob.* 5 (1968), 231–313.

Kempthorne, O., *An Introduction to Genetic Statistics*, John Wiley & Sons, New York, 1957, Chapters 5 and 6, pp. 72–139.

Kudo, A., A method for calculating the inbreeding coefficient, I, *Amer. J. Hum. Genet.* 14 (1962), 426–432.

Lewontin, R., Kirk, D., and Crow, J., Selective mating, assortative mating and inbreeding: definitions and implications, *Eugen. Quart.* 15 (1968), 141–143.

Li, C. C., *Population Genetics*, The University of Chicago Press, Chicago, 1955, Chapters 9–17, pp. 103–232.

Malécot, G., *Les Mathématiques de l'Hérédité*, Mason et Cie Paris, 1948.

Malécot, G., *Probabilités et Hérédité*, Presses Universitaires de France, Paris, 1966.

Pearl, R., A contribution toward an analysis of the problem of inbreeding, *Amer. Nat.* 47 (1913), 577–614; correction, 48 (1914), 57.

Robbins, R. B., Application of mathematics to breeding problems, II, *Genetics* 2 (1917), 489–504; 3 (1918), 73–92.

Sanghvi, L. D., Inbreeding, genes and phenotypes, *Amer. Nat.*, 89 (1955), 247–248.

Shikata, M., The generalized inbreeding coefficient in a Markov process, *Rep. Stat. Appl. Res. JUSE* 9 (1962), 127–136.

Shikata, M., A generalization of the inbreeding coefficient, *Biometrics* 21 (1965), 665–681.

Spuhler, J. N., Assortative mating with respect to physical characteristics, *Eugen. Quart.* 15 (1968), 128–140.

Steinberg, A. G., Methodology in human genetics (Part 2 of two parts), *Amer. J. Hum. Genet.* 11 (1959), 326–330.

Wahlund, S., Zusammensetzung von Populationen and Korrelationserscheinungen von Standpunkt der Vererbungslehre aus Betrachtet, *Hereditas* 11 (1928), 65–106.

Wentworth, E. M., and Remick, B. L., Some breeding properties of the generalized Mendelian population, *Genetics* 1 (1916), 608–616.

Workman, P. L., The maintenance of heterozygosity by partial negative assortative mating, *Genetics* 50 (1965), 1369–1382.

Wright, S., Systems of mating, *Genetics* **6** (1921), 111–178.
Wright, S., Coefficient of inbreeding and relationship, *Amer. Nat.* **56** (1922), 330–338.
Wright, S., The genetic structure of populations, *Ann. Eugen.* **15** (1950), 323–354.
Yasuda, N., An extension of Wahlund's principle to evaluate mating type frequency, *Amer. J. Hum. Genet.* **20** (1968), 1–23.

PROBLEMS

9.1. List all (fifteen) identity situations for two individuals, at a single auto-somal locus. (*Note:* "Identity situation" was defined in section 9.1.)

9.2. Let L be a father with coefficient of inbreeding F_L, and M be a mother with coefficient of inbreeding F_M. Let φ_{LM} be the coefficient of parentage between L and M. Let D be a daughter of $L \times M$ mating, and I be an offspring of $L \times D$ mating.

(*a*) Draw the pedigree.

(*b*) Show that the inbreeding coefficient of I is

$$F_I = \tfrac{1}{4}(1 + F_L + 2\varphi_{LM}).$$

(*c*) If S is a son of $L \times M$ mating and J is an offspring of $M \times S$ mating, what is the coefficient of inbreeding of J, that is, F_J?

9.3. Using the joint genotype distribution of sib-sib given in Table 7.12:

(*a*) Find the distribution of uniting gametes and show that the inbreeding coefficient for the offspring of brother-sister mating is $\tfrac{1}{4}$ (note that it is independent of gene frequencies; you can use $p = q = \tfrac{1}{2}$ to simplify the calculations).

(*b*) Find the joint genotype distribution of brother-sister in the offspring (i.e., second generation) of brother-sister mating [see Li (1955), Chapter 10].

(*c*) Using the results derived in (*a*) or (*b*), show that the coefficient of inbreeding in the second generation is $\tfrac{3}{8}$.

9.4. (*a*) Find the inbreeding coefficient of an individual K with the pedigree patterns shown in Figures 9(i) and 9(ii) on page 238.
(*Hint:* Note that the individuals in the loop, denoted by $\otimes$, are inbred.)

(*b*) Indicate (using dotted lines) the individuals which do not contribute to the inbreeding of K.

9.5. Show that the coefficient of inbreeding for double first cousins (i.e., individuals such that the parents of one of them are sib pairs of the parents of the other) is $\tfrac{1}{8}$. In your calculation use:

(*a*) The method of Malécot's coefficient of parentage.

(*b*) Wright's formula (9.18).

9.6. Suppose that an individual Z (in the pedigree shown in Figure 9(iii) on page 238) is hemophilic. Given that spouses (not indicated in the pedigree or indicated by dotted lines) are not hemophilic or carriers of hemophilia and that individual B is hemophilic, find the probability that:

(*a*) Individuals A_1, A_2, A_3 are carriers of hemophilia.

(*b*) Individual C is a carrier of hemophilia.

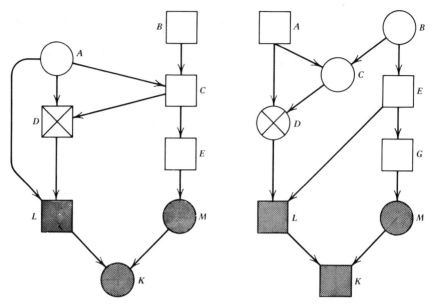

Figure 9(i)

Figure 9(ii)

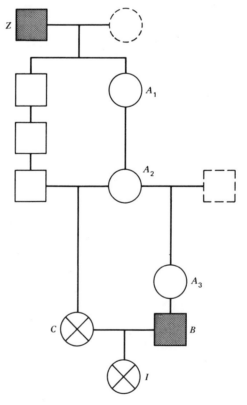

Figure 9(iii)

238

(c) If C marries B, their daughter I will be (1) a carrier of hemophilia; (2) hemophilic.

(d) Find the inbreeding coefficients of individuals C and I.

(e) What will be the answers to questions (a) through (d) if B is not a hemophilic?

9.7. Derive the solution for the recurrence formula (9.74) for parent-offspring mating as in Fig. 9.9.

9.8. Let an initial self-pollinated plant population (with $F_0 = 0$) have the genotype distribution

$$0.40AA + 0.50Aa + 0.10aa.$$

(a) Assuming that the population is considered to be fixed, if the proportion of heterozygotes is not more than 0.001 (i.e., 1 heterozygote in 1000 plants), find the number of generations needed for this population to be fixed.

(b) Find the genotype distribution under equilibrium.

(c) Let the initial population be

$$0.25AA + 0.50Aa + 0.25aa.$$

Answer questions (a) and (b) for this population. What general conclusions can you reach about the number of generations required for equilibrium and about the final structure of the fixed population, depending on the initial structure of a self-pollinated population?

CHAPTER 10

Natural Selection and Mutation

A. Selection

10.1. MECHANISMS OF SELECTION

In Chapter 4 we considered randomly mating populations and obtained equilibrium laws for such populations. To derive appropriate equilibrium formulae, we made essential assumptions that the fertilities of individuals and the viabilities of each gamete and each zygote are the same in every generation. These assumptions imply that the gene frequencies remain unchanged through generations.

The situation is not always so simple. One of the several causes which can change the gene frequencies in the next generation is *selection*. Certain genotypes have a smaller chance of surviving to reproduce themselves than others. This discrepancy may arise from the natural conditions in which a population breeds (natural selection), or some genotypes may be gradually eliminated by using a certain breeding system, in order to improve some economically valuable traits in plants and animals (artificial selection).

In this chapter, we restrict ourselves to the problems arising in natural selection and hence the word "natural" will often be omitted.

We may distinguish three different types of selection.

(i) Selection due to differential *viabilities* of genotypes at different stages of development. Selection can occur at the *gametic* stage, that is, certain types of gametes are subject to selective disadvantage before they unite into zygotes. Often selection occurs at the *zygotic* stage in the very early life before birth. But it can also take place at any stage of adult life.

(ii) Selection due to *differential fertility*. This can happen because certain genotypes are less fertile than others or because certain mating types are "incompatible" and fertilization does not occur.

(iii) *Familial* selection. By this, Haldane (1924), who introduced the term, meant that the offspring of a given mating type may have different viabilities, depending on the genotype of the individual offspring.

240

The endogenous forces acting in natural selection are often influenced by environmental conditions so that selection against a given genotype or gamete may not be the same in different populations.

Mathematical bases for the theory of natural selection and its measurements were developed in series of papers by Haldane (e.g., 1924–1931), Fisher (e.g., 1930), and Wright (e.g., 1931). More recently, several authors have contributed to the development of theoretical and practical analysis in this field. In particular, a great deal of work has been done by Li (e.g., 1955a, 1963, 1967a, 1967b), Kimura (e.g., 1956), Lewontin (e.g., 1958, 1967), and Crow (e.g., 1962).

We do not attempt to present in this chapter all the results and achievements in the theory of selection. The reader may find the key references at the end of this chapter, especially the expository paper by Li (1967b), helpful in further study.

Here we restrict ourselves to a few basic *deterministic* models, mostly for a single locus. We assume that the populations are large (theoretically "infinite") and that the genes have reasonable (i.e., neither too large nor too small) frequencies, so that random effects on the frequencies of genotypes can be neglected. For simplicity, we also assume that the generations are discrete and nonoverlapping. However, the effect of overlapping generations on the results of selection is usually small and, in most cases, can be neglected [Haldane (1927)].

10.2. FITNESSES OF GENOTYPES

How can the effects of selection be evaluated, and how can they be measured? One term especially associated with problems of natural selection is the *fitness* of a *genotype*. This denotes, in a broad sense, the ability of a genotype to survive and to reproduce itself. Using this descriptive "definition" of fitness, one can say only that some genotypes are "more fit," on the average, because they live longer and/or because their fertility is higher than that of other genotypes. Such statements give, of course, only rough comparisons, not measures of fitnesses. The fitness of a genotype determines the contribution of this genotype to the number of offspring in the next generation—for this reason it is often called the *reproductive fitness*. One may say that it would be reasonable to measure the fitness of each individual by the number of offspring it produces during its life. Of course, fitness defined in this way applies only to individuals. Some individuals will have fitness 0; others will have fitness 1 or 2 or 3 or 4, etc. These values alone do not give sufficient information about the fitness of genotypes in a population. We shall, then, measure the fitness of a genotype by the *mean* (expected) number of offspring per individual of a given genotype.

Suppose that we start with some initial population of newborn individuals. Some of them will not reach adult life; other survive to maturity but do not produce viable offspring. Should they be counted? Yes, and the number of progeny assigned to these individuals will be zero. Others may die during the reproductive period—they will also be counted with the appropriate number of offspring actually born. We may define (reproductive) fitness in the following way.

DEFINITION 10.1. *The mean (expected) number of live births ever born per newborn of a given genotype in a certain population is the fitness of this genotype in this population.*

Example 10.1. Let the genotype *aa* define a recessive deleterious trait, which can be recognized at birth. Suppose that a cohort of 100 newborn individuals with genotype *aa* was observed and at the end of reproductive life it was found that the *total* number of liveborn progeny of this cohort was 216. What is the fitness of the genotype *aa*?

According to our Definition 10.1, it is $(216/100) = 2.16$. It should be noticed that we counted only the total progeny, no matter whether all 100 or only 70 or only 50 individuals became parents.

Although the definition sounds simple, there are several difficulties in applying it in practical calculations. We will mention only a few of them.

1. If the trait is rare, it is difficult to start with a reasonable size of sample of the genotypes under consideration.
2. The fitness of a genotype may depend entirely on the genetic constitution of an individual with this genotype, but may also be a function of mating type. For instance, if mating is *aa* × *aa*, the offspring *aa* may die *in utero*, whereas from mating *aa* × *Aa* the offspring *aa* may be born. Interaction of parental genotypes such as in blood groups Rh+ and Rh− is another example of the possibility of the death of an offspring *in utero* while both Rh+ and Rh− are fully viable alone.
3. As a consequence of factors 1 and 2, the fitness in each generation can be a function of gene frequencies.
4. The onset of some characters occurs in later adult life.

Taking into account such situations, some authors prefer the following definition.

DEFINITION 10.1*a. The fitness of a genotype is measured by the mean (expected) number of its progeny, the parental and the offspring generations being counted at the same stage of development.*

Even using this definition, it is often difficult to evaluate the fitness with sufficient accuracy, especially in human populations.

The "absolute" fitness of a genotype, expressed as the average number of progeny, is not by itself very informative; is a fitness of 2 for a genotype *aa*, for instance, good or poor? The answer depends on the corresponding values for other genotypes in a population. Hence *ratios* of fitnesses are more useful than fitnesses themselves.

Example 10.2.   Let us consider the following pattern of genotype fitnesses:

$$\text{Genotype:}\quad AA\quad Aa\quad aa$$
$$\text{Fitness:}\quad\ \ 4\quad\ 3\quad\ 2.$$

For convenience (to avoid the subscripts), if there are only two alleles at a locus, we denote them by A and a. But this notation does not necessarily mean that they are, respectively, dominant and recessive. It is convenient to assume the fitness of one genotype as being 1.

(i) Suppose that we take the fitness of *aa* equal to 1. Then the pattern:

$$AA\qquad Aa\qquad aa$$
$$\tfrac{4}{2}=2\quad \tfrac{3}{2}=1.5\quad 1$$

shows the same ratios, $2:1.5:1 = 4:3:2$. The fitnesses 2 and 1.5 can be considered as *relative* fitnesses with respect to the fitness of the genotype *aa*.

(ii) Similarly assuming the fitness of the genotype *Aa* equal to 1, we find the relative fitnesses of *AA* and *aa* to be $4/3 = 1.3333$ and $2/3 = 0.6667$, respectively. The ratios

$$\tfrac{4}{3}:1:\tfrac{2}{3} = 4:3:2$$

are still the same. In (i) the unit was 2 offspring per genotype; in (ii), 3 offspring per genotype.

We also may express the relative quantities as proportions of the total, so that they add up to 1. In our Example 10.2, the total is $4 + 3 + 2 = 9$, so that the relative fitnesses could be

$$AA\qquad\qquad Aa\qquad\qquad aa$$
$$\tfrac{4}{9}=0.4444\quad \tfrac{3}{9}=0.3333\quad \tfrac{2}{9}=0.2222.$$

The ratios $\tfrac{4}{9}:\tfrac{3}{9}:\tfrac{2}{9} = 4:3:2$ are still the same, but $\tfrac{4}{9}+\tfrac{3}{9}+\tfrac{2}{9}=1$. We may, then, use the following definition.

DEFINITION 10.2.   (*Relative*) *fitnesses of genotypes in a population are quantities proportional to the mean (expected) numbers of offspring which these genotypes contribute to the next generation.*

In selection, usually fitness 1 is assigned to a certain genotype and the fitnesses of the remaining genotypes are evaluated in relation to this one.

We will be interested mostly in the ratios of genotypes, rather than in their absolute numbers, and the results will not be affected so far as the ratios are concerned. For convenience, we will omit the term "relative" in the text that follows and use only the term "fitness." Sometimes the synonym *adaptive value* will also be used.

10.3. A SINGLE AUTOSOMAL LOCUS WITH MULTIPLE ALLELES. CHANGES IN GENE FREQUENCIES DUE TO SELECTION

10.3.1. The Expected Fitness of a Population

Let us first discuss an example.

Example 10.3. Suppose that we consider a population, in regard to a single locus, of the following pattern:

Genotype:	AA	Aa	aa
Probability:	0.35	0.40	0.25
Fitness:	3	1	2.

Since the proportion of AA genotypes is 0.35, the expected contribution of genotypes AA to the offspring of the next generation is $3 \times 0.35 = 1.05$. Similarly, we have contributions $1 \times 0.40 = 0.40$ for Aa and $2 \times 0.25 = 0.50$ for aa genotypes. The expected total contribution to the offspring of next generation is, therefore, $1.05 + 0.40 + 0.50 = 1.95$. If we consider the fitnesses as values of a random variable with probabilities equal to genotype probabilities, then 1.95 is just the expected (mean) fitness of the initial population.

Thus, after selection, the proportions of the contributions to the next generations are $(1.05/1.95) = 0.5385$ of AA, $(0.40/1.95) = 0.2051$ of Aa, and $(0.50/1.95) = 0.2564$ of aa. In other words, if selection operates over the population (represented in the form of genotype array) $0.35AA + 0.40Aa + 0.25aa$ with adaptive values 3, 1, and 2, respectively, the expected effect on the distribution in the next generation will be the same as if each genotype contributes the same amount to the offspring, with the population structure $0.5385AA + 0.2051Aa + 0.2564aa$. For this reason, the last expression is sometimes called the *effective parental population*.

Let us now consider a general case of a single autosomal locus with s multiple alleles, $A_1, A_2, \ldots, A_s$, and gene frequencies, $p_1, p_2, \ldots, p_s$, with $\Sigma p_i = 1$. We will discuss a deterministic model for changes in gene frequencies due to selection.

We assume that the selection forces act in both sexes in the same way, so that there are no differences between fitnesses for males and females. Let us denote the fitness of genotype $A_i A_j$ by W_{ij}. For convenience, we also assume $A_i A_j = A_j A_i$ and $W_{ij} = W_{ji}$. The case $W_{ij} \neq W_{ji}$ (i.e., when selective forces act in different ways on maternal and paternal gametic contributions to the zygote) is straightforward, but is more troublesome since the genotypes $A_i A_j$ and $A_j A_i$ must be treated separately.

Let $\Pr\{A_i A_i\} = h_{ii}$ and $\Pr\{A_i A_j\} = 2h_{ij}$, with $\Sigma\Sigma h_{ij} = 1$, represent the genotype frequencies in the population. We now consider a population of the following pattern:

$$\left.\begin{array}{llllllllll}
\text{Genotype:} & A_1 A_1 & A_2 A_2 & \cdots & A_s A_s & A_1 A_2 & A_1 A_3 & \cdots & A_{s-1} A_s \\
\text{Probability:} & h_{11} & h_{22} & \cdots & h_{ss} & 2h_{12} & 2h_{13} & \cdots & 2h_{s-1,s} \\
\text{Fitness:} & W_{11} & W_{22} & \cdots & W_{ss} & W_{12} & W_{13} & \cdots & W_{s-1,s}
\end{array}\right\} \quad (10.1)$$

The fitnessess W_{ij} can be considered as values of a random variable, W, defined over the probability space given in (10.1).

Bearing in mind that $h_{ij} = h_{ji}$ and $W_{ij} = W_{ji}$, we find that the expected value of W, $\overline{W}$, say, is

$$E(W) = \overline{W} = \sum_{i=1}^{s} \sum_{j=1}^{s} W_{ij} h_{ij}, \qquad (10.2)$$

and its variance, $\sigma_W{}^2$, is

$$\text{Var}(W) = \sigma_W{}^2 = \sum_{i=1}^{s} \sum_{j=1}^{s} (W_{ij} - \overline{W})^2 h_{ij} = \sum_{i=1}^{s} \sum_{j=1}^{s} W_{ij}^2 h_{ij} - \overline{W}^2. \qquad (10.3)$$

DEFINITION 10.3. *The quantity $\overline{W}$ defined in (10.2) is called the mean (expected) fitness of the whole population with regard to a single autosomal locus.*

We notice that the frequency p_i of the gene A_i in this population is

$$p_i = \sum_{j=1}^{s} h_{ij}, \qquad i = 1, 2, \ldots, s. \qquad (10.4)$$

Let us now restrict ourselves to genotypes which include the allele A_i. The (conditional) distribution of the allele A_i in this reduced sample space will be as follows:

$$\begin{array}{lllllllll}
\text{Genotype:} & A_1 A_i & A_2 A_i & \cdots & A_{i-1} A_i & A_i A_i & A_i A_{i+1} & \cdots & A_i A_s \\
\text{Probability:} & \dfrac{h_{1i}}{p_i} & \dfrac{h_{2i}}{p_i} & \cdots & \dfrac{h_{i-1,i}}{p_i} & \dfrac{h_{ii}}{p_i} & \dfrac{h_{i,i+1}}{p_i} & \cdots & \dfrac{h_{is}}{p_i}.
\end{array} \quad (10.5)$$

DEFINITION 10.4.   *The quantity*

$$\overline{W}_i = \frac{1}{p_i} \sum_{j=1}^{s} W_{ij} h_{ij} \tag{10.6}$$

is called the mean (expected) fitness of the allele A_i in the population.

From (10.6) we have

$$\sum_{j=1}^{s} W_{ij} h_{ij} = \overline{W}_i p_i. \tag{10.6a}$$

Substituting (10.6a) into (10.2), we obtain

$$\overline{W} = \sum_{i=1}^{s} \sum_{j=1}^{s} W_{ij} h_{ij} = \sum_{i=1}^{s} \overline{W}_i p_i. \tag{10.7}$$

This means that the mean fitness of the population is a weighted mean of the mean fitnesses of the alleles at a given locus. The weights are frequencies of the corresponding alleles.

After selection, the corresponding "effective" frequencies, h'_{ij}, say, of the genotypes in the parental population are

$$h'_{ij} = \frac{W_{ij}}{\overline{W}} h_{ij}, \qquad i = 1, 2, \ldots, s, \tag{10.8}$$

and the "effective" parental genotypic array is

$$\sum_{i=1}^{s} \sum_{j=1}^{s} h'_{ij} A_i A_j = \frac{1}{\overline{W}} \sum_{i=1}^{s} \sum_{j=1}^{s} W_{ij} h_{ij} A_i A_j. \tag{10.9}$$

10.3.2.   Changes in Gene Frequencies

The "effective" gene frequencies [i.e., the gene frequencies calculated from the genotypic array (10.9)], p'_i, say, are

$$p'_i = \sum_{j=1}^{s} h'_{ij} = \frac{1}{\overline{W}} \sum_{j=1}^{s} W_{ij} h_{ij} = \frac{\overline{W}_i}{\overline{W}} p_i, \qquad i = 1, 2, \ldots, s. \tag{10.10}$$

Assuming that the generations are nonoverlapping, (10.10) represent the gene frequencies for the next generation.

The change in the gene frequency, Δp_i, say, due to selection is

$$\Delta p_i = p'_i - p_i = \frac{\overline{W}_i}{\overline{W}} p_i - p_i = \frac{\overline{W}_i - \overline{W}}{\overline{W}} p_i, \qquad i = 1, 2, \ldots, s. \tag{10.11}$$

For a single locus, the gamete frequencies are identical with the gene frequencies. Thus the condition for equilibrium, that the gamete frequencies remain constant through generations, is equivalent to the condition that the gene frequencies remain constant or that

$$\Delta p_i = \frac{\overline{W}_i - \overline{W}}{\overline{W}} p_i = 0, \qquad i = 1, 2, \ldots, s. \qquad (10.12)$$

Assuming all $p_i > 0$, the solution of the set of s equations (9.12) is

$$\overline{W}_i - \overline{W} = 0, \qquad i = 1, 2, \ldots, s, \qquad (10.13)$$

or

$$W_i = \overline{W}, \qquad i = 1, 2, \ldots, s. \qquad (10.13a)$$

We will return in section 10.6 to the problem of equilibrium under selection.

10.4. RANDOMLY MATING POPULATION UNDER SELECTION

For convenience and without loss of generality, let us assume that an initial population has a genotypic array

$$\sum_{i=1}^{s} p_i^2 A_i A_i + 2 \sum_{i<j} p_i p_j A_i A_j = \sum_{i=1}^{s} \sum_{j=1}^{s} p_i p_j A_i A_j.$$

Then, substituting $h_{ij} = p_i p_j$ in formula (10.2) and taking into account (10.7), we have

$$\overline{W} = \sum_{i=1}^{s} \sum_{j=1}^{s} W_{ij} p_i p_j = \sum_{i=1}^{s} \overline{W}_i p_i, \qquad (10.14)$$

with

$$\overline{W}_i = \sum_{j=1}^{s} W_{ij} p_j = \frac{1}{2} \frac{\partial \overline{W}}{\partial p_i}, \qquad i = 1, 2, \ldots, s. \qquad (10.15)$$

The quantity $\overline{W}$ in (10.14) defines the mean fitness at birth (or, more generally, *before* the onset of selection) of the *initial* population.

If the population mates at random, the genotypic array for the next generation is

$$\sum_{i=1}^{s} \sum_{j=1}^{s} p_i' p_j' A_i A_j = \frac{1}{\overline{W}^2} \sum_{i=1}^{s} \sum_{j=1}^{s} \overline{W}_i \overline{W}_j p_i p_j A_i A_j, \qquad (10.16)$$

On the assumption that W_{ij}'s are constant, the mean fitness, $\overline{W}'$, say, at the beginning of the *next* generation is

$$\overline{W}' = \sum_{i=1}^{s} \sum_{j=1}^{s} W_{ij} p_i' p_j' = \frac{1}{\overline{W}^2} \sum_{i=1}^{s} \sum_{j=1}^{s} W_{ij} \overline{W}_i \overline{W}_j p_i p_j. \qquad (10.17)$$

In sections 10.5–10.7 we shall derive some important results in natural selection for *randomly mating populations*. It is useful to formulate the assumptions under which these results hold:

> (i) The population is considered in regard to a single autosomal locus with multiple alleles.
> (ii) The fitnesses are constant from generation to generation.
> (iii) The fitnesses are the same in both sexes.
> (iv) The population is mating at random.
> (v) The generations are nonoverlapping.

(10.18)

In future, we will refer to these assumptions as "assumptions (10.18)." Theorems 10.1 and 10.2 in section 10.5 are based on these assumptions. In sections 10.8–10.12 we shall discuss the situations for which assumptions (10.18) do not all hold and the manner in which this affects the statements of Theorems 10.1 and 10.2.

10.5. THE RATE OF INCREASE OF THE MEAN FITNESS PER GENERATION. FUNDAMENTAL THEOREM OF NATURAL SELECTION

10.5.1. Change in the Mean Fitness per Generation

We now wish to find in *which direction the mean fitness changes per generation* due to selection if assumptions (10.18) hold. The model which we will consider is *discrete* with "jumps" of one generation.

THEOREM 10.1. *If assumptions* (10.18) *hold, the mean fitness increases in each successive generation until the stationary conditions* ($\Delta p_i = 0$, $i = 1, 2, \ldots, s$) *are obtained.*

To prove this, we shall evaluate the sign of $\Delta \overline{W} = \overline{W}' - \overline{W}$ for two successive generations, where $\overline{W}'$ and $\overline{W}$ are defined in (10.17) and (10.14), respectively. When $s > 2$, the algebra is rather tedious, so we restrict ourselves here to the case of two alleles, A and a. The proof for $s > 2$ can be found, for instance, in a paper by Mulholland and Smith (1959).

Proof for $s = 2$. Let us consider an initial population $p^2 AA + 2pq Aa + q^2 aa$ with fitnesses W_{11}, W_{12}, and W_{22} of genotypes AA, Aa, and aa, respectively. The mean fitness (10.14) can now be written as

$$\overline{W} = \overline{W}(p) = W_{11}p^2 + 2W_{12}p(1-p) + W_{22}(1-p)^2 = \overline{W}_1 p + \overline{W}_2(1-p),$$

(10.19)

with

$$\overline{W}_1 = W_{11}p + W_{12}(1-p), \quad \overline{W}_2 = W_{12}p + W_{22}(1-p). \quad (10.20)$$

After selection, the gene frequencies for the next generation are

$$p' = \frac{\overline{W}_1}{\overline{W}} p, \quad q' = 1 - p' = \frac{\overline{W}_2}{\overline{W}} (1 - p) = \frac{\overline{W}_2}{\overline{W}} q. \tag{10.21}$$

Hence,

$$\Delta p = p' - p = \frac{\overline{W}_1 - \overline{W}}{\overline{W}} p = \frac{\overline{W}_1 - \overline{W}_1 p - \overline{W}_2(1 - p)}{\overline{W}} p$$

$$= \frac{\overline{W}_1(1 - p) - \overline{W}_2(1 - p)}{\overline{W}} p = p(1 - p) \frac{\overline{W}_1 - \overline{W}_2}{\overline{W}}$$

$$= pq \frac{\overline{W}_1 - \overline{W}_2}{\overline{W}}. \tag{10.22}$$

Here $\overline{W}_1 - \overline{W}_2$ is the mean *effect* of the fitness due to substitution of the gene A for the gene a.

The mean fitness at the beginning of the *next* generation is

$$\overline{W}' = \overline{W}(p') = W_{11}p'^2 + 2W_{12}p'(1 - p') + W_{22}(1 - p')^2. \tag{10.23}$$

We now expand the function $\overline{W}(p')$ as a Taylor series about the point $p' = p$. This is a quadratic function of p', and expansion includes only terms up to the second derivative. We have

$$\overline{W}(p') = \overline{W}(p) + \frac{d\overline{W}}{dp} \Delta p + \frac{d^2\overline{W}}{dp^2} \frac{(\Delta p)^2}{2}, \tag{10.24}$$

where, for simplicity, we have denoted

$$\frac{d\overline{W}}{dp'}\bigg|_{p'=p} \text{ by } \frac{d\overline{W}}{dp}, \quad \text{and} \quad \frac{d^2\overline{W}}{dp'^2}\bigg|_{p'=p} \text{ by } \frac{d^2\overline{W}}{dp^2}.$$

We now find these derivatives. We have

$$\frac{d\overline{W}}{dp} = 2[W_{11}p + W_{12} - 2W_{12}p - W_{22}(1 - p)]$$

$$= 2\{[W_{11}p + W_{12}(1 - p)] - [W_{12}p + W_{22}(1 - p)]\}$$

$$= 2(\overline{W}_1 - \overline{W}_2); \tag{10.25}$$

$$\frac{d^2\overline{W}}{dp^2} = 2(W_{11} + W_{22} - 2W_{12}). \tag{10.26}$$

Substituting (10.23), (10.25), and (10.26), respectively, into (10.24) and subtracting $\overline{W}(p)$ from both sides of (10.24), we obtain

$$\Delta \overline{W} = \overline{W}(p') - \overline{W}(p) = \frac{2p(1 - p)(\overline{W}_1 - \overline{W}_2)^2}{\overline{W}}$$

$$+ (W_{11} + W_{22} - 2W_{12})\left[\frac{p(1 - p)(\overline{W}_1 - \overline{W}_2)}{\overline{W}}\right]^2$$

$$= 2p(1 - p)(\overline{W}_1 - \overline{W}_2)^2 \cdot \frac{2\overline{W} + p(1 - p)(W_{11} + W_{22} - 2W_{12})}{2\overline{W}^2} \tag{10.27}$$

Taking into account (10.19), we can write formula (10.27) in the form

$$\Delta \overline{W} = 2p(1 - p)(\overline{W}_1 - \overline{W}_2)^2 \cdot \frac{\overline{W} + [W_{11}p + W_{22}(1 - p)]}{2\overline{W}^2} \geqslant 0. \quad (10.27a)$$

In the nontrivial case, $p > 0$, $q > 0$, we see that the right-hand side of (10.27a) is always nonnegative. This means that the mean fitness will increase in each successive generation unless the stationary state $\{\Delta p = 0$ and hence $\Delta \overline{W} = 0)$ has been attained. We have then proved Theorem 10.1 for $s = 2$.

10.5.2. The Rate of Increase of the Mean Fitness

We now ask another question: *How much does the fitness increase per generation?*
We first notice that, if

$$W_{11} + W_{22} - 2W_{12} = 0, \quad (10.28)$$

formula (10.27) takes the form

$$\Delta \overline{W} = \frac{1}{\overline{W}} \cdot 2p(1 - p)(\overline{W}_1 - \overline{W}_2)^2. \quad (10.29)$$

We wish to find an interpretation of the term $2p(1 - p)(\overline{W}_1 - \overline{W}_2)^2$. Let us consider the fitnesses W_{ij} as "genotypic values" and suppose that they are *additive* functions of the gene fitnesses, $\overline{W}_i$, of the form

$$W_{11} = \overline{W} + 2\overline{W}_1, \quad W_{12} = \overline{W} + \overline{W}_1 + \overline{W}_2, \quad W_{22} = \overline{W} + 2\overline{W}_2. \quad (10.30)$$

We first notice that the functions given in (10.30) satisfy the condition (10.28).
We now recall the results of section 6.7. Putting $g_0 = \overline{W} + 2\overline{W}_2$ and $\beta = \overline{W}_1 - \overline{W}_2$ in (6.54), and replacing the variable G by W, we find, from (6.55) and (6.56), that

$$E(W) = \mu_W = \overline{W}, \quad \text{and} \quad \text{Var}(W) = \sigma^2_{Ad}(W) = 2p(1 - p)(\overline{W}_1 - \overline{W}_2)^2. \quad (10.31)$$

The variance $\sigma^2_{Ad}(W)$ represents here the *additive genetic variance* with respect the fitness. Thus, formula (10.29) can be written as

$$\Delta \overline{W} = \frac{1}{\overline{W}} \sigma^2_{Ad}(W). \quad (10.32)$$

We summarize the results in the following theorem, called the *fundamental theorem of natural selection.*

THEOREM 10.2. *If a population satisfies conditions* (10.18), *and additionally the fitnesses of the genotypes,* W_{ij}, *are additive functions of the fitnesses of genes,* $\overline{W}_i$, *as given in* (10.30), *then the rate of increase in mean fitness per generation is proportional to the additive genetic variance with respect to fitness.*

▼ 10.6. STABLE AND UNSTABLE EQUILIBRIA

10.6.1. Necessary Conditions for Equilibrium

We again consider a population with respect to a single locus with s alleles, satisfying the condition (10.18). We now wish to find values for gene frequencies, for which the mean fitness attains a stationary point so that $\Delta p_i = 0$, for $i = 1, 2, \ldots, s$. We denote these values by p_i^*, $i = 1, 2, \ldots, s$.

It will be convenient to represent the function $\overline{W}$, given by (10.14), as a quadratic form, that is,

$$\overline{W} = \overline{W}(\mathbf{p}) = \sum_{i=1}^{s} \sum_{j=1}^{s} W_{ij} p_i p_j = \mathbf{p}'\mathbf{W}\mathbf{p}, \tag{10.33}$$

where $\mathbf{p}' = (p_1, p_2, \ldots, p_s)$ is a $1 \times s$ vector of gene frequencies in the initial population, and

$$\mathbf{W} = \begin{bmatrix} W_{11} & W_{12} & \cdots & W_{1s} \\ W_{21} & W_{22} & \cdots & W_{2s} \\ \multicolumn{4}{c}{\dotfill} \\ W_{s1} & W_{s2} & \cdots & W_{ss} \end{bmatrix} \tag{10.34}$$

is a $s \times s$ matrix of the genotype fitnesses, with $W_{ij} = W_{ji}$, so that $\mathbf{W}$ is symmetrical, and we have $\mathbf{W} = \mathbf{W}'$ and $\mathbf{W}^{-1} = (\mathbf{W}^{-1})'$. (Note that now $\mathbf{p}'$, $\mathbf{W}'$, etc., are the transposes of the vector $\mathbf{p}$ and the matrix $\mathbf{W}$, respectively, and not the values of the parameters in the next generation.)

The stationary values for the p's will be obtained by solving the system of s equations $\Delta\mathbf{p} = 0$ or, equivalently [see (10.13a)], the system

$$\left.\begin{aligned} \overline{W}_1 &= W_{11}p_1 + W_{12}p_2 + \cdots + W_{1s}p_s = \overline{W}, \\ \overline{W}_2 &= W_{21}p_1 + W_{22}p_2 + \cdots + W_{2s}p_s = \overline{W}, \\ &\quad \cdots\cdots\cdots\cdots\cdots\cdots\cdots\cdots\cdots\cdots \\ \overline{W}_s &= W_{s1}p_1 + W_{s2}p_2 + \cdots + W_{ss}p_s = \overline{W}. \end{aligned}\right\} \tag{10.35}$$

System (10.35) can be written in the matrix form

$$\mathbf{W}\mathbf{p} = \overline{W}\mathbf{1}. \tag{10.35a}$$

where $\mathbf{1}' = (1, 1, \ldots, 1)$ is a $1 \times s$ unit vector.

Let us now find the *extrema* (minima or maxima) of $\overline{W}$, given in (10.33), as a function of $p_1, p_2, \ldots, p_s$. We notice that p's are subject to the constraint $\sum p_i = 1$. We then apply the method of Lagrange multipliers for finding extrema subject to side conditions. We introduce a function

$$M(\mathbf{p}) = \overline{W}(\mathbf{p}) - 2\phi(p_1 + p_2 + \cdots + p_s - 1) \qquad (10.36)$$

and solve the system of $s + 1$ equations

$$\left. \begin{array}{l} \dfrac{\partial M}{\partial p_i} = 0, \qquad i = 1, 2, \ldots, s, \\[2em] \displaystyle\sum_{i=1}^{s} p_i = 1. \end{array} \right\}$$

It is easy to see that this system takes the explicit form

$$\left. \begin{array}{l} W_{11}p_1 + W_{12}p_2 + \cdots + W_{1s}p_s = \phi, \\ W_{21}p_1 + W_{22}p_2 + \cdots + W_{2s}p_s = \phi, \\ \qquad \cdots\cdots\cdots\cdots\cdots\cdots\cdots\cdots\cdots \\ W_{s1}p_1 + W_{s2}p_2 + \cdots + W_{ss}p_s = \phi, \\ p_1 + p_2 + \cdots + p_s = 1, \end{array} \right\} \qquad (10.37)$$

or in matrix form

$$\left. \begin{array}{l} \mathbf{Wp} = \phi\mathbf{1}, \\ \mathbf{1'p} = 1. \end{array} \right\} \qquad \begin{array}{l} (10.38i) \\ (10.38ii) \end{array}$$

Solving (10.38i) with respect to p's, we obtain the vector of values at extremum, $\overset{\circ}{\mathbf{p}}$, say, that is,

$$\overset{\circ}{\mathbf{p}} = \mathbf{W}^{-1}\mathbf{1}\phi. \qquad (10.39)$$

Substituting (10.39) into (10.38ii), we have

$$\mathbf{1'W}^{-1}\mathbf{1}\phi = 1;$$

hence

$$\phi = \frac{1}{\mathbf{1'W}^{-1}\mathbf{1}}. \qquad (10.40)$$

Substituting (10.40) into (10.39), we finally obtain

$$\overset{\circ}{\mathbf{p}} = \frac{\mathbf{W}^{-1}\mathbf{1}}{\mathbf{1'W}^{-1}\mathbf{1}}. \qquad (10.41)$$

[This result is due to Tallis (1966).]

We also notice that, substituting (10.39) into (10.33), we obtain

$$\overline{W} = \overline{W}(\overset{\circ}{\mathbf{p}}) = \overset{\circ}{\mathbf{p}}'\mathbf{W}\overset{\circ}{\mathbf{p}} = \frac{\mathbf{1'W}^{-1}\mathbf{WW}^{-1}\mathbf{1}}{\mathbf{1'W}^{-1}\mathbf{1}}\phi = \phi. \qquad (10.42)$$

If we now compare the system of "equilibrium" equations (10.35) with the system for extrema (10.37), we notice that they are identical, with ϕ replaced by $\overline{W}$ evaluated at $\overset{o}{\mathbf{p}}$. Thus (10.41) gives the solution for both: the values of p's at which $\overline{W}$ attains an extremum and at which the gene frequencies are stationary (do not change). In other words, we have $\mathbf{p}^* = \overset{o}{\mathbf{p}}$. We summarize this in the following theorem.

THEOREM 10.3. *If assumptions (10.18) hold, the necessary condition for equilibrium is that the mean fitness of the population attains an extremum.*

In other words, the stationary gene frequencies p_i^*'s are identical with the corresponding $\overset{o}{p}$'s at which the function $\overline{W}$ given by (10.33) attains an extremum. The values of the gene frequencies at the equilibrium are

$$\mathbf{p}^* = \overset{o}{\mathbf{p}} = \frac{\mathbf{W}^{-1}\mathbf{1}}{\mathbf{1}'\mathbf{W}^{-1}\mathbf{1}}, \tag{10.43}$$

or in the explicit form

$$p_i^* = \overset{o}{p}_i = \frac{\sum\limits_{j=1}^{s} W^{ij}}{\sum\limits_{i=1}^{s}\sum\limits_{j=1}^{s} W^{ij}}, \qquad i = 1, 2, \ldots, s. \tag{10.43a}$$

It is interesting to notice that the mean fitness at equilibrium, $\overline{W}^*$, say, is

$$\overline{W}^* = \overline{W}(\overset{o}{\mathbf{p}}) = \frac{1}{\mathbf{1}'\overline{W}^{-1}\mathbf{1}} = \frac{1}{\sum\limits_{i=1}^{s}\sum\limits_{j=1}^{s} W^{ij}}, \tag{10.44}$$

which means that the mean fitness of the population under equilibrium is the reciprocal of the sum of the elements of the inverse matrix $\mathbf{W}^{-1}$.

10.6.2. Necessary and Sufficient Conditions for Stable Equilibrium

We shall now define what we call *stable* equilibrium and derive conditions, in terms of gene frequencies, which ensure that a population tends to stability.

Suppose that some factors disturb slightly the stationary state of a non-overlapping population which has attained equilibrium. If the population tends to return to the same stationary state, we shall call this equilibrium *stable*, if it tends to move away from this state, the equilibrium is *unstable*.

This can be expressed in terms of gene frequencies. Let p_i^* ($i = 1, 2, \ldots, s$) be the frequency of the gene A_i at the equilibrium, and p_i and p_i' its frequencies in two successive generations, respectively. If the "distance," $|p_i' - p_i^*|$, in the next generation is smaller than the "distance," $|p_i - p_i^*|$, in the previous generation, and this effect is observed in successive generations until the stationary state is attained again, then the equilibrium is *stable*. Otherwise, it will be called *unstable*.

We summarize this in the following definition.

DEFINITION 10.5.   *The equilibrium is stable at the point* $\mathbf{p} = \mathbf{p}^*$ *if, for all sufficiently small* $|p_i - p_i^*|$, *we have*

$$|p_i' - p_i^*| < |p_i - p_i^*|, \qquad i = 1, 2, \ldots, s, \qquad (10.45)$$

for all successive generations until the stationary point $\mathbf{p} = \mathbf{p}^*$ *is reached. If* (10.45) *is not satisfied, the equilibrium is unstable.*

It is worthwhile to notice that (10.45) implies that

$$\Delta p_i = p_i' - p_i < 0 \qquad \text{if } p_i > p_i^*,$$

and

$$\Delta p_i = p_i' - p_i > 0 \qquad \text{if } p_i < p_i^*, \qquad (i = 1, 2, \ldots, s) \qquad (10.45a)$$

but not conversely.

As was stated in Theorem 10.1, the mean fitness $\overline{W}(\mathbf{p})$ never decreases from one generation to the next. It follows that (10.45) will be satisfied if and only if p corresponds to maximum of $\overline{W}(\mathbf{p})$. Thus, stable equilibrium with respect to a single locus is reached when the mean fitness, $\overline{W}(\mathbf{p})$, attains a (local) maximum, while at a (local) minimum of $\overline{W}(\mathbf{p})$ the equilibrium is unstable. (See Figs. 10.1a and 10.1b.)

The mean fitness, $\overline{W}(\mathbf{p})$, as a function of p_i's represents a quadratic form (10.33). Thus the sufficient conditions for equilibrium to be stable are

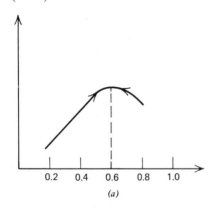

(a)

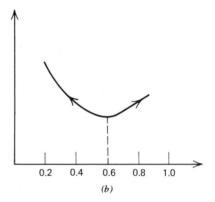

(b)

Stable equilibrium at $p^* = 0.6$ Unstable equilibrium at $p^* = 0.6$

$\Delta p < 0$ if $p > p^*$ $\Delta p > 0$ if $p > p^*$

$\Delta p > 0$ if $p < p^*$ $\Delta p < 0$ if $p < p^*$

Figure 10.1.

equivalent to those for the maximum of a quadratic form (see Appendix II).

First, we notice that to obtain $\mathbf{W}^{-1}$ the matrix $\mathbf{W}$ must be nonsingular, that is, the determinant $|\mathbf{W}| \neq 0$.

Second, to obtain nontrivial solutions for p_i's (i.e., $p_i > 0$ for each i) the numerator and the denominator of (10.43a) must have the same sign, that is, either

$$\left. \begin{aligned} \sum_{j=1}^{s} W^{ij} > 0, \\ \sum_{i=1}^{s} \sum_{j=1}^{s} W^{ij} > 0, \end{aligned} \right\} \qquad (10.46a)$$

or

$$\left. \begin{aligned} \sum_{j=1}^{s} W^{ij} < 0, \\ \sum_{i=1}^{s} \sum_{j=1}^{s} W^{ij} < 0, \end{aligned} \right\} \qquad (10.46b)$$

where $i = 1, 2, \ldots, s$. This can be reduced to

$$\sum_{j=1}^{s} W^{ij} > 0, \qquad (10.47a)$$

or

$$\sum_{j=1}^{s} W^{ij} < 0, \qquad (10.47b)$$

where $i = 1, 2, \ldots, s$, respectively.

Third, since $\sum p_i = 1$, we have to modify slightly the matrix of the quadratic form. We put

$$p_s = 1 - \sum_{j=1}^{s-1} p_j. \qquad (10.48)$$

Substituting (10.48) into (10.33), we obtain

$$\overline{W} = \sum_{i=1}^{s-1} \sum_{j=1}^{s-1} W_{ij} p_i p_j + W_{ss} \left(1 - \sum_{j=1}^{s} p_j \right)^2 + \sum_{i=1}^{s} W_{is} p_i \left(1 - \sum_{j=1}^{s-1} p_j \right). \qquad (10.49)$$

The second partial derivative, $(\partial^2 \overline{W} / \partial p_k \, \partial p_l)$, is

$$\frac{\partial^2 \overline{W}}{\partial p_k \, \partial p_l} = 2[(W_{kl} + W_{ss}) - (W_{ks} + W_{ls})] = 2t_{kl}, \quad \text{say.} \qquad (10.50)$$

According to the standard theory for evaluating the maximum of a quadratic form (see Appendix II), the matrix of second derivatives should be negative definite. This is equivalent to requiring that the leading principal minor, of order r, of the matrix $\mathbf{T} = (t_{kl})$, has the sign $(-1)^r$ for $r = 1, 2, \ldots, s - 1$. We have the following theorem.

THEOREM 10.4. *Necessary and sufficient conditions for a population satisfying* (10.18) *to attain a stable equilibrium are as follows:*

(i) *the determinant* $|\mathbf{W}| \neq 0$;

(ii) $\sum\limits_{j=1}^{s} W^{ij} \neq 0, i = 1, 2, \ldots, s$;

(iii) *the leading principal minor* $|\mathbf{T}_r|$, *of the order* r, *of the* $(s-1) \times (s-1)$ *matrix* $\mathbf{T}$, *whose elements* t_{kl} *are defined as*

$$t_{kl} = (W_{kl} + W_{ss}) - (W_{ks} + W_{ls}), \qquad k, l = 1, 2, \ldots, s,$$

has the sign $(-1)^r$, *for* $r = 1, 2, \ldots, s-1$.

Example 10.4. Let $s = 3$, $\mathbf{p'} = (0.25, 0.40, 0.35)$, and

$$\mathbf{W} = \begin{pmatrix} 1 & 2 & 2 \\ 2 & 1 & 2 \\ 2 & 2 & 1 \end{pmatrix},$$

which means that the fitness of each homozygote A_iA_i is $W_{ii} = 1$, and the fitness of each heterozygote A_iA_j is $W_{ij} = 2$, for $i, j = 1, 2, \ldots, s, i \neq j$.

(a) Find the gene frequencies at equilibrium.

We have $|\mathbf{W}| = 5$ and

$$\mathbf{W}^{-1} = \tfrac{1}{5} \begin{bmatrix} -3 & 2 & 2 \\ 2 & -3 & 2 \\ 2 & 2 & -3 \end{bmatrix}.$$

We have

$$\sum_{j=1}^{3} W^{1j} = \sum_{j=1}^{3} W^{2j} = \sum_{j=1}^{3} W^{3j} = \tfrac{1}{5}$$

so that

$$\sum_{i=1}^{s} \sum_{j=1}^{s} W^{ij} = \tfrac{3}{5}.$$

From (10.43a) we obtain

$$p_1^* = p_2^* = p_3^* = \tfrac{1}{3}.$$

The trivial equilibrium is, of course, $(1, 0, 0)$ or $(0, 1, 0)$ or $(0, 0, 1)$. Assuming $p_1 = 0$, we obtain another equilibrium at $(0, \tfrac{1}{2}, \tfrac{1}{2})$. Similarly, for $p_2 = 0$, we have equilibrium at $(\tfrac{1}{2}, 0, \tfrac{1}{2})$ and, for $p_3 = 0$, at $(\tfrac{1}{2}, \tfrac{1}{2}, 0)$.

(b) Calculate $\bar{W}_i$ $(i = 1, 2, 3)$ and $\bar{W}$ for the initial population, and $\bar{W}_i^*$ $(i = 1, 2, 3)$ and $\bar{W}^*$ for the population under equilibrium.

For the initial population we find, from (10.15) and (10.14), the values of the $\overline{W}_i$'s and $\overline{W}$, respectively:

$$\overline{W}_1 = 0.25 \times 1 + 0.40 \times 2 + 0.35 \times 2 = 1.75,$$

$$\overline{W}_2 = 0.25 \times 2 + 0.40 \times 1 + 0.35 \times 2 = 1.60,$$

$$\overline{W}_3 = 0.25 \times 2 + 0.40 \times 2 + 0.35 \times 1 = 1.65,$$

$$\overline{W} = 0.25 \times 1.75 + 0.40 \times 1.60 + 0.35 \times 1.65 = 1.6550.$$

For the population at equilibrium, we find

$$\overline{W}_1^* = \overline{W}_2^* = \overline{W}_3^* = \overline{W}^* = \tfrac{1}{3}(1 + 2 + 2) = \tfrac{5}{3} = 1.6667.$$

(c) Is the equilibrium, with three positive p's, stable? We now calculate the elements of the matrix $\mathbf{T}$:

$$t_{11} = (W_{11} + W_{33}) - (W_{13} + W_{13}) = (1 + 1) - (2 + 2) = -2,$$

$$t_{12} = t_{21} = (W_{12} + W_{33}) - (W_{13} + W_{23}) = (2 + 1) - (2 + 2) = -1,$$

$$t_{22} = (W_{22} + W_{33}) - (W_{23} + W_{23}) = (1 + 1) - (2 + 2) = -2.$$

The 2×2 matrix $\mathbf{T}$ is

$$\mathbf{T} = \begin{bmatrix} -2 & -1 \\ -1 & -2 \end{bmatrix}.$$

Here

$$|\mathbf{T}_1| \, t_{11} = -2 < 0, \quad |\mathbf{T}_2| = |\mathbf{T}| = \begin{vmatrix} -2 & -1 \\ -1 & -2 \end{vmatrix} = 3 > 0.$$

The conditions of Theorem 10.3 are satisfied, and therefore the equilibrium is stable.

10.7. SINGLE LOCUS WITH TWO ALLELES

The case $s = 2$ is of special interest. Let p and q be the frequencies of genes A and a, respectively, in the initial population, and p^* and q^* the corresponding stationary gene frequencies. The matrix of fitnesses and its inverse are

$$\mathbf{W} = \begin{bmatrix} W_{11} & W_{12} \\ W_{12} & W_{22} \end{bmatrix}, \quad \mathbf{W}^{-1} = \frac{1}{|\mathbf{W}|} \begin{bmatrix} W_{22} & -W_{12} \\ -W_{12} & W_{11} \end{bmatrix}, \tag{10.51}$$

where $|\mathbf{W}| = W_{11}W_{22} - W_{12}^2$. If $W_{12} \neq \sqrt{W_{11}W_{22}}$, then $|\mathbf{W}| \neq 0$.

The gene frequencies in the next generation are

$$p' = \frac{W_{11}p + W_{12}q}{\overline{W}} \, p, \quad q' = \frac{W_{22}q + W_{12}p}{\overline{W}} \, q. \tag{10.52}$$

The gene frequencies at equilibrium are

$$p^* = \frac{W_{22} - W_{12}}{W_{11} + W_{22} - 2W_{12}}, \quad q^* = \frac{W_{11} - W_{12}}{W_{11} + W_{22} - 2W_{12}}. \tag{10.53}$$

Here the matrix $\mathbf{T}$ reduces to one term, t_{11}, so that we have

$$\mathbf{T} = t_{11} = W_{11} + W_{22} - 2W_{12}. \tag{10.54}$$

Comparing (10.54) with (10.26), we notice that $\mathbf{T} = t_{11} = \frac{1}{2}(d^2\overline{W}/dp^2)$.

Applying Theorem 10.3 and requiring p^* and q^* to be positive, we find that stable equilibrium is attained when $W_{11} - W_{12} < 0$, $W_{22} - W_{12} < 0$ (this also implies that $W_{11} + W_{22} - 2W_{12} < 0$). This is equivalent to $W_{11} < W_{12} > W_{22}$, which means that stable equilibrium is reached when the heterozygotes are at a *selective advantage*.

The gene frequencies p^*, q^* are also positive if $W_{11} - W_{12} > 0$, $W_{22} - W_{12} > 0$ (which implies $W_{11} + W_{22} - 2W_{12} > 0$), or $W_{11} > W_{12} < W_{22}$. In this case equilibrium is unstable (since $\frac{1}{2}(d^2W/dp^2) = W_{11} + W_{22} - 2W_{12} > 0$ and $\overline{W}$ attains a minimum).

It is worthwhile to notice that, for $s > 2$, selective advantages of heterozygotes are not necessary for stable equilibrium (see Problem 10.1). On the other hand, if all homozygotes have the same fitnesses, and each heterozygote has the same selective advantage over a homozygote, then the equilibrium will be stable [Tallis (1966)]. (See also Problem 10.6.)

Special attention should be paid to the cases in which $\frac{1}{2}(d^2\overline{W}/dp^2) = W_{11} + W_{22} - 2W_{12} = 0$. Then formulae (10.53) do not apply, and each situation should be considered individually. Here we will present one simple model of this case.

10.7.1. A Model, When *aa* is Lethal and *Aa* Sublethal

Suppose that the fitnesses of the genotypes AA, Aa, and aa are $W_{11} = 2$, $W_{12} = 1$, and $W_{22} = 0$, respectively. We then have $W_{11} + W_{22} - 2W_{12} = 0$. This can happen when the gene a is lethal in the recessive condition, aa, and is sublethal in the heterozygous condition, Aa. From (10.20) we calculate $\overline{W}_1 = 2p + (1 - p) = 1 + p$; $\overline{W}_2 = p + 0 = p$. From (10.14), $\overline{W} = (1 + p)p + p(1 - p) = 2p$. From (10.21) we have

$$p' = \frac{1 + p}{2p} \cdot p = \tfrac{1}{2}(1 + p) \quad \text{and} \quad q' = \frac{p}{2p} \cdot q = \tfrac{1}{2}q.$$

Of course, these formulae apply to any two successive generations. Let $q^{(n-1)}$ and $q^{(n)}$ denote the frequencies of allele a after $n-1$ and n generations of selection, respectively. Then we have

$$q^{(n)} = \tfrac{1}{2}q^{(n-1)} = (\tfrac{1}{2})^{n-1}q, \tag{10.55}$$

where q is the frequency of allele a in the initial population. The equilibrium will be attained for

$$\lim_{n \to \infty} q^{(n)} = \lim_{n \to \infty} (\tfrac{1}{2})^{n-1}q = 0. \tag{10.56}$$

In other words, under equilibrium the population will be homogeneous, consisting entirely of homozygotes AA. The equilibrium, of course, will be stable.

10.8. RANDOMLY MATING POPULATION. VARIABLE FITNESSES, DEPENDING ON GENE FREQUENCIES

We now consider the situation for which assumptions (10.18) hold, *except for* (ii), that is, the fitnesses are not constant, but are functions of gene frequencies: $W_{ij} = W_{ij}(p_1, p_2, \ldots, p_s)$.

The "equilibrium" equations (10.35) are still valid, but, since the W_{ij}'s are now functions of p_i's, the solutions depend on the particular selection pattern.

We now find the extrema of the function $M(\mathbf{p})$ defined in (10.36). We have to solve the system of equations

$$\left.\begin{aligned} \frac{1}{2}\frac{\partial M}{\partial p_i} &= \sum_{j=1}^{s} W_{ij}p_j + \sum_{i=1}^{s}\sum_{j=1}^{s}\frac{\partial W_{ij}}{\partial p_i}p_ip_j - \phi = 0, \quad i = 1, 2, \ldots, s, \\ \sum_{i=1}^{s} p_i &= 1. \end{aligned}\right\} \tag{10.57}$$

It can be seen that this system is not equivalent to (10.37) unless

$$\sum_{i=1}^{s}\sum_{j=1}^{s}\frac{\partial W_{ij}}{\partial p_i}p_ip_j = 0.$$

This means that the stable equilibrium does not necessarily correspond to the maximum fitness. The solution of (10.57) depends, of course, on the selection model being considered. As we will see from Example 10.5, it is possible to construct a model in which stable equilibrium corresponds to a minimum, and not to a maximum, of $\overline{W}$.

Example 10.5. Sacks (1967) considered the following selection pattern:

Genotype: AA Aa aa

Probability: p^2 $2p(1-p)$ $(1-p)^2$

Fitness: $c + \dfrac{1}{p^2}$ $-c + \dfrac{1}{p(1-p)}$ $c + \dfrac{1}{(1-p)^2}$.

To have W_{ij} nonnegative, we must have $-1 \leqslant c \leqslant 4$. We calculate $\overline{W}$ from (10.14), that is,

$$\begin{aligned}
\overline{W} &= (cp^2 + 1) - [2cp(1-p) - 2] + c(1-p)^2 + 1 \\
&= 4 + c[p^2 - 2p(1-p) + (1-p^2)] = 4 + c(2p-1)^2 \\
&= 4 + 4c(\tfrac{1}{2} - p)^2.
\end{aligned}$$

When $p = \tfrac{1}{2}$, $\overline{W}$ is maximum if $c < 0$, and minimum if $c > 0$. We now calculate from (10.20)

$$\overline{W}_1 = \left(cp + \frac{1}{p}\right) + \left[-c(1-p) + \frac{1}{p}\right] = \frac{1}{p}\,(2cp^2 - cp + 2)$$

$$= \frac{2}{p}\,[-cp(\tfrac{1}{2} - p) + 1].$$

Hence

$$\Delta p = \frac{\overline{W}_1 - \overline{W}}{\overline{W}}\,p = \frac{2[-cp(\tfrac{1}{2} - p) + 1 - 2p - 2cp(\tfrac{1}{2} - p)^2]}{\overline{W}}$$

$$= \frac{4(\tfrac{1}{2} - p)[1 - cp(1-p)]}{\overline{W}}.$$

Since $[1 - cp(1-p)] \geqslant 0$, equilibrium ($\Delta p = 0$) occurs for $p = q = \tfrac{1}{2}$. We notice that the sign of Δp depends on the sign of $(\tfrac{1}{2} - p)$. We have

$$\Delta p < 0 \text{ if } p > \tfrac{1}{2}, \quad \text{and} \quad \Delta p > 0 \text{ if } p < \tfrac{1}{2}.$$

Thus, for $-1 \leqslant c \leqslant 4$, the equilibrium $p = \tfrac{1}{2}$ is stable, but in particular, when $0 < c \leqslant 4$, $\overline{W}$ attains, at $p = \tfrac{1}{2}$, its minimum.

10.9. DIFFERENT FITNESSES IN THE TWO SEXES

So far we have assumed in our models that the fitnesses in males and females are the same. But this is not always the case. If fitnesses in both sexes do not differ very much, the preceding results hold approximately. If there are substantial differences, however, new models have to be derived.

In such cases, after selection, the gene frequencies in males and females will change in different proportions and must be evaluated separately. After random mating, the genotype frequencies, before selection, can be evaluated in the usual manner. Again taking into account different sex fitnesses, the new gene frequencies, in males and females, have to be evaluated. The equilibrium conditions for an autosomal locus depend on the selection pattern and do not necessarily require equal gene frequencies in males and females.

Space does not permit us to go into detail concerning models appropriate for an autosomal locus; some such models can be found in a paper by Li (1967b) and in a book by Wright (1969). In the next section, however, we will consider an X-linked locus, where genotypes have different fitnesses, so that the reader can obtain some idea of the techniques for the autosomal locus case, with different fitnesses in the two sexes.

10.10. SELECTION AT THE X-LINKED LOCUS

Let us consider an X-linked locus A, with alleles A, a. Suppose that selection operates at this locus in a different way in males and females, so that fitnesses are not the same in the two sexes. We assume that the population mates at random. Let us consider the following model of a population *before* selection occurs:

	Males		Females		
Genotype:	A	a	AA	Aa	aa
Probability:	m_1	m_2	$m_1 f_1$	$m_1 f_2 + m_2 f_1$	$m_2 f_2$	(10.58)
Fitness:	V_1	V_2	W_{11}	W_{12}	$W_{22}.$

We recall the notations of section 4.8, where m_1, m_2 (with $m_1 + m_2 = 1$) are the *maternal* gamete frequencies, and f_1, f_2 (with $f_1 + f_2 = 1$) are the *paternal* gamete frequencies. We also remind the reader that for an X-linked locus the genotype frequencies in male offspring are equal to the maternal gamete frequencies so that, in fact, we need only to find the equilibrium conditions for females.

After selection is completed, the *paternal* gamete frequencies for the *next* generation of females, f_1', f_2', say, are

$$f_1' = \frac{V_1 m_1}{\bar{V}}, \quad f_2' = \frac{V_2 m_2}{\bar{V}}, \tag{10.59}$$

with

$$\bar{V} = V_1 m_1 + V_2 m_2. \tag{10.60}$$

The *maternal* gamete frequencies for the next generation, m_1', m_2', say, are

$$\left.\begin{aligned}
m_1' &= \frac{2W_{11}m_1f_1 + W_{12}(m_1f_2 + m_2f_1)}{2\overline{W}} \\
m_2' &= \frac{2W_{22}m_2f_2 + W_{12}(m_1f_2 + m_2f_1)}{2\overline{W}},
\end{aligned}\right\} \tag{10.61}$$

where

$$\overline{W} = W_{11}m_1f_1 + W_{12}(m_1f_2 + m_2f_1) + W_{22}m_2f_2. \tag{10.62}$$

If a generation is in equilibrium, the gene frequencies in maternal and paternal populations should be the same, respectively, as in the preceding generation, that is, we must have

$$m_1' = m_1, \quad m_2' = m_2 \quad \text{and} \quad f_1' = f_1, \quad f_2' = f_2. \tag{10.63}$$

Taking the ratio $m_1'/m_2' = m_1/m_2$ and substituting (10.61) for m_1' and m_2', we obtain

$$\frac{2W_{11}m_1f_1 + W_{12}(m_1f_2 + m_2f_1)}{2W_{22}m_2f_2 + W_{12}(m_1f_2 + m_2f_1)} = \frac{m_1}{m_2}. \tag{10.64}$$

But since, at equilibrium, also $f_1' = f_1, f_2' = f_2$, then, substituting (10.59) into (10.64) and taking into account the fact that $m_2 = 1 - m_1$, we obtain

$$2W_{11}V_1m_1 + W_{12}(V_1 + V_2)(1 - m_1) = 2W_{22}V_2(1 - m_1) + W_{12}(V_1 + V_2)m_1. \tag{10.65}$$

Solving (10.65) for m_1, we obtain the frequency under equilibrium for the maternal gene A, m_1^*, say,

$$m_1^* = \frac{W_{11}V_1 - \frac{1}{2}W_{12}(V_1 + V_2)}{(W_{11}V_1 + W_{22}V_2) - W_{12}(V_1 + V_2)}, \tag{10.66a}$$

and hence

$$m_2^* = 1 - m_1^* = \frac{W_{22}V_2 - \frac{1}{2}W_{12}(V_1 + V_2)}{(W_{11}V_1 + W_{22}V_2) - W_{12}(V_1 + V_2)}. \tag{10.66b}$$

Substituting these values into (10.59), we obtain the values for the paternal gamete frequencies at equilibrium, f_1^*, f_2^*, say, that is,

$$f_1^* = \frac{V_1}{\overline{V}^*}m_1^*, \quad f_2^* = \frac{V_2}{\overline{V}^*}m_2^*, \tag{10.67}$$

where

$$\overline{V}^* = V_1m_1^* + V_2m_2^*. \tag{10.68}$$

Note that the maternal and paternal gene frequencies at equilibrium are not in general the same, unless $V_1 = V_2$.

We now wish to find the conditions for *stable* equilibrium. We notice that

(10.66a) can be written in the form

$$m_1^* = \frac{W_{22}V_2 - \frac{1}{2}W_{12}(V_1 + V_2)}{[W_{11}V_1 - \frac{1}{2}W_{12}(V_1 + V_2)] + [W_{22}V_2 - \frac{1}{2}W_{12}(V_1 + V_2)]}. \quad (10.69)$$

In order to obtain positive nontrivial solutions (i.e., $0 < m_1^* < 1$), $W_{22}V_2 - \frac{1}{2}W_{12}(V_1 + V_2)$ and $W_{11}V_1 - \frac{1}{2}W_{12}(V_1 + V_2)$ should have the same sign, that is, either

$$W_{11}V_1 < \tfrac{1}{2}W_{12}(V_1 + V_2) > W_{22}V_2 \quad (10.70)$$

or

$$W_{11}V_1 > \tfrac{1}{2}W_{12}(V_1 + V_2) < W_{22}V_2. \quad (10.71)$$

Let us consider a randomly mating population with regard to an autosomal locus given in the scheme

Genotype:	AA	Aa	aa	
Probability:	p^2	$2pq$	q^2	(10.72)
Fitness:	$W_{11}V_1$	$\frac{1}{2}W_{12}(V_1 + V_2)$	$W_{22}V_2.$	

It is easy to see that the mean fitness of this population, $W_{11}V_1 p_1^2 + W_{12}(V_1 + V_2)pq + W_{22}V_2 q^2$, attains an extremum at the point $p^* = m_1^*$ [where m_1^* is given by (10.69)]. Using the results of section 10.7, we find that condition (10.70) corresponds to stable, and (10.71) to unstable, equilibrium.

We summarize the results in the following theorem.

THEOREM 10.5. *For X-linked genes, A and a, equilibrium is attained when the maternal gene frequencies, m_1^*, m_2^*, are as given in (10.66), and the paternal gene frequencies, f_1^*, f_2^*, are as given in (10.67). In the general case, $f_1^* \neq m_1^*$ and $f_2^* \neq m_2^*$ unless $V_1 = V_2$. The equilibrium is stable if conditions (10.70) are satisfied and unstable if (10.71) are satisfied.*

In our discussion, we have introduced the mean fitness for males, $\bar{V}$, given by (10.60), and for females, $\bar{W}$, given by (10.62). We may introduce the mean fitness for the whole population:

$$\bar{U} = \bar{U}(m_1, f_1) = \tfrac{1}{2}(\bar{V} + \bar{W})$$
$$= \tfrac{1}{2}[V_1 m_1 + V_2(1 - m_1)] + \tfrac{1}{2}\{W_{11}m_1 f_1 + W_{12}[m_1(1 - f_1) + (1 - m_1)f_1]$$
$$+ W_{22}(1 - m_1)(1 - f_1)\}. \quad (10.73)$$

Function (10.72) attains an extremum when

$$\left. \begin{aligned} \frac{\partial \bar{U}}{\partial m_1} &= \tfrac{1}{2}[(V_1 - V_2) + W_{11}f_1 + W_{12}(1 - 2f_1) - W_{22}(1 - f_1)] = 0, \\ \frac{\partial \bar{U}}{\partial f_1} &= \tfrac{1}{2}[W_{11}m_1 + W_{12}(1 - 2m_1) - W_{22}(1 - m_1)] = 0. \end{aligned} \right\} \quad (10.74)$$

Solving these two equations, we find the points, $\overset{o}{m}_1$ and $\overset{o}{f}_1$, for which $\bar{U}$ attains an extremum. These are

$$\overset{o}{m}_1 = \frac{W_{22} - W_{11}}{W_{11} + W_{22} - 2W_{12}}, \tag{10.75}$$

$$\overset{o}{f}_1 = \frac{(W_{22} - W_{11}) + (V_2 - V_1)}{W_{11} + W_{22} - 2W_{12}}. \tag{10.76}$$

We notice that, in general,

$$\overset{o}{m}_1 \neq m_1^*, \quad \overset{o}{f}_1 \neq f_1^*, \qquad \text{unless} \quad V_1 = V_2.$$

Example 10.6. Let $W_{11} = W_{12} = V_1 = 1$ and $W_{22} = V_2 = 0$, that is, there is complete selection against recessives. If we substitute these values into (10.69), the denominator becomes zero, so that (10.69) cannot be used. Let us calculate the paternal and maternal gene frequencies for the next generation in the usual manner.

Substituting V's and W's into (10.60) and (10.62), respectively, we obtain

$$\bar{V} = m_1, \quad \bar{W} = m_1 f_1 + m_1 f_2 + m_2 f_1 = 1 - m_2 f_2.$$

After selection, the paternal gene frequencies, f_1' and f_2', are, from (10.59),

$$f_1' = \frac{m_1 V_1}{\bar{V}} = 1 \quad \text{and} \quad f_2' = 0,$$

so that, after selection is completed, the male recessives are eliminated.

The maternal gene frequencies for the next generation, m_1' and m_2', are [from (10.61)]

$$m_1' = \frac{2m_1 f_1 + m_1 f_2 + m_2 f_1}{2(1 - m_2 f_2)} = \frac{1}{2} \cdot \frac{m_1 + f_1}{1 - m_2 f_2},$$

$$m_2' = \frac{1}{2} \cdot \frac{m_1 f_2 + m_2 f_1}{1 - m_2 f_2} = \frac{1}{2} \cdot \frac{m_2 + f_2 - 2m_2 f_2}{1 - m_2 f_2}.$$

Taking into account that $f_1' = 1$ and $f_2' = 0$, we have for the second generation

$$m_2'' = \tfrac{1}{2} m_2',$$

and for the nth generation

$$m_2^{(n)} = \tfrac{1}{2} m_2^{(n-1)} = (\tfrac{1}{2})^{n-1} m_2' = \left(\frac{1}{2}\right)^n \frac{m_1 + f_1}{1 - m_2 f_2}.$$

As $n \to \infty$, $m_2^{(n)} \to 0$, so that $m_1^{(n)} \to 1$, and at equilibrium the total population (of males and females) becomes homozygous with genotype AA.

▼ 10.11. SELECTION AT TWO AUTOSOMAL LOCI. RANDOMLY MATING POPULATION

We now discuss a situation for which all of assumptions (10.18) hold, except that the trait under consideration is controlled by *two* autosomal loci. Now, the equilibrium conditions will be defined in terms of *gamete*, *not* of *gene*, frequencies (compare Definition 4.2a).

Let us consider two *linked* autosomal loci, A and B, each with two alleles A, a and B, b, respectively, and recombination fraction λ. We recall the results of sections 4.5.1 and 4.7.2, which we now apply in finding equilibrium conditions under selection pressure.

Let the gamete output to the initial randomly mating population be

$$
\begin{array}{lcccc}
\text{Gamete:} & AB & Ab & aB & ab \\
& (\gamma_1) & (\gamma_2) & (\gamma_3) & (\gamma_4) \\
\text{Probability:} & g_1 & g_2 & g_3 & g_4,
\end{array}
\tag{10.77}
$$

with $\sum g_i = 1$.

We write the genotype compositions in the form of a 4×4 matrix $\mathbf{\Gamma} = \mathbf{\gamma\gamma'}$, as given explicitly in Table 4.1a, and their distribution in the form of a 4×4 matrix $\mathbf{G} = \mathbf{gg'}$, as given in Table 4.1b. Let

$$
\mathbf{W} =
\begin{bmatrix}
W_{11} & W_{12} & W_{13} & W_{14} \\
W_{21} & W_{22} & W_{23} & W_{24} \\
W_{31} & W_{32} & W_{33} & W_{34} \\
W_{41} & W_{42} & W_{43} & W_{44}
\end{bmatrix}
\tag{10.78}
$$

be the matrix of genotype fitnesses, and assume that these are constant through generations. In general, we define the *mean fitness* of the population with respect to two loci as

$$
\overline{W} = \mathbf{g'Wg} = \sum_{i=1}^{4} \sum_{j=1}^{4} W_{ij} g_i g_j.
\tag{10.79}
$$

We may also define the mean fitness, $\overline{W}_i$, of the gamete γ_i as

$$
\overline{W}_i = \frac{1}{2} \frac{1}{g_i} \left(\sum_{j=1}^{4} W_{ij} g_i g_j + \sum_{j=1}^{4} W_{ji} g_j g_i \right) = \sum_{j=1}^{4} \frac{W_{ij} + W_{ji}}{2} g_j = \frac{1}{2} \frac{\partial \overline{W}}{\partial g_i},
\tag{10.80}
$$

for $i = 1, 2, 3, 4$.

In particular, when $W_{ij} = W_{ji}$, we have

$$
\overline{W}_i = \sum_{j=1}^{4} W_{ij} g_j, \qquad i = 1, 2, \ldots, s.
\tag{10.81}
$$

Substituting (10.80) into (10.79), we have

$$\overline{W} = \sum_{i=1}^{4} \overline{W}_i g_i. \tag{10.82}$$

It is easy to see that these formulae resemble those of section 10.4, with gene frequencies replaced by gamete frequencies. *After* selection, gamete frequencies can be evaluated by using the segregation matrices, $\mathbf{C}_k$, given in (4.31) and applying formulae similar to those in (4.32), but replacing the matrix $\mathbf{G}$ by the matrix of "effective" genotype frequencies, that is, by

$$\frac{1}{\overline{W}}[\mathbf{W} \;\square\; \mathbf{G}] = \frac{1}{\overline{W}}(W_{ij}g_ig_j).$$

These, of course, will be the gamete frequencies for the next generation.

Thus, for example, the frequency g'_1 of the gamete AB, for the next generation, will be

$$\begin{aligned}
g'_1 &= \frac{1}{\overline{W}}\{[W_{11}g_1g_1 + \tfrac{1}{2}W_{12}g_1g_2 + \tfrac{1}{2}W_{13}g_1g_3 + \tfrac{1}{2}(1 - \lambda)W_{14}g_1g_4] \\
&\quad + [\tfrac{1}{2}W_{21}g_2g_1 + \tfrac{1}{2}\lambda W_{23}g_2g_3] + [\tfrac{1}{2}W_{31}g_3g_1 + \tfrac{1}{2}W_{32}g_3g_2] \\
&\quad + \tfrac{1}{2}(1 - \lambda)W_{41}g_4g_1\} \\
&= \frac{1}{\overline{W}}\left\{\tfrac{1}{2}g_1\left[\sum_{j=1}^{4}W_{1j}g_j + \sum_{i=1}^{4}W_{i1}g_i\right]\right. \\
&\quad \left. - \lambda\left[\frac{W_{14} + W_{41}}{2}g_1g_4 - \frac{W_{23} + W_{32}}{2}g_2g_3\right]\right\} \\
&= \frac{\overline{W}_1}{\overline{W}}g_1 - \frac{\lambda}{\overline{W}}[W_{14}g_1g_4 - W_{23}g_2g_3], \tag{10.83}
\end{aligned}$$

since $W_{ij} = W_{ji}$, and using (10.81).

Putting
$$Q = W_{14}g_1g_4 - W_{23}g_2g_3, \tag{10.84}$$

we write (10.83) in the form

$$g'_1 = \frac{\overline{W}_1}{\overline{W}}g_1 - \frac{\lambda}{\overline{W}}Q = \frac{1}{2\overline{W}}\frac{\partial \overline{W}}{\partial g_1}g_1 - \frac{\lambda}{\overline{W}}Q. \tag{10.85a}$$

In a similar way we find

$$g'_2 = \frac{\overline{W}_2}{\overline{W}}g_2 + \frac{\lambda}{\overline{W}}Q = \frac{1}{2\overline{W}}\frac{\partial \overline{W}}{\partial g_2}g_2 + \frac{\lambda}{\overline{W}}Q, \tag{10.85b}$$

$$g'_3 = \frac{\overline{W}_3}{\overline{W}}g_3 + \frac{\lambda}{\overline{W}}Q = \frac{1}{2\overline{W}}\frac{\partial \overline{W}}{\partial g_3}g_3 + \frac{\lambda}{\overline{W}}Q, \tag{10.85c}$$

$$g'_4 = \frac{\overline{W}_4}{\overline{W}}g_4 - \frac{\lambda}{\overline{W}}Q = \frac{1}{2\overline{W}}\frac{\partial \overline{W}}{\partial g_4}g_4 - \frac{\lambda}{\overline{W}}Q. \tag{10.85d}$$

The conditions for stationary points (stable or unstable equilibrium) are $g_i' - g_i = \Delta g_i = 0$, leading to the equations

$$\Delta g_i = \frac{\overline{W}_i - \overline{W}}{\overline{W}} g_i + (-1)^i \frac{\lambda}{\overline{W}} Q = 0, \qquad i = 1, 2, 3, 4. \qquad (10.86)$$

On the other hand, following the arguments of section 10.6, we find that the function $\overline{W}$ given by (10.79) with g_i's subject to the constraint $\sum g_i = 1$ attains an extremum when

$$\frac{1}{2} \frac{\partial \overline{W}}{\partial g_i} \frac{1}{\overline{W}} - \overline{W} = \frac{\overline{W}_i - \overline{W}}{\overline{W}} = 0, \qquad i = 1, 2, 3, 4. \qquad (10.87)$$

We see that (10.86) and (10.87) are not the same unless $Q = 0$. In other words;

Equilibrium with respect to two loci under selection is not necessarily attained at the extremum of the mean fitness of the population with respect to these loci.

Of particular interest might be the situation in which not only $W_{ij} = W_{ji}$, but also $W_{14} = W_{41} = W_{23} = W_{32}$, that is, the fitnesses of all double hetero-zygotes are the same. This can happen, for instance, when the selection is at the zygote rather than the gamete stage. In this case,

$$g_i' = \frac{\overline{W}_i}{\overline{W}} g_i + (-1)^i \frac{\lambda}{\overline{W}} W_{14}(g_1 g_4 - g_2 g_3)$$

$$= \frac{\overline{W}_i}{\overline{W}} g_i + (-1)^i \frac{\lambda}{\overline{W}} W_{14} \Delta_0, \qquad i = 1, 2, \dots, 4, \qquad (10.88)$$

where

$$\Delta_0 = \begin{vmatrix} g_1 & g_2 \\ g_3 & g_4 \end{vmatrix}. \qquad (10.89)$$

The equilibrium conditions are

$$\Delta g_i = \frac{\overline{W}_i - \overline{W}}{\overline{W}} g_i + (-1)^i \frac{\lambda}{\overline{W}} \Delta_0 = 0, \qquad i = 1, 2, 3, 4. \qquad (10.90)$$

If $\Delta_0 = 0$ (or, equivalently, $g_1 g_4 = g_2 g_3$), the equilibrium conditions are identical with the extremum conditions for $\overline{W}$. In this case the stable equilibrium will occur when $\overline{W}$ attains a maximum, and the unstable equilibrium when $\overline{W}$ attains a minimum. The quantity Δ_0 is sometimes called a *coefficient of epistatic disequilibrium* [Moran (1964)].

Example 10.7. Let us consider the situation in which $W_{14} = W_{41} = W_{23} = W_{32} = 2$, $W_{22} = W_{33} = 0$, and the fitnesses of the remaining geno-types are each equal to 1. This means that there is selective advantage of double heterozygotes $AaBb$, complete lethality of single recessives $aaBB$

and $AAbb$, and other genotypes have fitnesses equal to 1. We also assume that loci are independent, that is, $\lambda = \frac{1}{2}$. From (10.81), we have

$$\overline{W}_1 = g_1 + g_2 + g_3 + 2g_4 = 1 + g_4, \quad \overline{W}_2 = g_1 + 2g_3 + g_4 = 1 - g_2 + g_3,$$

$$\overline{W}_3 = g_1 + 2g_2 + g_4 = 1 + g_2 - g_3, \quad \overline{W}_4 = 2g_1 + g_2 + g_3 + g_4 = 1 + g_1.$$

$$\overline{W} = (1 + g_4)g_1 + (1 - g_2 + g_3)g_2 + (1 + g_2 - g_3)g_3 + (1 + g_1)g_4$$

$$= (g_1 + g_2 + g_3 + g_4) + g_1g_4 + g_1g_4 - (g_2 - g_3)^2$$

$$= 1 + 2g_1g_4 - (g_2 - g_3)^2.$$

We calculate from (10.88)

$$g_1' = \frac{(1 + g_4)g_1 - \Delta_0}{\overline{W}} = \frac{g_1 + (g_1g_4 - \Delta_0)}{\overline{W}},$$

and similarly

$$g_4' = \frac{(1 + g_1)g_4 - \Delta_0}{\overline{W}} = \frac{g_4 + (g_1g_4 - \Delta_0)}{\overline{W}}.$$

If the population is in equilibrium, then for these two generations we must have $g_1' = g_1$ and $g_4' = g_4$, or $g_1'/g_4' = g_1/g_4$, or

$$\frac{g_1'}{g_4'} = \frac{g_1 + (g_1g_4 - \Delta_0)}{g_4 + (g_1g_4 - \Delta_0)} = \frac{g_1}{g_4}.$$

The solution is either $g_1g_4 - \Delta_0 = 0$, which leads to $g_2g_3 = 0$ (trivial case) or $g_1 = g_4 \neq \frac{1}{2}$ (nontrivial).

In a similar way we obtain

$$\frac{g_2'}{g_3'} = \frac{(1 - g_2)g_2 + (g_2g_3 + \Delta_0)}{(1 - g_3)g_3 + (g_2g_3 + \Delta_0)} = \frac{g_2}{g_3}.$$

The nontrivial solution will be for $g_2 = g_3 \neq \frac{1}{2}$.

Since $\sum g_i = 1$, under equilibrium, we shall have $2g_1 + 2g_2 = 1$ or $g_2 = \frac{1}{2} - g_1$. The average fitness under equilibrium is $\overline{W} = 1 + 2g_1^2$, and $\overline{W}_1 = 1 + g_1$. We now evaluate the frequency of gamete AB, g_1^*, say, under equilibrium. Substituting $\overline{W}$ and $\overline{W}_1$ into (10.85a), we obtain

$$g_1 = \frac{(1 + g_1)g_1 - g_1^2 + g_2^2}{1 + 2g_1^2} = \frac{g_1 + (\frac{1}{2} - g_1)^2}{1 + 2g_1^2}.$$

This leads to the equation

$$(1 + 2g_1^2)g_1 = g_1 + (\tfrac{1}{2} - g_1)^2 \quad \text{or} \quad 2g_1^3 - g_1^2 + g_1 - \tfrac{1}{4} = 0,$$

which is satisfied for $g_1 = g_1^* = 0.28492$. This is also the value of g_4^*. We have $g_2^* = g_3^* = \frac{1}{2} - g_1^* = 0.21508$. It can be shown [Li (1967b)] that in this selection pattern the mean fitness can increase or decrease, depending on the values of gene frequencies in the initial population.

Several other patterns of selection can be constructed. Quite a number of so-called symmetrical selection patterns, constructed by several authors, have been summarized by Li (1967*b*).

10.12. SELECTION AND INBREEDING

We now consider the situation for which assumptions (10.18) hold except that the population is not entirely randomly mating, but rather is practicing inbreeding with the average coefficient of inbreeding F. We restrict ourselves to a single locus with only two alleles, A and a, having frequencies p and q, respectively. Using the results of section 9.7, we have the following genotype frequencies:

$$
\begin{aligned}
AA:\quad & h_{11} = p^2 + Fpq = (1 - F)p^2 + Fp, \\
Aa:\quad & 2h_{12} = 2(1 - F)pq, \\
aa:\quad & h_{22} = q^2 + Fpq = (1 - F)q^2 + Fq.
\end{aligned}
\tag{10.91}
$$

The structure of the population given in (10.91) is equivalent to a mixture of two subpopulations in the following proportions: a $(1 - F)$ fraction of a randomly mating population $(p^2AA + 2pqAa + q^2aa)$, and an F fraction of a self-mating population $(pAA + qaa)$.

Let W_{ij} be the fitnesses of genotypes in the whole population. Using formula (10.6), we find

$$
\begin{aligned}
\overline{W}_1 &= \frac{1}{p}\{[W_{11}(1 - F)p^2 + W_{11}Fp] + W_{12}(1 - F)pq\} \\
&= (1 - F)(W_{11}p + W_{12}q) + FW_{11} \\
&= (1 - F)\overline{W}_{1(R)} + FW_{11},
\end{aligned}
\tag{10.92}
$$

where $\overline{W}_{1(R)} = W_{11}p + W_{12}q$ denotes the mean fitness of the gene A in the randomly mating subpopulation. Similarly,

$$
\overline{W}_2 = (1 - F)(W_{22}q + W_{12}p) + FW_{22} = (1 - F)\overline{W}_{2(R)} + FW_{22},
\tag{10.93}
$$

where $\overline{W}_{2(R)} = W_{22}q + W_{12}p$ is the mean fitness of the gene a in the randomly mating subpopulation. Thus

$$
\begin{aligned}
\overline{W} &= \overline{W}_1 p + \overline{W}_2 q = (1 - F)[\overline{W}_{1(R)}p + \overline{W}_{2(R)}q] + F(W_{11}p + W_{22}q) \\
&= (1 - F)\overline{W}_{(R)} + F\overline{W}_{(F)},
\end{aligned}
\tag{10.94}
$$

where

$$
\overline{W}_{(R)} = \overline{W}_{1(R)}p + \overline{W}_{2(R)}q = W_{11}p^2 + 2W_{12}pq + W_{22}q^2
\tag{10.95}
$$

is the mean fitness in the randomly mating $(1 - F)$ fraction, and

$$\overline{W}_{(F)} = W_{11}p + W_{22}q \tag{10.96}$$

is the mean fitness in the inbred F fraction of the whole population. The gene frequency p' in the next generation is, from (10.10),

$$p' = \frac{\overline{W}_1}{\overline{W}}\, p = \frac{(1 - F)\overline{W}_{1(R)} + F\overline{W}_{11}}{(1 - F)\overline{W}_{(R)} + F\overline{W}_{(F)}}\, p. \tag{10.97}$$

Under equilibrium, $p' = p$, (and so $\overline{W}_1 = \overline{W}$), that is,

$$(1 - F)\overline{W}_{1(R)} + FW_{11} = (1 - F)[\overline{W}_{1(R)}p + \overline{W}_{2(R)}q] \\ + F(W_{11}p + W_{22}q)$$

or

$$(1 - F)\overline{W}_{1(R)}q + FW_{11}q = (1 - F)W_{2(R)}q + FW_{22}q$$

or

$$(1 - F)[W_{11}p + W_{12}(1 - p)] + FW_{11} \\ = (1 - F)[W_{12}p + W_{22}(1 - p)] + FW_{22}$$

or

$$(1 - F)(W_{11} + W_{22} - 2W_{12})p = (1 - F)(W_{22} - W_{12}) + F(W_{22} - W_{11}). \tag{10.98}$$

Hence the frequency of the gene A at equilibrium, p^*, is

$$p^* = \frac{(1 - F)(W_{22} - W_{12}) + F(W_{22} - W_{11})}{(1 - F)(W_{11} + W_{22} - 2W_{12})} \tag{10.99a}$$

and

$$q^* = 1 - p^* = \frac{(1 - F)(W_{11} - W_{12}) + F(W_{11} - W_{22})}{(1 - F)(W_{11} + W_{22} - 2W_{12})}, \tag{10.99b}$$

provided that

$$W_{11} + W_{22} - 2W_{12} \neq 0 \tag{10.100}$$

and

$$0 \leqslant F < 1. \tag{10.101}$$

In order to have both p^* and q^* positive fractions, the numerator and the denominator in (10.99a) must have the same sign, and the modulus of the numerator has to be smaller than the modulus of the denominator. This leads to the following condition on F:

$$F < \min\left(\frac{W_{11} - W_{12}}{W_{22} - W_{12}}, \frac{W_{22} - W_{12}}{W_{11} - W_{12}}\right). \tag{10.102}$$

Suppose that one of the two following situations occur

$$W_{11} < W_{12} > W_{22} \tag{10.103a}$$

or

$$W_{11} > W_{12} < W_{22}. \tag{10.103b}$$

It can be shown [Li (1955a)] that equilibrium will be stable when, in addition to conditions (10.100)–(10.102), relation (10.103a) holds, that is, when the heterozygotes are at a selective advantage.

The case in which $W_{11} + W_{22} - 2W_{12} = 0$ should be treated separately (see Problem 10.9).

We might ask this question: does the theory presented in section 10.6 apply to populations practicing inbreeding?

Using straightforward algebra, it is not difficult to show that function (10.94) attains its extremum at

$$\overset{0}{p} = \frac{2(1 - F)(W_{22} - W_{12}) + F(W_{22} - W_{11})}{2(1 - F)(W_{11} + W_{22} - 2W_{12})}, \tag{10.104}$$

so that

$$p^* \neq \overset{0}{p}. \tag{10.105}$$

This means that:

The equilibrium of the population practicing inbreeding does not correspond to the extremum of its mean fitness.

Example 10.8. Let $W_{11} = 1$, $W_{12} = 3$, $W_{22} = 0$, $F = \frac{1}{2}$, and $p = q = \frac{1}{2}$.

We wish to find the gene frequencies (*a*) in the next generation and (*b*) at stable equilibrium.

(*a*) From (10.92) and (10.93) we have

$$\overline{W}_1 = \tfrac{1}{2}(1 \cdot \tfrac{1}{2} + 3 \cdot \tfrac{1}{2}) + \tfrac{1}{2} \cdot 1 = \tfrac{3}{2} \quad \text{and} \quad \overline{W}_2 = \tfrac{1}{2} \cdot \tfrac{3}{2} = \tfrac{3}{4},$$

respectively. Thus, from (10.94)

$$\overline{W} = \tfrac{1}{2}(\tfrac{3}{2} + \tfrac{3}{4}) = \tfrac{9}{8}.$$

From (10.97) we obtain

$$p' = \frac{\overline{W}_1}{\overline{W}} p = \frac{3 \cdot 8}{2 \cdot 9} \cdot \frac{1}{2} = \frac{2}{3}, \qquad q' = \frac{1}{3}.$$

(*b*) We have

$$F < \frac{W_{11} - W_{12}}{W_{22} - W_{12}} = \frac{2}{3} \quad \text{and} \quad W_{11} < W_{12} > W_{22}$$

so that equilibrium will be stable. The equilibrium gene frequencies are, from (10.99a),

$$p^* = \frac{\tfrac{1}{2}(-2) + \tfrac{1}{2}(-1)}{\tfrac{1}{2}(-5)} = \frac{3}{5}, \qquad q^* = \frac{2}{5}.$$

B. Mutation

10.13. DISTRIBUTION OF MUTANTS IN A POPULATION

Another factor which affects gene frequencies in a population is mutation. By *mutation* we understand a sudden heritable change in the organism. The change can occur in a single gene (e.g., A may mutate into a, $A \to a$, or even into a new gene, a') and is then called a *gene mutation*. Changes can also occur in a part or the whole of a chromosome. Here we restrict ourselves to gene mutation at a *single locus*.

The mutation rate per gamete per generation, μ, *at a single locus* is the probability that in any given generation a given gamete (i.e., gene) will undergo a heritable change into a different allelomorphic form. It has been estimated that in man μ is of the order 10^{-4} to 10^{-5}, while in bacteria it is of the order 10^{-8} to 10^{-9}. Natural mutations are spontaneous and occur independently in time. Let $2N$ be the number of gametes in a population. The average number of mutant gametes is, then, $\lambda = 2N\mu$. Let x be the number of mutant gametes in this population. The distribution of x is binomial, $b(x; \mu; 2N)$. Since μ is very small and $2N$ large, it can be approximated by a Poisson distribution

$$p(x) = \frac{(2N\mu)^x}{x!} e^{-2N\mu}. \tag{10.106}$$

10.14. CHANGES IN GENE FREQUENCIES DUE TO MUTATION

Let us consider a single locus with two alleles, A and a, and gene frequencies p_0 and q_0, respectively, in the initial population. We shall be interested in how the gene frequencies change per generation if mutation occurs at a constant rate.

10.14.1. One-Way Mutation $A \to a$ with Rate μ

If A mutates into a at a constant rate, μ, then in the first generation (*before* the next mutation occurs) the frequency p_1, say, of the gene A will be

$$p_1 = (1 - \mu)p_0.$$

At the beginning of the second generation, we will have

$$p_2 = (1 - \mu)p_1 = (1 - \mu)^2 p_0,$$

and so on. Finally, at the beginning of the nth generation,

$$p_n = (1 - \mu)p_{n-1} = (1 - \mu)^n p_0 \qquad (10.107)$$

and

$$q_n = 1 - p_n = 1 - (1 - \mu)^n(1 - q_0). \qquad (10.107a)$$

When μ is small, we will have $(1 - \mu)^n \doteq e^{-n\mu}$ [since $(1 - \mu)^n = e^{n\ln(1-\mu)}$, and $\ln(1 - \mu) \doteq -\mu$, for small μ]. Thus

$$p_n \doteq p_0 e^{-n\mu} \qquad (10.108)$$

and

$$q_n \doteq 1 - (1 - q_0)e^{-n\mu}. \qquad (10.108a)$$

10.14.2. Reverse Mutations

Suppose that both genes mutate, $A \to a$ with mutation rate μ, and $a \to A$ with mutation rate ν. Then in the first generation we will have

$$p_1 = (1 - \mu)p_0 + \nu(1 - p_0) = \nu + (1 - \mu - \nu)p_0;$$

in the second generation,

$$p_2 = \nu + (1 - \mu - \nu)p_1 = \nu + \nu(1 - \mu - \nu) + (1 - \mu - \nu)^2 p_0;$$

and so on. Finally,

$$\begin{aligned} p_n &= \nu + (1 - \mu - \nu)p_{n-1} = \nu + \nu(1 - \mu - \nu) + \nu(1 - \mu - \nu)^2 + \cdots \\ &\quad + \nu(1 - \mu - \nu)^{n-1} + (1 - \mu - \nu)^n p_0 \\ &= \nu[(1 - \mu - \nu) + (1 - \mu - \nu)^2 + \cdots + (1 - \mu - \nu)^{n-1}] \\ &\quad + (1 - \mu - \nu)^n p_0. \end{aligned} \qquad (10.109)$$

The expression in square brackets is a geometric series, and its sum is

$$\frac{1 - (1 - \mu - \nu)^n}{1 - (1 - \mu - \nu)} = \frac{1 - (1 - \mu - \nu)^n}{\mu + \nu}.$$

Hence

$$\begin{aligned} p_n &= \frac{\nu}{\mu + \nu}[1 - (1 - \mu - \nu)^n] + (1 - \mu - \nu)^n p_0 \\ &= \frac{\nu}{\mu + \nu} - \left(\frac{\nu}{\mu + \nu} - p_0\right)(1 - \mu - \nu)^n. \end{aligned} \qquad (10.110)$$

For small $\mu + \nu$, $(1 - \mu - \nu)^n \doteq e^{-n(\mu+\nu)}$, so that

$$p_n \doteq \frac{\nu}{\mu + \nu} - \left(\frac{\nu}{\mu + \nu} - p_0\right)e^{-n(\mu+\nu)}, \qquad (10.110a)$$

and similarly

$$q_n = \frac{\mu}{\mu + \nu} - \left(\frac{\mu}{\mu + \nu} - q_0\right)(1 - \mu - \nu)^n \qquad (10.111)$$

or

$$q_n \doteq \frac{\mu}{\mu + \nu} - \left(\frac{\mu}{\mu + \nu} - q_0\right)e^{-n(\mu+\nu)}. \qquad (10.111a)$$

We have the recurrence formula

$$p_n = \nu + (1 - \mu - \nu)p_{n-1}.$$

Thus

$$\Delta p_n = p_n - p_{n-1} = \nu - (\mu + \nu)p_n. \qquad (10.112)$$

Equilibrium will be obtained when $\Delta p_n = 0$, that is, for

$$p^* = \frac{\nu}{\mu + \nu} \quad \text{and} \quad q^* = \frac{\mu}{\mu + \nu}. \qquad (10.113)$$

Substituting (10.113) into (10.110a), we obtain

$$p_n - p^* \doteq (p_0 - p^*)e^{-n(\mu+\nu)}.$$

Hence

$$-n(\mu + \nu) \doteq \ln \frac{p_n - p^*}{p_0 - p^*}$$

or

$$n(\mu + \nu) \doteq 2.3026 \log_{10} \frac{p_0 - p^*}{p_n - p^*}. \qquad (10.114)$$

Similarly, we have

$$n(\mu + \nu) \doteq 2.3026 \log_{10} \frac{q_0 - q^*}{q_n - q^*}. \qquad (10.114a)$$

From (10.114a) we can calculate how many generations are necessary in order to increase q by the amount $q_n - q_0$ (or to decrease p from p_0 to p_n). A numerical example is given in Problem 10.11.

10.15. THE PROBABILITY OF EXTINCTION OF A MUTANT GENE

Suppose that we start with a homogeneous population in which each individual is AA. Suppose also that one gene, A, mutates into a, so there will be one individual Aa. Even if Aa survives and produces offspring, the gene a can be lost in the first generation. Assuming that the number of families with s offspring in a large population has an approximately Poisson distribution with $\lambda = 2$ (for convenience, to keep the population size constant), Fisher (1930) [see also Li (1955b), Chapter 18] showed that the probability, γ_1, say, that the mutant will be lost in the first generation is

$$\gamma_1 = e^{-1} = 0.3679, \tag{10.115}$$

and the probability, γ_2, that it will be lost within two generations, is

$$\gamma_2 = e^{-(1-\gamma_1)} = e^{-(1-0.3679)} = e^{-0.6321} = 0.5315.$$

By means of the same procedure, the probability of extinction of the mutant within n generations can be evaluated from the formula

$$\gamma_n = e^{-(1-\gamma_{n-1})}. \tag{10.116}$$

Fisher (1930) calculated that, for $n = 127$, $\gamma_{127} = 0.9847$. As n becomes large, $\gamma_n \to 1$, so that finally the mutant should disappear.

10.16. JOINT EFFECTS OF MUTATION AND SELECTION

Let us assume that mutation, with constant rates μ and v for $A \to a$ and $a \to A$, respectively, occurs in each generation. We also assume that selection takes place with fitnesses W_{11}, W_{12}, and W_{22} of genotypes AA, Aa, and aa, respectively.

The model showing how the gene frequencies change in the next generation depends on how the two factors, mutation and selection, operate jointly during the generation time. We will outline only two simple models to demonstrate how different assumptions can lead to different results.

Let p and q be the gene frequencies in the population of the parents *before* mutation or selection occurs.

10.16.1. A Model When Mutation Occurs at the Gametic Stage and Selection at the Zygotic Stage

Suppose that mutation, $A \to a$, occurs in the gametes *before selection* takes place. This means that, for instance, the frequency of gene A will change, after mutation, from p, to p_1, say, where

$$p_1 = (1 - \mu)p \tag{10.117}$$

and

$$q_1 = 1 - p_1 = q + \mu p = \mu + (1 - \mu)q. \tag{10.117a}$$

If selection operates on the zygotes (or during adult life) *after* mutation has occurred, then the gene frequency for the next generation, p', say, will be evaluated according to formula (10.52), where for p we substitute p_1, that is,

$$p' = \frac{W_{11}p_1 + W_{12}(1 - p_1)}{\overline{W}(p_1)}, \qquad (10.118)$$

with

$$\overline{W}(p_1) = W_{11}p_1{}^2 + 2W_{12}p_1(1 - p_1) + W_{22}(1 - p_1)^2. \qquad (10.119)$$

Substituting (10.117) and (10.117a) into (10.118), we obtain p' as a function of p, $p' = \varphi(p)$, say. The evaluation of p', $\Delta p = p' - p$, and the equilibrium value p^* may encounter some difficulties if the expression for $\varphi(p)$ does not have explicit form. Some graphical methods for problems of this kind were proposed by Cannings (1969).

10.16.2.   A Model When Mutation and Selection Can Occur at the Same Stage of Development

It can happen that mutation and selection operate simultaneously in such a way that their effects on the change in gene frequencies are approximately additive. If unit time is taken equal to one generation, the change in gene frequency will be approximately equal to the sum of the changes due to mutation and selection, that is,

$$\Delta p = p' - p \doteq \mu p + \frac{\overline{W}_1 - \overline{W}}{\overline{W}} p, \qquad (10.120)$$

where $\overline{W}_1$ is the mean fitness of the allele A, and $\overline{W}$ is the mean fitness of the total initial population. The equilibrium solutions for this model and the one discussed in section 10.16.1 are not the same. More complicated models are required when migration is introduced.

SUMMARY

The basic concept in selection, the "reproductive fitness" of a genotype, has been defined. It has been shown that, for a randomly mating population with constant fitnesses of the genotypes defined for a single autosomal locus, the mean fitness, $\overline{W}$, of the population always increases until stable equilibrium (that is, when $\overline{W}$ is maximum) is attained. If either the population does not mate at random, or fitnesses are not constant from generation to generation, or we consider more loci, the theorem that $\overline{W}$ always increases is not necessarily valid. Several specific situations and examples have been considered, and the conditions for stable equilibrium in such instances have been discussed. Simple problems concerning the effect of gametic mutation and of mutation and selection jointly on the changes in gene frequencies have been outlined.

REFERENCES

Bennett, J. H., The existence and stability of a selectively balanced polymorphism at the sex-linked locus, *Australian J. Biol. Sci.* **11** (1958), 598–602.

Cannings, C., Equilibrium convergence and stability at a sex-linked locus under natural selection, *Genetics* **56** (1967), 613–618.

Cannings, C., A graphical method for the study of complex genetical systems, with reference to equilibria, *Biometrics* **25** (1969), 747–754.

Conference on Genetics, *Mutation*, ed. by W. J. Shull, The University of Michigan Press, Ann Arbor, 1962.

Conference on Genetics, *Genetic Selection in Man*, ed. by W. J. Shull, The University of Michigan Press, Ann Arbor, 1963.

Crow, J. F., Mutation in man. In: *Progress in Medical Genetics*, Vol. 1, ed. by A. Steinberg, Grune & Stratton, New York, 1961, pp. 1–26.

Crow, J. F., Population genetics: selection. In: *Methodology in Human Genetics*, ed. by W. J. Burdette, Holden-Day, Inc., San Francisco, 1962, pp. 53–75.

Fisher, R. A., *The Genetic Theory of Natural Selection*, Clarendon Press, 1930 (second edition, 1962).

Fisher, R. A., Average excess and average effect of a gene substitution, *Ann. Eugen.* **11** (1941), 53–63.

Haldane, J. B. S., A mathematical theory of natural and artificial selection, *Trans. Camb. Phil. Soc.* **23** (1924), 19–41; *Proc. Camb. Phil. Soc.* **23** (1926), 363–372, 607–615; (1927), 838–844; **27** (1931), 137–142.

Haldane, J. B. S., The effect of variation of fitness, *Amer. Nat.* **71** (1937), 337–349.

Karlin, S., Equilibrium behavior of population genetic models with non-random mating, I. Preliminaries and special mating system, *J. Appl. Prob.* **5** (1968), 231–313.

Kimura, M., Rules for testing stability of a selective polymorphism, *Proc. Nat. Acad. Sci.* **42** (1956), 336–340.

Kirkman, H. N., Properties of X-linked alleles during selection, *Amer. J. Hum. Genet.* **18** (1966), 424–432.

Kojima, K., and Kelleher, F. M., Changes of mean fitnesses in random mating population where epistasis and linkage are present, *Genetics* **46** (1961), 527–540.

Lewontin, R. C., A general method for investigating equilibrium of a gene frequency in a population, *Genetics* **43** (1958), 419–434.

Lewontin, R. C., Population genetics. In: *Annual Review of Genetics*, Vol. 1, 1967, 37–70.

Li, C. C., The stability of an equilibrium and the average fitness of a population, *Amer. Nat.* **89** (1955a), 281–295.

Li, C. C., *Population Genetics*, University of Chicago Press, Chicago, Illinois, 1955b.

Li, C. C., Equilibrium under differential selection in the sexes. *Evolution* **17** (1963), 493–496.

Li, C. C., Fundamental theorem of natural selection, *Nature* **214** (1967a), 505–506.

Li, C. C., Genetic equilibrium under selection, *Biometrics* **24** (1967b), 397–484.

Mandel, S. P. H., Stable equilibrium at a sex-linked locus, *Nature* **183** (1959a), 1347–1348.

Mandel, S. P. H., The stability of a multiple allelic system, *Heredity* **13** (1959*b*), 289–302.

Moran, P. A. P., On the non-existence of adaptive topographies, *Ann. Hum. Genet. Lond.* **27** (1964), 383–393.

Mulholland, H. P., and Smith, C. A. B., An inequality arising in genetical theory, *Amer. Math. Monthly* **66** (1959), 673–683.

Penrose, L. S., The meaning of "fitness" in human populations, *Ann. Eugen.* **14** (1949), 301–304.

Penrose, L. S., Smith, S. M., and Sprott, D. A., On the stability of allelic systems with special reference to hemoglobins A, S and C, *Ann. Hum. Genet. Lond.* **21** (1956), 90–93.

Reed, T. E., The definition of relative fitness of individuals with specific genetic traits, *Amer. J. Hum. Genet.* **11** (1959), 147–155.

Sacks, J. M., A stable equilibrium with minimum average fitness, *Genetics* **56** (1967), 705–708.

Tallis, G. M., Equilibria under selection for *k* alleles, *Biometrics* **22** (1966), 121–127.

Wright, S., *Evolution and the Genetics of Populations. The Theory of Gene Frequencies*, The University of Chicago Press, Chicago 1969.

PROBLEMS

10.1. The data below [from Allison, *Ann. Hum. Genet. Lond.* **21** (1956), 67–89] represent genotype frequencies and their fitnesses with respect to hemoglobin locus with alleles Hb_1^A (A hemoglobin, normal) Hb_1^S (S hemoglobin, sickle cell), and Hb_1^C (C hemoglobin, mixture of A and S), in a certain adult African population. Assume that selection takes place at the zygotic or early life phase, and that the adult population is in equilibrium.

(*a*) Find the distribution of genotypes at the zygotic stage *before* selection.

(*b*) Prove that the equilibrium is stable.

(For brevity, we denote the allele Hb_1^A by A, Hb_1^S by S, and Hb_1^C by C.)

Genotype:	AA	SS	CC	AS	AC	SC
Probability (after selection):	0.6906	0.0019	0.0019	0.1912	0.1096	0.0048
Fitness:	0.9764	0.1921	0.5495	1.1381	1.1026	0.4072.

10.2. Consider a randomly mating population of the following pattern:

Genotype:	AA	Aa	aa
Probability:	$\frac{1}{4}$	$\frac{1}{2}$	$\frac{1}{4}$
Fitness:	2	1	0.

(Note that $W_{11} + W_{22} - 2W_{12} = 0$.) As has been shown in section 10.7.1, equilibrium will be attained when the population becomes homozygous with the genotype AA.

Calculate how many generations are necessary in order to obtain 99% of the population with genotype AA.

10.3. Suppose that the initial, randomly mating population has frequencies $p = q = \frac{1}{2}$ of genes A and a. Let the fitnesses of the genotypes be as follows:

$$\text{(i)} \quad W_{11} = W_{22} = 1, \ W_{12} = 0;$$
$$\text{(ii)} \quad W_{11} = W_{22} = 1, \ W_{12} = 2;$$
$$\text{(iii)} \quad W_{11} = W_{22} = 1, \ W_{12} = \tfrac{1}{2};$$
$$\text{(iv)} \quad W_{11} = W_{22} = 0, \ W_{12} = 1.$$

(a) Find the rate of increase in the mean fitness for the next generation for each of these populations.

(b) Calculate the gene frequencies at equilibrium for cases (i), (ii), and (iii). Which of these equilibria are stable, and which unstable?

10.4. Let us consider a single locus with $s = 3$ alleles, A_1, A_2, A_3, and gene frequencies $p_1 = 0.5$, $p_2 = 0.3$, $p_3 = 0.2$, respectively. Let the fitnesses of the genotypes in a randomly mating population be $W_{ii} = 1$, $W_{ij} = 2$, with $i, j = 1, 2, 3$, $i \neq j$.

(a) Calculate the gene frequencies in the next two successive generations.

(b) Calculate the changes in the mean fitness in the next two generations.

(c) Find the gene frequencies at equilibrium. Is the equilibrium, with three positive p's, stable or unstable?

10.5. Fill in the appropriate entries in the table below for the patterns of selection given in the first column of this table (from Conference on Genetics (1963), p. 30.)

Initial distribution and fitnesses					Equilibrium	
AA p^2	Aa $2pq$	aa q^2	Selection against	Δq	q^*	Stable or unstable
$1 - c$	1	$1 - \alpha c$	Homozygotes AA and aa	$\dfrac{pqc(p - \alpha q)}{1 - c(p^2 + \alpha q^2)}$	$\dfrac{1}{1 - \alpha}$	Stable
1	1	$1 - c$				
1	$1 - c$	$1 - 2c$				
1	$1 - c$	1				

We assume $0 < c < 1$ and $0 < \alpha \leqslant (1/c)$.

10.6. Let s be the number of alleles at a single autosomal locus. Let the fitnesses of genotypes be $W_{ii} = 1$, with $i = 1, 2, \ldots, s$, and $W_{ij} = 1 + \gamma$, with $\gamma > 0$, $i, j = 1, 2, \ldots, s$, $i \neq j$. Prove that stable equilibrium will be attained at $p_i^* = (1/s)$, $i = 1, 2, \ldots, s$ [G. M. Tallis, *Biometrics* **22** (1966), 121–127].

10.7. Let the selection pattern, with fitnesses depending on gene frequencies, be as follows:

Genotype:	AA	Aa	aa
Probability:	p^2	$2p(1-p)$	$(1-p)^2$
Fitness:	$ct^2 + \left(\dfrac{b}{p}\right)^2$	$bt\left[-c + \dfrac{1}{p(1-p)}\right]$	$cb^2 + \left[\dfrac{t}{1-p}\right]^2.$

Show that equilibrium occurs for $p^* = [b/(b+t)]$ provided that b and t have the same sign. Prove that the equilibrium is stable. [J. M. Sacks, *Genetics* **56** (1967), 705–708.]

10.8. Let a be a rare recessive, X-linked gene with frequencies $m = 0.05$ of the maternal and $f = 0.01$ of the paternal gametes, in the initial population. Let the corresponding fitnesses of genotypes (as defined in section 10.10) be, in males, $V_1 = 1.0$, $V_2 = 0.9$, and, in females, $W_{11} = 1.0$, $W_{12} = 1.0$, and $W_{22} = 0.5$.

(a) Calculate m^* (i.e., the frequency of the maternal gene a at equilibrium).

(b) Is the equilibrium stable or unstable?

(c) Let $\bar{U}$ be the mean fitness for the whole population as defined in (10.73). Calculate the maximum value of $\bar{U}$, $\overset{\circ}{U}$, say, and $\bar{U}^*$, that is, the value of $\bar{U}$ for $m = m^*$, and $f = f^*$. Compare these values.

10.9. Suppose that the frequencies of A and a genes in the initial population are p_0 and q_0, respectively. Let F be the inbreeding coefficient in this population, and let the corresponding fitnesses of the genotypes be $W_{11} = 2$, $W_{12} = 1$, $W_{22} = 0$.

Derive the recurrence formulae for the gene frequencies in the nth generation. [Note that $W_{11} + W_{22} - 2W_{12} = 0$, and formulae (10.99) for p^* and q^* cannot be applied.]

▼ **10.10.** Let U_1, U_2, U_3 be the fitnesses of the genotypes AA, Aa, aa, and V_1, V_2, V_3 be the fitnesses of the genotypes BB, Bb, bb, respectively. We assume that $U_1 < U_2 > U_3$ and $V_1 < V_2 > V_3$, and that the loci A and B are independent.

There are nine different genotypes with respect to two loci, obtained from the genotype array $(AA + Aa + aa)$ $(BB + Bb + bb)$. Let W_{ij} be the fitness of a joint genotype with respect to two loci. We wish to consider two models:

(i) $W_{ij} = U_i + V_j$ ("additive" model);

(ii) $W_{ij} = U_i V_j$ ("multiplicative" model).

Let p_1^*, q_1^*, and p_2^*, q_2^* be the equilibrium gene frequencies at locus A and at locus B separately.

Prove that stable equilibrium with respect to two loci jointly is attained when the frequencies of gametes AB, Ab, aB, and ab are $p_1^* p_2^*$, $p_1^* q_2^*$, $q_1^* p_2^*$, and $q_1^* q_2^*$ respectively [P. A. P. Moran, *Ann. Hum. Genet. Lond.* **32** (1968), 183–190].

10.11. Let $p_0 = 0.90$ and $q_0 = 10$ be the frequencies of A and a genes in the initial population. Let $\mu = 5 \cdot 10^{-5}$ be the mutation rate for $A \to a$, and $\nu = 2 \cdot 10^{-6}$ be the reverse mutation rate (i.e., $a \to A$). Assuming no selection and infinite population, calculate:

(a) The gene frequencies at equilibrium.

(b) The gene frequencies in the fifth generation.

(c) The number of generations required to increase q by 0.005.

10.12.   Let μ be the mutation rate $A \to a$, ν be the mutation rate $a \to A$, and p and q be the frequencies of genes A and a, respectively. Suppose that the fitnesses of the genotypes AA, Aa, and aa are $W_{11} = W_{12} = 1$ and $W_{22} = 1 - c$, respectively, that is, selection is against recessives. Find the conditions for equilibrium, assuming:

(a) mutation and selection occur at the same stage of reproduction;

(b) mutation occurs *before* selection.

PART II

Experimental Genetics.

Estimation of Genetic Parameters

and Testing Genetic Hypotheses

In Chapters 2–10 we considered hypothetical ("infinite") populations and derived some useful probability models which describe, in mathematical forms, general laws of transmission of inheritance. We introduced genetic parameters such as gene frequency, recombination fraction in linkage, fitness, and inbreeding coefficient.

As has been pointed out, living populations are always finite and the laws of population genetics hold only approximately if a population is large.

Usually the genetic parameters for the whole, even a finite, population are unknown. They can be *estimated* from observations on part of a population taken randomly or by using especially designed experiments. Very often not only the parameters but also the mode of inheritance are unknown. In such cases, one can make *hypotheses*, that is, construct some models and try to fit the experimental data to these models.

Statistical methods used in estimation and in testing hypotheses depend on how the data have been collected, that is, on the experimental designs, which reflect different sampling techniques. We shall distinguish here between independent and dependent samples, data obtained by random selection from one and two generations (Chapters 11, 12, 13, 14, 19) and data collected by different ascertainments of families in human populations (Chapters 17 and 18). Special emphasis is laid on the chi-square analysis of categorical data, since it has a broad application in studies of the inheritance of qualitative traits (Chapters 12 and 13).

Several classical, and also some recently developed, methods for problems such as estimating gene frequencies in blood groups (Chapter 14), detecting linkage in experimental populations (Chapter 15), studying modes of inheritance of heritable diseases in human beings by the methods of segregation analysis (Chapters 17 and 18) and the completely new field of histocompatibility testing in tissue transplantation (Chapter 19), *inter alia*, are presented.

CHAPTER 11

Estimation of Genetic Parameters.
General Theory

11.1. GENERAL PROBLEM OF ESTIMATION

In several preceding chapters we discussed probability distributions of random variables associated with characters or events in populations. Genotype and phenotype distributions of populations under different mating systems were especially widely discussed. We introduced some characteristics which determine the genetic properties, genetic structure, or mode of inheritance. These are called *genetic parameters*. The most common, and also most important, genetic parameters are: (*a*) gene frequencies; (*b*) prevalence and/or incidence of rare (usually abnormal) traits; (*c*) recombination fraction in linkage; (*d*) inbreeding coefficient; (*e*) penetrance, viability, fitness; and (*f*) number of alleles and/or loci controlling a certain character.

In deriving our models we have assumed that our mating populations are so large ("infinite") that the parameters are expected values, not subject to random fluctuations. In practice, however, we can deal only with finite populations of reasonable sizes, which can be considered as samples from "infinite" populations. Thus we have to *estimate* our parameters from observed data and test whether our model fits the data. This chapter is entirely concerned with estimation. Let us first take a simple example.

Example 11.1. Let us consider a population under equilibrium of the type $p^2AA + 2pqAa + q^2aa$. Suppose that all three genotypes are phenotypically distinguishable. Let x be the number of genes A in the genotype;* that is, it can take values $x = 0, 1, 2$, depending on whether a genotype is aa, Aa, or AA, respectively. Suppose that we wish to estimate the frequency, q, of gene a.

To do this, we observe n individuals and assign to each the number 0 or

* Note that now we use lower case letters for both random variables and observed values.

1 or 2, depending on its genotype. Let r_1, r_2, and r_3 (with $r_1 + r_2 + r_3 = n$) be the numbers of AA, Aa, and aa genotypes, respectively.

(i) One way of estimating q is to use the observed proportion of the aa genotypes. Since q^2 is the expected value or r_3/n, it is reasonable to take the *estimated q*, which we denote here by $\tilde{q}$, as $\tilde{q} = \sqrt{r_3/n}$.

(ii) Another way of estimating q is to *count the number of genes a* in Aa and aa, and in estimating q, to use the ratio

$$\hat{q} = \frac{r_2 + 2r_3}{2n}.$$

These two functions of the sample, $\tilde{q}$ and $\hat{q}$, can be used to estimate the frequency of gene a. We shall discuss in section 12.5 which of these functions is more appropriate. We now consider a more general situation.

Let x be a random variable with probability distribution (or density function) $p(x; \theta)$, where θ is a parameter of the distribution (note that here θ denotes any parameter of the distribution; in particular cases the symbol μ will be used for mean, σ^2 for variance, P for binomial proportion, etc.). Let $x_1, x_2, \ldots, x_n$ be a random sample from a population having the distribution $p(x; \theta)$. We have denoted the random sample, as well as the values obtained in a particular sample, by lower case letters — the reader should by now be able to distinguish between these two situations. Let $\hat{\theta} = \hat{\theta}(x_1, x_2, \ldots, x_n)$ be a function* of a random sample (i.e., a statistic).

DEFINITION 11.1.   *Any function of a random sample constructed for the purpose of estimating a parameter of a distribution is called an estimator of this parameter.*

DEFINITION 11.2.   *The value of the estimator of a parameter calculated from an observed sample is called an estimate of this parameter.*

We notice the difference between estimator and estimate: *estimator* is a random variable; *estimate* is a particular value which this variable can take. We again use the same notation for estimator and estimate, bearing in mind when it denotes a random variable, $\hat{\theta}$, and when a particular value of $\hat{\theta}$. Some other symbols, such as $\tilde{\theta}$ or $\check{\theta}$, or Latin letters are used for estimators. In particular, the sample mean, $\bar{x}$, is used as an estimator of a population mean, μ, and s^2, defined in (8.75), often serves as an estimator of population variance σ^2.

A number of different statistics can be used as estimators of the same parameter; the choice of estimator depends on several circumstances which

---

* To emphasize that $\hat{\theta}$ is a function of the sample size n, some authors use the notation $\hat{\theta}_n$ instead of $\hat{\theta}$. We shall omit the subscript n, since it might cause some ambiguity in the later text.

will be discussed in a number of examples. We shall here consider a few statistical properties which may serve as criteria for classifying estimators as "better" or "worse."

11.2. PROPERTIES OF A "GOOD" ESTIMATOR

11.2.1. Consistency

Consistency is a property whereby the accuracy of an estimate increases when the sample size increases.

DEFINITION 11.3. *An estimator, $\hat{\theta}$, is called consistent if the probability that it differs from the true value, θ, by no more than ε, where $\varepsilon > 0$ is an arbitrary positive small number, tends to 1 as the sample size, n, increases; that is, if*

$$\lim_{n \to \infty} \Pr\{|\hat{\theta} - \theta| < \varepsilon\} = 1. \tag{11.1}$$

This means that, as n increases, the estimate, $\hat{\theta}$, becomes more likely to be close (within a small fixed distance, $\pm\varepsilon$) to the true parameter, θ. This is an *asymptotic* property of an estimator. It applies to "sufficiently large" samples. A sufficient pair of conditions for an estimator to be consistent is

$$\lim_{n \to \infty} E(\hat{\theta}) = \theta; \quad \lim_{n \to \infty} \text{Var}(\hat{\theta}) = 0. \tag{11.2}$$

11.2.2. Unbiasedness

An estimator, $\hat{\theta}$, as a random variable, has a certain distribution in repeated samples of size n. In a particular sample, the calculated value may deviate more or less from θ, but we hope that, on the average, it will give the true value. *Unbiasedness* is a property which ensures that, on the average, the estimator is correct.

DEFINITION 11.4. *An estimator, $\hat{\theta}$, is called unbiased if its expected value is equal to the true value, θ, that is,*

$$E(\hat{\theta}) = \theta. \tag{11.3}$$

Any estimator, $\hat{\theta}$, for which

$$E(\hat{\theta}) = \theta + b(\theta) \quad \text{with } b(\theta) \neq 0 \tag{11.4}$$

is called *biased*; the quantity $b(\theta)$ is called the *bias*.

By analogy with chemical or biochemical experiments, bias corresponds to the "systematic error" or the "error of the method." A chemist may use a certain method for which the results obtained can be very close to each other in repeated experiments but, on the average, do not give the correct answer. Similar situations may occur to a statistician in constructing an estimator. Of course, it is not always worthwhile to bother obtaining an exactly unbiased estimator; usually when the sample size increases

$$\lim_{n \to \infty} E(\hat{\theta}) = \theta, \tag{11.5}$$

so that $\hat{\theta}$ is *asymptotically unbiased*.

Example 11.2.   It can be shown that

$$E(\bar{x}) = \mu,$$

that is, $\bar{x}$ is an unbiased estimator of μ. But the sample second central moment,

$$\tilde{\sigma}^2 = \frac{1}{n} \sum_{i=1}^{n} (x_i - \bar{x})^2$$

is a biased estimator of σ^2, because

$$E(\tilde{\sigma}^2) = \frac{n-1}{n} \sigma^2 = \sigma^2 - \frac{1}{n} \sigma^2,$$

the bias being $b(\sigma^2) = -(1/n)\sigma^2$. If we take the "adjusted" estimator,

$$\frac{n}{n-1} \tilde{\sigma}^2 = s^2 = \frac{1}{n-1} \sum_{i=1}^{n} (x_i - \bar{x})^2,$$

then s^2 is an unbiased estimator of σ^2, because now

$$E(s^2) = \sigma^2.$$

For this reason we defined s^2, in section 8.16.2, as the sample variance. Of course, for $n \to \infty$, we have for both estimators

$$\lim_{n \to \infty} E(\tilde{\sigma}^2) = \lim_{n \to \infty} E(s^2) = \sigma^2,$$

that is, both $\tilde{\sigma}^2$ and s^2 are asymptotically unbiased.

It can also be shown that an unbiased estimator of the covariance between two variables, x and y, is the sample covariance

$$s_{xy} = \frac{1}{n-1} \sum_{i=1}^{n} (x_i - \bar{x})(y_i - \bar{y}).$$

Proofs of these well-known results can be found in almost any book on statistics.

It should be noticed that, even if $\hat{\theta}$ is an unbiased estimator of θ, the function $g(\hat{\theta})$ is not necessarily an unbiased estimator of $g(\theta)$ and, in fact, usually it is not. For example, s^2 is an unbiased estimator of the variance σ^2, but s is not an unbiased estimator of the standard deviation σ.

11.2.3. Efficiency

An estimator, $\hat{\theta}$, being a random variable, has not only an expected value but also a variance. If an estimator is unbiased, that is, if $E(\hat{\theta}) = \theta$, then $E[(\hat{\theta} - \theta)^2]$ defines the variance of $\hat{\theta}$. But if $E(\hat{\theta}) \neq \theta$, that is, if $\hat{\theta}$ is a biased estimator. the expression $E[(\hat{\theta} - \theta)^2]$ is called the *mean square error* (MSE) of $\hat{\theta}$.

Efficiency is a property describing the reliability of an estimator, expressed in terms of its sampling variation. We now restrict ourselves to unbiased estimators. Within this class, for a given sample $(x_1, x_2, \ldots, x_n)$, if there is an estimator with the smallest variance, it is called a *minimum variance unbiased* estimator or, briefly, an *MVU estimator*.

DEFINITION 11.5. *An MVU estimator is called an efficient estimator.*

DEFINITION 11.6. *Let $\hat{\theta} = \hat{\theta}(x_1, x_2, \ldots, x_n)$ be an efficient estimator and $\tilde{\theta} = \tilde{\theta}(x_1, x_2, \ldots, x_n)$ be another unbiased estimator from the same sample. The efficiency of $\tilde{\theta}$ is defined as*

$$\text{Eff}(\tilde{\theta}) = \frac{\text{Var}(\hat{\theta})}{\text{Var}(\tilde{\theta})}. \tag{11.6}$$

We notice that

$$0 < \text{Eff}(\tilde{\theta}) \leqslant 1. \tag{11.7}$$

For finite sample sizes, n, the MVU estimator (or even any unbiased estimator) very often does not exist. But quite often, as n tends to infinity ($n \to \infty$), the value of $n \, \text{Var}(\hat{\theta})$ tends to its *lower bound* (if it exists), so that $\hat{\theta}$ is *asymptotically MVU*, that is, $\hat{\theta}$ is *asymptotically efficient* [for explanation see section 11.3, and, in particular, formulae (11.16), (11.21), and (11.22)].

Suppose that $\hat{\theta}_1$ and $\hat{\theta}_2$ are two unbiased (arbitrary) estimators of θ. The *relative efficiency* of $\hat{\theta}_2$ with respect to $\hat{\theta}_1$ is defined as

$$\text{Rel eff}(\hat{\theta}_2) = \frac{\text{Var}(\hat{\theta}_1)}{\text{Var}(\hat{\theta}_2)}. \tag{11.8}$$

For biased estimators, comparisons are often made by using the mean square errors in the same way as variances were used above.

Example 11.3. Let P be the "true" proportion of individuals possessing a certain trait C, in a very large ("infinite") population. Suppose that a simple random sample of n_1 individuals is observed and that, among these, r_1 individuals possess trait C. We estimate P by the sample proportion $\hat{P}_1 = r_1/n_1$, say. As we can easily see, P_1 is a binomial sample proportion, so we have $E(\hat{P}_1) = P$ and $\text{Var}(\hat{P}_1) = PQ/n_1$, where $Q = 1 - P$.

Suppose that the sampling is independently repeated, and a sample of size n_2, with r_2 individuals possessing trait C, is obtained. We can obtain an estimator from the second sample, $\hat{P}_2 = r_2/n_2$. We shall have $E(\hat{P}_2) = P$ and $\text{Var}(\hat{P}_2) = PQ/n_2$. We find that the relative efficiency of $\hat{P}_2$ with respect to $\hat{P}_1$ is

$$\text{Rel eff}(\hat{P}_2) = \frac{\text{Var}(\hat{P}_1)}{\text{Var}(\hat{P}_2)} = \frac{n_2}{n_1}.$$

If $n_2 > n_1$, then $\hat{P}_2$ is more efficient than $\hat{P}_1$, and conversely if $n_2 < n_1$.

We may now use both samples to estimate P. How should we combine these results?

(i) Suppose that we take an ordinary arithmetic mean, that is, $\tilde{P} = \frac{1}{2}(\hat{P}_1 + \hat{P}_2)$. We shall have

$$E(\tilde{P}) = \frac{1}{2}[E(\hat{P}_1) + E(\hat{P}_2)] = P,$$

and

$$\text{Var}(\tilde{P}) = \frac{1}{4}[\text{Var}(\hat{P}_1) + \text{Var}(\hat{P}_2)] = \frac{PQ}{4} \cdot \frac{n_1 + n_2}{n_1 n_2}.$$

(ii) We may also pool the two samples and estimate P as

$$\hat{P} = \frac{r_1 + r_2}{n_1 + n_2} = \frac{n_1}{n_1 + n_2} \hat{P}_1 + \frac{n_2}{n_1 + n_2} \hat{P}_2.$$

The estimator $\hat{P}$ is usually called a *pooled* estimator. We find

$$E(\hat{P}) = \frac{n_1}{n_1 + n_2} E(\hat{P}_1) + \frac{n_2}{n_1 + n_2} E(\hat{P}_2) = P \quad \text{and} \quad \text{Var}(\hat{P}) = \frac{PQ}{n_1 + n_2}.$$

We notice that both estimators, $\tilde{P}$ and $\hat{P}$, are unbiased. Also they are consistent since, as $n_1 \to \infty$, $n_2 \to \infty$, both variances tend to zero. We now calculate the relative efficiency of $\tilde{P}$ with respect to $\hat{P}$, obtaining

$$\text{Rel eff}(\tilde{P}) = \frac{\text{Var}(\hat{P})}{\text{Var}(\tilde{P})} = \frac{PQ}{n_1 + n_2} \cdot \frac{4n_1 n_2}{PQ(n_1 + n_2)} = \frac{4n_1 n_2}{(n_1 + n_2)^2} \leqslant 1 \text{ always.}$$

Thus, for $n_1 \neq n_2$, the arithmetic mean of two sample proportions, P, is less efficient than the proportion calculated from pooled data, which is identical with the *weighted* proportion, $\hat{P}$.

The result obtained for sample proportions applies also to sample means from any kind of distribution; the properly *weighted* mean of two (or more) sample means is a more efficient estimator than their arithmetic mean.

11.3. THE MAXIMUM LIKELIHOOD METHOD OF ESTIMATION. ONE PARAMETER

A number of statistical methods can be used in estimating parameters from samples. As was mentioned in section 11.1, genetic parameters are concerned mostly with proportions such as the prevalence of a certain hereditary disease in a population, gene frequencies, or the recombination fraction in linkage. An easy and quite often obvious method is to take a simple random sample of size n and to *count* the number of individuals with this disease in the sample, or the number of genes in the sample, or the number of recombinants in the offspring of intercrosses or backcrosses. If r is the number of cases in a sample of size n, the proportion (r/n) is usually a "good" estimator of the expected proportion in the population. But there are several situations for which counting is not a simple procedure, especially when more than one parameter must be estimated.

It is desirable, then, to present a method (or methods) which can be applied in a rather large class of problems, and which provides estimators that can be expected to have the properties of "good" estimators, if not precisely, then at least approximately.

One such method, well known in statistics, is called the *method of maximum likelihood*. We shall first present this method for the situation in which only one parameter, θ, has to be estimated.

Let $x_1, x_2, \ldots, x_n$ be a random sample of size n drawn from a population in which the character x has the probability function $p(x; \theta)$. Depending on the sampling procedure, one can derive an appropriate probability function of obtaining a sample $(x_1, x_2, \ldots, x_n)$ which, in general, will be represented by a joint probability function, $p(x_1, x_2, \ldots, x_n; \theta)$. in n-dimensional sample space. Of particular importance and convenience is the case in which $(x_1, x_2, \ldots, x_n)$ represents a simple random sample of *independent* observations, so that their joint probability function is $p(x_1; \theta)p(x_2; \theta) \cdots p(x_n; \theta)$ [compare section 8.13.1].

DEFINITION 11.7. *The probability function evaluated for sample values* $(x_1, x_2, \ldots, x_n)$,

$$L(x_1, x_2, \ldots, x_n; \theta) = p(x_1, x_2, \ldots, x_n; \theta), \qquad (11.9)$$

is called the likelihood function of this sample.

In particular, when the sample is a simple random sample of independent observations, the likelihood function is of the form

$$L(x_1, x_2, \ldots, x_n; \theta) = p(x_1; \theta)\, p(x_2; \theta) \cdots p(x_n; \theta) = \prod_{i=1}^{n} p(x_i; \theta), \quad (11.10)$$

where $p(x; \theta)$ is the probability function of x in the population.

When the sample values are given, the *likelihood* is usefully regarded as *a function of θ* with x's fixed [briefly, $L(\theta)$], although, regarded as a probability function, it is a function of the x's with θ fixed.

The definition and further results are also valid when the probability function, $p(x; \theta)$, of a discrete random variable is replaced by the density function, $f(x; \theta)$, of a continuous variable.

We now restrict our consideration to the most common case of a simple random sample, and therefore we shall deal with the likelihood function defined in (11.10). (In the following text the word "simple" will often be omitted.)

DEFINITION 11.8.   *The estimator $\hat{\theta}$ of the parameter θ, which maximizes the likelihood function $L(x_1, x_2, \ldots, x_n; \theta)$ with respect to θ, is called the maximum likelihood estimator (briefly, the ML estimator) of θ.*

To find the ML estimator, $\hat{\theta}$, say, we usually have to solve the equation

$$\frac{\partial L}{\partial \theta} = 0 \tag{11.11}$$

(see Appendix II) for θ with the condition

$$\frac{\partial^2 L}{\partial \theta^2}\bigg|_{\theta=\hat{\theta}} < 0.$$

Since log L attains its maximum for the same value of θ as L and it is usually easier to deal with log L, we take the logarithm of L, defined in (11.10), that is,

$$\log L = \log p(x_1; \theta) + \log p(x_2; \theta) + \cdots + \log p(x_n; \theta)$$

$$= \sum_{i=1}^{n} \log p(x_i; \theta). \tag{11.12}$$

Differentiating both sides with respect to θ, we obtain

$$\frac{\partial \log L}{\partial \theta} = \sum_{i=1}^{n} \frac{\partial \log p(x_i; \theta)}{\partial \theta}. \tag{11.13}$$

Therefore, the likelihood equation (11.11) is equivalent to

$$\frac{\partial \log L}{\partial \theta} = \sum_{i=1}^{n} \frac{\partial \log p(x_i; \theta)}{\partial \theta} = 0 \tag{11.14}$$

(unless the maximum is reached at a boundary value of θ).

Since the ML estimator $\hat{\theta} = \hat{\theta}(x_1, x_2, \ldots, x_n)$ is a random variable, we would also like to find its variance. Although it is not always easy to find an exact formula for Var($\hat{\theta}$), it can be proved [see, for instance, Kendall and Stuart (1961), Chapter 17], under rather wide general conditions, that for any unbiased estimator

$$\text{Var}(\hat{\theta}) \geqslant -\frac{1}{E\left(\dfrac{\partial^2 \log L}{\partial \theta^2}\right)} = \frac{1}{E\left[\left(\dfrac{\partial \log L}{\partial \theta}\right)^2\right]}. \tag{11.15}$$

If the estimator is biased, that is, if $b(\theta) \neq 0$, we have

$$\text{Var}(\theta) \geqslant -\frac{\left[1 + \dfrac{\partial b(\theta)}{\partial \theta}\right]^2}{E\left(\dfrac{\partial^2 \log L}{\partial \theta^2}\right)} = \frac{\left[1 + \dfrac{\partial b(\theta)}{\partial \theta}\right]^2}{E\left[\left(\dfrac{\partial \log L}{\partial \theta}\right)^2\right]}. \tag{11.16}$$

Of course, if (11.10) holds and all $p(x_i; \theta)$, $i = 1, 2, \ldots, n$, are identical, then all

$$E\left[\frac{\partial^2 \log (p(x_i; \theta))}{\partial \theta^2}\right]$$

are the same, and

$$E\left(\frac{\partial^2 \log L}{\partial \theta^2}\right) = nE\left(\frac{\partial^2 \log p}{\partial \theta^2}\right) = -nE\left[\left(\frac{\partial \log p}{\partial \theta}\right)^2\right]. \tag{11.17}$$

Expression (11.15) is known as the *Cramér-Rao inequality*. The quantity

$$-\frac{1}{E\left(\dfrac{\partial^2 \log L}{\partial \theta^2}\right)} = -\frac{1}{nE\left(\dfrac{\partial^2 \log p}{\partial \theta^2}\right)} = \frac{1}{nE\left[\left(\dfrac{\partial \log p}{\partial \theta}\right)^2\right]} \tag{11.18}$$

is called the *lower bound of the variance* (briefly, the LBV) of an estimator of θ. The equality in (11.15) holds only under special conditions, namely, if and only if $(\partial \log L)/\partial \theta$ can be expressed as a multiple of $\hat{\theta} - \theta$, that is, if

$$\frac{\partial \log L}{\partial \theta} = k(\theta)(\hat{\theta} - \theta), \tag{11.19}$$

where k is independent of $\hat{\theta}$ but can depend on θ.

If (11.19) holds, then

$$\text{Var}(\hat{\theta}) = -\frac{1}{E\left(\dfrac{\partial^2 \log L}{\partial \theta^2}\right)} = -\frac{1}{nE\left(\dfrac{\partial^2 \log p}{\partial \theta^2}\right)}, \tag{11.20}$$

and $\hat{\theta}$ must be an efficient, that is, an MVU, estimator.

The variance of $\hat{\theta}$ always decreases as the sample size, n, increases. It can be shown (again under wide conditions) that, as $n \to \infty$,

$$\lim_{n \to \infty} [n \, \text{Var}(\hat{\theta})] = -\frac{1}{E\left(\dfrac{\partial^2 \log p}{\partial \theta^2}\right)} = \frac{1}{E\left[\left(\dfrac{\partial \log p}{\partial \theta}\right)^2\right]}. \tag{11.21}$$

In other words, even if the ML estimator, $\hat{\theta}$, does not satisfy condition (11.19), when the sample size, n, is sufficiently large, its variance can still be calculated from the approximate formula

$$\text{Var}(\hat{\theta}) = \sigma_{\hat{\theta}}^2 \doteq -\frac{1}{E\left(\dfrac{\partial^2 \log L}{\partial \theta^2}\right)} = -\frac{1}{nE\left(\dfrac{\partial^2 \log p}{\partial \theta^2}\right)}$$

$$= \frac{1}{E\left[\left(\dfrac{\partial \log L}{\partial \theta}\right)^2\right]} \tag{11.22}$$

which, in fact, represents its lower bound, defined in (11.15). The ML estimator, $\hat{\theta}$, is asymptotically an MVU estimator (compare section 11.2).

Binomial Distribution.   Let r be the number of "successes" in n independent trials, where the probability of a "success" in a single trial is P. Thus r is a binomial variable* with probability distribution

$$b(r; P, n) = \binom{n}{r} P^r (1 - P)^{n-r}.$$

Suppose that P is unknown and we wish to estimate it from the "successes" observed in n independent trials. Then $b(r; P, n)$ is the likelihood function for this experimental design, that is,

$$L = \binom{n}{r} P^r (1 - P)^{n-r}.$$

Taking logarithms of both sides, we have

$$\log L = \log \binom{n}{r} + r \log P + (n - r)\log(1 - P).$$

Differentiating $\log L$ with respect to P, we obtain

$$\frac{\partial \log L}{\partial P} = \frac{r}{P} + \frac{n - r}{1 - P}. \tag{11.23}$$

Equalizing (11.23) to zero, we obtain

$$\hat{P} = \frac{r}{n}. \tag{11.24}$$

Expression (11.23) can be written in the form

$$\frac{\partial \log L}{\partial P} = \frac{1}{nP(1 - P)}\left(\frac{r}{n} - P\right) = \frac{1}{nP(1 - P)}(\hat{P} - P) = k(P)(\hat{P} - P).$$

We see that the ML estimator, $\hat{P}$, given by (11.24) satisfies condition (11.19),

* Note that now r denotes the random variable as well as its observed value.

so $\hat{P}$ is the MVU estimator of P. Let us find $\text{Var}(\hat{P})$. We have

$$\frac{\partial^2 \log L}{\partial P^2} = -\frac{r}{P^2} - \frac{n-r}{(1-P)^2}.$$

Since $E(r) = nP$, then

$$E\left(\frac{\partial^2 \log L}{\partial P^2}\right) = -\frac{n}{P(1-P)},$$

and so

$$\text{Var}(\hat{P}) = \frac{P(1-P)}{n}. \tag{11.25}$$

Of course, we should notice that, since $\hat{P} = (r/n)$ is a binomial proportion, its variance is $P(1-P)/n$ *exactly* [see formula (8.45)]. The derivation of (11.25) was carried out only to demonstrate the maximum likelihood method.

The result can be summarized as follows:

The ML estimator of the binomial parameter P is the observed proportion of "successes" in n independent trials, $\hat{P} = (r/n)$, and its (exact) variance is $\text{Var}(\hat{P}) = P(1-P)/n$. The estimator $\hat{P}$ is the MVU estimator.

11.4. ASYMPTOTIC PROPERTIES OF MAXIMUM LIKELIHOOD ESTIMATORS

1. It can be shown that ML estimators are *consistent*, that is,

$$\lim_{n \to \infty} P\{|\hat{\theta} - \theta| < \varepsilon\} = 1. \tag{11.26}$$

Most of the ML estimators calculated from a small sample are *biased*. If an ML estimator is biased, that is, $E(\hat{\theta}) = \theta + b(\theta)$, then $E[(\hat{\theta} - \theta)^2]$ is the mean square error rather than the variance.

2. It can be shown, however, that, as $n \to \infty$, the ML estimators are *asymptotically unbiased*, that is [see formula (11.5)],

$$\lim_{n \to \infty} E(\hat{\theta}) = \theta.$$

Even if an ML estimator is unbiased, its variance may not always attain the Cramér-Rao lower bound.

3. It can be shown, however, that, as $n \to \infty$, the ML estimator, $\hat{\theta}$, is *asymptotically efficient*; that is, as $n \to \infty$, $n \, \text{Var}(\hat{\theta})$ tends to its lower bound and the variance is *approximately* evaluated from formula (11.22), that is,

$$\text{Var}(\hat{\theta}) \doteq -\frac{1}{E\left(\dfrac{\partial^2 \log L}{\partial \theta^2}\right)} = \frac{1}{E\left[\left(\dfrac{\partial \log L}{\partial \theta}\right)^2\right]}.$$

Of course, if the ML estimator is only asymptotically unbiased, (11.22) is still valid if n is large enough, but the approximation is not so good as in the case of an unbiased estimator. But for large n, the error can be negligible.

4. Another very useful property of ML estimators is that they are *approximately normally* distributed, namely, the statistic

$$Z = \frac{\hat{\theta} - \theta}{\sqrt{\text{Var}(\hat{\theta})}} \doteq \sqrt{n}\,(\hat{\theta} - \theta)\sqrt{-E\left(\frac{\partial^2 \log p}{\partial \theta^2}\right)} \tag{11.27}$$

is asymptotically distributed as $N(0, 1)$. This property is very useful in applications.

Example 11.4. The main blood group locus in rats is called *Ag-B*. Several alleles are associated with this locus, and it appears that in certain crosses significant deviations from Mendelian ratios have been observed. Ramseier and Palm (1967) obtained $n = 200$ serological tests from the offspring of intercrosses of two strains of rats, Lewis (Ag-B^1Ag-B^1) and DA (Ag-B^4Ag-B^4). (For simplicity, we will use only the symbol B and will omit Ag.) The results of crosses $B^1B^4 \times B^1B^4$ were as follows:

Genotype:	B^1B^1	B^1B^4	B^4B^4
Observed frequency:	58	129	13

which, at first glance, differ markedly from the Mendelian ratios $1:2:1$. Several explanations (hypotheses) of this phenomenon can be proposed, and this problem will be discussed in section 13.7.

Let us here *assume* that the mechanism which modified the Mendelian ratios was due to *selective disadvantage* of homozygote B^4B^4 at the embryonic stage. Let the survival ratios of zygotes be $B^1B^1:B^1B^4:B^4B^4 = 1:1:\phi$, with $0 < \phi < 1$. Thus the genotype ratios in F_2 will be $B^1B^1:B^1B^4:B^4B^4 = \frac{1}{4}:\frac{2}{4}:\frac{1}{4}\phi = 1:2:\phi$. Hence the expected proportions (distributions) of the corresponding genotypes are

$$\begin{array}{ccc} B^1B^1 & B^1B^4 & B^4B^4 \\[4pt] \dfrac{1}{3+\phi} & \dfrac{2}{3+\phi} & \dfrac{\phi}{3+\phi} \end{array}.$$

Suppose that n individuals from F_2 were selected at random and tested for their antigens. Let r_1, r_2, r_3, with $r_1 + r_2 + r_3 = n$, be the observed numbers of genotypes B^1B^1, B^1B^4, B^4B^4, respectively. Thus r_1, r_2, r_3, considered as random variables, have the multinomial distribution with parameters

$$P_1 = \frac{1}{3+\phi}, \quad P_2 = \frac{2}{3+\phi}, \quad P_3 = \frac{\phi}{3+\phi}, \quad \text{with} \quad P_1 + P_2 + P_3 = 1.$$

The likelihood function is

$$L = \frac{n!}{r_1! r_2! r_3!} \left(\frac{1}{3 + \phi}\right)^{r_1} \left(\frac{2}{3 + \phi}\right)^{r_2} \left(\frac{\phi}{3 + \phi}\right)^{r_3},$$

and its logarithm is

$$\log L = \log \frac{n!}{r_1! r_2! r_3!} + r_1 \log 1 + r_2 \log 2 + r_3 \log \phi - n \log(3 + \phi).$$

Differentiating with respect to ϕ and equating the first derivative to zero. we have

$$\frac{\partial L}{\partial \phi} = \frac{r_3}{\phi} - \frac{n}{3 + \phi} = 0.$$

Hence $\hat{\phi} = [3r_3/(n - r_3)]$. In our example, $n = 200$, $r_3 = 13$, and $n - r_3 = 187$, so that $\hat{\phi} = (39/187) = 0.2086$.

We note that

$$E(\hat{\phi}) = E\left(\frac{3r_3}{n - r_3}\right) \neq \frac{3E(r_3)}{n - E(r_3)} = \frac{3P}{1 - P}.$$

Assuming that $\Pr\{r_3 = n\} > 0$, the expected value of $\hat{\phi}$ is infinite. However, if we exclude the case $r_3 = n$, the *conditional* expected value of $\hat{\phi}$ is *approximately* equal to $3P/(1 - P)$ for a large sample.

We now find the approximate variance, using formula (11.22). We have

$$\frac{\partial^2 \log L}{\partial \phi^2} = -\frac{r_3}{\phi^2} + \frac{n}{(3 + \phi)^2}.$$

Since

$$E(r_3) = nP = n \frac{\phi}{3 + \phi},$$

then

$$-E\left(\frac{\partial^2 \log L}{\partial \phi^2}\right) = \frac{n}{\phi(3 + \phi)} - \frac{n}{(3 + \phi)^2} = \frac{3n}{\phi(3 + \phi)^2}.$$

Hence

$$\mathrm{Var}(\hat{\phi}) \doteq \frac{\phi(3 + \phi)^2}{3n}.$$

Since in our example the true ϕ is unknown, we calculate the estimated variance by substituting $\hat{\phi}$ for ϕ. We obtain

$$\mathrm{Est\,Var}(\hat{\phi}) = \frac{0.2086 \times 3.2086^2}{600} = 0.003579, \quad \text{and} \quad \mathrm{Est}\ \sigma = 0.0598.$$

11.5. FUNCTIONS OF THE ML ESTIMATORS

It is very often easier to find the ML estimator of a certain function of θ than of θ itself. Let $\tau = \tau(\theta)$ be a monotonic function of θ such that the inverse function $\theta = \theta(\tau)$ and $(d\theta/d\tau)$ exist. It is very easy to see that:

If $\hat{\tau}$ is an ML estimator of τ, then $\hat{\theta} = \theta(\hat{\tau})$ is an ML estimator of θ.

Applying formula (8.91), we notice that $\hat{\theta}$ is only asymptotically unbiased even if $\hat{\tau}$ is unbiased.

From (8.93) we obtain the approximate variance of $\hat{\theta}$:

$$\mathrm{Var}(\hat{\theta}) \doteq \left(\frac{d\theta}{d\tau}\right)^2 \mathrm{Var}(\hat{\tau}), \qquad (11.28)$$

where

$$\mathrm{Var}(\hat{\tau}) \doteq - \frac{1}{E\left(\dfrac{\partial^2 \log L}{\partial \tau^2}\right)} = \frac{1}{E\left[\left(\dfrac{\partial \log L}{\partial \tau}\right)^2\right]}.$$

On the other hand, we have

$$\frac{\partial \log L}{\partial \theta} = \frac{\partial \log L}{\partial \tau} \cdot \frac{d\tau}{d\theta},$$

and so the large-sample formula for the variance of $\hat{\theta}$, given in (11.22), takes the form

$$\mathrm{Var}(\hat{\theta}) \doteq - \frac{1}{E\left(\dfrac{\partial^2 \log L}{\partial \theta^2}\right)} = \frac{1}{\left(\dfrac{d\tau}{d\theta}\right)^2} \cdot \frac{1}{E\left[\left(\dfrac{\partial \log L}{\partial \tau}\right)^2\right]} = \left(\frac{d\theta}{d\tau}\right)^2 \mathrm{Var}(\hat{\tau}). \quad (11.29)$$

We note that this is the same as the approximate relation (11.28).

We can thus formulate a practical rule for calculating the ML estimator of θ and its approximate variance if obtaining the ML estimator of $\tau = \tau(\theta)$ is more convenient:

(i) Calculate the ML estimator of τ, $\hat{\tau}$ say, by a standard method; the estimator $\hat{\theta} = \theta(\hat{\tau})$ will be the ML estimator of θ. In particular, when τ is the binomial parameter P, that is, $\tau(\theta) = P(\theta)$, estimate P by the proportion of successes in n independent trials, $\hat{P} = (r/n)$, and solve the equation $P(\hat{\theta}) = \hat{P}$ with respect to $\hat{\theta}$.

(ii) For a "sufficiently large" sample, calculate the approximate variance, $\mathrm{Var}(\hat{\theta})$, using the "statistical differential method," from formula (11.28) [or (11.29)].

It is left to the reader to show that, by applying formula (11.29) to example 11.4, we obtain results identical with those obtained by the standard method.

11.6. CONFIDENCE INTERVALS

In preceding sections we discussed the possibilities of estimating a parameter θ by a *single value* $\hat{\theta}$ obtained from a sample. Even if $\hat{\theta}$ has all the properties of a "good" estimator, we still have to be careful in using $\hat{\theta}$ instead of θ unless we have a rather high probability that the deviation, $\hat{\theta} - \theta$, is relatively unimportant for our purpose.

Suppose that the probability that $\hat{\theta} - \theta$ is contained between two limits, d_1 and d_2, is $1 - \alpha$. The limits d_1 and d_2 are some functions of α, that is, $d_1 = d_1(\alpha)$, $d_2 = d_2(\alpha)$, and so we have

$$\Pr\{d_1(\alpha) \leqslant \hat{\theta} - \theta \leqslant d_2(\alpha)\} = 1 - \alpha. \tag{11.30}$$

Relation (11.30) can be rewritten in the form

$$\Pr\{\hat{\theta} - d_2(\alpha) \leqslant \theta \leqslant \hat{\theta} - d_1(\alpha)\} = \Pr\{l_1(\hat{\theta}, \alpha) \leqslant \theta \leqslant l_2(\hat{\theta}, \alpha)\} = 1 - \alpha.$$
$$\cdot (11.31)$$

The quantities $l_1(\hat{\theta}, \alpha) = \hat{\theta} - d_2(\alpha)$ and $l_2(\hat{\theta}, \alpha) = \hat{\theta} - d_1(\alpha)$ can be calculated from sample values. A few important definitions and remarks should be made.

1. The limits $l_1 = l_1(\hat{\theta}, \alpha)$ and $l_2 = l_2(\hat{\theta}, \alpha)$ are called the *lower* and the *upper confidence limits*, respectively, and the interval (l_1, l_2) is termed a $100(1 - \alpha)\%$ *confidence interval;* its length is equal to $l_2 - l_1$. We notice that the confidence limits, $l_1(\hat{\theta}, \alpha)$ and $l_2(\hat{\theta}, \alpha)$, are *random variables* since they depend on $\hat{\theta}$, which itself is a random variable.

2. The probability $1 - \alpha$ is called the *confidence coefficient*. It means that, in a very large ("infinite") number of evaluated confidence intervals, the expected proportion of those which include θ, is $1 - \alpha$. It should be emphasized that it is *not the probability that, given a particular sample*, the calculated limits, l_1 and l_2, include θ.

3. For a given sample size n and a given probability $1 - \alpha$, d_1 and d_2 depend on n and α. If d_1, d_2, and n are fixed, $1 - \alpha$ can be determined. If d_1, d_2 and $1 - \alpha$ are fixed, the required n can be evaluated.

How can confidence limits be constructed? If we know the distribution of the estimator $\hat{\theta}$, the values of d_1 and d_2, for given n and $1 - \alpha$, can be evaluated by solving (11.31). We demonstrate this procedure by constructing confidence limits for μ of normal distribution.

11.6.1. Confidence Interval for the Population Mean of a Normal Distribution

Let x be a random variable normally distributed with mean μ and variance σ^2. Suppose that σ^2 is *known*, and we estimate μ using the sample mean $\bar{x}$.

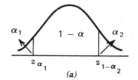 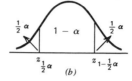 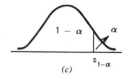

Figure 11.1

We know that $\bar{x} \frown N[\mu, (\sigma^2/n)]$. For any two values, $\alpha_1 (<\frac{1}{2})$ and $\alpha_2 (<\frac{1}{2})$, such that $\alpha_1 + \alpha_2 = \alpha$ we have

$$\Pr\left\{z_{\alpha_1} \leqslant \frac{\bar{x} - \mu}{\sigma/\sqrt{n}} \leqslant z_{1-\alpha_2}\right\} = \Pr\left\{\bar{x} - z_{1-\alpha_2}\frac{\sigma}{\sqrt{n}} \leqslant \mu \leqslant \bar{x} - z_{\alpha_1}\frac{\sigma}{\sqrt{n}}\right\}$$

$$= 1 - \alpha. \tag{11.32}$$

Thus, we can construct several confidence intervals satisfying (11.32) (see Figs. 11.1a, b, c). It can be shown that, when $\alpha_1 = \alpha_2 = \frac{1}{2}\alpha$, the confidence interval will have some "optimal" properties (e.g., being the shortest in this case; Fig. 11.1b). We then confine ourselves to such confidence intervals. Since now $z_{\frac{1}{2}\alpha} = -z_{1-\frac{1}{2}\alpha}$, our condition (11.32) takes the form

$$\Pr\left\{\bar{x} - z_{1-\frac{1}{2}\alpha}\frac{\sigma}{\sqrt{n}} \leqslant \mu \leqslant \bar{x} + z_{1-\frac{1}{2}\alpha}\frac{\sigma}{\sqrt{n}}\right\} = 1 - \alpha. \tag{11.33}$$

The $100(1 - \alpha)\%$ confidence limits will be then calculated from the formulae

$$l_1 = \bar{x} - z_{1-\frac{1}{2}\alpha}\frac{\sigma}{\sqrt{n}} \; ; \quad l_2 = \bar{x} + z_{1-\frac{1}{2}\alpha}\frac{\sigma}{\sqrt{n}} . \tag{11.34}$$

The choice of $1 - \alpha$ is arbitrary. It is customary to use 90%, 95%, and 99% confidence intervals.

Example 11.5. Let $n = 10$, $\sigma^2 = 1.6$, and $\bar{x} = 25.62$. We wish to find 95% confidence limits. We have $1 - \alpha = 0.95$; thus $\frac{1}{2}\alpha = 0.025$, $z_{0.975} = 1.96$, and $\sigma/\sqrt{n} = \sqrt{0.16} = 0.4$. Hence

$$z_{1-\frac{1}{2}\alpha}\frac{\sigma}{\sqrt{n}} = 1.96 \times 0.4 = 0.78,$$

and

$$l_1 = 25.62 - 0.78 = 24.84, \quad l_2 = 25.62 + 0.78 = 26.40.$$

The 95% confidence interval is $24.84 < \mu < 26.40$. We may regard this as meaning that with 95% confidence we "believe" that the interval $(24.84, 26.40)$ includes the true mean. It should be noted that the statement $\Pr\{24.84 \leqslant \mu \leqslant 26.40\} = 0.95$ is now *not true*. This probability is 1 if μ is between these limits and 0 otherwise. Now the two limits are no longer random variables, so relation (11.33) does not hold.

When σ^2 is *unknown*, a confidence interval can be constructed using the t-distribution, that is, from the formula,

$$\Pr\left\{\bar{x} - t_{1-\frac{1}{2}\alpha;n-1}\frac{s}{\sqrt{n}} \leqslant \mu \leqslant \bar{x} + t_{1-\frac{1}{2}\alpha;n-1}\frac{s}{\sqrt{n}}\right\} = 1 - \alpha. \quad (11.35)$$

Here the value of t depends not only on α but also on the degrees of freedom, $n - 1$.

11.6.2. Application of Large-Sample Theory in Calculating Confidence Intervals

The genetic parameters in which we are chiefly interested do not usually represent the parameters of normal distributions. Moreover, the exact distribution of the estimator is quite often unknown. However, assuming that the sample is large, we can take advantage of property (11.27), that is, that the ML estimator, $\hat{\theta}$, is asymptotically normally distributed with mean θ and variance σ_{θ}^2, given by (11.22). This means that the relation

$$\Pr\{\hat{\theta} - z_{1-\frac{1}{2}\alpha}\sigma_\theta \leqslant \theta \leqslant \hat{\theta} + z_{1-\frac{1}{2}\alpha}\sigma_\theta\} \doteq 1 - \alpha \quad (11.36)$$

holds for large n and σ_θ given by (11.22). The $100(1 - \alpha)\%$ confidence limits can be calculated from the formulae

$$l_1 \doteq \hat{\theta} - z_{1-\frac{1}{2}\alpha}\sigma_\theta; \quad l_2 \doteq \hat{\theta} + z_{1-\frac{1}{2}\alpha}\sigma_\theta. \quad (11.37)$$

Example 11.6. We wish to find a 95% confidence interval for the parameters discussed in Example 11.4. Here the true value of ϕ is unknown, so we can use only an estimated variance of $\hat{\phi}$, Est $\sigma_{\hat{\phi}}^2 = \tilde{\sigma}_{\hat{\phi}}^2 = 0.0598$ as calculated in Example 11.4. We have $z_{0.975}\tilde{\sigma}_{\hat{\phi}} \doteq 1.96 \times 0.0598 = 0.1172$. Hence $l_1 \doteq 0.2086 - 0.1172 = 0.0914$; $l_2 \doteq 0.2086 + 0.1172 = 0.325$. We believe with about 95% confidence that the interval $(0.0914, 0.3258)$ includes the true value, ϕ.

11.6.3. Confidence Interval for a Binomial Proportion

(i) An estimator of the binomial parameter P is the observed proportion of "successes" in n independent trials, that is, $\hat{P} = r/n$, where r is the number of "successes." Although the exact distribution of $\hat{P}$ is also of binomial form, exact $100(1 - \alpha)\%$ confidence limits cannot be calculated since the distribution is discrete. Tables or graphs are available to find values r_1 and r_2, such that $\Pr\{r_1 \leqslant r \leqslant r_2 \mid P\} \doteq 1 - \alpha$.

(ii) For sufficiently large samples, the common practice is to apply normal theory as presented in section 11.6.2. In this case, *approximate* $100(1 - \alpha)\%$ confidence limits are

$$\hat{P} - z_{1-\frac{1}{2}\alpha}\sqrt{\frac{PQ}{n}} \leqslant P \leqslant \hat{P} + z_{1-\frac{1}{2}\alpha}\sqrt{\frac{PQ}{n}} . \qquad (11.38)$$

A better approximation is usually obtained by using the continuity correction, $\pm(1/2n)$, that is,

$$\left(\hat{P} - \frac{1}{2n}\right) - z_{1-\frac{1}{2}\alpha}\sqrt{\frac{PQ}{n}} \leqslant P \leqslant \left(\hat{P} + \frac{1}{2n}\right) + z_{1-\frac{1}{2}\alpha}\sqrt{\frac{PQ}{n}} . \qquad (11.38a)$$

It can be seen that the limits for P depend on P itself. But (11.38) or (11.38a) can be rearranged so that P does not occur in the limits. This requires solution of some quadratic equation [see Brownlee (1960), Chapter 3].

Another possibility is to replace the variance PQ/n by the estimated variance, $\hat{P}\hat{Q}/n$. This makes the confidence coefficient $1 - \alpha$ even more uncertain, but we shall have at least some idea of the order of accuracy of our estimator.

(iii) A useful confidence interval can be obtained by using the F-distribution. If r_0 is the observed number of "successes," the $100(1 - \alpha)\%$ confidence limits for the binomial proportion, P, can be evaluated as follows:

$$l_1 = \frac{r_0}{r_0 + (n - r_0 + 1)F_{1-\frac{1}{2}\alpha}}, \quad l_2 = \frac{(r_0 + 1)F_{1-\frac{1}{2}\alpha}}{n - r_0 + (r_0 + 1)F_{1-\frac{1}{2}\alpha}}, \qquad (11.39)$$

where $F_{1-\frac{1}{2}\alpha}$ for the *lower* limit, l_1, is obtained from tables for $\nu_1 = 2(n - r_0 + 1)$, $\nu_2 = 2r_0$; and for the *upper* limit, l_2, from tables for $\nu_1 = 2(r_0 + 1)$, $\nu_2 = 2(n - r_0)$. [For details of these formulae see Brownlee (1960), Chapter 3.] Application of formula (11.39) is demonstrated in Problem 19.1.

11.7. SCORES AND AMOUNT OF INFORMATION

We now introduce a terminology first employed by R. A. Fisher.

DEFINITION 11.9. *The first derivative of the logarithm of the likelihood function with respect to θ is called the total score. For a given sample, the score can be considered as a function of θ. Denoting this score by $U(\theta)$ (or U_θ), we have*

$$U(\theta) = \frac{\partial \log L}{\partial \theta} = \sum_{i=1}^{n} \frac{\partial \log p(x_i; \theta)}{\partial \theta} . \qquad (11.40)$$

DEFINITION 11.10. *The quantity*

$$u(x_i; \theta) = \frac{\partial \log p(x_i; \theta)}{\partial \theta} \qquad (11.41)$$

is the individual score of a particular observation, x_i.

An ML estimator may be obtained by equating the total score to zero. Of course, individual scores, evaluated for $\theta = \hat{\theta}$, are not all zeros, but they must satisfy the condition that their sum or the *average score*, for $\theta = \hat{\theta}$, is equal to zero.

An estimator $\hat{\theta}$ utilizes certain "information" about the true parameter, θ. The less $\hat{\theta}$ varies, the more precise is the information about θ.

DEFINITION 11.11. *The inverse of the minimum variance of $\hat{\theta}$ (termed also the invariance) is called the expected total amount of information about θ obtainable from the sample through $\hat{\theta}$, and is denoted by $I(\theta)$ (or I_0).*

In practice, we usually take the inverse of the lower bound for variance, that is,

$$I(\theta) \doteq \frac{1}{\mathrm{Var}(\hat{\theta})} = -E\left(\frac{\partial^2 \log L}{\partial \theta^2}\right) = -nE\left(\frac{\partial^2 \log p}{\partial \theta^2}\right) = -E\left(\frac{\partial U}{\partial \theta}\right). \qquad (11.42)$$

If n is large, we may use $\hat{\theta}$ instead of θ and calculate the *estimated* total amount of information, that is,

$$I(\hat{\theta}) \doteq \frac{\partial^2 \log L}{\partial \hat{\theta}^2} = -n\frac{\partial^2 \log p}{\partial \hat{\theta}^2} = -\frac{\partial U}{\partial \hat{\theta}}, \qquad (11.42a)$$

where $(\partial^2 \log L/\partial \hat{\theta}^2)$ is the second derivative of $\log L$ with respect to θ evaluated at $\theta = \hat{\theta}$.

DEFINITION 11.12. *The expected average amount of information per single observation, $i(\theta)$ (or i_0), is*

$$i(\theta) = \frac{1}{n} I(\theta) = -E\left(\frac{\partial^2 \log p}{\partial \theta^2}\right) = E\left[\left(\frac{\partial \log p}{\partial \theta}\right)^2\right], \qquad (11.43)$$

The estimated (from the observed sample) average amount of information per observation is

$$i(\hat{\theta}) = \frac{1}{n} I(\hat{\theta}) = -\frac{\partial^2 \log p}{\partial \hat{\theta}^2} = \left(\frac{\partial \log p}{\partial \hat{\theta}}\right)^2. \qquad (11.43a)$$

11.7.1. Efficiency of Experimental Designs

Let $\hat{\theta}_1$ and $\hat{\theta}_2$ be two estimators of the same parameter, θ, obtained from two different experimental designs, Exp. 1 and Exp. 2, say. Let $i_1 = i_1(\theta)$

and $i_2 = i_2(\theta)$ be the average amounts of information per observation obtained from Exp. 1 and Exp. 2, respectively. If $(i_2/i_1) < 1$, this means that the average amount of information per observation from Exp. 1 is larger than that from Exp. 2. Suppose that we wish the total amounts of information from each experiment to be the same, that is, $I_1 = I_2$. Since $I_1 = n_1 i_1$ and $I_2 = n_2 i_2$, the condition $I_1 = I_2$ is equivalent to

$$\frac{i_2}{i_1} = \frac{n_1}{n_2}.$$ (11.44)

In other words, to obtain the same total amount of information from Exp. 2 as can be obtained from Exp. 1 we have to increase the sample size, n_2, in Exp. 2 in the ratio (i_1/i_2), that is,

$$n_2 = \frac{i_1}{i_2} n_1.$$ (11.45)

We might regard ratio (11.44) as the *relative efficiency of Exp. 2 with respect to Exp. 1* in estimating the same parameter, θ.

Example 11.7. Suppose that the "selective disadvantage," ϕ, of the allele Ag-B⁴, determined in Example 11.4, has also to be estimated from backcrosses $B^1 B^4 \times B^4 B^4$. Using the same argument as in Example 11.4, we find that the number, r, of genotypes $B^4 B^4$ in n independent observations on the offspring of backcrosses has a binomial distribution with $P = [\phi/(1 + \phi)]$. The ML estimator of ϕ is $\hat{\phi} = [r/(n - r)]$, and its approximate variance is $\mathrm{Var}(\hat{\phi}) = [\phi(1 + \phi)^2/n]$.

Let us call the experimental design with intercrosses — Exp.1, and that on backcrosses — Exp.2. The corresponding expected average amounts of information per observation are: $i_1 = \frac{1}{3}\phi(3 + \phi)^2$, $i_2 = \phi(1 + \phi)^2$, respectively. We have

$$\frac{i_2}{i_1} = \frac{3(1 + \phi)^2}{(3 + \phi)^2}.$$

For $\phi = 0$, $(i_2/i_1) = \frac{1}{3}$, and for $\phi = 1$, $(i_2/i_1) = \frac{3}{4}$, so we have $\frac{1}{3} < (i_2/i_1) < \frac{3}{4}$. This means that more information is always obtained from intercrosses than from the same number of backcrosses. In extreme cases, for $\phi = 0$ the information from one intercross is equivalent to the information from three backcrosses, and for $\phi = 1$ this ratio is $3:4$.

11.7.2. An Iterative Method of Obtaining an ML Estimator

In a number of situations the explicit solution of the likelihood equation

$$\frac{\partial \log L}{\partial \theta} = U(\theta) = 0$$

cannot be obtained by any standard method, and an iterative method must be used.

Let θ_1 be a trial value which makes $U(\theta)$ rather small. We call θ_1 the first approximation to θ. We expand $U(\theta)$ in a Taylor series at the point θ_1, and confine ourselves to the first two terms. We then have to solve the numerical equation

$$U(\theta) \doteq U(\theta_1) + (\theta - \theta_1)\frac{\partial U}{\partial \theta}\bigg|_{\theta=\theta_1} = 0 \qquad (11.46)$$

or

$$-(\theta - \theta_1)\frac{\partial U}{\partial \theta}\bigg|_{\theta=\theta_1} \doteq U(\theta_1). \qquad (11.46a)$$

We note (from 11.42) that $-E(\partial U/\partial \theta) = I(\theta)$. But, of course, $I(\theta)$, is unknown, and for a large sample we evaluate the amount of information for the value θ_1, that is, we have

$$\frac{\partial U}{\partial \theta}\bigg|_{\theta=\theta_1} = I(\theta_1). \qquad (11.47)$$

We now write (11.46a) in the form

$$\theta - \theta_1 \doteq \frac{U(\theta_1)}{I(\theta_1)}. \qquad (11.48)$$

Hence, as the *second* approximation of $\hat{\theta}$, we take

$$\hat{\theta} \doteq \theta_2 = \theta_1 + \frac{U(\theta_1)}{I(\theta_1)}. \qquad (11.49)$$

Repeating the same operations on θ_2, we find the *third* approximation for $\hat{\theta}$, θ_2, say:

$$\hat{\theta} \doteq \theta_3 = \theta_2 + \frac{U(\theta_2)}{I(\theta_2)}. \qquad (11.50)$$

The process can be repeated several times until the desired precision is obtained. This method can be tedious, but there should be no difficulty when an electronic computer is available.

▼11.8. ESTIMATION OF SEVERAL PARAMETERS BY THE METHOD OF MAXIMUM LIKELIHOOD

11.8.1. Functionally Independent Parameters

Suppose that there are s parameters, $\theta_1, \theta_2, \ldots, \theta_s$ or, in vector form, $\boldsymbol{\theta}' = (\theta_1, \theta_2, \ldots, \theta_s)$, which must be estimated. We assume that $\theta_1, \theta_2, \ldots, \theta_s$

are *functionally independent**. This assumption is essential for further consideration.

With simple random sample of n observations, $x_1, x_2, \ldots, x_n$, the likelihood function can be regarded as a function of s variables, $\theta_1, \theta_2, \ldots, \theta_s$, that is,

$$L(\theta_1, \theta_2, \ldots, \theta_s) = L(\mathbf{\theta}) = \prod_{t=1}^{n} p(x_t; \theta_1, \theta_2, \ldots, \theta_s), \qquad (11.51)$$

where $p(x; \theta_1, \theta_2, \ldots, \theta_s)$ is the probability distribution of a random variable x in the population.

The first derivative of $\log L$ with respect to θ_i is the total score for θ_i, that is,

$$\frac{\partial \log L}{\partial \theta_i} = U_i(\theta_1, \theta_2, \ldots, \theta_s) = U_i(\mathbf{\theta}), \qquad (11.52)$$

where $i = 1, 2, \ldots, s$.

Now the total amount of information obtained from the sample about $(\theta_1, \theta_2, \ldots, \theta_s)$ can be expressed in a form of the $s \times s$ *expected information matrix*, $\mathbf{I}(\mathbf{\theta})$, whose elements are defined by

$$I_{ij}(\theta_1, \theta_2, \ldots, \theta_s) = I_{ij}(\mathbf{\theta}) = -E\left(\frac{\partial^2 \log L}{\partial \theta_i\, \partial \theta_j}\right), \qquad (11.53)$$

where $i, j = 1, 2, \ldots, s$. In particular, for $i = j$ we have

$$I_{ii}(\theta_1, \theta_2, \ldots, \theta_s) = I_{ii}(\mathbf{\theta}) = -E\left(\frac{\partial^2 \log L}{\partial \theta_i^2}\right). \qquad (11.54)$$

The information matrix takes the form

$$\mathbf{I}(\mathbf{\theta}) = \begin{bmatrix} I_{11}(\mathbf{\theta}) & I_{12}(\mathbf{\theta}) & \cdots & I_{1s}(\mathbf{\theta}) \\ I_{21}(\mathbf{\theta}) & I_{22}(\mathbf{\theta}) & \cdots & I_{2s}(\mathbf{\theta}) \\ \cdots\cdots\cdots\cdots\cdots\cdots \\ \cdots\cdots\cdots\cdots\cdots\cdots \\ I_{s1}(\mathbf{\theta}) & I_{s2}(\mathbf{\theta}) & \cdots & I_{ss}(\mathbf{\theta}) \end{bmatrix}. \qquad (11.55)$$

The inverse of the information matrix gives the approximate *variance-covariance matrix* of the estimators $\mathbf{\hat{\theta}}' = (\hat{\theta}_1, \hat{\theta}_2, \ldots, \hat{\theta}_s)$, $V(\mathbf{\hat{\theta}})$, say, that is,

$$V(\mathbf{\hat{\theta}}) \doteq \mathbf{I}^{-1}(\mathbf{\theta}). \qquad (11.56)$$

In other words, the variance of the estimator $\hat{\theta}_i$ is the element $I^{ii}(\mathbf{\theta})$, and the covariance of two estimators $\hat{\theta}_i, \hat{\theta}_j$ is the element $I^{ij}(\mathbf{\theta})$ of the inverse matrix $\mathbf{I}^{-1}(\mathbf{\theta})$. In order to obtain the ML estimators, $\hat{\theta}_1, \hat{\theta}_2, \ldots, \hat{\theta}_s$, we now have

---

* For definition of functional dependence see formula (5.45).

to solve a *set* of s equations with respect to $\theta_1, \theta_2, \ldots, \theta_s$, that is,

$$\frac{\partial \log L}{\partial \theta_i} = U_i(\theta_1, \theta_2, \ldots, \theta_s) = 0, \tag{11.57}$$

for $i = 1, 2, \ldots, s$.

Usually it is not very easy to obtain exact solutions, and iterative methods often have to be used, as in the case of one parameter. We will only outline such a method. We take s "trial" values, $\boldsymbol{\theta}'_1 = (\theta_{11}, \theta_{21}, \ldots, \theta_{s1})$, say, as *first* approximations. Using a Taylor expansion about the point $\boldsymbol{\theta} = \boldsymbol{\theta}_1$ and taking into account (11.57), we obtain a *second* approximation for the estimator $\hat{\theta}_i$ ($i = 1, 2, \ldots, s$), θ_{i2}, say, from the formula

$$\hat{\theta}_i \doteq \theta_{i2} = \theta_{i1} + \sum_{j=1}^{s} I^{ij}(\boldsymbol{\theta}_1) U_j(\boldsymbol{\theta}_1)$$

$$= \theta_{i1} + U_i(\boldsymbol{\theta}_1) \mathrm{Var}(\hat{\theta}_i)|_{\boldsymbol{\theta}=\boldsymbol{\theta}_1} + \sum_{j \neq i}^{s} U_j(\boldsymbol{\theta}_1) \mathrm{Cov}(\hat{\theta}_i, \hat{\theta}_j)|_{\boldsymbol{\theta}=\boldsymbol{\theta}_1}, \tag{11.58}$$

for $i = 1, 2, \ldots, s$. The procedure, which is repeated until the desired precision is obtained, is, of course, very tedious, and for $s > 2$ it is almost essential to use an electronic computer.

11.8.2. Parameters Subject to Constraint

It should be emphasized that the condition of functional independence of the parameters $\theta_1, \theta_2, \ldots, \theta_s$ is essential. Suppose now that they are subject to the constraint $\psi(\theta_1, \theta_2, \ldots, \theta_s) = 0$. In such cases we maximize the function $\log L + \lambda\psi$, that is, we solve for $\theta_1, \theta_2, \ldots \theta_s$ and λ the set of $s + 1$ equations

$$\left.\begin{array}{l} \dfrac{\partial \log L}{\partial \theta_i} + \lambda \dfrac{\partial \psi}{\partial \theta_i} = U_i(\theta_1, \theta_2, \ldots, \theta_s) + \lambda \dfrac{\partial \psi}{\partial \theta_i} = 0, \quad i = 1, 2, \ldots, s, \\ \psi(\theta_1, \theta_2, \ldots, \theta_s) = 0. \end{array}\right\} \tag{11.59}$$

In general, if there are s parameters subject to f constraints, only $s - f$ parameters need be estimated.

In most problems which we encounter in mathematical genetics, the parameters are subject to *linear* constraints. For example, in the case of multiple alleles at a single locus the gene frequencies must add up to 1. If $p_1, p_2, \ldots, p_s$ denote the gene frequencies, we have $p_1 + p_2 + \cdots + p_s = 1$. We can then write

$$p_s = 1 - \sum_{j=1}^{s-1} p_j,$$

In this way we reduce the number of parameters to $s - 1$, and these are now functionally independent. The general theory discussed in section 11.8.1 can then be applied to these $s - 1$ functionally independent parameters. Some applications will be shown in the next chapter.

11.9. CONFIDENCE REGIONS

When there are s (functionally independent) parameters $\boldsymbol{\theta}' = (\theta_1, \theta_2, \ldots, \theta_s)$, it is sometimes useful to consider them jointly and to find a confidence region for them.

Let $\Omega_\boldsymbol{\theta}$ be the set of all possible values for $\boldsymbol{\theta}$, and $\hat{\boldsymbol{\theta}}' = (\hat{\theta}_1, \hat{\theta}_2, \ldots, \hat{\theta}_s)$ be a vector of estimators. Note that $\hat{\boldsymbol{\theta}} \in \Omega_\boldsymbol{\theta}$ ("$\hat{\boldsymbol{\theta}}$ belongs to $\Omega_\boldsymbol{\theta}$"). Furthermore, let $D(\hat{\boldsymbol{\theta}})$ be a certain subset in $\Omega_\boldsymbol{\theta}$ (depending on $\hat{\boldsymbol{\theta}}$), such that the true values $\boldsymbol{\theta}' = (\theta_1, \theta_2, \ldots, \theta_s)$ are covered by $D(\hat{\boldsymbol{\theta}})$ with probability $1 - \alpha$, that is,

$$\Pr\{\boldsymbol{\theta} \in D(\hat{\boldsymbol{\theta}})\} = 1 - \alpha. \tag{11.60}$$

The set $D(\hat{\boldsymbol{\theta}})$ is called a *confidence region* for $\boldsymbol{\theta}$. The probability $1 - \alpha$ is called the *(joint) confidence coefficient.*

It is not always easy to find $D(\hat{\boldsymbol{\theta}})$. The methods employed in the theory of finding general solutions for $D(\hat{\boldsymbol{\theta}})$ require knowledge of the joint distribution of the estimators $\hat{\theta}_1, \hat{\theta}_2, \ldots, \hat{\theta}_s$, and are beyond the scope of this book.

In special cases, when the estimators $\hat{\theta}_1, \hat{\theta}_2, \ldots, \hat{\theta}_s$ are *statistically** independent, it is convenient to use a rectangular parallelepiped in s dimensions as a confidence region obtained by the method of simultaneous intervals.

Let $1 - \gamma$ be the probability that a single confidence interval $[l_1(\hat{\theta}_i), l_2(\hat{\theta}_i)]$ contains the parameter θ_i, $i = 1, 2, \ldots, s$ (for simplicity, we take the same γ for each θ_i). The joint probability that every confidence interval $[l_1(\hat{\theta}_i), l_2(\hat{\theta}_i)]$ contains the corresponding true value θ_i, that is, the probability that the s-dimensional point $\boldsymbol{\theta}' = (\theta_1, \theta_2, \ldots, \theta_s)$ lies in the s-dimensional parallelepiped defined by $l_1(\hat{\theta}_i) \leqslant \theta_i \leqslant l_2(\hat{\theta}_i)$, $i = 1, 2, \ldots, s$, is $(1 - \gamma)^s$. If $1 - \alpha$ is the joint confidence coefficient, then we have

$$(1 - \gamma)^s = 1 - \alpha.$$

Hence

$$1 - \gamma = \sqrt[s]{1 - \alpha} \tag{11.61}$$

or

$$\gamma = 1 - \sqrt[s]{1 - \alpha}. \tag{11.61a}$$

For example, if $s = 2$ and the joint confidence coefficient is $1 - \alpha = 0.95$, for each parameter we should have $1 - \gamma = \sqrt{0.95} \doteq 0.975$. In other

* For statistical independence of random variables see formula (5.44).

words, in order to have the joint probability that the rectangle defined as $l_1(\hat{\theta}_1) \leqslant \theta_1 \leqslant l_2(\hat{\theta}_2)$ and $l_1(\hat{\theta}_2) \leqslant \theta_2 \leqslant l_2(\hat{\theta}_2)$ covers the point (θ_1, θ_2), equal to 0.95, we have to evaluate the separate confidence intervals for each parameter, with the confidence coefficients equal to 0.975.

11.10. PLANNING OF EXPERIMENTS. REQUIRED SAMPLE SIZE

The problems which arise in mathematical genetics are, broadly speaking, of two kinds:

1. Investigations of the mode of inheritance of several traits and characters are problems of constructing genetic hypotheses and testing these hypotheses using data from experiments. We shall discuss the statistical aspects of these problems in Chapter 13.

2. On the assumption that the mode of inheritance is known, the problem is to estimate genetic parameters. This has been discussed in the present chapter.

In both kinds of problems two important questions arise: how to plan the experiments so that they are most economical and most informative, and how large an experiment should be in order to obtain satisfactory precision with a high probability. There are no unique answers to these questions because they depend on the special conditions of each problem.

We now outline a few general features which may be of special importance and interest when problems of estimating genetic parameters are involved.

11.10.1. Choice of Experimental Designs

If we have a choice of experimental design, such as using intercrosses or backcrosses for estimation of the viability of gametes with linkage between two loci, it is always useful to examine the *efficiency* of such experimental techniques by calculating expected average amounts of information, as described in section 11.7, and to use the most efficient.

If we have a choice of more than one experimental design, it is desirable to use a number of different designs, even if some are less efficient than others. The data obtained not only can be used in estimation, but may also throw light on our assumptions. If the data from different experiments give significantly different estimates, perhaps the assumptions should be revised. Data from different kinds of experiments are of special importance when we wish to discriminate between two (or more) hypotheses (see Chapters 13, 16, 17, and 18).

11.10.2. Sampling Techniques

It is necessary to emphasize again the importance of the method by which a sample is obtained. The estimation methods discussed in this chapter apply only to simple random samples. We especially draw the attention of the reader to sampling methods for related pairs, such as mother-child or sib-sib. These points will be discussed in section 12.9.

11.10.3. Required Sample Size

If we want to make some statement about the genetic parameters in the population on the basis of their estimates, these estimates should be of a certain desired accuracy with a rather high probability.

(i) Let θ be a parameter, and $\hat{\theta}$ its estimator. Suppose that we wish to estimate θ with such precision that the probability that $\hat{\theta}$ differs from θ by no more than a certain value, d, say, is not less than $1 - \alpha$, where $1 - \alpha$ is high (0.95 or 0.99, say). In other words, we want

$$\Pr\{|\hat{\theta} - \theta| \leqslant d\} \geqslant 1 - \alpha. \tag{11.62}$$

Assuming that $\hat{\theta}$ is approximately normally distributed, we have

$$\Pr\{|\hat{\theta} - \theta| \leqslant z_{1-\frac{1}{2}\alpha}\sigma_{\hat{\theta}}\} \doteq 1 - \alpha.$$

Hence

$$z_{1-\frac{1}{2}\alpha}\sigma_{\theta} \leqslant d. \tag{11.63}$$

The standard deviation, σ_{θ}, is usually inversely proportional to $\sqrt{n}$. Let $\sigma_{\theta} = (A/\sqrt{n})$. Thus we have

$$z_{1-\frac{1}{2}\alpha} \frac{A}{\sqrt{n}} \leqslant d$$

and hence

$$n \geqslant \left(\frac{z_{1-\frac{1}{2}\alpha}}{d}\right)^2 A^2. \tag{11.64}$$

In other words, to satisfy (11.62) the sample size, n, should be greater than the integer part of the right-hand side of (11.64). In particular, if θ is a binomial parameter, $\theta = P$, say, then

$$\sigma_{\hat{P}} = \sqrt{\frac{P(1 - P)}{n}} = \frac{\sqrt{P(1 - P)}}{\sqrt{n}}.$$

Here $A = \sqrt{P(1 - P)}$. Substituting into (11.64), we obtain

$$n \geqslant \left(\frac{z_{1-\frac{1}{2}\alpha}}{d}\right)^2 P(1 - P). \tag{11.65}$$

Of course, P is usually unknown, but often it can be estimated from preliminary investigations. In such cases the calculated lower limit for n is only a "rough" approximation. (Whatever the value of P, $\frac{1}{4}(z_{1-\frac{1}{2}\alpha}/d)^2$ is a safe lower bound for n.)

Example 11.8. We apply the results discussed in section 11.10.3 to the problem of Example 11.4. We ask this question: how large should the sample size be if we wish to have a probability not less than 0.95 that the estimate of parameter ϕ is subject to an error not less than 0.05?

We apply formula (11.64). In Example 11.4 we obtained

$$\mathrm{Var}(\hat{\phi}) \doteq \frac{\phi(3 + \phi)^2}{3n} ;$$

hence

$$A^2 \doteq \frac{\phi(3 + \phi)^2}{3} .$$

Since the true ϕ is unknown, we can only estimate A^2, using $\hat{\phi} = 0.2086$. We have $A^2 \doteq 0.7159$. Taking $z_{0.975} = 1.96$ and $d = 0.05$, we obtain from (11.65)

$$n \geqslant \left(\frac{1.96}{0.05}\right)^2 \times 0.7159 \doteq 1100.$$

(ii) If there is *more than one* parameter, we should use the theory of joint confidence regions. For practical purposes, we might calculate n for each parameter separately with different requirements for the accuracy of each estimator. The maximal value of n's so obtained can be used as the (approximate) required sample size.

(iii) When the trait is rare, we might wonder how long the sampling must be continued in order to have a high probability, not less than $1 - \alpha$, say, that *at least one* individual possessing this trait is present in the sample. If P is the expected prevalence of this trait, then the number, X, of individuals possessing this trait in a sample of size n has a binomial distribution. Thus

$$\mathrm{Pr}\{X \geqslant 1\} = 1 - \mathrm{Pr}\{X = 0\} = 1 - (1 - P)^n.$$

To find n, we have to solve the inequality

$$1 - (1 - P)^n \geqslant 1 - \alpha$$

or

$$(1 - P)^n \leqslant \alpha;$$

hence

$$n \geqslant \frac{\log \alpha}{\log(1 - P)}. \tag{11.66}$$

In a similar way we may calculate the required sample size to satisfy the condition that *at least r* individuals possessing this trait are present in the sample. The solution for n will not be so simple in this case.

Finally, we should also take into account the economic aspects of collecting data — the cost and labor of different experimental designs and techniques should be balanced against the value of the information obtained and against the precision of the estimates.

SUMMARY

The general theory of estimation of the distribution parameters from a sample has been outlined. The desirable properties of estimators, such as consistency, unbiasedness, and efficiency, have been defined. The maximum likelihood method of obtaining estimators from a simple random sample of independent observations has been presented. The solving of maximum likelihood equations is usually tedious; an iterative method has been outlined for this purpose. This method is usually used for computer programs. Finally, on the basis of the theory presented in this chapter, some suggestions for planning genetic experiments have been made.

REFERENCES

Bailey, N. T. J., *Introduction to the Mathematical Theory of Genetic Linkage*, Clarendon Press, Oxford, 1961.

Brownlee, K. A., *Statistical Theory and Methodology in Science and Engineering*, John Wiley & Sons, New York, 1960.

Cramér, H., *Mathematical Methods of Statistics*, Princeton University Press, Princeton, N.J., 1946.

Kendall, M. G., and Stuart, A., *The Advanced Theory of Statistics*, Vol. 2, Hafner Publishing Co., New York, 1961.

Pfanzagl, J., Confidence intervals and regions. In: *International Encyclopedia of the Social Sciences*, The Macmillan Company and Free Press, New York, 1968, Vol. 5, pp. 150–157.

Ramseier, H., and Palm, J., Further studies of histocompatibility loci in rats, *Transplantation* 5 (1967), 721–729.

PROBLEMS

11.1. There is evidence that the mechanism which controls tissue rejection is basically immunogenetic. Let A_1 and A_2 be two codominant alleles at locus A which control two "transplantation" antigens. We will also denote these antigens by A_1 and A_2, respectively. If a donor of tissue possesses an antigen which is lacking in a recipient, the recipient's cells produce antibodies against this foreign antigen and the tissue graft will be rejected after a certain time. Thus recipient A_1A_1 rejects skin from donors A_1A_2 and A_2A_2 but accepts skin from A_1A_1; recipient A_1A_2 accepts grafts from A_1A_1, A_1A_2, and A_2A_2; etc. If more than one locus controls the acceptance or rejection of transplanted tissue (histocompatibility), the graft will be accepted only if the donor is compatible with the recipient with respect to *all* histocompatibility loci (for more details see Chapter 19).

Suppose that two homozygous strains of rats, Pt_1 and Pt_2, say, differ with respect to n independent histocompatibility loci, and we wish to estimate n. Suppose also that the $F_1 = Pt_1 \times Pt_2$ and $F_2 = F_1 \times F_1$ generations have been produced and skin grafts exchanged.

Let R denote a recipient and D a donor of the skin graft; $D \to R$ means that the skin from donor D has been placed on recipient R. The following grafting experiments ($D \to R$) have been designed:

$$\text{(i) } Pt_1 \to F_2; \quad \text{(ii) } F_1 \to F_2; \quad \text{and (iii) } F_2 \to F_2.$$

(*a*) Show that the expected proportions of accepted grafts in each experiment are as follows: (i) $(\frac{3}{4})^n$; (ii) $(\frac{1}{2})^n$; (iii) $(\frac{5}{8})^n$, respectively.

(*b*) Suppose that N grafts have been applied in each experiment. Find the maximum likelihood estimators and their approximate variances in each case (note that n is a function of the binomial parameter in each experiment).

(*c*) Calculate the relative efficiencies of experiments (ii) and (iii) with respect to (i).

11.2. The following data were obtained from crosses of strains Lewis and B.N. in grafting B.N. $\to F_2$. The number of grafts performed, N, was 329, and the number of grafts accepted, r was 6. [Billingham et al. *Proc. Nat. Acad. Sci.* **48** (1962), 138–147].

(*a*) Find the estimate and 95% confidence interval for the number of histocompatibility loci, n, with respect to which these strains differ.

(*b*) Assuming that the "true" value of n is 14, calculate how many grafts should be carried out in experiments (ii) and (iii) of Problem 11.1 in order to obtain the same efficiency as in experiment (i), for which $N = 329$.

(*c*) Assuming that $n = 14$, calculate how many grafts should be performed in each experiment to obtain *at least one* graft for which the probability of not being rejected is at least 0.90.

11.3. Find approximate $100(1 - \alpha)\%$ confidence intervals for Snyder's ratios, $S_{10} = [q/(1 + q)]$ and $S_{11} = [q^2/(1 + q)^2]$, using the property of asymptotic normality of maximum likelihood estimators.

CHAPTER 12

Estimation of Parameters Associated with Multinomial Distributions

12.1. INTRODUCTION

In Chapter 11 we considered a single random variable x with probability distribution $p(x; \theta)$, where θ can be, in the general case, a vector of s functionally independent parameters. We showed how to estimate these parameters, using the observations from a simple random sample.

In a more general case there are k random variables with joint distribution $p(\mathbf{x}; \theta)$. If x is replaced by a k-dimensional vector $\mathbf{x}$, all results of Chapter 11 derived for the univariate case become valid for multivariate cases with only slight modifications.

However, we shall not discuss here the general multivariate case but shall confine ourselves to *multinomial* distributions, which are very common and very important in genetic problems. The results presented here, although derived from the general theory, are so useful and convenient that it seems worthwhile to consider them in a separate chapter.

12.2. MAXIMUM LIKELIHOOD ESTIMATORS OF MULTINOMIAL PARAMETERS

Suppose that n observations fall into k classes. Let r_l be the number of individuals in the lth class, and P_l be the expected proportion in the lth class for $l = 1, 2, \ldots, k$, with the conditions $\sum r_l = n$ and $\sum P_l = 1$, respectively. Therefore, the random variables* $r_1, r_2, \ldots, r_k$ have the joint multinomial distribution

$$L = \Pr\{r_1, r_2, \ldots, r_k; P_1, P_2, \ldots, P_k\} = n! \prod_{l=1}^{k} \frac{P_l^{r_l}}{r_l!} \qquad (12.1)$$

* Now $r_1, r_2, \ldots, r_k$ denote random variables as well as their observed values.

314

[compare formula (8.54a)]. The probability given in (12.1), regarded as a function of $P_1, P_2, \ldots, P_k$, represents the likelihood function. Since $\sum P_l = 1$, only $k - 1$ parameters are independent. Substituting

$$P_k = 1 - \sum_{j=1}^{k-1} P_j \tag{12.2}$$

into (12.1) and taking logarithms of both sides, we obtain

$$\log L = \log(n!) + \sum_{j=1}^{k} \log(r_j!) + \sum_{j=1}^{k-1} r_j \log P_j$$
$$+ \left[\left(n - \sum_{j=1}^{k-1} r_j \right) \log \left(1 - \sum_{j=1}^{k-1} P_j \right) \right]. \tag{12.3}$$

Differentiating (12.3) with respect to P_l, we obtain

$$\frac{\partial \log L}{\partial P_l} = \frac{r_l}{P_l} - \frac{n - \sum_{j=1}^{k-1} r_j}{1 - \sum_{j=1}^{k-1} P_j}, \qquad l = 1, 2, \ldots, k - 1. \tag{12.4}$$

Equating this to zero, we obtain

$$\frac{r_l}{P_l} = \frac{r_k}{P_k}, \qquad l = 1, 2, \ldots, k - 1. \tag{12.4a}$$

Hence, the maximum likelihood estimators, $\hat{P}_1, \hat{P}_2, \ldots, \hat{P}_k$, are proportional to $r_1, r_2, \ldots, r_k$. Using the condition $\sum \hat{P}_l = 1$, we find

$$\hat{P}_l = \frac{r_l}{n}, \qquad l = 1, 2, \ldots, k. \tag{12.5}$$

We obtain the following result: *the ML estimators of multinomial parameters are the observed proportions in the corresponding classes.* This is a well-established practical procedure apart from its connection with the maximum likelihood method. Of course, we know the *exact* variance-covariance matrix, **V**. We recall section 8.11, where we obtained

$$\text{Var}(\hat{P}_l) = \frac{P_l(1 - P_l)}{n}, \quad \text{Cov}(\hat{P}_l, \hat{P}_m) = -\frac{P_l P_m}{n}, \qquad l, m = 1, 2, \ldots, k;$$
$$l \neq m.$$

12.3. THE ML ESTIMATOR OF A FUNCTION OF MULTINOMIAL PROPORTIONS. SINGLE PARAMETER

In Example 11.1 we obtained two estimators for the gene frequency, q, at the locus in equilibrium $p^2 AA + 2pq Aa + q^2 aa$. We now wish to find the maximum likelihood estimator of q. First, we notice that $p = 1 - q$ and that, in fact, only one parameter, q, say, need be estimated. The genotype array, given above, can be written as $(1 - q)^2AA + 2q(1 - q)Aa + q^2aa$. In a sample of size n with frequencies r_1 of AA, r_2 of Aa, and r_3 of aa, and with $r_1 + r_2 + r_3 = n$, the variables r_1, r_2, r_3 have multinomial proportions $P_1 = (1 - q)^2$, $P_2 = 2q(1 - q)$, $P_3 = q^2$, that is, the multinomial proportions are functions of a single parameter q, $P_l = P_l(q)$, $l = 1, 2, 3$, say. Of course, one can apply the standard procedure described in section 11.3, but it is also useful to use, for scores and amounts of information derived for a general multinomial distribution, the especially convenient formulae presented below.

Let us consider a k-nomial distribution as defined in section 12.2. Suppose that the expected multinomial proportions are certain functions of a parameter θ, that is, $P_l = P_l(\theta)$, $l = 1, 2, \ldots, k$, and we are interested in estimating θ. The likelihood function (12.1) can be now written in the form

$$L(r_1, r_2, \ldots, r_k; \theta) = n! \prod_{l=1}^{k} \frac{P_l^{r_l}(\theta)}{r_l!}. \tag{12.6}$$

We note that $E(r_l) = nP_l(\theta)$. Taking logarithms of both sides and differentiating with respect to θ, we obtain the total score:

$$U(\theta) = \frac{\partial \log L}{\partial \theta} = \sum_{l=1}^{k} \frac{r_l}{P_l(\theta)} \frac{\partial P_l}{\partial \theta}. \tag{12.7}$$

We shall call the function

$$u(l; \theta) = \frac{1}{P_l(\theta)} \frac{\partial P_l}{\partial \theta} \tag{12.8}$$

the *individual score* corresponding to any single observation in the lth class. All individual scores in the lth class are the same; thus the score corresponding to the lth class is

$$r_l u(l; \theta) = \frac{r_l}{P_l(\theta)} \frac{dP_l}{d\theta}, \qquad l = 1, 2, \ldots, k, \tag{12.9}$$

and the total score is

$$U(\theta) = \sum_{l=1}^{k} r_l u(l; \theta). \tag{12.10}$$

Equating (12.10) to zero and solving with respect to θ, we obtain the ML estimator of θ.

We now evaluate the expected total amount of information. We have

$$\begin{aligned} I(\theta) = E\left(\frac{\partial U}{\partial \theta}\right) &= E\left\{ \sum_{l=1}^{k} \left[\frac{r_l}{P_l^2(\theta)} \left(\frac{\partial P_l}{\partial \theta}\right)^2 - \frac{\partial^2 P_l}{\partial \theta^2} \cdot \frac{r_l}{P_l(\theta)} \right] \right\} \\ &= \sum_{l=1}^{k} \frac{E(r_l)}{P_l^2(\theta)} \left(\frac{\partial P_l}{\partial \theta}\right)^2 - \sum_{l=1}^{k} \frac{\partial^2 P_l}{\partial \theta^2} \cdot \frac{E(r_l)}{P_l(\theta)} \\ &= n\left\{ \sum_{l=1}^{k} \frac{1}{P_l(\theta)} \left(\frac{\partial P_l}{\partial \theta}\right)^2 - \frac{\partial^2}{\partial \theta^2} \left[\sum_{l=1}^{k} P_l(\theta) \right] \right\}. \end{aligned} \tag{12.11}$$

But, since $\sum P_l(\theta) = 1$, then

$$\frac{\partial^2}{\partial \theta^2}\left[\sum_{l=1}^{k} P_l(\theta)\right] = 0,$$

so finally we obtain

$$I(\theta) = n \sum_{l=1}^{k} \frac{1}{P_l(\theta)}\left(\frac{\partial P_l}{\partial \theta}\right)^2. \tag{12.12}$$

The expected *average* amount of information about θ per observation is

$$i(\theta) = \frac{1}{n} I(\theta) = \sum_{l=1}^{k} \frac{1}{P_l(\theta)}\left(\frac{\partial P_l}{\partial \theta}\right)^2. \tag{12.13}$$

Formula (12.12) can be written in another form. Since $E(r_l) = nP_l(\theta)$, then $n = E(r_l)/P_l(\theta)$. Substituting into (12.12), we obtain

$$I(\theta) = \sum_{l=1}^{k} E(r_l)\left[\frac{1}{P_l(\theta)}\frac{\partial P_l}{\partial \theta}\right]^2 = \sum_{l=1}^{k} E(r_l)u^2(l; \theta). \tag{12.14}$$

In practice, θ and so also $P_l(\theta)$ and $E(r_l)$ are usually unknown, and we replace them by estimated values. Thus the *estimated* total amount of information, $I(\hat{\theta})$, say, is

$$I(\hat{\theta}) = \sum_{l=1}^{k} r_l u^2(l; \hat{\theta}), \tag{12.15}$$

and the estimated average amount per observation, $i(\hat{\theta})$, is

$$i(\hat{\theta}) = \frac{1}{n} I(\hat{\theta}) = \frac{1}{n}\sum_{l=1}^{k} r_l u^2(l; \hat{\theta}). \tag{12.16}$$

Formulae (12.7)–(12.16) are all useful in practical applications; of course, we always have a free choice whether to use these formulae or to follow the standard method described in section 11.3.

12.4. ESTIMATION OF GENE FREQUENCIES IN RANDOMLY MATING NATURAL POPULATIONS. GENERAL REMARKS

Among several genetic parameters, gene frequencies appear to be of the widest application in studies of the structure, dynamics, and evolution of natural (in particular, of human) populations. Several problems in medical genetics, anthropology, the theory of evolution, and other fields can be investigated by calculating gene frequencies and interpreting their relevance to a particular phenomenon. It is, then, of great importance that the statistical methods employed in estimating gene frequencies be used correctly. These methods reflect the sampling techniques, and it seems that this point is not always recognized by practical geneticists.

Two main ways of studying gene frequencies in human populations may be distinguished:

(a) Sampling from the whole population of one or two generations.
(b) Analysis of family data.

In this chapter we shall discuss only problems relevant to (a); Chapter 18 is especially concerned with the handling of data obtained from family records.

Several sampling techniques can be designed in order to obtain a *simple random sample*. In the following sections we shall consider a few possibilities. At present, we confine ourselves to a single locus with two alleles so that, in fact, only the gene frequency of one allele (i.e., a single parameter) has to be estimated. Problems of gene frequencies of multiple alleles will be discussed in Chapter 14. Also, we assume here no selection, mutation, or migration.

12.5. ML ESTIMATORS OF GENE FREQUENCIES FROM THE NUMBERS OF GENOTYPES OR PHENOTYPES OBSERVED IN A SIMPLE RANDOM SAMPLE FROM A POPULATION

The theorem on genetic equilibrium for randomly mating populations, derived in Chapter 4, applies to distinct generations when a certain "generation time" is assigned to each breeding population. This can be effectively the case in annual populations of plants and can be arranged in certain animal populations but never holds for human populations, which are "mixtures" of overlapping generations. However, assuming that the mixture consists of two (or more) generations in fixed proportions and that each generation is in equilibrium, we may consider the whole population as being in equilibrium with the same genotype (and hence also gene) frequencies (compare the results of section 6.9).

Therefore, to estimate gene frequencies in a population we may take a sample from one generation or from the whole population. In the latter case, we must realize that to obtain a simple random sample each individual has to be selected independently "on its own." If, for instance, whole families are typed for certain characters (e.g., blood groups), such samples will be correlated (and hence not independent).

Experiment 1. Suppose that the heterozygote Aa is distinguishable from both homozygotes, AA and aa. Then, in estimating the frequency q of gene a, we use the distribution of genotypes. Table 12.1 gives the evaluation of scores and amounts of information in order to obtain the maximum likelihood estimator of q.

TABLE 12.1 Scores and expected amounts of information based on genotypes (Exp. 1)

Class (genotype)	Class number, l	Class frequency, r_l	$P_l(q)$	$\dfrac{\partial P_l}{\partial q}$	$\dfrac{1}{P_l(q)}\dfrac{\partial P_l}{\partial q}$	$\dfrac{1}{P_l(q)}\left(\dfrac{\partial P_l}{\partial q}\right)^2$
AA	1	r_1	$(1-q)^2$	$-2(1-q)$	$-\dfrac{2}{1-q}$	4
Aa	2	r_2	$2q(1-q)$	$2(1-2q)$	$\dfrac{1-2q}{q(1-q)}$	$\dfrac{2(1-2q)^2}{q(1-q)}$
aa	3	r_3	q^2	$2q$	$\dfrac{2}{q}$	4
Total		n	1			$\dfrac{2}{q(1-q)}$

The total score for Experiment 1, $U_1(q)$, say, is

$$U_1(q) = -\frac{2r_1}{1-q} + \frac{r_2(1-2q)}{q(1-q)} + \frac{2r_3}{q} = \frac{(r_2 + 2r_3) - 2nq}{q(1-q)}. \quad (12.17)$$

Solving $U_1(q) = 0$, we obtain

$$\hat{q}_1 = \frac{r_2 + 2r_3}{2n}. \quad (12.18)$$

The ML estimator, $\hat{q}_1$ based on genotypes, is, then, identical with the estimator obtained in Example 11.1 by gene counting. We notice that

$$E(\hat{q}_1) = \frac{E(r_2) + 2E(r_3)}{2n} = \frac{n[2q(1-q) + 2q^2]}{2n} = q,$$

so that $\hat{q}_1$ is unbiased. Thus $\hat{q}_1$ is an MVU estimator of q. The expected average amount of information is given in the last row and column of Table 12.1, that is,

$$i_1(q) = \frac{2}{q(1-q)},$$

and the variance of $\hat{q}_1$ is

$$\mathrm{Var}(\hat{q}_1) = \frac{1}{ni_1(q)} = \frac{q(1-q)}{2n}. \quad (12.19)$$

Experiment 2. If the heterozygote Aa is not distinguishable from the homozygote AA, then only two classes, dominants (AA or Aa) and recessives

TABLE 12.2   Scores and expected amounts of information based on phenotypes (Exp. 2)

Class (phenotype)	Class number, l	Class frequency, r_l	$P_l(q)$	$\dfrac{\partial P_l}{\partial q}$	$\dfrac{1}{P_l(q)}\dfrac{\partial P_l}{\partial q}$	$\dfrac{1}{P_l(q)}\left(\dfrac{\partial P_l}{\partial q}\right)^2$
AA or Aa	1	r_1	$1 - q^2$	$-2q$	$-\dfrac{2q}{1 - q^2}$	$\dfrac{4q^2}{1 - q^2}$
aa	2	r_2	q^2	$2q$	$\dfrac{2}{q}$	4
Total		n				$\dfrac{4}{1 - q^2}$

(aa), are recognizable. Table 12.2 gives the scores and amounts of information based on phenotypes.

Here the total score, $U_2(q)$, say, is

$$U_2(q) = \frac{2(r_2 - nq^2)}{q(1 - q^2)}. \tag{12.20}$$

Solving the equation $U_2(q) = 0$, we obtain the estimator $\hat{q}_2 = \sqrt{r_2/n}$, which is identical with the estimator of q obtained in Example 11.1 based on recessives, and is only asymptotically unbiased (compare section 11.5).

For Experiment 2 we have

$$i_2(q) = \frac{4}{1 - q^2}, \quad \text{and} \quad \text{Var}(\hat{q}_2) \doteq \frac{1 - q^2}{4n}. \tag{12.21}$$

The relative efficiency of Experiment 2 with respect to Experiment 1 is approximately

$$\text{Rel eff(Exp. 2)} \doteq \frac{i_2(q)}{i_1(q)} = \frac{4}{1 - q^2} \cdot \frac{q(1 - q)}{2} = \frac{2q}{1 + q} \leqslant 1. \tag{12.22}$$

We notice that the relative efficiency depends on q. For large q it is close to 1; for $q < 0.1$ it decreases rapidly as q decreases.

Example 12.1. Plato et al. (1964) published the following data on haptoglobin in Cyprus:

Haptoglobin type:	Hp1-1	Hp1-2	Hp2-2	
Observed frequency:	10	68	112	$(n = 190)$.

Assuming equilibrium, we estimate the frequency q of allele Hp² from formula (12.18), $\hat{q} = (68 + 224)/380 = 0.7684$. Since the true value of q is unknown, we calculate an approximate variance, replacing q by $\hat{q}$ in (12.19). We obtain $\mathrm{Var}(\hat{q}) \doteq 0.000468320$ and so $\sigma_{\hat{q}} \doteq 0.0216$. We now find the 99% confidence interval, using approximate formula (11.37). We have $z_{0.995} = 2.58$; therefore $z_{.995}\sigma_{\hat{q}} \doteq 2.58 \times 0.0216 = 0.0557$. Hence

$$l_1 \doteq 0.7684 - 0.0557 = 0.7127; \quad l_2 \doteq 0.7684 + 0.0557 = 0.8241.$$

Thus, we "believe" with 99% confidence that the interval (0.7127, 0.8241) includes the true gene frequency q.

12.6. SIMPLE RANDOM SAMPLE FROM A "TRUNCATED" POPULATION

If the gene a is lethal in the homozygous state, the homozygotes aa will be missing. If the heterozygotes Aa are distinguishable from the homozygotes AA, the distribution of genotypes will be truncated. We have

$$p^2 + 2pq = (1 - q)^2 + 2q(1 - q) = (1 - q)(1 + q) = 1 - q^2.$$

Thus, the frequency of AA is $p^2/(1 - q^2) = (1 - q)/(1 + q)$, and the frequency of Aa is $2pq/(1 - q)^2 = 2q/(1 + q)$. Now the genotypes, in a sample of size n, have a binomial distribution and the ML estimator can be found from the observed frequency, r, of heterozygotes Aa, that is,

$$\frac{2\hat{q}}{1 + \hat{q}} = \frac{r}{n} \quad \text{or} \quad \hat{q} = \frac{r}{2n - r}.$$

It should be noted, however, that the sample must be obtained from *one generation only* and will be valid for just this generation, unless equilibrium has been reached. Otherwise, the elimination of homozygotes aa changes the gene frequencies in the next generation.

12.7. SAMPLING FROM ONE GENERATION WHEN INFORMATION ABOUT PARENTAL GENOTYPES IS AVAILABLE

When a homozygote aa is not very common and heterozygote Aa is indistinguishable from homozygote AA, an extremely large sample is required in order to observe a few aa individuals. It is then more practical to select a sample from families in which at least one parent is aa (for rare characters the matings $aa \times aa$ are very unlikely).

Snyder's ratio, $S_{10} = q/(1 + q)$, can be used as the binomial parameter if the sampling procedure can be represented by independent Bernoulli trials. As was said in section 7.13, S_{10} gives the expected proportion of aa children in a very large ("infinite") population of children of parents $A-$ × aa, combined (mixed) together. If the population of $A-$ × aa parents is really large, a simple random sample can be obtained by "mixing" children together and selecting from them a random sample of size n. Usually such populations are small, however, so that the practical and safe procedure is to choose families in which one parent is observed to be aa and another $A-$, and then to select from each family *one* child. If in such a sample of n children there are r of aa, the proportion r/n will be an ML estimator of S_{10}, that is,

$$\hat{S}_{10} = \frac{\hat{q}}{1 + \hat{q}} = \frac{r}{n}, \tag{12.23}$$

and

$$\text{Var}(\hat{S}_{10}) = \frac{S_{10}(1 - S_{10})}{n} = \frac{q}{n(1 + q)^2}. \tag{12.24}$$

We can estimate $\hat{q}$ from Snyder's ratio. From (12.23) we find

$$\hat{q} = \frac{r}{n - r} = \frac{\hat{S}_{10}}{1 - \hat{S}_{10}} \tag{12.25}$$

and

$$\text{Var}(\hat{q}) \doteq \left(\frac{d\hat{q}}{d\hat{S}_{10}}\right)^2_{\hat{q}=q} \text{Var}(\hat{S}_{10}) = \frac{q(1 + q)^2}{n}. \tag{12.26}$$

If the number of families is small, we cannot pool the data. Also, selecting one child per family usually gives a poor estimate of S_{10} (and so of q). In such cases methods presented in Chapter 18 should be used.

12.8. SAMPLING FROM TWO GENERATIONS

Snyder's ratio S_{10} can now be estimated in a different way. Suppose that q is estimated from a sample of size N, say, taken from the population of parents. Let R be the number of aa individuals in this sample. Thus, the estimator of q, $\hat{q}'$, say, is

$$\hat{q}' = \sqrt{\frac{R}{N}}, \tag{12.27}$$

and its variance is approximately

$$\text{Var}(\hat{q}') \doteq \frac{1 - q^2}{4N} \tag{12.28}$$

[see formula (12.21)].

Suppose that we now estimate S_{10}, using $\hat{q}'$ calculated from (12.27). Denoting this estimator by $\hat{S}'_{10}$, we obtain

$$\hat{S}'_{10} = \frac{\hat{q}'}{1 + \hat{q}'},$$
(12.29)

and its approximate variance is

$$\mathrm{Var}(\hat{S}'_{10}) \doteq \left(\frac{d\hat{S}'_{10}}{d\hat{q}'}\right)^2_{\hat{q}'=q} \mathrm{Var}(\hat{q}') = \frac{1 - q}{4N(1 + q)^3}.$$
(12.30)

We may compare $\hat{S}'_{10}$ given by (12.29) with $\hat{S}_{10}$ calculated from (12.23) when a sample of n children is used. If both N and n are sufficiently large and are obtained by a method of simple random sampling, $\hat{S}_{10}$ and $\hat{S}'_{10}$ should be close to each other.

12.9. SAMPLING FROM A POPULATION OF RELATED PAIRS

It is sometimes convenient to use records obtained from *pairs of related* individuals. When a child is tested for a certain trait, quite often the mother is present at the same time and can be tested as well. There are some characters for which the onset depends not only on age but also on environmental conditions. Sib pairs are suitable subjects in such situations since they are usually about the same age and live in similar conditions. In such cases we recall the joint genotype distributions of parent-child combinations or sib-sib pairs discussed in sections 7.5 and 7.11 respectively, and use the joint genotype (or phenotype) proportions as multinomial parameters in the random sample of n pairs.

It is essential always to bear in mind the importance of the sampling procedure used to select related pairs from the basic population of such pairs.

Mother-Child Combinations. Mothers represent the "base" population in one dimension, and all their children the base population in the other dimension. The sampling procedure might be as follows: a child is selected at random from the population of children, and his mother identified — such a pair undergoes testing. It can happen that more than one child per family is selected, but each child must be selected at *random* from the whole pool of children. If the population is not very large, it would be better to select a *family only once* and to examine the *mother* and *one* randomly selected child from this family. If all the children in each of n families are tested, methods appropriate for family study should be employed.

Sib-Sib Pairs. The "base" population is composed of all possible pairs of siblings. For a simple random sample, a child is selected at random and then a sibling is chosen at random from the remaining sibs in the same family. Again, it can happen that more than one pair from a family is included in the sample. Since we are usually dealing with populations of reasonable sizes, the safest procedure would be to select randomly only *one pair* of siblings from each (randomly chosen) family. When all possible pairs in each of n families are tested, methods appropriate to family study should again be used.

12.10. MAXIMUM LIKELIHOOD SCORES FOR SAMPLES OF MOTHER-CHILD PAIRS

We will now outline the method of maximum likelihood when the test records for mother-child pairs are available. The techniques presented here apply to simple random samples of mother-child pairs from a joint population, as described above, that is, they apply to samples of random pairs, *one pair per family.*

12.10.1. Mother-Child Genotype Distribution

Suppose that the character under consideration is such that the heterozygote can be distinguished from both homozygotes and that the tests were applied to the mother and a randomly selected child. Recalling the joint parent-child genotype distribution given in Table 7.4, and replacing p by $1 - q$, we notice that the $3 \times 3 = 9$ parent-child combination of genotypes, in the sample of n pairs, will have a multinomial distribution with parameters corresponding to the joint probabilities given in Table 7.4. In fact, we have only seven classes, since the combinations $A_i A_i$, $A_j A_j$ $(i \neq j)$, $i, j = 1, 2$, do not occur. The results are given in Table 12.3.

We have

$$U(q) = \sum_{l=1}^{7} r_l \frac{1}{P_l(q)} \frac{dP_l}{dq} = \frac{-Aq + B}{q(1 - q)}, \tag{12.31}$$

where

$$A = 3n - r_1; \quad B = r_2 + r_3 + r_4 + 2(r_5 + r_6). \tag{12.32}$$

Taking $S(q) = 0$, we obtain the ML estimator of q:

$$\hat{q} = \frac{B}{A} = \frac{r_2 + r_3 + r_4 + 2(r_5 + r_6)}{3n - r_1}.$$

The approximate variance of $\hat{q}$ is

$$\mathrm{Var}(\hat{q}) \doteq \frac{1}{ni(q)} = \frac{q(1 - q)}{n(q^2 - q + 3)}. \tag{12.33}$$

TABLE 12.3 Scores and expected amounts of information based on joint genotypes of mother-child combinations

Class Mother	Class Child	Class number, l	Class frequency, r_l	$P_l(q)$	$\dfrac{\partial P_l}{\partial q}$	$\dfrac{1}{P_l(q)} \cdot \dfrac{\partial P_l}{\partial q}$	$\dfrac{1}{P_l(q)}\left(\dfrac{\partial P_l}{\partial q}\right)^2$
A_1A_1	A_1A_1	1	r_1	$(1-q)^3$	$-3(1-q)^2$	$-\dfrac{3}{1-q}$	$9(1-q)$
	A_1A_2	2	r_2	$q(1-q)^2$	$(1-q)(1-3q)$	$\dfrac{1-3q}{q(1-q)}$	$\dfrac{(1-3q)^2}{q}$
A_1A_2	A_1A_1	3	r_3	$q(1-q)^2$	$(1-q)(1-3q)$	$\dfrac{1-3q}{q(1-q)}$	$\dfrac{(1-3q)^2}{q}$
	A_1A_2	4	r_4	$q(1-q)$	$1-2q$	$\dfrac{1-2q}{q(1-q)}$	$\dfrac{(1-2q)^2}{q(1-q)}$
	A_2A_2	5	r_5	$q^2(1-q)$	$q(2-3q)$	$\dfrac{2-3q}{q(1-q)}$	$\dfrac{(2-3q)^2}{1-q}$
A_2A_2	A_1A_2	6	r_6	$q^2(1-q)$	$q(2-3q)$	$\dfrac{2-3q}{q(1-q)}$	$\dfrac{(2-3q)^2}{1-q}$
	A_2A_2	7	r_7	q^3	$3q^2$	$\dfrac{3}{q}$	$9q$
Total			n				$i(q) = \dfrac{q^2-q+3}{q(1-q)}$

325

Example 12.2. Hirschfeld (1959) studied a factor, called Gc, of α_2-globulins, which can be detected electrophoretically in human sera. Genetic studies have confirmed the inheritance of the Gc system as an autosomal two-allelic locus without dominance. This factor is related to haptoglobin; by analogy with haptoglobin types, the alleles are called Gc^1 and Gc^2, and the genotypes are denoted by Gc1-1, Gc2-1, and Gc2-2.

Hirschfeld and Heiken (1963) applied this factor to medicolegal problems of paternity. We shall use their data to demonstrate the application of mother-child distribution in estimating gene frequency. Let q denote the frequency of the gene Gc^2 and $p = 1 - q$ the frequency of the gene Gc^1. The sample of $n = 142$ women and a child of each of them were tested with the following results:

Child Mother	Gc1-1	Gc2-1	Gc2-2	Total
Gc1-1	$60(r_1)$	$21(r_2)$	0	81
Gc2-1	$21(r_3)$	$32(r_4)$	$1(r_5)$	54
Gc2-2	0	$6(r_6)$	$1(r_7)$	7
Total	81	59	2	142

We calculate $A = 3 \times 142 - 60 = 366$, $B = 21 + 21 + 32 + 2(1 + 6) = 88$. Hence

$$\hat{q} = \frac{B}{A} = \frac{88}{366} = 0.2404.$$

From studies of 2259 unrelated individuals the frequency of the gene Gc^2 has been found to be $\hat{q} = 0.2541$. We use it as the "true" gene frequency and then evaluate the approximate variance from (12.33). We have

$$\text{Var}(\hat{q}) \doteq \frac{0.2541 \times 0.7459}{142 \times 2.81046681} = \frac{0.18953319}{399.08628702} = 0.00047492.$$

12.10.2. Mother-Child Phenotype Distribution

When there is dominance, we can use the joint phenotype distribution as given in Table 7.5b and construct the table of scores as in Table 12.4. The total score is

$$U(q) = \frac{A - Bq - Cq^2 + 3nq^3}{q(1 - q)(1 + q - q^2)}, \tag{12.34}$$

TABLE 12.4 Scores and expected amount of information based on joint phenotypes of mother-child combinations

Class Mother	Child	Class number, l	Class frequency, r_l	$P_l(q)$	$\dfrac{\partial P_l}{\partial q}$	$\dfrac{1}{P_l(q)}\dfrac{\partial P_l}{\partial q}$	$\dfrac{1}{P_l(q)}\left(\dfrac{\partial P_l}{\partial q}\right)^2$
AA or Aa {	AA or Aa	1	r_1	$(1-q)(1+q-q^2)$	$-q(4-3q)$	$-\dfrac{q(4-3q)}{(1-q)(1+q-q^2)}$	$\dfrac{q^2(4-3q)}{(1-q)(1+q-q^2)}$
	aa	2	r_2	$q^2(1-q)$	$q(2-3q)$	$\dfrac{2-3q}{q(1-q)}$	$\dfrac{(2-3q)^2}{1-q}$
aa {	AA or Aa	3	r_3	$q^2(1-q)$	$q(2-3q)$	$\dfrac{2-3q}{q(1-q)}$	$\dfrac{(2-3q)^2}{1-q}$
	aa	4	r_4	q^3	$3q^2$	$\dfrac{3}{q}$	$9q$
Total			n				

$$i(q) = \dfrac{8 - 7q + 2q^2}{(1-q)(1+q-q^2)}$$

327

where

$$A = 2(n - r_1) + r_4; \quad B = r_2 + r_3; \quad C = 4n + r_2 + r_3 + 2r_4. \quad (12.35)$$

Equating the numerator of (12.34) to zero, we have to solve a cubic equation in q. For this purpose we can use the iterative method described in section 11.7.2. The expected amount of information (from the last column of Table 12.4) is

$$I(q) = n \cdot i(q) = n \frac{8 - 7q + 2q^2}{(1 - q)(1 + q - q^2)}. \quad (12.36)$$

Result (12.36) was obtained by R. A. Fisher (1940). [In his original paper, he gave $I(\theta)$, where $\theta = q^2$.]

Similar procedures can be applied to random samples of other related pairs. Using, for instance, the genotype distribution of full-sib pairs (Table 7.12), we can derive the scores and amount of information for such pairs. Similarly, we can calculate these quantities when the phenotype distribution of full sibs is considered.

It can be seen from the discussion in the present section and preceding ones that the use of data from related pairs has both advantages and disadvantages. It is usually a drawback if the data were not collected in an independent manner. In estimation problems it is desirable to use samples from unrelated individuals (if possible, of course). If, in addition, data from related pairs are available, they should be used preferably in testing genetic hypotheses (see Chapter 13).

12.11. COMBINATION OF DATA

It quite often happens that more than one set of data, from different *independent* experiments, is available for estimating a parameter. For example, for laboratory animals there may be data from intercrosses, backcrosses, an F_3 generation, etc. If the assumptions regarding the mode of inheritance are true, such data should be more informative than information from a single experiment alone. It is desirable, then, to combine the data and evaluate an estimator which utilizes all the information obtainable from these experiments. We will sometimes call this estimator a *combined* or *pooled* estimator and denote it by $\hat{\theta}$.

Suppose that we wish to estimate the parameter θ and that m sets of data, each from a different and independent experiment, are available. Let $L_g(\theta)$ denote the likelihood function for the sample of size n_g in the gth experiment, for $g = 1, 2, \ldots, m$. Since the experiments are independent, the overall likelihood function is

$$L(\theta) = \prod_{g=1}^{m} L_g(\theta), \qquad (12.37)$$

and its logarithm is

$$\log L(\theta) = \sum_{g=1}^{m} \log L_g(\theta). \qquad (12.38)$$

Let us assume that the samples in each experiment were obtained by simple random sampling. Thus, from (11.12) we have

$$\log L_g(\theta) = \sum_{i=1}^{n_g} \log p_g(x_i; \theta), \qquad g = 1, 2, \ldots, m.$$

If $u(x_{ig}; \theta)$ denotes an individual score with respect to the ith observation in the gth sample $U_g(\theta)$ the total score in the gth experiment, and $U(\theta)$ the overall score, it is easy to see that

$$U(\theta) = \sum_{g=1}^{m} U_g(\theta) = \sum_{g=1}^{m} \sum_{i=1}^{n_g} u(x_{ig}; \theta). \qquad (12.39)$$

Thus we have the following result.

The scores from different independent sets of data (concerning the same parameter, θ) *are additive.*

Solving the equation $U(\theta) = 0$ with respect to θ, we obtain a pooled estimator, $\hat{\theta}$, from the combined data. The property of additivity applies also to the total amount of information. If

$$I_g(\theta) = -n_g E\left(\frac{\partial^2 \log p_g}{\partial \theta^2}\right)$$

denotes the expected amount of information in the gth experiment, the expected overall amount of information is

$$I(\theta) = \sum_{g=1}^{m} I_g(\theta) = -\sum_{g=1}^{m} n_g E\left(\frac{\partial^2 \log p_g}{\partial \theta^2}\right), \qquad (12.40)$$

that is, *the amounts of information are additive, too.*

Of special interest is the situation in which the data in each experiment arise from *multinomial* distributions. Let $P_{lg}(\theta)$ denote the proportion and r_{lg} the number of individuals in the lth class of the gth experiment; also let k_g be the number of classes and n_g the sample size in the gth experiment, with the conditions

$$\sum_{l=1}^{k_g} P_{lg} = 1 \quad \text{and} \quad \sum_{l=1}^{k_g} r_{lg} = n_g$$

for each $g = 1, 2, \ldots, m$. Then the overall score is

$$U(\theta) = \sum_{g=1}^{m} \sum_{l=1}^{k_g} \frac{r_{lg}}{P_{lg}(\theta)} \cdot \frac{\partial P_{lg}}{\partial \theta} = \sum_{g=1}^{m} U_g(\theta), \qquad (12.41)$$

where

$$U_g(\theta) = \sum_{l=1}^{k_g} \frac{r_{lg}}{P_{lg}(\theta)} \frac{\partial P_{lg}}{\partial \theta}, \qquad g = 1, 2, \ldots, m, \qquad (12.42)$$

is the total score in the gth experiment.

The expected overall amount of information available from all m samples is

$$I(\theta) = \sum_{g=1}^{m} n_g \sum_{l=1}^{k_g} \frac{1}{P_{lg}(\theta)} \left(\frac{\partial P_{lg}}{\partial \theta} \right)^2 = \sum_{g=1}^{m} I_g(\theta), \qquad (12.43)$$

where

$$I_g(\theta) = n_g \sum_{l=1}^{k_g} \frac{1}{P_{lg}(\theta)} \left(\frac{\partial P_{lg}}{\partial \theta} \right)^2, \qquad g = 1, 2, \ldots, m, \qquad (12.44)$$

is the expected total amount of information from the gth experiment.

The expected overall average amount of information per observation is

$$i(\theta) = \frac{1}{N} I(\theta) = \frac{1}{N} \sum_{g=1}^{m} n_g \sum_{l=1}^{k_g} \frac{1}{P_{lg}(\theta)} \left(\frac{\partial P_{lg}}{\partial \theta} \right) = \frac{1}{N} \sum_{g=1}^{m} n_g i_g(\theta), \qquad (12.45)$$

where $N = \sum n_g$ is the total number of observations, and $i_g(\theta)$ is the expected average amount of information in the gth experiment. As we see, $i(\theta)$ is a weighted mean of $i_g(\theta)$'s.

Example 12.3. We will use some data quoted by Mather (1935) on segregation in autotetraploid tomatoes. Let R be a dominant and r be its recessive allele for a certain trait in tetraploid tomatoes. It appears that the viability of the gametes rr is much lower than that of the gametes RR and Rr. Suppose that the survival ratios for these gametes are $RR:Rr:rr = 1:1:\theta$, with $0 < \theta < 1$.

Two sets of data are available: from Experiment 1, in which self-crosses $Rrrr \times Rrrr$ were performed; and from Experiment 2, in which backcrosses $Rrrr \times rrrr$ were performed.

Experiment 1. The ratios of genotype frequencies for the offspring of self-crosses are $RRrr:Rrrr:rrrr = \frac{1}{4}:\frac{2}{4}\theta:\frac{1}{4}\theta^2 = 1:2\theta:\theta^2$. If, for simplicity, we denote the phenotype including gene R (i.e., $RRrr$ or $Rrrr$) by R and the recessive $rrrr$ by r, the phenotypic ratios are $R:r = (1 + 2\theta):\theta^2$, and the phenotype distribution is

$$\Pr\{R\} = \frac{1 + 2\theta}{(1 + \theta)^2}, \quad \Pr\{r\} = \frac{\theta^2}{(1 + \theta)^2}, \quad \text{with} \quad \Pr\{R\} + \Pr\{r\} = 1.$$

Experiment 2. In a similar way, we find the phenotype distribution of the offspring of backcrosses

$$\Pr\{R\} = \frac{1}{1 + \theta}, \quad \Pr\{r\} = \frac{\theta}{1 + \theta}.$$

Let us denote

$$\vartheta = \frac{\theta}{1 + \theta} \, .$$

Thus ϑ is the expected proportion of gametes rr which survive in each experiment. The accompanying table summarizes the results, in terms of ϑ, for both experiments. The last row includes the observed numbers of the two kinds of phenotypes in samples of $n_1 = 826$ from offspring of self-crosses and $n_2 = 115$ from offspring of backcrosses, respectively.

	Exp. 1		Exp. 2	
Phenotype:	R	r	R	r
Expected prop., $P_{lg}(\vartheta)$:	$1 - \vartheta^2$	ϑ^2	$1 - \vartheta$	ϑ
Observed numbers, r_{lg}:	605	221	48	67.

The model presented in Table 12.2 applies to Experiment I. Replacing q by ϑ in (12.20), and denoting the score for Exp. 1 by $U_1(\vartheta)$ we obtain

$$U_1(\vartheta) = \frac{2(221 - 826\vartheta^2)}{\vartheta(1 - \vartheta^2)} \, ,$$

which yields

$$\hat{\vartheta}_1 = \sqrt{\frac{221}{826}} = 0.5173.$$

From (12.21) we obtain

$$I_1(\vartheta) = \frac{4n_1}{1 - \vartheta^2} = \frac{3304}{1 - \vartheta^2} \, .$$

It is easy to show that for Exp. 2 we have

$$U_2(\vartheta) = \frac{67 - 115\vartheta}{\vartheta(1 - \vartheta)} \, ,$$

which yields the estimate

$$\hat{\vartheta}_2 = \frac{67}{115} = 0.5826, \quad \text{and} \quad I_2(\vartheta) = \frac{115}{\vartheta(1 - \vartheta)} \, .$$

The question of whether these estimators differ significantly will be discussed in Chapter 13. Let us *assume* here that the true parameter ϑ is the same in both experiments and calculate the estimator, $\hat{\hat{\vartheta}}$, say, from combined data. We have

$$U(\vartheta) = U_1(\vartheta) + U_2(\vartheta) = \frac{2(221 - 826\vartheta^2)}{\vartheta(1 - \vartheta^2)} + \frac{67 - 115\vartheta}{\vartheta(1 - \vartheta)}$$

$$= - \frac{1767\vartheta^2 + 48\vartheta - 509}{\vartheta(1 - \vartheta^2)} \, .$$

Solving $U(\vartheta) = 0$, that is, $1767\vartheta^2 + 48\vartheta - 509 = 0$, we find that the positive root of this quadratic equation is $\hat{\hat{\vartheta}} = 0.5233$.

The expected total amount of information from the combined data is

$$I(\vartheta) = I_1(\vartheta) + I_2(\vartheta) = \frac{3304}{1 - \vartheta^2} + \frac{115}{\vartheta(1 - \vartheta)} = \frac{3419\vartheta + 115}{\vartheta(1 - \vartheta^2)}.$$

Substituting for ϑ its combined estimator $\hat{\hat{\vartheta}} = 0.5233$, we find the approximate (estimated) variance:

$$\text{Est Var}(\hat{\hat{\vartheta}}) = \frac{1}{I(\hat{\hat{\vartheta}})} = \frac{1}{5010.9806} = 0.0001996.$$

▼ 12.12.   ML ESTIMATORS OF FUNCTIONS OF MULTINOMIAL PROPORTIONS. SEVERAL PARAMETERS

Let $\theta_1, \theta_2, \ldots, \theta_s$ be s functionally independent parameters. Suppose that the multinomial parameters are functions of θ's, that is,

$$P_l = P_l(\theta_1, \theta_2, \ldots, \theta_s) = P_l(\boldsymbol{\theta}), \qquad l = 1, 2, \ldots, k. \tag{12.46}$$

It can be shown by a method similar to that used in section 12.3 that the score with respect to θ_i, $U_i(\boldsymbol{\theta})$, is

$$U_i(\boldsymbol{\theta}) = \frac{\partial \log L}{\partial \theta_i} = \sum_{l=1}^{k} \frac{r_l}{P_l(\boldsymbol{\theta})} \frac{\partial P_l}{\partial \theta_i}, \qquad i = 1, 2, \ldots, s. \tag{12.47}$$

The individual score in the lth class, with respect to θ_i, $u_i(l; \boldsymbol{\theta})$, is

$$u_i(l; \boldsymbol{\theta}) = \frac{1}{P_l(\boldsymbol{\theta})} \frac{\partial P_l}{\partial \theta_i}, \qquad i = 1, 2, \ldots, s. \tag{12.48}$$

Thus the total score with respect to θ_i,

$$U_i(\boldsymbol{\theta}) = \sum_{l=1}^{k} r_l u_i(l; \boldsymbol{\theta}), \qquad i = 1, 2, \ldots, s, \tag{12.49}$$

is the weighted sum of individual scores. It resembles formula (12.10) for a score of a single parameter.

The (i, j)th element of the expected information matrix, $I_{ij}(\boldsymbol{\theta})$, is

$$I_{ij}(\boldsymbol{\theta}) = -E\left(\frac{\partial^2 \log L}{\partial \theta_i \, \partial \theta_j}\right) = n \sum_{l=1}^{k} \frac{1}{P_l(\boldsymbol{\theta})} \frac{\partial P_l}{\partial \theta_i} \cdot \frac{\partial P_l}{\partial \theta_j}$$

$$= \sum_{l=1}^{k} E(r_l) u_i(l; \boldsymbol{\theta}) u_j(l; \boldsymbol{\theta}), \tag{12.50}$$

for $i, j = 1, 2, \ldots, s \ (i \neq j)$.

If $i = j$, we have

$$I_{ii}(\boldsymbol{\theta}) = n \sum_{l=1}^{k} \frac{1}{P_l(\boldsymbol{\theta})} \left(\frac{\partial P_l}{\partial \theta_i}\right)^2 = \sum_{l=1}^{k} E(r_l)u_i^2(l; \boldsymbol{\theta}), \quad i = 1, 2, \ldots, s. \quad (12.51)$$

Replacing $E(r_l)$ by r_l, and $\boldsymbol{\theta}$ by $\hat{\boldsymbol{\theta}}$, we obtain the elements of the *estimated* information matrix $I(\hat{\boldsymbol{\theta}})$, that is,

$$I_{ij}(\hat{\boldsymbol{\theta}}) = \sum_{l=1}^{k} r_l u_i(l; \hat{\boldsymbol{\theta}}) u_j(l; \hat{\boldsymbol{\theta}}), \quad i, j = 1, 2, \ldots, s, \quad i \neq j. \quad (12.52)$$

For $i = j$ we have

$$I_{ii}(\hat{\boldsymbol{\theta}}) = \sum_{l=1}^{k} r_l u_i^2(l; \hat{\boldsymbol{\theta}}), \quad i = 1, 2, \ldots, s. \quad (12.53)$$

The following important and useful properties should be noted.

(*i*) *Estimability.* If k is the number of multinomial parameters (i.e., only $k - 1$ are functionally independent), then the parameters $\theta_1, \theta_2, \ldots, \theta_s$ are *estimable* if and only if $s \leqslant k - 1$.

It should be noted that the condition $s \leqslant k - 1$ for estimability is necessary when we consider only a single experiment. If, however, the combined data are from m (in the general case, different) experiments, the condition for estimability is

$$s \leqslant \sum_{g=1}^{m} (k_g - 1),$$

where k_g is the number of classes in the gth experiment.

(*ii*) *Special Cases.* The solutions are especially easy to obtain for $s = k - 1$. In such cases, the maximum likelihood estimators of $\theta_1, \theta_2, \ldots, \theta_s$ are obtained by equating the expressions for multinomial parameters (being functions of θ's) to the observed proportions in each class, that is, by solving any set of $s = k - 1$ equations of the type

$$P_l(\theta_1, \theta_2, \ldots, \theta_s) = \frac{r_l}{n}, \quad l = 1, 2, \ldots, k, \quad (12.54)$$

with respect to $\theta_1, \theta_2, \ldots, \theta_s$.

(*iii*) *Variances and Covariances.* The estimators $\hat{\theta}_1, \hat{\theta}_2, \ldots, \hat{\theta}_s (s \leqslant k - 1)$ are certain functions of observed multinomial proportions, $\hat{P}_1, \hat{P}_2, \ldots, \hat{P}_k$, that is,

$$\hat{\theta}_i = \hat{\theta}_i(\hat{P}_1, \hat{P}_2, \ldots, \hat{P}_k), \quad i = 1, 2, \ldots, s.$$

Therefore, using the asymptotic formula (8.95) for variance, we obtain

$$\mathrm{Var}(\hat{\theta}_i) \doteq \sum_{l=1}^{k} \left(\frac{\partial \theta_i}{\partial P_l}\right)^2 \mathrm{Var}(\hat{P}_l) + 2 \sum_{l < m} \sum \frac{\partial \theta_i}{\partial P_l} \frac{\partial \theta_i}{\partial P_m} \mathrm{Cov}(\hat{P}_l, \hat{P}_m)$$

$$= \sum_{l=1}^{k} \left(\frac{\partial \theta_i}{\partial P_l}\right)^2 \frac{P_l(1 - P_l)}{n} - \sum_{l < m} \sum \frac{\partial \theta_i}{\partial P_l} \frac{\partial \theta_i}{\partial P_m} \frac{P_l P_m}{n}$$

$$i = 1, 2, \ldots, s. \quad (12.55)$$

Here $(\partial \theta_i / \partial P_l)$ denotes the derivative of $\hat{\theta}_i$ taken with respect to $\hat{P}_l$ and calculated at the point $(\hat{P}_1, \hat{P}_2, \ldots, \hat{P}_k) = (P_1, P_2, \ldots, P_k)$, or, briefly,

$$\left(\frac{\partial \hat{\theta}_i}{\partial \hat{P}_l}\right)\Bigg|_{\hat{\mathbf{P}} = \mathbf{P}} = \frac{\partial \theta_i}{\partial P_l}, \quad i = 1, 2, \ldots, s; \quad l = 1, 2, \ldots, k. \quad (12.56)$$

Similarly, it can be shown that the covariance between two estimators $\hat{\theta}_i$ and $\hat{\theta}_j$, $\mathrm{Cov}(\hat{\theta}_i, \hat{\theta}_j)$, is approximately

$$\mathrm{Cov}(\hat{\theta}_i, \hat{\theta}_j) \doteq \sum_{l=1}^{k} \frac{\partial \theta_i}{\partial P_l} \frac{\partial \theta_j}{\partial P_l} \mathrm{Var}(\hat{P}_l) + 2 \sum_{l < m} \sum \frac{\partial \theta_i}{\partial P_l} \frac{\partial \theta_j}{\partial P_m} \mathrm{Cov}(\hat{P}_l, \hat{P}_m)$$

$$= \sum_{l=1}^{k} \frac{\partial \theta_i}{\partial P_l} \frac{\partial \theta_j}{\partial P_l} \frac{P_l(1 - P_l)}{n} - 2 \sum_{l < m} \sum \frac{\partial \theta_i}{\partial P_l} \frac{\partial \theta_j}{\partial P_m} \frac{P_l P_m}{n}, \quad (12.57)$$

for $i, j = 1, 2, \ldots, s$; $(i \neq j)$.

The results obtained in this section will be applied in Chapter 14 to the estimation of gene frequencies at multiallelic blood group loci. In the next section, we present an application of these results in estimating the proportions of admixtures in "hybrid" populations.

▼ 12.13.   ESTIMATION OF ADMIXTURES IN A "HYBRID" POPULATION

Almost every human population is a mixture of several ethnic groups which have contributed in different ways and at different periods of time to its social and anthropological structure. Here we restrict our analysis to populations of rather simple structures. We assume that a population called a *hybrid* population is composed of $c + 1$ contributing populations, termed *base* populations. When such a hybrid population lives under certain conditions and no in- or out-migration is observed, it is called an *isolate*. For example, there are certain isolates in the United States (North Carolina) which appear to be of triracial origin: White, Negro, and Indian (see Example 12.4).

The contributions of base populations are usually expressed in terms of proportions of gene frequencies of known characters, very often in terms of blood group gene frequencies. We derive a probability model which can be applied in estimating the proportions of base populations in a hybrid population, under the following assumptions:

(a) The hybrid population is an isolate.

(b) The hybrid population is in equilibrium with respect to the loci under consideration.

(c) The gene frequencies in the base populations are known.

We first consider a single locus with s alleles, and $c + 1$ base populations. We introduce the following notations: q_{it} is the (known) gene frequency of the allele A_t ($t = 1, 2, \ldots, s$) in the ith base population ($i = 1, 2, \ldots, c + 1$); h_t is the (unknown) gene frequency of the allele A_t in the hybrid population; α_i is the proportion of admixture (the "admixture coefficient") of the ith base population in the hybrid population, with

$$\sum_{j=1}^{c+1} \alpha_j = 1.$$

Using a sample of n individuals, we wish to find the ML estimators of $\alpha_1, \alpha_2, \ldots, \alpha_{c+1}$. Since only c parameters are functionally independent, we take

$$\alpha_{c+1} = 1 - \sum_{j=1}^{c} \alpha_j.$$

The gene frequency, h_t, in the hybrid population can be expressed as a linear function of the gene frequencies q_{it} as follows:

$$h_t = \alpha_1 q_{1t} + \alpha_2 q_{2t} + \cdots + \left(1 - \sum_{j=1}^{c} \alpha_j\right) q_{c+1,t}, \qquad t = 1, 2, \ldots, s; \quad (12.58)$$

or

$$h_t^* = \alpha_1 q_{1t}^* + \alpha_2 q_{2t}^* + \cdots + \alpha_c q_{ct}^*, \qquad (12.59)$$

where

$$h_t^* = h_t - q_{c+1,t} \quad \text{and} \quad q_{it}^* = q_{it} - q_{c+1,t}, \qquad (12.60)$$

for $t = 1, 2, \ldots, s; i = 1, 2, \ldots, c$.

Let k be the number of phenotypes at this locus, and let

$$P_l = P_l[h_1(\alpha_1, \ldots, \alpha_c), \ldots, h_s(\alpha_1, \ldots, \alpha_c)], \qquad l = 1, 2, \ldots, s, \quad (12.61)$$

represent the expected proportion of individuals with the lth phenotype in the hybrid population. Also let r_l be the number of individuals with the lth phenotype, $l = 1, 2, \ldots, k$, in a sample of size n from the hybrid population, and $\Sigma r_l = n$. We wish to estimate $\alpha_1, \alpha_2, \ldots, \alpha_c$.

The likelihood function of this sample with respect to the parameters $\alpha_1, \alpha_2, \ldots, \alpha_c$, — briefly, $L(\alpha_1, \alpha_2, \ldots, \alpha_c) = L(\boldsymbol{\alpha})$ — is proportional to $P_1^{r_1} P_2^{r_2} \cdots P_k^{r_k}$. Therefore

$$\log L(\boldsymbol{\alpha}) = \text{const} + \sum_{l=1}^{k} r_l \log P_l(\boldsymbol{\alpha}),$$

where $P_l(\boldsymbol{\alpha})$ is a function of the α's defined in (12.61). Thus, the score with respect to α_i, $U_i(\boldsymbol{\alpha})$, is

$$U_i(\boldsymbol{\alpha}) = U_i(\alpha_1, \alpha_2, \ldots, \alpha_c)$$

$$= \sum_{l=1}^{k} r_l \sum_{t=1}^{s} \frac{1}{P_l(\boldsymbol{\alpha})} \frac{\partial P_l}{\partial h_t} \cdot \frac{\partial h_t}{\partial \alpha_i}, \qquad i = 1, 2, \ldots, c. \tag{12.62}$$

We notice that

$$\frac{\partial h_t}{\partial \alpha_i} = q_{it} - q_{c+1,t} = q_{it}^*, \qquad i = 1, 2, \ldots, c; \quad t = 1, 2, \ldots, s. \tag{12.63}$$

Substituting (12.63) into (12.62), we obtain

$$U_i(\boldsymbol{\alpha}) = \sum_{l=1}^{k} r_l \sum_{t=1}^{s} \frac{1}{P_l(\boldsymbol{\alpha})} \frac{\partial P_l}{\partial h_t} q_{it}^*, \qquad i = 1, 2, \ldots, c. \tag{12.64}$$

The individual score with respect to α_i in the lth class, $u_i(l; \boldsymbol{\alpha})$, is

$$u_i(l; \boldsymbol{\alpha}) = \sum_{t=1}^{s} \frac{1}{P_l(\boldsymbol{\alpha})} \frac{\partial P_l}{\partial h_t} q_{it}^*, \qquad i = 1, 2, \ldots, c. \tag{12.65}$$

Thus the total score with respect to α_i can be written in the form

$$U_i(\boldsymbol{\alpha}) = \sum_{l=1}^{k} r_l u_i(l; \boldsymbol{\alpha}), \qquad i = 1, 2, \ldots, c. \tag{12.66}$$

[Compare formula (12.49).] Solving the system of equations

$$U_i(\boldsymbol{\alpha}) = 0, \qquad i = 1, 2, \ldots, c, \tag{12.67}$$

we obtain the maximum likelihood estimators $\hat{\alpha}_1, \hat{\alpha}_2, \ldots, \hat{\alpha}_c$.

It can be shown that the (ij)th element of the expected information matrix is

$$I_{ij}(\boldsymbol{\alpha}) = -E\left(\frac{\partial^2 \log L}{\partial \alpha_i \, \partial \alpha_j}\right) = \sum_{l=1}^{k} E(r_l) u_i(l; \boldsymbol{\alpha}) u_j(l; \boldsymbol{\alpha}), \quad i, j = 1, 2, \ldots, c; \quad i \neq j. \tag{12.68}$$

In particular, for $i = j$

$$I_{ii}(\boldsymbol{\alpha}) = \sum_{l=1}^{k} E(r_l)u_i^2(l; \boldsymbol{\alpha}), \qquad i = 1, 2, \ldots, c. \qquad (12.69)$$

[Compare formulae (12.50) and (12.51) respectively].

The corresponding *estimators* of elements of the information matrix are

$$I_{ij}(\hat{\boldsymbol{\alpha}}) = \sum_{l=1}^{k} r_l u_i(l; \hat{\boldsymbol{\alpha}})u_j(l; \hat{\boldsymbol{\alpha}}), \qquad (12.68a)$$

and

$$I_{ii}(\hat{\boldsymbol{\alpha}}) = \sum_{l=1}^{k} r_l u_i^2(l; \hat{\boldsymbol{\alpha}}). \qquad (12.69a)$$

It should be noted that $\alpha_1, \alpha_2, \ldots, \alpha_c$ are *estimable* if and only if $c < s \leqslant k - 1$. In particular, when $c = s - 1$, it may sometimes be easier to calculate first the ML estimators of gene frequencies in the hybrid population, $\hat{h}_1$, $\hat{h}_2, \ldots, \hat{h}_s$. The ML estimators of $\alpha_1, \alpha_2, \ldots, \alpha_c$ will then be obtained by equating any $s - 1$ expected gene frequencies to their observed proportions in the sample, that is, by solving $s - 1$ equations of the type

$$\alpha_1 q_{1t}^* + \alpha_2 q_{2t}^* + \cdots + \alpha_c q_{ct}^* = \hat{h}_t^*, \qquad t = 1, 2, \ldots, s - 1, \quad (12.70)$$

where $\hat{h}_t^* = \hat{h}_t - q_{c+1,t}$.

It should also be mentioned that when $2 < c < s$, equations (12.67) become rather complicated and an iterative method has to be used. [In such cases a computer program will be helpful. A program in FORTRAN IV for the IBM 7040 computer has already been written; see Krieger et al. (1965), p. 120.]

The general model can be extended to m *independent* loci, provided that

$$c \leqslant \sum_{g=1}^{m}(k_g - 1), \qquad (12.71)$$

where k_g ($g = 1, 2, \ldots, m$) is the number of *phenotypes* at the gth locus.

The overall scores and amounts of information can be obtained by adding the scores or the amounts of information, respectively, from individual experiments.

We now demonstrate only a simple case with $c = s - 1 = 2$.

Example 12.4. Pollitzer (1964) studied a certain relatively isolated population in the state of North Carolina, the members of which call themselves "Indians." In fact, however, the population is a "hybrid" of triracial origin: Whites (mostly of English origin), Charleston Negroes, and Cherokee Indians. Several blood group systems have been typed, mostly of school children.

A sample of size 1273 was examined. All the data available should be used

TABLE 12.5   Observed frequencies of phenotypes in the
sample of 1273 individuals from the "hybrid" isolate

Blood group	Class, l	Observed number, r_l	Proportion
O	1	643	0.5051
A	2	450	0.3535
B	3	136	0.1068
AB	4	44	0.0346
Total		1273	1.0000

in estimating the admixture coefficients in this isolate. In our example, however, we restrict ourselves to the data on the *ABO* blood group system, given in Tables 12.5 and 12.6.

The gene frequencies, h_t ($t = 1, 2, 3$), were estimated by means of Bernstein's method (see Chapter 14). Since we now have $c = s - 1 = 2$, we evaluate the estimators of α_1 and α_2 by solving the following equations:

$$q_{11}^*\alpha_1 + q_{21}^*\alpha_2 = \hat{h}_1^*$$
$$q_{21}^*\alpha_1 + q_{22}^*\alpha_2 = \hat{h}_2^*$$

that is

$$0.290\alpha_1 + 0.263\alpha_2 = 0.264$$
$$0.239\alpha_1 + 0.119\alpha_2 = 0.200.$$

We obtain $\hat{\alpha}_1 = 0.7473$ and $\hat{\alpha}_2 = 0.1798$; hence $\hat{\alpha}_3 = 1 - \hat{\alpha}_1 - \hat{\alpha}_2 = 0.0729$. Therefore, using the *ABO* blood group system, we estimate that this isolate is composed of about 75% Whites, 18% Negroes, and 7% Indians. The estimated information matrix and its inverse (i.e., the variance-covariance matrix) can be calculated using (12.68a) and (12.69a). (See Problem 12.4.)

12.14. MINIMUM CHI-SQUARE ESTIMATORS

In section 11.4 we discussed asymptotic properties such as consistency, unbiasedness, efficiency, and normality of maximum likelihood estimators.

TABLE 12.6   Gene frequencies in the base populations and estimated gene frequencies in the sample from "hybrid" isolate

Population (i) Allele (t)	(1) English Whites q_{1t}	(2) Charleston Negroes q_{2t}	(3) Cherokee Indians q_{3t}	"Hybrid" isolate $(\hat{h}_t)$	$q_{1t}^* = q_{1t} - q_{3t}$	$q_{2t}^* = q_{2t} - q_{3t}$	$\hat{h}_t^* = \hat{h}_t - q_{3t}$
O (1)	0.683	0.710	0.973	0.709	−0.290	−0.263	−0.264
A (2)	0.257	0.137	0.018	0.218	0.239	0.119	0.200
B (3)	0.060	0.153	0.009	0.073			

There are other estimators, in addition to the ML estimators, which possess these asymptotic properties. The class of such estimators is called the class of *best asymptotically normal* (briefly, BAN) estimators. It includes ML estimators as a particular subclass.

Another subclass of BAN estimators consists of so-called *minimum chi-square* (or, briefly, MCS) estimators. The method of minimum chi-square can be applied when the parameters which must be estimated are functions of *multinomial* parameters, whereas the maximum likelihood method is of general use, whatever the population distributions may be.

We recall the notation used in section 12.12. The k multinomial parameters, $P_1, P_2, \ldots, P_k$, are functions of s functionally independent parameters, $\theta' = (\theta_1, \theta_2, \ldots, \theta_s)$, that is,

$$P_l = P_l(\theta_1, \ldots, \theta_s) = P_l(\theta), \qquad l = 1, 2, \ldots, k,$$

and r_l denotes the number of individuals in the lth class in the sample of size n. If n is large enough and $nP_l \geqslant 3$, the statistic

$$X^2 = \sum_{l=1}^{k} \frac{[r_l - nP_l(\theta)]^2}{nP_l(\theta)} \tag{12.72}$$

is distributed approximately as χ^2 with $k - 1$ degrees of freedom (see section 8.19). We wish to find the set of estimators, $\check{\theta}_1, \check{\theta}_2, \ldots, \check{\theta}_s$, for which the function X^2 given in (12.72) is *minimum*.

Let us represent (12.72) in a rather simpler form. Expanding the binomial in (12.72), we obtain

$$X^2 = \sum_{l=1}^{k} \frac{r_l^2}{nP_l(\theta)} - 2\sum_{l=1}^{k} r_l + n\sum_{l=1}^{k} P_l.$$

Since

$$\sum_{l=1}^{k} r_l = n \quad \text{and} \quad \sum_{l=1}^{k} P_l = 1,$$

we obtain

$$X^2 = \frac{1}{n}\sum_{l=1}^{k} \frac{r_l^2}{P_l(\theta)} - n. \tag{12.73}$$

To minimize (12.73) with respect to $\theta_1, \theta_2, \ldots, \theta_s$ we have to solve a set of s equations:

$$\frac{\partial(X^2)}{\partial \theta_i} = -\frac{1}{n}\sum_{l=1}^{k} \left[\frac{r_l}{P_l(\theta)}\right]^2 \cdot \frac{\partial P_l}{\partial \theta_i} = 0, \qquad i = 1, 2, \ldots, s. \tag{12.74}$$

The parameters, $\theta_1, \theta_2, \ldots, \theta_s$, are estimable provided $s < k$.

The estimators, $\check{\theta}_1, \check{\theta}_2, \ldots, \check{\theta}_s$, obtained by this method are BAN estimators.

It should be noted that this method is valid also for combined data from independent experiments. If m denotes the number of experiments, n_g the sample size, k_g the number of multinomial classes in the gth experiments, $P_{lg}(\theta)$ the expected proportion, and r_{lg} the number of individuals in the lth class in the gth experiment, then the statistic

$$X^2 = \sum_{g=1}^{m} n_g \sum_{l=1}^{k_g} \frac{[r_{lg} - n_g P_{lg}(\theta)]^2}{n_g P_{lg}(\theta)} \tag{12.75}$$

is approximately distributed as χ^2 with

$$\nu = \sum_{g=1}^{m} (k_g - 1)$$

degrees of freedom.

To minimize (12.75) with respect to $\theta_1, \ldots, \theta_s$ we have to solve a system of s equations of the type

$$\frac{\partial(X^2)}{\partial \theta_i} = -\sum_{g=1}^{m} \frac{1}{n_g} \sum_{l=1}^{k_g} \left(\frac{r_{lg}}{P_{lg}(\theta)}\right)^2 \frac{\partial P_{lg}}{\partial \theta_i} = 0, \qquad i = 1, 2, \ldots, s. \tag{12.76}$$

The s parameters are estimable if $s \leqslant \sum (k_g - 1)$.

It can be shown that if n is very large, equation (12.75) or (12.76) becomes identical with the maximum likelihood equations.

12.14.1. Modified Minimum Chi-Square Estimators

The MCS estimators can be preferred to ML estimators only when equation (12.75) or (12.76) is easier to solve than the ML equations, which is not always the case. Some simplifications are often obtained when the expected values, $E(r_l) = nP_l(\theta)$, in the denominator of (12.72) are replaced by the observed values, r_l (assuming $r_l \neq 0$), for $l = 1, 2, \ldots, k$. We now wish to minimize

$$X^2_{(\text{mod})} = \sum_{l=1}^{k} \frac{[r_l - nP_l(\theta)]^2}{r_l} = n^2 \sum_{l=1}^{k} \frac{[P_l(\theta)]^2}{r_l} - n \tag{12.77}$$

with respect to $\theta_1, \theta_2, \ldots, \theta_s$. We then have to solve a system of s equations of the type

$$\frac{\partial(X^2_{(\text{mod})})}{\partial \theta_i} = 2n^2 \sum_{l=1}^{k} \frac{P_l(\theta)}{r_l} \cdot \frac{\partial P_l}{\partial \theta_i} = 0, \qquad i = 1, 2, \ldots, s. \tag{12.78}$$

The estimators obtained by minimizing (12.77) are called *modified minimum chi-square* (briefly, MMCS) estimators. For large samples they are also BAN estimators [Neyman (1949)].

The same method can also be applied to combined data from m independent experiments. By analogy to (12.76), in order to obtain MMCS estimators from combined data we have to solve a system of s equations of the type

$$\frac{1}{2}\frac{\partial(X^2_{(\mathrm{mod})})}{\partial\theta_i} = \sum_{g=1}^{m} n_g{}^2 \sum_{l=1}^{k_g} \frac{P_{lg}(\boldsymbol{\theta})}{r_{lg}} \cdot \frac{\partial P_{lg}}{\partial\theta_i} = 0, \qquad i = 1, 2, \ldots, s. \quad (12.79)$$

Some applications of the method of minimum χ^2 will be demonstrated in later chapters.

SUMMARY

This chapter has been devoted chiefly to maximum likelihood estimators of parameters which are functions of multinomial proportions. Scores and expected amounts of information have been presented in especially convenient forms. The ML estimators of gene frequencies at a single two-allelic locus have been derived for different kinds of experimental designs. The emphasis has been on the appropriate sampling procedures, especially when selecting a random sample of related pairs. Also, the problem of estimating proportions in a "hybrid" population has been outlined. Minimum chi-square estimators and their relation to maximum likelihood estimators have been briefly discussed.

REFERENCES

Bailey, N. T. J., *Introduction to the Mathematical Theory of Genetic Linkage*, Clarendon Press, Oxford, 1961.

Cotterman, C. W., Estimation of gene frequencies in non-experimental populations. In: *Statistics and Mathematics in Biology*, ed. by O. Kempthorne, Hafner Publishing Co., New York, 1954, pp. 449–465.

Cramér, H., *Mathematical Methods of Statistics*, Princeton University Press, Princeton, N.J., 1946.

Fisher, R. A., The estimation of the proportion of recessives from tests carried out on a sample not wholly unrelated, *Ann. Eugen. Lond.* **10** (1940), 160–170.

Hirschfeld, J., Immuno-electrophoretic demonstration of qualitative differences in human sera and their relation to haptoglobins, *Acta Path. Microbiol. Scand.* **47** (1959), 160–168.

Hirschfeld, J., and Heiken, A., Application of the Gc system in paternity cases, *Amer. J. Hum. Genet.* **15** (1963), 19–23.

Kempthorne, O., *An Introduction to Genetic Statistics*, John Wiley & Sons, New York, 1957.

Kendall, M. G., and Stuart, A., *The Advanced Theory of Statistics*, Vol. 2, Hafner Publishing Co., New York, 1961.

Krieger, H., Morton, N. E., Mi, M. P., Azevedo, E., Freire-Maia, A., and Yasuda, N., Racial admixture in northeastern Brazil, *Ann. Hum. Genet. Lond.* **29** (1965), 113–125.

Mather, K., The combination of data, *Ann. Eugen. Lond.* **6** (1935), 399–410.

Neyman, J., Contribution to the theory of the X^2 test, *Proc. First Berkeley Symp. Math. Statist. and Prob.*, University of California Press, Berkeley, 1949, p. 239.

Plato, C. C., Rucknagel, D. L., and Gershowitz, H., Studies of the distribution of glucose-6-phosphate dehydrogenase deficiency, thalassemia and other genetic traits in the coastal and mountain villages of Cyprus, *Amer. J. Hum. Genet.* **16** (1964), 267–283.

Pollitzer, W. S., Analysis of tri-racial isolate, *Hum. Biol.* **36** (1964), 362–373.

PROBLEMS

12.1. The locus *Gm* is one of the two loci which control gamma globulins present in human sera and showing antigenic properties. The three most common alleles at this locus are denoted by Gm^a, Gm^{ax}, and Gm^b.* Let p, q, and t, with $p + q + t = 1$, be their respective frequencies.

The standard reagents used in the laboratory are such that only certain substances, called a, b, and x factors, can be serologically detected. Therefore the genotypes Gm^aGm^{ax} and $Gm^{ax}Gm^{ax}$ are indistinguishable.

Let $+$ be the presence and $-$ be the absence of the factor in the phenotype. Thus there are five phenotypes, and their distribution, assuming equilibrium, is as follows:

l	1	2	3
Phenotype:	$Gm(a + b - x-)$	$Gm(a + b - x+)$	$Gm(a + b + x-)$
Distribution:	p^2	$q^2 + 2pq$	$2pt$

l	4	5
Phenotype:	$Gm(a + b + x+)$	$Gm(a - b + x-)$
Distribution:	$2qt$	t^2 .

Let us, for convenience, number the phenotypes 1, 2, 3, 4, 5 as above. Let r_l be the observed number of the lth phenotype in a sample of size n, with $\sum r_l = n$.

(*a*) Show that the maximum likelihood equations are given by

$$\frac{\partial \log L}{\partial p} = U_p = \frac{2r_1 + r_3}{p} + \frac{2r_2}{q + 2p} - \frac{r_3 + r_4 + 2r_5}{1 - p - q} = 0,$$

$$\frac{\partial \log L}{\partial q} = U_q = \frac{r_2 + r_4}{q} + \frac{r_2}{q + 2p} - \frac{r_3 + r_4 + 2r_5}{1 - p - q} = 0,$$

and $t = 1 - p - q$.

(*b*) Find the estimated and expected information matrices $I(\hat{q}, \hat{t})$ and $I(q, t)$, respectively.

* The new nomenclature of these alleles is Gm^1, $Gm^{1,2}$, and $Gm^{3,5,13,14}$. For convenience, we retain the symbols used in the earlier (before 1966) literature.

12.2. The following data are given by Steinberg et al. [*Amer. J. Hum. Genet.* **13** (1961), 205–213] for the sample $n = 303$ of five observed phenotypes at the *Gm* locus in U.S. Whites (for notation, see Problem 12.1.)

$$\text{Phenotype:} \quad 1 \quad 2 \quad 3 \quad 4 \quad 5$$
$$\text{Observed frequency:} \quad 14 \quad 11 \quad 87 \quad 42 \quad 149 \,.$$

(*a*) Calculate the ML estimates $\hat{p}$, $\hat{q}$, and $\hat{t}$.

(*b*) Calculate the scores for the phenotypic classes.

(*c*) Calculate the estimated information matrix, $\mathbf{I}(\hat{q}, \hat{t})$.

(*d*) Find the estimated variance-covariance matrix, $\mathbf{V}(\hat{q}, \hat{t}) \doteq \mathbf{I}^{-1}(\hat{q}, \hat{t})$.

[*Hint:* Invert $\mathbf{I}(\hat{q}, \hat{t})$ or apply approximate formulae (12.55) and (12.57), bearing in mind that r_l/n ($l = 1, 2, \ldots, 5$) are multinomial proportions.]

12.3. In Example 12.4 we estimated the coefficients of admixture, α_1 and α_2, in a hybrid population.

(*a*) Derive the solutions for α_1 and α_2 in the general form (in terms of q's and $\hat{h}$'s) of the likelihood equations:

$$\alpha_1 q_{11}^* + \alpha_2 q_{21}^* = \hat{h}_1^*,$$
$$\alpha_1 q_{21}^* + \alpha_2 q_{22}^* = \hat{h}_2^*,$$

where

$$q_{1t}^* = q_{1t} - q_{3t}, \quad q_{2t}^* = q_{2t} - q_{3t}, \quad \hat{h}_t^* = \hat{h}_t - q_{3t},$$

for $t = 1, 2$.

(*b*) Find $\text{Var}(\hat{\alpha}_1)$, $\text{Var}(\hat{\alpha}_2)$, and $\text{Cov}(\hat{\alpha}_1, \hat{\alpha}_2)$. (*Hint:* Note that $\hat{h}_1$, $\hat{h}_2$, and $\hat{h}_3$, with $\sum \hat{h}_t = 1$, are functions of binomial proportions r_l/n, where r_l are phenotype frequencies as given in Table 12.5.) Write down the variance-covariance matrix, $\mathbf{V} = \mathbf{V}(\hat{\alpha}_1, \hat{\alpha}_2)$.

(*c*) Find the expected information matrix from the approximate relationship $\mathbf{I}^{-1}(\alpha_1, \alpha_2) \doteq \mathbf{V}(\hat{\alpha}_1, \hat{\alpha}_2)$.

12.4. Calculate the elements of the estimated information matrix $\mathbf{I}(\hat{\alpha}_1, \hat{\alpha}_2)$, using the results of Example 12.4 (i.e., $\hat{\alpha}_1 = 0.7473$, $\hat{\alpha}_2 = 0.1798$).

12.5. Derive the formulae for modified minimum chi-square (MMCS) estimators for Problem 12.1, and calculate the MMCS estimates using the data given in Problem 12.2. Compare these with the ML estimates.

12.6. Assuming genetic equilibrium, derive the maximum likelihood estimators of the gene frequencies for the X-linked locus with alleles A (dominant) and a (recessive), using the following samples of males and females:

	Males			Females			
Genotype:	A	a	Total	AA	Aa	aa	Total
Observed number:	m_1	m_2	m	n_1	n_2	n_3	n .

(*a*) Use only the sample of males.

(*b*) Use only the sample of females.

(*c*) Use combined samples.

12.7.   Find the variance of the estimator of the frequency of the gene A obtained from combined samples in Problem 12.6.

12.8.   Apply the results to the Xg-blood group locus with alleles Xg^a (dominant) and Xg (recessive or "silent"), using the following data (from Race and Sanger, *The Blood Groups in Man*, F. A. Davis Co., Philadelphia, 1968, p. 524): males — $m_1 = 268$, $m_2 = 132$, $m = 400$; females — $n_1 = 122$, $n_2 = 119$, $n_3 = 29$, $n = 270$.

CHAPTER 13

Testing Genetic Hypotheses

13.1. GENETIC AND STATISTICAL HYPOTHESES

A statement or supposition about a phenomenon occurring under certain conditions we will call a *scientific hypothesis*. In particular, *genetic hypotheses* are concerned with the structure and transmission of heredity material. *Statistical hypotheses*, on the other hand, are associated with distributions of random variables.

In principle, scientific hypotheses are disprovable. Using appropriate experiments, one may show that the data are consistent with his hypothesis and so he may accept it. Nevertheless, he must keep in mind that with increasing knowledge this hypothesis can be modified or even disproved.

In this chapter, we will be concerned with methods of constructing criteria (experimental designs and tests) by which we can reject or not reject (i.e., "provisionally" accept) genetic hypotheses. Genetic hypotheses can be "translated" into statistical language. Hence, in fact, we shall present here a brief theory of testing statistical hypotheses, with emphasis on their relevance to genetic problems. For demonstration, we use a simple example about binomial proportions.

Example 13.1. Suppose that a plant-breeder observes the flower color of a new variety of a certain plant species and makes a supposition that it is inherited as a simple Mendelian dominant. This is equivalent to the statement that the expected proportion, P, of colored flowers is $\frac{3}{4}$. In other words, his genetic hypothesis on dominance implies that the observations fall into two categories:

$$x = \begin{cases} 1 & \text{if a plant has colored flowers,} \\ 0 & \text{otherwise,} \end{cases}$$

and that in a sample of n independently observed plants the number of colored flowers, r, will be binomially distributed with the expected proportion $P = \frac{3}{4}$.

DEFINITION 13.1. *A statement about the distribution(s) of random variables is called a statistical hypothesis.*

A hypothesis can be made about the form of a distribution (e.g., "x is normally distributed") or about the parameters of the distribution (e.g., "the expected value of a normally distributed character is μ_0") or about both the form and the parameter(s) of a distribution (as in Example 13.1).

DEFINITION 13.2. *A hypothesis which specifies the distribution (i.e., the form and the parameters) is called a simple hypothesis; a hypothesis which is not simple is called a composite hypothesis.*

For example, for given n, "r is binomially distributed with $P = \frac{3}{4}$" is a simple hypothesis, whereas "r is binomially distributed with $P > \frac{3}{4}$" is a composite hypothesis, as it is associated with a family of distributions with $\frac{3}{4} < P < 1$.

DEFINITION 13.3. *The hypothesis which one tests is called the null hypothesis and is often denoted by H_0. Any other hypothesis, different from H_0, is called an alternative hypothesis; this can be denoted by H_1, H_2, or, generally, by H.*

In our Example 13.1, the null hypothesis is $H_0: P = \frac{3}{4}$. If one suspects that the character might be controlled by two loci with mutual epistasis, the alternative will be a simple hypothesis $H_1: P = \frac{13}{16}$. If, however, one has the figure $\frac{3}{4}$ in mind only as the most likely value of P and has no clear idea what other specific hypotheses could be, his alternative is $H: P \neq \frac{3}{4}$.

Usually the null hypothesis is stated in such a way that there is no effect, that is, no deviation from a certain (standard) model. The alternative hypothesis, on the other hand, suggests that there is some effect or deviation from the model defined by H_0.

13.2. STATISTICAL TEST

13.2.1. Errors of Two Kinds

To decide with certainty whether a hypothesis is true or false, one would have to examine the entire population. In many problems such a population is very large. Thus the decision is made (with some uncertainty) on the basis of a sample. Using a given set of data, we have a choice of two decisions: to accept or to reject the null hypothesis.

DEFINITION 13.4. *A statistical procedure by which one decides to reject or to accept a statistical hypothesis is called a statistical test.*

Suppose that in our Example 13.1 a sample of $n = 200$ plants was observed, and it was found that $r = 154$ plants had colored flowers, while $n - r = 46$ had white flowers. It is natural to use the sample proportion, $\hat{P} = r/n = \frac{154}{200} = 0.77$, in this case, and to compare it to $P_0 = \frac{3}{4} = 0.75$. The sample proportion $\hat{P}$ will be called a *test statistic*. Its distribution is that of a binomial proportion, and if our H_0 is true, the binomial parameter is $P_0 = 0.75$. This question arises: is the observed value $\hat{P} = 0.77$ sufficiently close to $P_0 = 0.75$ that we would be willing to accept H_0, or is it so far away that H_0 should be rejected?

In performing a test we may make a correct decision or commit one of two kinds of errors:

(*a*) *Reject H_0 when it is* true. This error is called an *error of the first kind*. The probability of committing an error of the first kind is usually denoted by α; it is the expected proportion (in an "infinite" series of repeated tests) of rejecting the true H_0 (sometimes α is called briefly "the error of the first kind").

(*b*) *Accept H_0 when it is* false. This error is called an *error of the second kind*. The probability of committing an error of the second kind will be denoted by β (sometimes β is called briefly "the error of the second kind").

It is clear that the consequences of these two errors are different. Although it is desirable to keep both types to a minimum, for a given n both probabilities cannot be controlled simultaneously. It is customary to control α and to minimize β as far as possible with the given value of α.

13.2.2. Critical Region, Significance Test

Suppose that in our Example 13.1 we decide to reject H_0: $P = \frac{3}{4}$ (in general, H_0: $P = P_0$) if the observed $\hat{P}$ falls *outside* the interval between two values, P_L (the *lower* limit) and P_U (the *upper* limit), with P_L and P_U satisfying the relation

$$\Pr\{\hat{P} < P_L \mid H_0\} + \Pr\{\hat{P} > P_U \mid H_0\} = \alpha. \qquad (13.1)$$

However, since the observed proportion $\hat{P}$ has a binomial distribution, which is discrete, relation (13.1) cannot be exactly satisfied. Therefore we replace it by

$$\Pr\{\hat{P} < P_L \mid H_0\} + \Pr\{\hat{P} > P_U \mid H_0\} \leqslant \alpha, \qquad (13.1a)$$

with the left-hand side in (13.1a) as close as possible to the given value of α.

DEFINITION 13.5. *The probability, α, of rejecting H_0 when it is true is called the significance level of the test.*

The choice of the value of α is rather arbitrary, depending on how often we are willing to reject (incorrectly) the true H_0. The values $\alpha = 0.10$, 0.05, and 0.01 are commonly used.

DEFINITION 13.6. *The region of rejection of the null hypothesis, H_0, associated with the significance level α, is called the critical region for H_0; we will denote it generally by $w = w(\alpha)$. The complementary region, $\bar{w} = \bar{w}(\alpha)$ (i.e., non-w), is called the acceptance region for H_0.*

DEFINITION 13.7. *A test procedure using a critical region associated with a given significance level, α, is called a significance test.*

In our example, the critical region is defined by $\hat{P} < P_L$ or $\hat{P} > P_U$, where P_L and P_U can be evaluated by solving (13.1a). The acceptance region is $P_L \leqslant \hat{P} \leqslant P_U$. The values P_L and P_U are called the *critical values*. For a given α, there can be several pairs (P_L, P_U) satisfying (13.1a). It is desirable to select P_L and P_U in such a manner that, for a given α, the probability, β, of accepting H_0 when it is false is as small as possible.

13.3. THE POWER OF THE TEST

We now introduce the concept of the power of a test, which will be helpful in selecting the most appropriate critical region.

DEFINITION 13.8. *The probability of rejecting H_0 when H_1 is true is called the power of the test with respect to H_1. In our notation, it is equal to $1 - \beta$.*

Suppose that, for given α, two tests have been constructed, using different critical regions based either on the same or on different test statistics. The one which has greater power with respect to the same alternative, H_1, is said to be *more powerful* than the other.

DEFINITION 13.9. *If among the class of tests (for the same H_0 against a simple alternative, H_1), each with the same significance level α, there is one such a test which has greatest power, $1 - \beta$, then this test is called the most powerful test with respect to H_1.*

Keeping H_0, α, and the test criterion the same but changing H_1, we obtain different values of the power. These constitute the *power function* of the test.

It may happen that the most powerful test with respect to H_1 has this optimal property with respect to the whole class of alternatives (i.e., with respect to the elements of a certain composite alternative, H).

DEFINITION 13.10. *If the alternative hypothesis H is composite and a test is most powerful with respect to each of its components (i.e., each simple hypothesis in H), then the test is called the uniformly most powerful test (briefly, the UMP test).*

We notice that the UMP test maximizes $1 - \beta$ (or minimizes β) when α and n are given. If such a test exists, it will naturally be the most desirable one to use in testing H_0.

Returning to our hypothesis H_0: $P = P_0$, we wish now to select our two values P_L and P_U in such a way that $1 - \beta$ will be maximized if possible. It is clear that the construction of this "best" critical region depends on the composite alternative. We shall give only the appropriate formulae; the proof that this will be the "best" (in an approximate sense) test is beyond the scope of this book.

First, we notice that our sample $n = 200$ is sufficiently large to use normal approximation. Therefore, if P_0 is the true binomial proportion and $\hat{P} = r/n$ is the sample proportion, $\mathrm{Var}(\hat{P}) = \sigma_{\hat{P}}^2 = P_0(1 - P_0)/n$, and the statistic $Z = (\hat{P} - P_0)/\sigma_{\hat{P}}$ is approximately normally distributed with $\mu = 0$ and $\sigma^2 = 1$.

(i) Suppose that the alternative is H: $P < P_0$ (left-hand, one-sided alternative). It appears that the "best" selection of the critical region will be to find P_L and P_U such that

$$\Pr\{\hat{P} < P_L \mid H_0\} \doteq \alpha, \quad \Pr\{\hat{P} > P_U \mid H_0\} = 0. \tag{13.2}$$

Using normal approximation, we replace (13.2) by

$$\left.\begin{aligned}
\Pr\{Z < z_\alpha\} &= \Pr\left\{\frac{\hat{P} - P_0}{\sigma_{\hat{P}}} < -z_{1-\alpha}\right\} \doteq \alpha, \\[2mm]
\Pr\{Z > z_1\} &= \Pr\left\{\frac{\hat{P} - P_0}{\sigma_{\hat{P}}} > z_1\right\} = 0,
\end{aligned}\right\} \tag{13.3}$$

or, equivalently,

$$\left.\begin{aligned}
\Pr\left\{\hat{P} < P_0 - z_{1-\alpha}\sqrt{\frac{P_0(1 - P_0)}{n}}\right\} &= \alpha, \\[2mm]
\Pr\left\{\hat{P} > P_0 + z_1\sqrt{\frac{P_0(1 - P_0)}{n}}\right\} &= 0.
\end{aligned}\right\} \tag{13.3a}$$

Hence (noting that $z_1 = \infty$) we obtain

$$P_L = P_0 - z_{1-\alpha}\sqrt{\frac{P_0(1 - P_0)}{n}}, \quad P_U = \infty. \tag{13.4}$$

This means that there is only one critical point, P_L. Thus the approximate

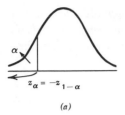 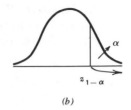 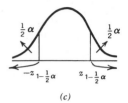

$$z_\alpha = -z_{1-\alpha} \qquad\qquad z_{1-\alpha} \qquad\qquad -z_{1-\frac{1}{2}\alpha} \qquad z_{1-\frac{1}{2}\alpha}$$

(a)   (b)   (c)

Figure 13.1

critical region is

$$\hat{P} < P_0 - z_{1-\alpha}\sqrt{\frac{P_0(1-P_0)}{n}} \tag{13.5}$$

(see Fig. 13.1a). The region of acceptance is

$$\hat{P} \geqslant P_0 - z_{1-\alpha}\sqrt{\frac{P_0(1-P_0)}{n}}. \tag{13.6}$$

If the observed value $\hat{P}$ falls into the critical region, we reject $H_0: P = P_0$; if it falls into the acceptance region, we may accept H_0. However, if we have in mind another simple hypothesis, $H_1: P = P_1$, say, we have to evaluate the power of the test against this alternative. This problem will be discussed in sections 13.14 and 13.15.

(ii) Similarly, if the alternative is $H_0: P > P_0$ (right-hand, one-sided), the approximate critical region is

$$P > P_0 + z_{1-\alpha}\sqrt{\frac{P_0(1-P_0)}{n}}, \tag{13.7}$$

and the acceptance region is

$$P \leqslant P_0 + z_{1-\alpha}\sqrt{\frac{P_0(1-P_0)}{n}} \tag{13.8}$$

(see Fig. 13.1b).

It appears that these two one-sided tests are very nearly uniformly most powerful tests. However, if the alternative is two-sided, $H: P \neq P_0$, such nearly UMP test does exist. A good procedure, however, is to select P_L and P_U in such a way that

$$\Pr\{\hat{P} < P_L \mid H_0\} = \Pr\{\hat{P} > P_U \mid H_0\} \doteq \tfrac{1}{2}\alpha. \tag{13.9}$$

This leads to the approximate critical region

$$\hat{P} < P_0 - z_{1-\frac{1}{2}\alpha}\sqrt{\frac{P_0(1-P_0)}{n}} \quad\text{or}\quad \hat{P} > P_0 + z_{1-\frac{1}{2}\alpha}\sqrt{\frac{P_0(1-P_0)}{n}} \tag{13.10}$$

or, briefly,

$$|\hat{P} - P_0| > z_{1-\frac{1}{2}\alpha}\sqrt{\frac{P_0(1 - P_0)}{n}},$$ (13.10a)

and the acceptance region is

$$P_0 - z_{1-\frac{1}{2}\alpha}\sqrt{\frac{P_0(1 - P_0)}{n}} \leqslant \hat{P} \leqslant P_0 + z_{1-\frac{1}{2}\alpha}\sqrt{\frac{P_0(1 - P_0)}{n}}$$ (13.11)

or, briefly,

$$|\hat{P} - P_0| \leqslant z_{1-\frac{1}{2}\alpha}\sqrt{\frac{P_0(1 - P_0)}{n}},$$ (13.11a)

(see Fig. 13.1c).

Since a (discrete) binomial distribution has here been approximated by a (continuous) normal, it is sometimes useful to introduce a continuity correction, $\pm 1/2n$. This is subtracted from P_0 on the left-hand side, and added to P_0 on the right-hand side. For instance, a better approximation for determining the two-sided critical region seems to be

or

$$\left.\begin{array}{l} \hat{P} < \left(P_0 - \dfrac{1}{2n}\right) - z_{1-\frac{1}{2}\alpha}\sqrt{\dfrac{P_0(1 - P_0)}{n}} \\[4mm] \hat{P} > \left(P_0 + \dfrac{1}{2n}\right) + z_{1-\frac{1}{2}\alpha}\sqrt{\dfrac{P_0(1 - P_0)}{n}} \end{array}\right\}$$ (13.12)

or, equivalently,

$$\left(|\hat{P} - P_0| - \frac{1}{2n}\right) > z_{1-\frac{1}{2}\alpha}\sqrt{\frac{P_0(1 - P_0)}{n}}.$$ (13.12a)

When n is large, the continuity correction changes the results very little and can be neglected.

Let us apply these results to our data in Example 13.1. We wish to test $H_0: P = \frac{3}{4}$. First, suppose that we have no special alternative in mind. Thus our composite alternative will be $H: P \neq \frac{3}{4}$. Let us choose $\alpha = 0.05$. We construct a two-sided test, using normal approximation. We have

$$z_{1-\frac{1}{2}\alpha} = z_{0.975} = 1.96, \quad \sigma_{\hat{P}}^2 = \frac{P_0(1 - P_0)}{n} = \frac{0.75 \times 0.25}{200} = 0.00093750,$$

and so $\sigma_{\hat{P}} = 0.0306$. Thus

$$z_{1-\frac{1}{2}\alpha}\sqrt{\frac{P_0(1 - P_0)}{n}} = 1.96 \times 0.0306 = 0.0600.$$

Using the continuity correction $\pm 1/400 = 0.0025$, we obtain the approximate critical points:

$$P_L = (0.75 - 0.0025) - 0.0600 = 0.6875,$$

$$P_U = (0.75 + 0.0025) + 0.0600 = 0.8125.$$

The observed proportion $P = 0.77$ is between these limits, so we do not reject H_0: $P = \frac{3}{4}$. Without the continuity correction, the critical points are 0.69 and 0.81—our decision (not to reject H_0) has not been changed. Should we, then, accept H_0?

First, it may be useful to find the *least* significant level, α_0, say, at which H_0 will be rejected, that is, to evaluate the probability

$$\Pr\{|Z| > \overset{\circ}{z}\} = \alpha_0, \tag{13.13}$$

where $\overset{\circ}{z} = (|\hat{P} - P_0|)/\sqrt{P_0(1 - P_0/n}$ [or, using the continuity correction, $\overset{\circ}{z} = [|\hat{P} - P_0| - (1/2n)]/\sqrt{P_0(1 - P_0)/n}$ is the value of z evaluated for the observed $\hat{P}$. In our example, $\overset{\circ}{z} = (0.77 - 0.75)/0.0306 = 0.65$. Thus $\Pr\{|Z| > 0.65\} = 0.5156 \doteq 0.52$. This is well above the conventional level, $\alpha = 0.05$, or even $\alpha = 0.10$, and is often regarded as favorable evidence for the validity of H_0. We may then provisionally accept H_0: $P = \frac{3}{4}$ (i.e., the hypothesis that the flower color is inherited as a simple dominant), but with the reservation that we are prepared to change our decision if we later obtain some evidence against H_0. For instance, using the same procedure, it is easy to show that the hypothesis H_1: $P = \frac{13}{16}$ will not be rejected either. Discrimination between these two hypotheses requires a specially designed experiment. Some formal statistical methods for dealing with this problem will be discussed in sections 13.14 and 13.16.

It may be of some interest to mention that, for testing H_0: $P = P_0$ against a two-sided alternative, H: $P \neq P_0$, another equivalent test statistic can be used. We note that $Z = (\hat{P} - P_0)/\sigma_{\hat{P}}$ is approximately distributed as $N(0, 1)$. Thus

$$Z^2 = \frac{(\hat{P} - P_0)^2}{\sigma_{\hat{P}}^2} \tag{13.14}$$

is approximately distributed as χ^2 with 1 degree of freedom. If α is the significance level, an approximate critical region can be evaluated from the condition

$$\Pr\{Z^2 > \chi^2_{1-\alpha;1}\} = \Pr\left\{\frac{(\hat{P} - P_0)^2}{\sigma_{\hat{P}}} > \chi^2_{1-\alpha;1}\right\} \doteq \alpha. \tag{13.15}$$

This means that, if

$$Z^2 = \frac{(P - P_0)^2}{\sigma_{\hat{P}}^2} = \frac{(\hat{P} - P_0)^2 n}{P_0(1 - P_0)} > \chi_{1-\alpha;1}^2, \qquad (13.16)$$

we reject H_0; otherwise we do not.

In our example, $Z^2 = (0.77 - 0.75)^2/0.00093750 = 0.4267$. Taking $\alpha = 0.05$, we find $\chi_{0.95;1}^2 = 3.84$. Since $0.4267 < 3.84$, we do not reject H_0.

The Z^2 can be improved by introducing the continuity correction, that is, by using Z_{Cor}^2 ("corrected"):

$$Z_{\text{Cor}}^2 = \frac{\left(|\hat{P} - P_0| - \dfrac{1}{2n}\right)^2}{\sigma_{\hat{P}}^2} \qquad (13.14a)$$

instead of Z^2.

The test based on Z^2 and defined in (13.14) is commonly called a "χ^2-test," since it utilizes the approximate χ^2-distribution. It should be emphasized that the Z^2-test is equivalent to the Z-test defined in (13.10) for the *two-sided* alternative. If the alternative is one-sided, the Z^2-test cannot be used.

When n is *small* ($n < 10$, say) and P_0 differs substantially from $\frac{1}{2}$ (is less than 0.01 or greater than 0.99, say), the normal approximation is rather poor and a test based on the *exact* binomial distribution should be used. [For details of this test see Brownlee (1960), section 3.6, pp. 117–118.]

13.4. CONSTRUCTION OF A SIGNIFICANCE TEST FOR A PARAMETRIC HYPOTHESIS

We now summarize, in general terms, the important steps in constructing a significance test for testing a *parametric* hypothesis, that is, a hypothesis about a parameter, θ, say, of a certain distribution.

(i) Formulate assumptions, the null hypothesis $H_0: \theta = \theta_0$, and the alternative H (e.g., $H: \theta > \theta_0$ or $H: \theta \neq \theta_0$). Make a clear distinction between assumptions and hypotheses. Remember that the hypotheses should be stated *before* the sample is obtained.

(ii) Take a sample and choose a statistic, $\hat{\theta}$, say, with known distribution $p(\hat{\theta}; \theta)$ as the test criterion. If $p(\hat{\theta}; \theta)$ is unknown and the sample is large, it is likely that normal approximation can be used. This requires a knowledge of only the (possibly approximate) expected value and variance of $\hat{\theta}$ as a function of θ. If σ_θ^2 is the variance of $\hat{\theta}$, when $\theta = \theta_0$, then under the null hypothesis, $H_0: \theta = \theta_0$, the statistic $Z = (\hat{\theta} - \theta_0)/\sigma_\theta$ is approximately distributed as $N(0, 1)$.

(iii) Choose the significance level α and define a critical region $w = w(\alpha)$ by the condition

$$\Pr\{\hat\theta \in w \mid H_0\} = \alpha. \tag{13.17}$$

[We read (13.17) as follows: "The probability that $\hat\theta$ belongs to the region w, given that H_0 is true, is equal to α."] For the one-sided alternative (e.g., $H: \theta > \theta_0$) the critical region is also one-sided [e.g., for the right-hand alternative $\Pr\{\hat\theta - \theta > d(\alpha)\} \doteq \alpha$], and for the two-sided alternative, $H: \theta \neq \theta_0$, the critical region is also two-sided, $\Pr\{|\hat\theta - \theta_0| > d(\tfrac{1}{2}\alpha)\} \doteq \alpha$. In particular, for a test based on normal approximation, the right-hand, one-sided critical region is defined by $\Pr\{\hat\theta - \theta > z_{1-\alpha}\sigma_\theta\} \doteq \alpha$, and the two-sided critical region by $\Pr\{|\hat\theta - \theta_0| > z_{1-\frac{1}{2}\alpha}\sigma_\theta\} \doteq \alpha$.

(iv) Make a decision. If the observed value of $\hat\theta$ falls into the critical region, w, reject H_0. The statement "The difference between the hypothetical value θ_0 and the estimate $\hat\theta$ is *significant*" is often used instead of saying, "We reject H_0." If $\hat\theta$ falls into the acceptance region, $\bar{w}$, we say, "The difference between $\hat\theta$ and θ_0 is *nonsignificant*." If sufficient additional (usually not statistical) information supporting H_0 exists, we may "provisionally" accept H_0. On the other hand, if we have some specific simple alternative(s) in mind, the power of the test should be evaluated (see section 13.13).

(v) Take advantage of the fact that, for testing the same hypothesis H_0, different test statistics can be used. Select (if possible) the test statistic which gives the most powerful test. In our example, the sample proportion, $\hat{P}$, is the "best" test statistic.

(vi) Apply the above arguments also in testing a *joint* hypothesis about s parameters, $\theta_1, \theta_2, \ldots, \theta_s$. Replacing θ by an $s \times 1$ vector $\boldsymbol{\theta}$, and using the *joint* distribution $p(\hat{\boldsymbol{\theta}}; \boldsymbol{\theta})$, we can evaluate the critical region by following the same steps as for a single parameter, θ. When $p(\hat{\boldsymbol{\theta}}; \boldsymbol{\theta})$ is unknown, we can often use a multivariate normal approximation. Some details on this matter will be given in section 13.6.

13.5. CHI-SQUARE TESTS OF SIGNIFICANCE USEFUL IN GENETICS. GENERAL REMARKS

Most of the rest of this chapter will be devoted to applications of approximate χ^2-distributions in testing hypotheses which arise frequently in genetic problems. Especially useful is the X^2-statistic defined for multinomial variates (see section 8.19). All of the formulae which we will derive are only approximately valid, and we assume that the sample is "sufficiently large." In practice, this means that, for instance, for multinomial distributions the *expected* number of observations in each class, nP_i, say, should be greater than 3. In sections 13.6–13.13 we shall assume that the conditions for "large-sample theory" are satisfied, and we shall present methods for constructing the four main types of significance tests employing the approximate χ^2-distribution:

(a) Tests about specified values of parameters under consideration.
(b) Tests of goodness of fit.
(c) Tests of independence (more often called tests for association).
(d) Tests of homogeneity (more often called tests for heterogeneity).

We shall give a few examples to demonstrate the techniques and the interpretation of the results.

13.6. APPROXIMATE SIGNIFICANCE TESTS ABOUT PARAMETERS. LARGE-SAMPLE THEORY

13.6.1. A Single Parameter. Testing $H_0: \theta = \theta_0$

In the general case, when testing $H_0: \theta = \theta_0$, it may be difficult to find an appropriate statistic, with known distribution, which can be used as test criterion. For large samples, however, it is often convenient to use a BAN estimator of θ, $\check{\theta}$, say, since it has the property of asymptotic normality. As a test criterion, then, we can use

$$Z = \frac{\check{\theta} - \theta_0}{\sqrt{\mathrm{Var}(\check{\theta})}}, \qquad (13.18)$$

which is approximately normally distributed as $N(0, 1)$. The equivalent statistic for a two-sided test is

$$Z^2 = \frac{(\check{\theta} - \theta_0)^2}{\mathrm{Var}(\check{\theta})}. \qquad (13.19)$$

When $\theta = \theta_0$, this is approximately distributed as χ^2 with 1 d.f. Application of this result was demonstrated in section 13.3 for the particular case of binomial proportion (to data given in Example 13.1). Especially convenient is the situation in which the maximum likelihood estimator, $\hat{\theta}$, say, is used. Let $U(\theta_0)$ and $I(\theta_0)$ be the score and the amount of information evaluated at $\theta = \theta_0$, respectively. Expanding $U(\hat{\theta})$ into a Taylor series up to the second term and taking into account that $U(\hat{\theta}) = 0$, we obtain

$$\hat{\theta} \doteq \theta_0 + \frac{U(\theta_0)}{I(\theta_0)} \qquad (13.20)$$

or

$$\hat{\theta} - \theta_0 \doteq \frac{U(\theta_0)}{I(\theta_0)}. \qquad (13.20a)$$

If H_0 is true, we also have

$$\text{Var}(\hat{\theta}) \doteq \frac{1}{I(\theta_0)}. \qquad (13.21)$$

Thus for the ML estimator, statistic (13.18) takes the form

$$Z = \frac{\hat{\theta} - \theta_0}{\sqrt{\text{Var}(\hat{\theta})}} = (\hat{\theta} - \theta_0)\sqrt{I(\theta_0)} = \frac{U(\theta_0)}{\sqrt{I(\theta_0)}}, \qquad (13.22)$$

and the statistic Z^2 takes the form

$$Z^2 = \frac{(\hat{\theta} - \theta_0)^2}{\text{Var}(\hat{\theta})} = (\hat{\theta} - \theta_0)^2 I(\theta_0) = \frac{[U(\theta_0)]^2}{I(\theta_0)}. \qquad (13.23)$$

▼ 13.6.2.   **Several Parameters. Testing the Joint Hypothesis** $H_0\colon \boldsymbol{\theta} = \boldsymbol{\theta}_0$

Suppose that s *functionally independent* parameters, $\boldsymbol{\theta}' = (\theta_1, \theta_2, \ldots, \theta_s)$, are involved in a certain genetic problem. We may be interested in testing a hypothesis about a single parameter, $H_0\colon \theta_i = \theta_{i0}$, say. In such cases, the results of section 13.6.1 can be applied to each individual parameter.

More often we are interested in testing the *joint* hypothesis that the parameters $\boldsymbol{\theta}' = (\theta_1, \theta_2, \ldots, \theta_s)$ take, respectively, a set of values $\boldsymbol{\theta}_0' = (\theta_{10}, \theta_{20}, \ldots, \theta_{s0})$, *simultaneously*, against the alternative that at least one of the parameters does not take the specified value. This can be written in an abbreviated form, using matrix notation, as $H_0\colon \boldsymbol{\theta} = \boldsymbol{\theta}_0$ and $H\colon \boldsymbol{\theta} \neq \boldsymbol{\theta}_0$. (Notice that we are concerned here only with two-sided alternatives.) Let us assume that all the s parameters are estimable from a given experimental design.

In the general case of testing the joint hypothesis, the joint distribution of the estimators should be known. Very often, however, such a distribution is not available, and then for BAN estimators the multivariate normal approximation can be used. The details of construction of the test are beyond the scope of this book, but the reader familiar with matrix algebra may find useful the formulae given below.

Let $\boldsymbol{\check{\theta}}' = (\check{\theta}_1, \check{\theta}_2, \ldots, \check{\theta}_s)$ be BAN estimators of $\boldsymbol{\theta}' = (\theta_1, \theta_2, \ldots, \theta_s)$, respectively, and $\mathbf{V} = \mathbf{V}(\boldsymbol{\check{\theta}})$ be the variance-covariance matrix of the estimators. If H_0 is true, the quadratic form

$$Z^2 = (\boldsymbol{\check{\theta}} - \boldsymbol{\theta}_0)'\mathbf{V}^{-1}(\boldsymbol{\check{\theta}} - \boldsymbol{\theta}_0) \qquad (13.24)$$

is approximately distributed as χ^2 with s d.f.

In particular, let $\hat{\boldsymbol{\theta}}' = (\hat{\theta}_1, \hat{\theta}_2, \ldots, \hat{\theta}_s)$ be the vector of the ML estimators and $\mathbf{U}'(\boldsymbol{\theta}_0) = [U_1(\boldsymbol{\theta}_0), U_2(\boldsymbol{\theta}_0), \ldots, U_s(\boldsymbol{\theta}_0)]$ be the vector of scores for θ_1, $\theta_2, \ldots, \theta_s$, respectively, evaluated at $\boldsymbol{\theta} = \boldsymbol{\theta}_0$. We also have $\mathbf{V}(\hat{\boldsymbol{\theta}}) \doteq \mathbf{I}^{-1}(\boldsymbol{\theta}_0)$. It can be shown that statistic (13.24) now takes the form

$$Z^2 = (\hat{\boldsymbol{\theta}} - \boldsymbol{\theta}_0)'\mathbf{I}(\boldsymbol{\theta}_0)(\hat{\boldsymbol{\theta}} - \boldsymbol{\theta}_0) = \mathbf{U}'(\boldsymbol{\theta}_0)\mathbf{I}^{-1}(\boldsymbol{\theta}_0)\mathbf{U}(\boldsymbol{\theta}_0) = \mathbf{U}'(\boldsymbol{\theta}_0)\mathbf{V}(\hat{\boldsymbol{\theta}})\mathbf{U}(\boldsymbol{\theta}_0).$$

$$(13.25)$$

If $Z^2 > \chi^2_{1-\alpha;s}$, we reject H_0; otherwise we do not. Some practical applications using (13.25) in testing joint hypotheses about parameters will be discussed briefly in section 13.11 and broadly in Chapter 18.

In sections 13.7–13.13 we will restrict ourselves to ML estimators. All results derived in these sections, however, are approximately valid also for BAN estimators. [Note in section 13.11 the expected information matrices, $\mathbf{I}_g(\boldsymbol{\theta}_0)$ and $\mathbf{I}(\boldsymbol{\theta}_0)$, have to be replaced by $\mathbf{V}_g^{-1}(\check{\boldsymbol{\theta}})$ and $\mathbf{V}^{-1}(\check{\boldsymbol{\theta}})$, respectively, if the BAN estimators are used.]

13.7. TESTS OF GOODNESS OF FIT

Statistical procedures for testing a hypothesis that a population from which a sample was obtained has a certain form of distribution are called *tests of goodness of fit*.

Let $F(\mathbf{x})$ denote the cumulative distribution of k random variables, $\mathbf{x}' = (x_1, x_2, \ldots, x_k)$. (*a*) We may wish to test a hypothesis that $F(\mathbf{x})$ takes a specified form, $F_0(\mathbf{x})$, say. In this case we are interested in testing a *simple hypothesis*, H_0, about the distribution. For instance, the statement that the height of people in a certain population has a normal distribution with $\mu = 69$ inches and $\sigma = 2$ inches is an example of such a hypothesis for $k = 1$. (*b*) We may be interested only in the *form* of the distribution; for instance, that the height is normally distributed. In such cases we have to estimate μ and σ and construct a test of goodness of fit for this *composite hypothesis*. We shall discuss these two cases separately.

Many tests of goodness of fit have been constructed, using different statistical methods. We restrict ourselves to the most common, based on the X^2-statistic introduced in section 8.19 and sometimes called "Pearson's χ^2-test." To apply this test, the observations have to be arranged into k mutually exclusive and exhaustive classes. Let P_l ($l = 1, 2, \ldots, k$) be the expected proportion associated with the lth class, with $\sum P_l = 1$. Let n be the sample size and r_l ($l = 1, 2, \ldots, k$) the number of cases observed in the lth class, with $\sum r_l = n$. The quantities $r_1, r_2, \ldots, r_k$, considered as random variables, are multinomially distributed with parameters $P_1, P_2, \ldots, P_k$, respectively (see section 8.11). We now construct tests for goodness of fit for simple and composite hypotheses, respectively.

13.7.1.　Test of Goodness of Fit When Multinomial Proportions Are Specified. Simple H_0

We shall distinguish two slightly different situations.

(i) Suppose that the parameters P_l are specified by H_0. For instance, H_0 might specify the phenotypic ratios of a certain trait as 9:3:3:1 (i.e., $P_1 = \frac{9}{16}, P_2 = \frac{3}{16}, P_3 = \frac{3}{16}, P_4 = \frac{1}{16}$). In general, suppose that we wish to test the hypothesis $H_0: P_l = P_{l0}$, $l = 1, 2, \ldots, k$, against the alternative that at least one P_l is not equal to its specified value P_{l0}.

If H_0 is true, the statistic

$$X^2 = \sum_{l=1}^{k} \frac{(r_l - nP_{l0})^2}{nP_{l0}} = n \sum_{l=1}^{k} \frac{(\hat{P}_l - P_{l0})^2}{P_{l0}} = \frac{1}{n} \sum_{l=1}^{k} \frac{r_l^2}{P_{l0}} - n, \quad (13.26)$$

where $\hat{P}_l = r_l/n$, is approximately distributed as χ^2 with $k - 1$ d.f. [To obtain the right-hand side in (13.26) see formula (12.73).]

The X^2 defined in (13.26) can be used for testing H_0. For given α, if $X^2 > \chi^2_{1-\alpha;k-1}$ we reject H_0; if $X^2 \leqslant \chi^2_{1-\alpha;k-1}$ we do not reject H_0 and we conclude that our data are consistent with this hypothesis.

Example 13.2.　In Example 11.4 we assumed that segregation ratios at the *Ag-B* locus in rats differ from the Mendelian ratios 1:2:1 in intercrosses $B^1B^4 \times B^1B^4$. Let us see whether there is statistical justification for this statement. Our null hypothesis will be that the mode of inheritance is the one leading to simple Mendelian ratios 1:2:1. In other words, we wish to fit the data by a multinomial distribution with parameters $P_1 = 0.25$, $P_2 = 0.50$, $P_3 = 0.25$. Using the $n = 200$ observations of Example 11.4, we have

Genotype:	B^1B^1	B^1B^4	B^4B^4
Observed (r_l):	58	129	13
Expected (nP_{l0}):	50	100	50 .

We calculate

$$X^2 = \frac{(58 - 50)^2}{50} + \frac{(129 - 100)^2}{100} + \frac{(13 - 50)^2}{50}$$

$$= 1.28 + 8.41 + 27.38 = 37.07.$$

Comparing this value with the value of χ^2 for $\nu = 3 - 1 = 2$ d.f., and $\alpha = 0.001$ we find that it is much greater than $\chi^2_{0.999;2} = 13.82$. Therefore we definitely reject the hypothesis that the data fit the simple Mendelian ratios $1:2:1$.

(ii) The multinomial proportions may be *functions* of $s(< k - 1)$ functionally independent parameters, $\boldsymbol{\theta}' = (\theta_1, \theta_2, \ldots, \theta_s)$, and these can be specified by $H_0: \theta_i = \theta_{i0}$, $i = 1, 2, \ldots, s$ (briefly, $H_0: \boldsymbol{\theta} = \boldsymbol{\theta}_0$). Calculating $P_l(\boldsymbol{\theta}_0)$ and substituting into (13.26), we obtain

$$X^2 = \sum_{l=1}^{k} \frac{[r_l - nP_l(\boldsymbol{\theta}_0)]^2}{nP_l(\boldsymbol{\theta}_0)} = n \sum_{l=1}^{k} \frac{[\hat{P}_l - P_l(\boldsymbol{\theta}_0)]^2}{P_l(\boldsymbol{\theta}_0)}, \qquad (13.27)$$

which is just the same as (13.26).

In the statistical literature, the observed number in the lth class, r_l, is often denoted by O_l, and the expected number, nP_{l0}, when H_0 is true, by E_l, so that (13.26) and (13.27) can each be written as

$$X^2 = \sum_{l=1}^{k} \frac{(O_l - E_l)^2}{E_l}. \qquad (13.28)$$

13.7.2. Test of Goodness of Fit When Multinomial Proporitons Depend on Unknown Parameters. Composite H_0

Sometimes the multinomial parameters are functions of some s functionally independent parameters, $\boldsymbol{\theta}' = (\theta_1, \theta_2, \ldots, \theta_s)$, which cannot be specified by H_0. These can be, for instance, gene frequencies or parameters which modify Mendelian ratios, such as recombination fraction in linkage, different fitnesses of genotypes, or incomplete penetrance. These parameters can only be estimated.

The null hypothesis which we now wish to test is that the mode of inheritance is defined by special functions of genetic parameters, that is, $H_0: P_l = P_l(\boldsymbol{\theta})$, $l = 1, 2, \ldots, k$, with θ's unspecified. It can be proved that in testing this hypothesis the X^2 given in (13.27), with $\boldsymbol{\theta}_0$ replaced by the vector of the ML estimates $\hat{\boldsymbol{\theta}}' = (\hat{\theta}_1, \hat{\theta}_2, \ldots, \hat{\theta}_s)$, can be used. But the number of degrees of freedom will be reduced by the number of estimated parameters. Thus, if H_0 is true, the statistic

$$X^2 = \sum_{l=1}^{k} \frac{[r_l - nP_l(\hat{\boldsymbol{\theta}})]^2}{nP_l(\hat{\boldsymbol{\theta}})} = n \sum_{l=1}^{k} \frac{[\hat{P}_l - P_l(\hat{\boldsymbol{\theta}})]^2}{P_l(\hat{\boldsymbol{\theta}})} \qquad (13.29)$$

is approximately distributed as χ^2 with $\nu = k - 1 - s$ d.f. If only c parameters are estimated, $\nu = k - 1 - c$.

Formula (13.29) is also often written in the form (13.28). Here $nP_l(\hat{\boldsymbol{\theta}})$ is also called an "expected" frequency; the quotation marks emphasize that, in fact, this is not the true expected value but rather its ML estimate evaluated under the null hypothesis.

Example 13.3. In Example 11.4 we made the assumption that the Mendelian ratios were disturbed because of the selective disadvantage of homozygotes B^4B^4. From that example we find that the theoretical distribution for this mode of inheritance should be

$$B^1B^1 \qquad\qquad B^1B^4 \qquad\qquad B^4B^4$$

$$P_1(\phi) = \frac{1}{3+\phi} \qquad P_2(\phi) = \frac{2}{3+\phi} \qquad P_3(\phi) = \frac{\phi}{3+\phi}, \qquad (i)$$

where ϕ $(0 < \phi < 1)$ is the relative viability of B^4B^4 with respect to B^1B^1 or B^1B^4. The probabilities given in (i) play the role of multinomial parameters $P_l(\phi)$, $l = 1, 2, 3$. We now wish to test the hypothesis that the mode of gene action is such that the segregation ratios in F_2 follow the model given in (i). Substituting for ϕ its ML estimate, $\hat{\phi} = 0.2086$ (see Example 11.4), we obtain $P_1(\hat{\phi}) = 0.3117$, $P_2(\hat{\phi}) = 0.6233$, $P_3(\hat{\phi}) = 0.0650$. The "expected" numbers of genotypes under H_0 are, then, $nP_1(\hat{\phi}) = 200 \times 0.3117 = 62.34$, $nP_2(\hat{\phi}) = 124.66$, $nP_3(\hat{\phi}) = 13.00$. The corresponding observed values are 58, 129, and 13. Thus, the calculated X^2-statistic is

$$X^2 = \frac{(58 - 62.34)^2}{62.34} + \frac{(129 - 124.66)^2}{124.66} + \frac{(13 - 13.00)^2}{13}$$

$$= 0.3021 + 0.1511 = 0.4532.$$

This value must be compared with the value of $\chi^2_{1-\alpha;\nu}$ with $\nu = 3 - 1 - 1 = 1$ d.f. If we choose $\alpha = 0.05$, we find $\chi^2_{0.95} = 3.84$, which is much greater than 0.4532. The nearest tabulated χ^2 is $\chi^2_{0.50} = 0.4549$. This means that even at the significance level $\alpha = 0.50$ we would not reject H_0 — a fact that, to some extent, convinces us that our hypothesis may be true. But, of course, these data might fit other hypotheses as well.

13.8. TESTS OF INDEPENDENCE (OR TESTS FOR ASSOCIATION)

Suppose that observations on individuals or results of some experiments are classified with respect to two different types of categories or characters or responses as in Table 13.1. For instance, classification C (rows) may correspond to k different phenotypes at the multiallelic locus A, and classification S to m different phenotypes at locus B. The symbol r_{lg} represents the number of individuals belonging to classes C_l and S_g simultaneously. The quantities r_{lg} (i.e., the frequencies r_{lg}) are sometimes called *categorical data*, and any classification of categorical data into a two (or more) way table as

TABLE 13.1   A $k \times m$ contingency table

C \\ S	1 (S_1)	2 (S_2)	$\cdots$	g (S_g)	$\cdots$	m (S_m)	Total
1 (C_1)	r_{11}	r_{12}	$\cdots$	r_{1g}	$\cdots$	r_{1m}	$n_{1.}$ (R_1)
2 (C_2)	r_{21}	r_{22}	$\cdots$	r_{2g}	$\cdots$	r_{2m}	$n_{2.}$ (R_2)
$\cdots$	$\cdots$	$\cdots$	$\cdots$	$\cdots$	$\cdots$	$\cdots$	$\cdots$
l (C_l)	r_{l1}	r_{l2}	$\cdots$	r_{lg}	$\cdots$	r_{lm}	$r_{l.}$ (R_l)
$\cdots$	$\cdots$	$\cdots$	$\cdots$	$\cdots$	$\cdots$	$\cdots$	$\cdots$
k (C_k)	r_{k1}	r_{k2}	$\cdots$	r_{kg}	$\cdots$	r_{km}	r_k (R_k)
Total	$n_{.1}$ (n_1)	$n_{.2}$ (n_2)	$\cdots$	$n_{.g}$ (n_g)	$\cdots$	$n_{.m}$ (n_m)	N

in Table 13.1 is called a $k \times m$ *contingency table.* We have

$$\sum_{l=1}^{k} r_{lg} = n_{.g}, \quad \sum_{g=1}^{m} r_{lg} = n_{l.}, \quad \sum_{l=1}^{k} \sum_{g=1}^{m} r_{lg} = N. \tag{13.30}$$

A $k \times m$ contingency table such as Table 13.1 represents observed frequencies, r_{lg}, from a total of N observations. We can associate with it a joint distribution such that $\Pr\{C_l S_g\} = P_{lg}$ ($l = 1, 2, \ldots, k; g = 1, 2, \ldots, m$) and marginal distributions with probabilities $\Pr\{C_l\} = P_{l.}$ ($l = 1, 2, \ldots, k$) for classification C and $\Pr\{S_g\} = P_{.g}$ ($g = 1, 2, \ldots, m$) for classification S. We have

$$\sum_{l=1}^{k} \sum_{g=1}^{m} P_{lg} = \sum_{l=1}^{k} P_{l.} = \sum_{g=1}^{m} P_{.g} = 1. \tag{13.31}$$

These probabilities can be arranged in a table which resembles Table 2.2. (We are using here capital letters, P_{lg}, instead of p_{lg}, to be consistent with other notations in this chapter.)

If the two classifications (or characters) are independent, then

$$P_{lg} = P_{l.} \times P_{.g}, \qquad l = 1, 2, \ldots, k; g = 1, 2, \ldots, m. \tag{13.32}$$

We now wish to construct a test of the hypothesis $H_0: P_{lg} = P_{l.} \times P_{.g}$ ($l = 1, 2, \ldots, k; g = 1, 2, \ldots, m$). Such a test will be called a *test of independence;* however, since we are usually looking for some kind of relationship, it will also be called a *test for association.* We will consider two situations.

13.8.1.   The Marginal Distributions are Specified

Suppose that we specify the marginal distributions, that is, $P_{l.} = P_{l.0}$ ($l = 1, 2, \ldots, k$) and $P_{.g} = P_{.g0}$ ($g = 1, 2, \ldots, m$). We wish to test the hypothesis that the C and S classifications are independent; that is, $H_0: P_{lg} = P_{l.0} \times P_{.g0}$ $(l = 1, 2, \ldots, k; g = 1, 2, \ldots, m)$ against $H: P_{lg} \neq P_{l.0} \times P_{.g0}$ for at least one (l, g). The observed frequencies, r_{lg} (denoted also by O_{lg}), are variates having a multinomial joint distribution with km classes and, if H_0 is true, with expected values $E(r_{lg}) = E_{lg} = NP_{l.0}P_{.g0}$. Thus the statistic

$$X^2 = \sum_{l=1}^{k} \sum_{g=1}^{m} \frac{(r_{lg} - NP_{l.0}P_{.g0})^2}{NP_{l.0}P_{.g0}} = \sum_{l=1}^{k} \sum_{g=1}^{m} \frac{(O_{lg} - E_{lg})^2}{E_{lg}} \tag{13.33}$$

is approximately distributed as χ^2 with $mk - 1$ d.f. when H_0 is true. If $X^2 > \chi^2_{1-\alpha; mk-1}$, then H_0 is rejected. This can be taken as evidence that the two characters are not independent, but it might also mean that our "specification" of the marginal distributions was not correct, or both.

13.8.2. The Marginal Distributions Are Unspecified

Very often the marginal distributions are *unknown*. It can be shown (see Problem 13.9) that the ML estimators of $P_{l.}$, $P_{.g}$, and P_{lg} (if H_0 is true) are, respectively,

$$\hat{P}_{l.} = \frac{n_{l.}}{N}; \quad \hat{P}_{.g} = \frac{n_{.g}}{N}; \quad \hat{P}_{lg} = \frac{n_{l.}}{N} \cdot \frac{n_{.g}}{N}, \tag{13.34}$$

for $l = 1, 2, \ldots, k; g = 1, ,2 \ldots, m$.

Now the analogue of the X^2 given in (13.33) is

$$X^2 = \sum_{g=1}^{m} \sum_{l=1}^{k} \frac{[r_{lg} - (n_{l.}n_{.g}/N)]^2}{n_{l.}n_{.g}/N} = \sum_{g=1}^{m} n_{.g} \sum_{l=1}^{k} \frac{[(r_{lg}/n_{.g}) - (n_{l.}/N)]^2}{n_{l.}/N}. \tag{13.35}$$

This is approximately distributed as χ^2 with $(mk - 1) - (k - 1) - (m - 1) = (k - 1)(m - 1)$ d.f. Formula (13.34) is also often written as

$$X^2 = \sum_{g=1}^{m} \sum_{l=1}^{k} \frac{(O_{lg} - E_{lg})^2}{E_{lg}},$$

but here the "expected" values, E_{lg}, are the ML estimators under H_0, that is, $E_{lg} = n_{l.} \times n_{.g}/N$.

A special kind of analysis, utilizing the X^2-statistic for association, will be applied in Chapter 15 to detect linkage in experimental populations.

13.9. TEST FOR ASSOCIATION AND THE SAMPLE CORRELATION COEFFICIENT FOR A 2×2 CONTINGENCY TABLE

Especially useful in many applications is the X^2-test for association with $k = m = 2$. By using straightforward algebraic manipulations, it can be

TABLE 13.2   A 2×2 contingency table

x \ y	1 (+)	0 (−)	Total
1 (+)	a	b	$a + b$
0 (−)	c	d	$c + d$
Total	$a + c$	$b + d$	N

shown that the X^2, with 1 d.f., is

$$X^2 = N \sum_{g=1}^{2} \sum_{l=1}^{2} \frac{(r_{lg} - n_{l.}n_{.g}/N)^2}{n_{l.}n_{.g}} = \frac{N(r_{11}r_{22} - r_{12}r_{21})^2}{n_{1.}n_{2.}n_{.1}n_{.2}}. \qquad (13.36)$$

For a 2×2 contingency table, it is customary to use the notation shown in Table 13.2. In this notation, $r_{11} = a, r_{12} = b, r_{21} = c, r_{22} = d, n_{1.} = a + b,$ $n_{2.} = c + d, n_{.1} = a + c, n_{.2} = b + d, N = a + b + c + d$. Thus formula (13.36) now takes the form

$$X^2 = \frac{N(ad - bc)^2}{(a + b)(c + d)(a + c)(b + d)}. \qquad (13.37)$$

It has been suggested that a better approximation can be obtained by using a continuity correction, $\pm \frac{1}{2}N$, called *Yates's correction*. The corrected X^2 (denoted here by X^2_{Cor}) is

$$X^2_{\text{Cor}} = \frac{N(|ad - bc| - \frac{1}{2}N)^2}{(a + b)(c + d)(a + c)(b + d)}. \qquad (13.37a)$$

Recent investigations [e.g., Grizzle (1967)] have shown that, if we wish to evaluate $\Pr\{X^2 \leq \chi^2_{1-\alpha}\} \doteq 1 - \alpha$, the corrected X^2 gives a closer approximation of $1 - \alpha$. However, in testing hypotheses, with given α, the uncorrected X^2 is more powerful.

A 2×2 contingency table may be considered as a finite two-dimensional sample space, in which the sample proportions are estimators of the true proportions in each cell. Let x be a random variable taking the value 1 if the response is observed (sometimes denoted by +) and the value 0 otherwise (sometimes denoted by −), and similarly for y. Then, applying the theory developed in Chapter 6, we can evaluate the estimators of means, variances, and covariance for x and y. We have

$$\hat{\mu}_x = 1 \times \frac{a + b}{N} + 0 \times \frac{c + d}{N} = \frac{a + b}{N};$$

$$\hat{\sigma}_x^2 = 1^2 \times \frac{a + b}{N} - \frac{(a + b)^2}{N^2} = \frac{a + b}{N}\left(1 - \frac{a + b}{N}\right) = \frac{(a + b)(c + d)}{N^2}.$$

Similarly,

$$\hat{\mu}_y = \frac{a+c}{N} \quad \text{and} \quad \hat{\sigma}_y^2 = \frac{(a+c)(b+d)}{N^2}.$$

An estimator of covariance is

$$\hat{\sigma}_{xy} = 1 \times 1 \times \frac{a}{N} = \frac{a+b}{N} \cdot \frac{a+c}{N} = \frac{Na - (a+b)(a+c)}{N^2}$$

$$= \frac{(a+b+c+d)a + (a+b)(a+c)}{N^2} = \frac{ad - bc}{N^2}.$$

The square of the sample correlation coefficient, $\hat{\rho}^2$, is

$$\hat{\rho}^2 = \frac{\hat{\sigma}_{xy}^2}{\hat{\sigma}_x^2 \hat{\sigma}_y^2} = \frac{(ad - bc)^2}{(a+b)(c+d)(a+c)(b+d)} = \frac{X^2}{N}. \tag{13.38}$$

Hence,

$$\hat{\rho} = \pm\sqrt{\frac{X^2}{N}}. \tag{13.39}$$

This is a useful result.

The sample correlation coefficient for a 2 × 2 contingency table is the square root (with the sign + if ad > bc, and − if ad < bc) of the X^2 divided by the sample size, N.

Thus X^2 given in (13.37) also tests the hypothesis that x and y are un-correlated. In fact, X^2 is used to detect the association; $\hat{\rho}$ measures, in some sense, the "strength" of association.

Example 13.4. The data below [Bodmer and Payne (1965)] represent the results of testing on the leucocyte antigens of 107 unrelated individuals, by the agglutination method. Two sera, called here 4a and 4c, were used.

4c \ 4a	+	−	Total
+	23	2	25
−	47	35	82
Total	70	37	107

In this table, + denotes that the cells react with a serum, that is, that agglutination occurs, while − indicates no agglutination. We obtain for the data in our example $X^2 = 10.19$, $X_{\text{Cor}}^2 = 8.71$. Both are highly significant ($\alpha \leqslant 0.005$). Thus we may conclude that there is a rather strong association. The correlation coefficient is $\hat{\rho} = \sqrt{10.19/107} = 0.3086$.

We notice that the combination $(4c+, 4a-)$ is very rare, whereas $(4c-, 4a+)$ is rather frequent. This suggests that $4a$ and $4c$ may be alleles at the same locus, but also that serum $4c$ includes practically all determinants which may be present in serum $4a$, so $4a$ may be dominant over $4c$. In fact, in this case the theoretical frequency in the cell $(4c+, 4a-)$ is zero.

13.10. TESTS OF HOMOGENEITY (OR FOR HETEROGENEITY). GENERAL CONCEPTS

Suppose that a population is described by classifying the individuals in it into k classes or categories. For instance, for a particular mode of inheritance we may classify individuals with respect to their phenotypes. In a similar way we may classify the results of a certain type of experiment into k categories or responses. Let us assume that such a classification is exhaustive, that is, includes all possible classes relevant to a problem under consideration. Let us call this "classification C" and denote the classes by $C_1, C_2, \ldots, C_k$. In the general population, some probabilities may be associated with these classes, such that P_l is the probability that an individual belongs to class C_l (with $\sum P_l = 1$), but these probabilities are *unknown*.

Suppose that we collect m sets of data, one from each of m populations (or from m independent experiments). These data can be arranged in the same way as in Table 13.1. Now classification C corresponds to k multinomial classes, while classification S represents m independent samples. To be consistent with our notation in section 12.11 we denote the size of the gth sample (experiment) by n_g; this is identical with $n_{.g}$ in Table 13.1. For the gth sample the random variables $r_{1g}, r_{2g}, \ldots, r_{kg}$ have the multinomial distribution with parameters $P_{1g}, P_{2g}, \ldots, P_{kg}$, respectively.

We wish to test the hypothesis that these parameters are the same for each population without specifying their values or that these populations are *homogeneous* with respect to classification C. The hypothesis tested is $H_0: P_{l1} = P_{l2} = \cdots = P_{lm}$ $(l = 1, 2, .., k)$. Let

$$\hat{P}_{lg} = \frac{r_{lg}}{n_g}, \qquad l = 1, 2, \ldots, k; \quad g = 1, 2, \ldots, m, \qquad (13.40)$$

be the ML estimator of the proportion in the gth experiment for the lth class. We now put

$$\sum_{g=1}^{m} r_{lg} = R_l, \qquad l = 1, 2, \ldots, k.$$

The quantity R_l is the total number of individuals in the lth class for all m

samples. This is identical with $n_{l.}$ in Table 13.1. We note that

$$\sum_{l=1}^{k} R_l = \sum_{g=1}^{m} n_g = N.$$

Let

$$\hat{P}_{l.} = \frac{R_l}{N} = \frac{\sum\limits_{g=1}^{m} r_{lg}}{\sum\limits_{g=1}^{m} n_g}, \qquad l = 1, 2, \ldots, k, \tag{13.41}$$

be the combined *ML* estimator of the proportion in the *l*th class from all *m* samples. If the samples are from homogeneous populations, each $\hat{P}_{lg}$ should be "close" to the corresponding $\hat{P}_{l.}$.

Using large-sample theory, we find that the statistic

$$X^2_{\text{Het}} = \sum_{g=1}^{m} \sum_{l=1}^{k} \frac{(r_{lg} - n_g \hat{P}_{l.})^2}{n_g \hat{P}_{l.}} = \sum_{g=1}^{m} n_g \sum_{l=1}^{k} \frac{(\hat{P}_{lg} - \hat{P}_{l.})^2}{\hat{P}_{l.}} \tag{13.42}$$

is approximately distributed as χ^2 with $m(k-1) - (k-1) = (m-1)(k-1)$ d.f. if the populations are homogeneous. Hence, if $X^2_{\text{Het}} > \chi^2_{1-\alpha;(m-1)(k-1)}$, we conclude that the populations are heterogeneous with respect to classification *C*. Although this is a test of homogeneity, we are usually interested in detecting also some kind of heterogeneity and therefore the test is sometimes called a *test for heterogeneity*.

The right-hand side of (13.42) can also be written in the form

$$X^2_{\text{Het}} = \sum_{g=1}^{m} n_g \sum_{l=1}^{k} \frac{(r_{lg}/n_g - R_l/N)^2}{R_l/N}. \tag{13.42a}$$

Taking into account that $n_g = n_{.g}$ and $R_l = n_{l.}$ from the preceding section, we see that (13.42a) is identical with (13.35). We have obtained an interesting result:

The X^2 for association in the $k \times m$ contingency table and the X^2 for heterogeneity for m populations, each classified with respect to k different classes, are identical.

▼**13.11. A GENERAL METHOD OF HOMOGENEITY TESTING BASED ON LARGE-SAMPLE THEORY**

Let $\mathbf{x}' = (x_1, x_2, \ldots, x_k)$ be k random variables, and $\boldsymbol{\theta}' = (\theta_1, \theta_2, \ldots, \theta_s)$ be s functionally independent parameters associated with a certain mode of inheritance. These parameters may represent, for example, segregation ratios or gene frequencies or fitnesses.

Suppose that we select m independent samples from m populations or, equivalently, perform m independent experiments. We wish to test the hypothesis, H_0, say, that these parameters are the same in all m populations or that these populations are *homogeneous with respect to the parameters* $\boldsymbol{\theta}' = (\theta_1, \theta_2, \ldots, \theta_s)$. We shall distinguish two situations: (*a*) that in which the parameters are specified, and (*b*) that in which they are not specified.

13.11.1. The Parameters θ are Specified

Let $H_0^{(a)}$ be the hypothesis that the parameters θ_i', s take the particular values θ_{i0} ($i = 1, 2, \ldots, s$) or, briefly, $H_0^{(a)} : \boldsymbol{\theta} = \boldsymbol{\theta}_0$. Let $\hat{\boldsymbol{\theta}}_g' = (\hat{\theta}_{1g}, \hat{\theta}_{2g}, \ldots, \hat{\theta}_{gs})$ be the vector of the ML estimators in the gth experiment. Let $\mathbf{U}_g'(\boldsymbol{\theta}_0) = (U_{1g}(\boldsymbol{\theta}_0), U_{2g}(\boldsymbol{\theta}_0), \ldots, U_{sg}(\boldsymbol{\theta}_0))$ be the $s \times 1$ vector of scores, and $\mathbf{I}_g(\boldsymbol{\theta}_0) = (I_{ijg}(\boldsymbol{\theta}_0))$ be the information matrix, both evaluated at $\boldsymbol{\theta} = \boldsymbol{\theta}_0$. Assuming that n is large, we find that the statistic

$$Z_g^2 = (\hat{\boldsymbol{\theta}}_g - \boldsymbol{\theta}_0)' \mathbf{I}_g(\boldsymbol{\theta}_0)(\hat{\boldsymbol{\theta}}_g - \boldsymbol{\theta}_0) = \mathbf{U}_g'(\boldsymbol{\theta}_0)\mathbf{I}_g^{-1}(\boldsymbol{\theta}_0)\mathbf{U}_g(\boldsymbol{\theta}_0) \qquad (13.43)$$

is approximately distributed as χ^2 with s d.f. if $H_0^{(a)}$ is true for the gth experiment ($g = 1, 2, \ldots, m$).

We may also be interested in a hypothesis, H_0', say, that the parameters $\boldsymbol{\theta}$ yield the same values $\boldsymbol{\theta}_0$ in all m populations or that these populations are *homogeneous* and fit a model defined by $H_0^{(a)}$.

Since the samples are independent, the statistic

$$Z_{\text{Total}}^2 = \sum_{g=1}^m Z_g^2 = \sum_{g=1}^m \mathbf{U}_g'(\boldsymbol{\theta}_0)\mathbf{I}_g^{-1}(\boldsymbol{\theta}_0)\mathbf{U}_g(\boldsymbol{\theta}_0) \qquad (13.44)$$

is approximately distributed as χ^2 with ms d.f. if $H_0^{(a)}$ is true for each population. This is an appropriate statistic for testing H_0'. If $Z_{\text{Total}}^2 \leqslant \chi_{1-\alpha;ms}^2$, we may accept H_0'. However, if $Z_{\text{Total}}^2 > \chi_{1-\alpha;ms}^2$, we reject H_0'. This can happen for at least two reasons: (1) $H_0^{(a)}$ is "true on the average," but some populations deviate from $H_0^{(a)}$; (2) $H_0^{(a)}$ is false, and no populations agree with $H_0^{(a)}$. In the second case these populations might still be homogeneous and fit a model defined by another hypothesis, $H_0^{(x)}$, say, but $H_0^{(x)}$ is unknown. A significant value of Z_{Total}^2 does not discriminate between these two kinds of heterogeneity, and further tests are necessary.

Let $\hat{\boldsymbol{\theta}}' = (\hat{\theta}_1, \hat{\theta}_2, \ldots, \hat{\theta}_s)$ be the pooled estimators obtained from the combined sample of m experiments (see section 12.11). Also let

$$\mathbf{U}(\boldsymbol{\theta}_0) = \sum_{g=1}^m \mathbf{U}_g(\boldsymbol{\theta}_0), \quad \mathbf{I}(\boldsymbol{\theta}_0) = \sum_{g=1}^m \mathbf{I}_g(\boldsymbol{\theta}_0) \qquad (13.45)$$

be the vector of total scores and the information matrix, respectively, evaluated at $\boldsymbol{\theta} = \boldsymbol{\theta}_0$. Thus the statistic Z_{Combined}^2 (briefly, Z_{Comb}^2), defined as

$$Z_{\text{Comb}}^2 = (\hat{\boldsymbol{\theta}} - \boldsymbol{\theta}_0)' \mathbf{I}(\boldsymbol{\theta}_0)(\hat{\boldsymbol{\theta}} - \boldsymbol{\theta}_0) = \mathbf{U}'(\boldsymbol{\theta}_0)\mathbf{I}^{-1}(\boldsymbol{\theta}_0)\mathbf{U}(\boldsymbol{\theta}_0), \qquad (13.46)$$

is approximately distributed as χ^2 with s d.f. if $H_0^{(a)}$ is "true on the average."

We now construct another statistic, which we call $Z^2_{\text{Difference}}$ (briefly, Z^2_{Diff}), defined as

$$Z^2_{\text{Diff}} = Z^2_{\text{Total}} - Z^2_{\text{Comb}}$$

$$= \sum_{g=1}^{m} \mathbf{U}'_g(\boldsymbol{\theta}_0)\mathbf{I}_g^{-1}(\boldsymbol{\theta}_0)\mathbf{U}_g(\boldsymbol{\theta}_0) - \mathbf{U}'(\boldsymbol{\theta}_0)\mathbf{I}^{-1}(\boldsymbol{\theta}_0)\mathbf{U}(\boldsymbol{\theta}_0). \qquad (13.47)$$

This will be a suitable statistic for testing a hypothesis, H_0, say, that the populations are homogeneous with respect to a model given by a hypothesis $H_0^{(x)}$ [including $H_0^{(a)}$, too], about the parameters $\boldsymbol{\theta}$. If H_0 is true, the statistic Z^2_{Diff} is approximately distributed as χ^2 with $ms - s = s(m - 1)$ d.f.

If Z^2_{Total} is significant, the other test statistics may be significant or non-significant, and it is necessary to consider the results of the three tests *jointly*.

(a) Z^2_{Total} significant, Z^2_{Comb} significant, Z^2_{Diff} significant. This means that our primary hypothesis, $H_0^{(a)}$, cannot be accepted and also that the data are not homogeneous whatever $H_0^{(x)}$ may be (i.e., H_0 is rejected).

(b) Z^2_{Total} significant, Z^2_{Comb} significant, Z^2_{Diff} nonsignificant. In this case $H_0^{(a)}$ is not acceptable, but the data are homogeneous and fit another model corresponding to some unknown $H_0^{(x)}$. Our further investigation should be to construct $H_0^{(x)}$.

(c) Z^2_{Total} significant, Z^2_{Comb} nonsignificant, Z^2_{Diff} significant. In this case $H_0^{(a)}$ is probably true, but particular samples exhibit deviations from $H_0^{(a)}$. (This might be due, for instance, to nonrandom sampling.)

It should be noticed that these three Z^2-statistics are based on large sample-theory, according to which the joint distributions of the ML estimators are approximately normal when n is large. Therefore knowledge of the distributions of $\mathbf{x}$ in each population is not required. In fact, these can be different, that is, our data might be collected from experiments of different designs (e.g., from backcrosses and intercrosses).

In particular, when there is only one parameter, θ (i.e., $s = 1$), Z^2_{Total} and Z^2_{Comb} take special forms:

$$Z^2_{\text{Total}} = \sum_{g=1}^{m} \frac{[U_g(\theta_0)]^2}{I_g(\theta_0)} \quad \text{with } m \text{ d.f.,} \qquad (13.48)$$

and

$$Z^2_{\text{Comb}} = \frac{\left[\sum_{g=1}^{m} U_g(\theta_0)\right]^2}{\sum_{g}^{m} I_g(\theta_0)} \quad \text{with } 1 \text{ d.f.,} \qquad (13.49)$$

respectively.

Applications of this theory will be demonstrated in Chapter 18, where data from families of different sizes are used for testing a hypothesis regarding a mode of inheritance of certain traits.

13.11.2. The Parameters are Unknown

In some problems only the mode of inheritance, that is, the form of the distribution of phenotypes is determined; the parameters cannot be specified. For instance, gene frequencies, linkage, and proportions of viable zygotes can only be estimated. In such cases, the Z^2-statistics defined in section 13.11.1 can be used, but with the vector $\boldsymbol{\theta}_0' = (\theta_{10}, \theta_{20}, \ldots, \theta_{s0})$ replaced by the vector of pooled estimators, $\hat{\boldsymbol{\theta}}' = (\hat{\theta}_1, \hat{\theta}_2, \ldots, \hat{\theta}_s)$, and the degrees of freedom decreased by the number of estimated parameters, s. Of course, now $Z_{\text{Comb}}^2 = 0$, so that

$$Z_{\text{Total}}^2 = Z_{\text{Diff}}^2 = \sum_{g=1}^{m}(\hat{\boldsymbol{\theta}}_g - \hat{\boldsymbol{\theta}})' \mathbf{I}_g(\hat{\boldsymbol{\theta}})(\hat{\boldsymbol{\theta}}_g - \hat{\boldsymbol{\theta}}) = \sum_{g=1}^{m} \mathbf{U}_g'(\hat{\boldsymbol{\theta}}) \mathbf{I}_g^{-1}(\hat{\boldsymbol{\theta}}) \mathbf{U}_g(\hat{\boldsymbol{\theta}}) \quad (13.50)$$

is approximately distributed as χ^2 with $ms - s = s(m - 1)$ d.f. Therefore this statistic is appropriate for testing goodness of fit "confounded" with testing for heterogeneity, and these two components cannot be separated.

If only c parameters are estimated and $s - c$ are specified, the degrees of freedom for Z_{Total}^2, Z_{Comb}^2, and Z_{Diff}^2 will be $m(s - c)$, $s - c$, and $(s - c)(m - 1)$, respectively.

It may happen that the parameters are not estimable from a single experiment but can be estimated from m experiments combined. In such cases, the results presented in sections 13.11.1 and 13.11.2 are valid, but the degrees of freedom must be appropriately adjusted (see Example 18.1).

Example 13.5. We now apply our results to the data discussed in Example 12.3 on segregation ratios in tetraploid tomatoes. The ML estimate of the parameter ϑ, determining the selective disadvantage of gamete rr, was obtained from combined experiments on selfing (Exp. 1) and backcrossing (Exp. 2). We found $\hat{\hat{\vartheta}} = 0.5233$. Since scores, $U_g(\hat{\hat{\vartheta}})$, and amounts of information, $I_g(\hat{\hat{\vartheta}})$, have already been calculated, we apply the heterogeneity test given in (13.48), substituting $\hat{\hat{\vartheta}} = 0.5233$ for ϑ_0. The results are shown in the table. (Note: The scores should add up to zero. The small disagreement is due to rounding error.)

If we take $\alpha = 0.05$, then $\chi_{0.95}^2 = 3.841$ with 1 d.f. Since $1.7859 < 3.841$, we conclude that our hypothesis about the selective disadvantage of the rr gamete is supported by these data.

Exp.	$U_g(\hat{\hat{\vartheta}})$	$I_g(\hat{\hat{\vartheta}})$	$U_g(\hat{\vartheta})$	$I_g(\hat{\vartheta})$	$[U_g(\hat{\vartheta})]^2/I_g(\hat{\vartheta})$
1	$\dfrac{2(221 - 826\hat{\hat{\vartheta}}^2)}{\hat{\hat{\vartheta}}(1 - \hat{\hat{\vartheta}}^2)}$	$\dfrac{3304}{1 - \hat{\hat{\vartheta}}^2}$	-27.3382	4549.9795	0.1643
2	$\dfrac{67 - 115\hat{\hat{\vartheta}}}{\hat{\hat{\vartheta}}(1 - \hat{\hat{\vartheta}})}$	$\dfrac{115}{\hat{\hat{\vartheta}}(1 - \hat{\hat{\vartheta}})}$	$+27.3414$	461.0011	1.6216
	Z^2_{Total} (with $\nu = 1$ d.f.)				1.7859

13.12. TESTS FOR HETEROGENEITY ABOUT MULTINOMIAL PARAMETERS SPECIFIED BY $H_0^{(a)}$

The theory presented in section 13.11 can be applied when the distribution of x's is of any form (continuous or discrete), and the m experiments can have different designs. When there is more than one parameter ($s > 1$), the calculations become laborious and usually computer programs are required.

However, if the data of each sample are represented by variables having multinomial distributions with the same general classification C (or, equivalently, the m sets of data are from the same type of experimental designs), the calculations can be facilitated by using a few types of X^2-statistics of relatively simple structures.

13.12.1. The Multinomial Parameters are Specified by $H_0^{(a)}$: $P_{lg} = P_{l0}$ ($l = 1, 2, \ldots, k; g = 1, 2, \ldots, m$)

Using the results of section 13.7 and following the arguments developed in section 13.11, we find that for testing H_0' [i.e., the hypothesis that the populations are homogeneous with respect to $H_0^{(a)}$] a suitable statistic with $m(k - 1)$ d.f. would be

$$X^2_{\text{Total}} = \sum_{g=1}^{m} X_g^2 = \sum_{g=1}^{m} \sum_{l=1}^{k} \frac{(r_{lg} - n_g P_{l0})^2}{n_g P_{l0}} = \sum_{g=1}^{m} n_g \sum_{l=1}^{k} \frac{(\hat{P}_{lg} - P_{l0})^2}{P_{l0}}, \quad (13.51)$$

where $\hat{P}_{lg} = r_{lg}/n_g$ is defined in (13.40).

For testing the hypothesis that $H_0^{(a)}$ is "true on the average," the statistic

$$X^2_{\text{Comb}} = \sum_{l=1}^{k} \frac{(R_l - N P_{l0})^2}{N P_{l0}} = N \sum_{l=1}^{k} \frac{(\hat{P}_{l.} - P_{l0})^2}{P_{l0}}, \quad (13.52)$$

with $k - 1$ d.f., will be suitable. [Here $\hat{P}_{l.} = R_l/N$ is defined by (13.41).]

One might calculate

$$X^2_{\text{Diff}} = X^2_{\text{Total}} - X^2_{\text{Comb}} = \sum_{g=1}^{m} n_g \sum_{l=1}^{k} \frac{(\hat{P}_{lg} - \hat{P}_{l.})^2}{P_{l0}} \qquad (13.53)$$

and wish to use it in testing for heterogeneity with respect to some unknown model defined by $H_0^{(x)}$. However, this statistic has P_{l0} in the denominator which is specified by $H_0^{(a)}$, and (13.52) is not relevant for testing for heterogeneity with respect to (unknown) $H_0^{(x)}$. In this case the correct statistic is X^2_{Het}, given by (13.42).

Interpretation of the joint results of tests based on X^2_{Total}, X^2_{Comb}, and X^2_{Het} will be similar to that given in the preceding section in regard to the joint significance (or nonsignificance) of Z^2_{Total}, Z^2_{Comb}, and Z^2_{Diff}, respectively.

However, we note that if we wish to test $H_0^{(a)}$ that the phenotypic ratios are $9:3:3:1$ (i.e., $P_1 = \frac{9}{16}$, $P_2 = \frac{3}{16}$, $P_3 = \frac{3}{16}$, $P_4 = \frac{1}{16}$) in m independent experimental intercrosses, it will be simpler to apply the X^2-statistics discussed in this section than the Z^2-statistics defined in section 3.11.

13.12.2. The Multinomial Proportions are Functions of $s(<k - 1)$ Functionally Independent Parameters, $\boldsymbol{\theta}' = (\theta_1, \theta_2, \ldots, \theta_s)$

Suppose that $P_l = P_l(\boldsymbol{\theta})$ and $\boldsymbol{\theta}$ are specified by $H_0^{(a)}: \boldsymbol{\theta} = \boldsymbol{\theta}_0$. Now $P_{l0} = P_l(\boldsymbol{\theta}_0)$. To obtain the required X^2 we substitute $P_l(\boldsymbol{\theta}_0)$ for P_{l0} into the appropriate formula for X^2. For instance,

$$X^2_{\text{Total}} = \sum_{g=1}^{m} n_g \sum_{l=1}^{k} \frac{[\hat{P}_{lg} - P_l(\boldsymbol{\theta}_0)]^2}{P_l(\boldsymbol{\theta}_0)}. \qquad (13.52a)$$

13.13. TESTS FOR HETEROGENEITY WHEN MULTINOMIAL PROPORTIONS ARE FUNCTIONS OF SOME PARAMETERS NOT SPECIFIED BY $H_0^{(a)}$. EQUILIBRIUM AND HETEROGENEITY

We now consider a situation like that in section 13.12.2 but with the parameters not specified by $H_0^{(a)}$. For any particular genetic model expressed in terms of the $H_0^{(a)}$ hypothesis, it is possible to construct a series of X^2-tests, which should utilize the information included in $H_0^{(a)}$.

Let $H_0^{(a)}$ be a hypothesis that the multinomial proportions P_l are special functions of θ's, that is, $H_0^{(a)}: P_l = P_l(\boldsymbol{\theta})$. However, the θ's are not specified but must be estimated. For instance, if a population is assumed to be in equilibrium, the genotype frequencies can be expressed as a quadratic expansion of the gamete array; the gene (and so the gamete) frequencies, are usually unknown and have to be estimated.

Suppose that we select m independent samples, each from a different population, and wish to analyze these data jointly. The following problems may be of some interest.

(i) We wish to test whether each population has the distribution determined by $H_0^{(o)}$.

Let $\hat{\boldsymbol{\theta}}_g' = (\hat{\theta}_{1g}, \hat{\theta}_{2g}, \ldots, \hat{\theta}_{sg})$ be the ML estimators of θ's in the gth sample of size n_g. If $H_0^{(a)}$ is true for the gth population, then the statistic

$$[X_g^{(a)}]^2 = \sum_{l=1}^{k} \frac{[r_{lg} - n_g P_l(\hat{\boldsymbol{\theta}}_g)]^2}{n_g P_l(\hat{\boldsymbol{\theta}}_g)} = n_g \sum_{l=1}^{k} \frac{[\hat{P}_{lg} - P_l(\hat{\boldsymbol{\theta}}_g)]^2}{P_l(\hat{\boldsymbol{\theta}}_g)} \tag{13.54}$$

is approximately distributed as χ^2 with $k - 1 - s$ d.f. We note that this is the X^2-statistic for testing goodness of fit, given in (13.29), applied to the gth sample.

If $H_0^{(a)}$ is true for each of the m populations, the statistic

$$[X^{(a)}]^2 = \sum_{g=1}^{m}[X_g^{(a)}]^2 = \sum_{g=1}^{m} \sum_{l=1}^{k} \frac{[r_{lg} - n_g P_l(\hat{\boldsymbol{\theta}}_g)]^2}{n_g P_l(\hat{\boldsymbol{\theta}}_g)} = \sum_{g=1}^{m} n_g \sum_{l=1}^{k} \frac{[\hat{P}_{lg} - P_l(\hat{\boldsymbol{\theta}}_g)]^2}{P_l(\hat{\boldsymbol{\theta}}_g)} \tag{13.55}$$

is distributed as χ^2 with $m(k - 1 - s)$ d.f. If $[X^{(a)}]^2 \geqslant \chi^2_{1-\alpha;m(k-1-s)}$, we reject $H_0^{(a)}$; this means either that the genetic model is incorrect (e.g., populations are not in equilibrium) or that some populations deviate from this model. Usually values of individual $[X_g^{(a)}]^2$'s give clues as to which situation is more likely.

(ii) We wish to test the hypothesis, H_0', that the m populations are homogeneous and fit the model defined by $H_0^{(a)}$ (e.g., are in equilibrium and have the same expected gene frequencies).

Let $\hat{\boldsymbol{\theta}}' = (\hat{\hat{\theta}}_1, \hat{\hat{\theta}}_2, \ldots, \hat{\hat{\theta}}_s)$ be the "pooled" estimators of θ's obtained from the combined m samples. The statistic

$$X_{\text{Total}}^2 = \sum_{g=1}^{m} \sum_{l=1}^{k} \frac{[r_{lg} - n_g P_l(\hat{\hat{\boldsymbol{\theta}}})]^2}{n_g P_l(\hat{\hat{\boldsymbol{\theta}}})} = \sum_{g=1}^{m} n_g \sum_{l=1}^{k} \frac{[\hat{P}_{lg} - P_l(\hat{\hat{\boldsymbol{\theta}}})]^2}{P_l(\hat{\hat{\boldsymbol{\theta}}})}, \tag{15.56}$$

with $m(k - 1) - s$ d.f., will be appropriate for testing H_0'. If X_{Total}^2 is non-significant, we may accept H_0', the hypothesis that the populations are homogeneous with respect to the genetic model specified by $H_0^{(a)}$. However, if X_{Total}^2 is significant, we have to interpret the results of tests $[X^{(a)}]^2$ and X_{Total}^2 jointly.

(a) If $[X^{(a)}]^2$ is significant, X_{Total}^2 nonsignificant, then the model specified by $H_0^{(a)}$ may be accepted, but the populations are heterogeneous (i.e., the populations are in equilibrium but have different gene frequencies).

(b) We may have $[X^{(a)}]^2$ significant, X_{Total}^2 significant. On the basis of the significance of $[X^{(a)}]^2$ we conclude that the model is incorrect; the significance of X_{Total}^2 may result from fitting an inappropriate model, but the populations may or may not be homogeneous with respect to the other model defined by $H_0^{(x)}$. To test for heterogeneity with respect to a model defined by $H_0^{(x)}$, the X_{Het}^2 given by (13.42), should be applied.

Of course, assuming that the model given by $H_0^{a)}$ is correct, the Z^2-test given by (13.50) is always the right one to apply.

Example 13.6. One of the two loci which determine gamma globulins in human serum is known as the *Gm* locus. There are a number of alleles at this locus, but the data presented in Table 13.3 [Vos et al. (1963)] are from two subpopulations in Pakistan in which only the three most common alleles, denoted as Gm^a,* Gm^{ax}, and Gm^b, were present. Let p, q, and t, with $p + q + t = 1$, be their respective frequencies.

Serologically only factors a, b, and x can be detected. Hence the genotypes Gm^aG^{ax} and $Gm^{ax}Gm^{ax}$ are indistinguishable. Using $+$ for the presence and $-$ for the absence of each factor, we can show the phenotypes (numbered, for convenience, as $l = 1, 2, \ldots, 5$), their distribution, and observed frequencies in the sample of size n as follows:

l:	1	2	3	4	5
	$Gm(a + b - x -)$	$Gm(a + b - x +)$	$Gm(a + b + x -)$	$Gm(a + b + x +)$	$Gm(a - b + x -)$
P_l:	p^2	$q^2 + 2pq$	$2pt$	$2qt$	t^2
r_l:	r_1	r_2	r_3	r_4	r_5.

In fact, only two parameters are functionally independent, so only $s = 2$ parameters need be estimated.

The ML estimates of gene frequencies, assuming equilibrium, can be calculated using the maximum likelihood equations (see Problem 12.2). In fact, we used here the gene-counting method described in Chapter 14, which is identical with the ML method (see also Problem 14.10). The values of individual and pooled estimates of gene frequencies are given in the right-hand part of Table 13.3.

(i) We may first analyze the data in the left-hand side of Table 13.3 as to whether they are homogeneous, without taking into account the supposed phenotype distribution. Then formula (13.42) will be used. We calculate $\hat{P}_{1.} = 36/312 = 0.1154$; $\hat{P}_{2.} = 0.0769$; $\hat{P}_{3.} = 0.4071$; $\hat{P}_{4.} = 0.0833$; $\hat{P}_{5.} = 0.3173$. Hence

$$X_{\text{Het}}^2 = \frac{(24 - 203 \times 0.1154)^2}{203 \times 0.1154} + \frac{(16 - 203 \times 0.0769)^2}{203 \times 0.1154}$$

$$+ \cdots + \frac{(12 - 109 \times 0.1154)^2}{109 \times 0.1154} + \cdots + \frac{(38 - 109 \times 0.3173)^2}{109 \times 0.3173}$$

$$= 0.0141 + 0.0097 + \cdots + 0.0266 + \cdots + 0.3371 = 1.8684.$$

See footnote on p. 342 about the recent nomenclature for the alleles at the *Gm* locus

TABLE 13.3 [a]Phenotype distributions and gene frequencies at the Gm locus in Pakistan

Subpopulation		Sample size, n_g	Phenotype frequencies at the Gm locus					Gene frequencies		
			$(a+b-x-)$	$(a+b-x+)$	$(a+b+x-)$	$(a+b+x+)$	$(a-b+x-)$	Gm^a	Gm^{ax}	Gm^{b}
g	Name		r_{1g}	r_{2g}	r_{3g}	r_{4g}	r_{5g}	$\hat{p}_g$	$\hat{q}_g$	$\hat{t}_g$
1	Punjabis	203	24	16	87	15	61	0.3681	0.0802	0.5517
2	Pathans	109	12	8	40	11	38	0.3257	0.0917	0.5826
	Total	312	36	24	127	26	99	0.3533	0.0842	0.5625
								$\hat{\hat{p}}$	$\hat{\hat{q}}$	$\hat{\hat{t}}$

[a] Note that it is more convenient here to represent the data in the form of a $m \times k$ table.

If the data are homogeneous, X_{Het}^2 given by (13.42) should be approximately distributed as χ^2 with $(2 - 1)(5 - 1) = 4$ d.f. Using χ^2 tables, we find that the significance level, α, corresponding to the value of $X_{\text{Het}}^2 = 1.8684$ is between 0.70 and 0.80, and therefore we might well accept the hypothesis H_0 that the data are obtained from homogeneous populations. However, this test does not inform us whether the hypothesis regarding equilibrium can be considered correct.

(ii) We wish to test whether each population is separately in equilibrium. The $[X^{(a)}]^2$ defined in (13.55) will be used for this purpose. Here $P_l(\boldsymbol{\theta}) = P_l(p, q, t)$, but since $p + q + t = 1$ there are only $s = 2$ gene frequencies which are functionally independent. Substituting the estimates $\hat{p}_1$, $\hat{q}_1$, and $\hat{t}_1$, from the first sample, for p, q, and t, respectively, we obtain the values $P_1(\hat{\boldsymbol{\theta}}_1) = P_1(\hat{p}_1, \hat{q}_1, \hat{t}_1) = 0.3680^2 = 0.1354$, $P_2(\hat{\boldsymbol{\theta}}_1) = P_2(\hat{p}_1, \hat{q}_1, \hat{t}_1) = (0.0802)^2 + 2 \times 0.3681 \times 0.0802 = 0.0655$, etc. Similarly, using the estimates $\hat{p}_2$, $\hat{q}_2$, $\hat{t}_2$ from the second sample, we calculate the values of $P_l(\hat{\boldsymbol{\theta}}_2) = P_l(\hat{p}_2, \hat{q}_2, \hat{t}_2)$. The values of $P_l(\hat{\boldsymbol{\theta}}_g)$, $l = 1, 2, \ldots, 5$; $g = 1, 2$, are given in the fourth column of Table 13.4. The calculation steps for $[X_1^{(a)}]^2$ and $[X_2^{(a)}]^2$, as well as the values of these $[X_g^{(a)}]^2$'s and their sum $[X^{(a)}]^2 = [X_1^{(a)}]^2 + [X_2^{(a)}]^2$, are given in the fifth column. Since the significance level for $[X^{(a)}]^2 = 1.9155$ is between 0.70 and 0.80, we may accept the hypothesis that both populations are in equilibrium with respect to the *Gm* locus.

(iii) Since $[X^{(a)}]^2$ and X_{Het}^2 are both nonsignificant, we may conclude that both populations are in equilibrium and that the gene frequencies do not differ significantly in the two populations.

Suppose, however, that we start our analysis by testing $H_0^{(a)}$ for each population, using the (13.55) test. We conclude that $H_0^{(a)}$ in regard to equilibrium can be accepted for each population. Then, for testing about homogeneity, if $H_0^{(a)}$ is assumed to be correct, the X_{Total}^2 given by (13.56) is more appropriate. From Table 13.3 we have the pooled estimates of gene frequencies obtained from combined data: $\hat{\hat{p}} = 0.3533$, $\hat{\hat{q}} = 0.0842$, $\hat{\hat{t}} = 0.5625$. The appropriate genotype frequencies, $P_l(\hat{\hat{\boldsymbol{\theta}}}) = P_l(\hat{\hat{p}}, \hat{\hat{q}}, \hat{\hat{t}})$, are as follows: $P_1(\hat{\hat{\boldsymbol{\theta}}}) = 0.1248$, $P_2(\hat{\hat{\boldsymbol{\theta}}}) = 0.0666$, $P_3(\hat{\hat{\boldsymbol{\theta}}}) = 0.3975$, $P_4(\hat{\hat{\boldsymbol{\theta}}}) = 0.0947$, $P_5(\hat{\hat{\boldsymbol{\theta}}}) = 0.3164$. The penultimate column in Table 13.4 shows the terms $[r_{lg} - n_g P_l(\hat{\hat{\boldsymbol{\theta}}})]^2/[n_g P_l(\hat{\hat{\boldsymbol{\theta}}})]$ and X_{Total}^2 with 6 d.f. It is not significant. We then assume the validity of the hypothesis that both populations are in equilibrium with the same gene frequencies.

One general remark should be made in regard to X^2-tests: it is desirable that the authors of papers including X^2-analyses indicate that X^2's have been calculated and state which hypotheses have been tested. It is sometimes difficult for the reader to understand the meanings of X^2-statistics published in reports of analysis of human data.

TABLE 13.4 Testing of fit in individual subpopulations and heterogeneity test

(1)	(2) g	(3) l	(4) $P_l(\hat{\hat{\theta}}_g)$	(5) $\dfrac{[r_{lg} - n_g P_l(\hat{\hat{\theta}}_g)]^2}{n_g P_l(\hat{\hat{\theta}}_g)}$	(6) $\dfrac{[r_{lg} - n_g P_l(\hat{\hat{\theta}})]^2}{n_g P_l(\hat{\hat{\theta}})}$	Significance level, α
Punjabis ($n_1 = 203$)	1	1	0.1355	0.4470	0.0703	
		2	0.0655	0.5497	0.4550	
		3	0.4062	0.2521	0.4930	
		4	0.0885	0.4895	0.9282	
		5	0.3044	0.0102	0.1624	
	Test of goodness of fit $[X_1^{(a)}]^2$ with $\nu = 2$			1.7465		$0.40 < \alpha < 0.50$
Pathans ($n_2 = 109$)	2	1	0.1061	0.0164	0.1889	
		2	0.0681	0.0449	0.0756	
		3	0.3795	0.0451	0.2555	
		4	0.1068	0.0353	0.0445	
		5	0.3394	0.0273	0.3577	
	Test of goodness of fit $[X_2^{(a)}]^2$ with $\nu = 2$			0.1690		$0.90 < \alpha < 0.95$
$[X^{(a)}]^2 = [X_1^{(a)}]^2 + [X_2^{(a)}]^2$ with $\nu = 4$				1.9155		$0.70 < \alpha < 0.80$
X_{Total}^2 with $\nu = 6$					3.0311	$0.80 < \alpha < 0.90$
X_{Het}^2 with $\nu = 4$					1.8684	$0.70 < \alpha < 0.80$

▼ 13.14. EVALUATION OF THE POWER OF THE TEST. DISCRIMINATION BETWEEN TWO SIMPLE HYPOTHESES

In Example 13.1 we tested the hypothesis $H_0: P = \frac{3}{4}$, and this hypothesis was not rejected. It can also be shown that, when the same $\alpha = 0.05$ (or even 0.01) is used, another hypothesis, $H_1: P = \frac{13}{16}$, will not be rejected, either. Geneticists can discriminate between these hypotheses by using special designs. In this situation, one may use crosses White × White. If white flowers are simple recessives (i.e., $aa \times aa$), the next generation will be only white (aa). However, if white flowers arise from mutual epistasis, some crosses might be $A-B- \times A-B-$, which would be able to produce offspring with the phenotype $aaB-$, representing colored flowers. Of course, lack of colored flowers does not contradict the hypothesis H_1: $P = \frac{13}{16}$, but for sufficiently large samples it would be unlikely that no colored flowers would be observed if H_1: $P = \frac{13}{16}$ is true.

However, there may be situations in which performance of such a "discriminating" experiment is impossible or impractical, and we have to make a decision by using statistical methods. In such situations the use of the *power* of the test will be helpful. We restrict ourselves here to parametric hypotheses. Let H_0: $\theta = \theta_0$ and H: $\theta > \theta_0$. Let α be the significance level, and $w = w(\alpha)$ be the critical region defined in (13.17). We recall Definition 13.8 of the power of a test. Let H_1: $\theta = \theta_1$ be the specified alternative, and $\hat{\theta}$ be an estimator of θ. The probability of rejecting H_0 when the alternative, H_1, is true is called the *power* of the test with respect to H_1. Denoting this power by $K(\theta_1)$, we have

$$K(\theta_1) = \Pr\{\hat{\theta} \in w(\alpha) \mid H_1\} = 1 - \beta. \qquad (13.57)$$

(We read: "the probability that $\hat{\theta}$ belongs to the critical region, $w(\alpha)$, given that H_1 is true, is equal to $1 - \beta$.")

Suppose that we perform a test at the given significance level α, for which $1 - \beta$ is large. Expression (13.57) tells us that if H_1 is true it is very likely that the observed statistic $\hat{\theta}$ falls into the critical region. Hence, if the power, $1 - \beta$, is large and the test statistic $\hat{\theta}$ is in the acceptance region, we have strong support to accept H_0 and reject H_1.

The evaluation of the power of our X^2-statistic requires a knowledge of the noncentral χ^2-distribution, which is beyond the scope of this book. Restricting ourselves to normal distributions, we will evaluate the power of a test about the mean and apply this to the evaluation of the approximate power of a test about a binomial proportion.

13.14.1. The Power of the Test for H_0: $\mu = \mu_0$ for a Normal Distribution

Suppose that we construct a test for H_0: $\mu = \mu_0$ against the one-sided (right-hand) alternative, H: $\mu > \mu_0$. We assume that the character, X, under consideration has a normal distribution, $X \frown N(\mu, \sigma^2)$, and that σ^2 is known. As a test statistic we use the sample mean, $\bar{x}$. If H_0 is true, the statistic

$$Z = \frac{\bar{x} - \mu_0}{\sigma/\sqrt{n}}$$

is distributed as $N(0, 1)$. Thus, for a given α, we have

$$\Pr\{Z > z_{1-\alpha}\} = \Pr\left\{\bar{x} > \mu_0 + z_{1-\alpha}\frac{\sigma}{\sqrt{n}}\right\} = \alpha. \qquad (13.58)$$

In other words, the critical point, $\bar{x}_c$, say, is $\bar{x}_c = \mu_0 + z_{1-\alpha}(\sigma/\sqrt{n})$, and the critical region is $\bar{x} > \bar{x}_c$.

Suppose now that we have in mind another simple hypothesis, H_1: $\mu = \mu_1 > \mu_0$, and want to evaluate the power of the test against H_1 as defined in (13.57). We have

$$K(\mu_1) = 1 - \beta = \Pr\{\bar{x} \in w \mid H_1\} = \Pr\{\bar{x} > \bar{x}_c \mid H_1\}$$

$$= \Pr\left\{\frac{\bar{x} - \mu_1}{\sigma/\sqrt{n}} > \frac{\bar{x}_c - \mu_1}{\sigma/\sqrt{n}}\right\} = \Pr\left\{Z > \frac{\mu_0 + z_{1-\alpha}(\sigma/\sqrt{n}) - \mu_1}{\sigma/\sqrt{n}}\right\}$$

$$= \Pr\left\{Z > z_{1-\alpha} - \frac{\mu_1 - \mu_0}{\sigma/\sqrt{n}}\right\} = \Pr\{Z < -z_{1-\alpha} + \delta\}$$

$$= \Phi(-z_{1-\alpha} + \delta), \tag{13.59}$$

where now

$$Z = \frac{\bar{x} - \mu_1}{\sigma/\sqrt{n}}, \quad \delta = \frac{\mu_1 - \mu_0}{\sigma/\sqrt{n}}, \tag{13.60}$$

and $\Phi(-z_{1-\alpha} + \delta)$ is the standard normal cumulative distribution function, evaluated at the point $z = -z_{1-\alpha} + \delta$.

In a similar way, it can be shown that for a left-hand, one-sided test (i.e., H_1: $\mu = \mu_1 < \mu_0$) the critical region is $\bar{x} < \mu_0 - z_{1-\alpha}(\sigma/\sqrt{n})$ and the power is

$$K(\mu_1) = 1 - \beta = \Phi(-z_{1-\alpha} - \delta). \tag{13.61}$$

For the two-sided alternative, the critical region is $|\bar{x} - \mu_0| > z_{1-\frac{1}{2}\alpha}(\sigma/\sqrt{n})$ and the (approximate) power is

$$K(\mu_1) = 1 - \beta = \Phi(-z_{1-\frac{1}{2}\alpha} + |\delta|) + \Phi(-z_{1-\frac{1}{2}\alpha} - |\delta|). \tag{13.62}$$

In practice, the second term on the right-hand side of (13.62) is often so small that it is sufficient to use the approximate power

$$K(\mu_1) = 1 - \beta \doteq \Phi(-z_{1-\frac{1}{2}\alpha} + |\delta|). \tag{13.62a}$$

13.14.2. The Approximate Power of the Test for Hypothesis about Binomial Proportion, $H_0: P = P_0$

If the alternative is one-sided, $H: P > P_0$, then, by the use of large-sample theory, the critical point is $P_c \doteq P_0 + z_{1-\alpha}\sqrt{P_0(1 - P_0)/n}$, and the approximate critical region is $\hat{P} > P_c$ [compare formula (13.7)].

Let $H_1: P = P_1 > P_0$ be a specific alternative. Following the same arguments as in the derivation of formula (13.59), we obtain

$$K(P_1) = 1 - \beta \doteq \Pr\left\{\frac{\hat{P} - P_1}{\sqrt{P_1(1 - P_1)/n}} > \frac{P_c - P_1}{\sqrt{P_1(1 - P_1)/n}}\right\}$$

$$= \Pr\left\{Z < -\frac{z_{1-\alpha}\sqrt{P_0(1 - P_0)} - (P_1 - P_0)\sqrt{n}}{\sqrt{P_1(1 - P_1)}}\right\}$$

$$= \Phi\left\{-z_{1-\alpha}\sqrt{\frac{P_0(1 - P_0)}{P_1(1 - P_1)}} + \delta\right\}, \tag{13.63}$$

where

$$Z = \frac{\hat{P} - P_1}{\sqrt{P_1(1 - P_1)/n}}, \quad \delta = \frac{P_1 - P_0}{\sqrt{P_1(1 - P_1)/n}}. \tag{13.64}$$

Similarly, for a left-hand, one-sided alternative, $H: P < P_0$, the approximate critical region is $\hat{P} < P_0 - z_{1-\alpha}\sqrt{P_0(1 - P_0)/n}$, and the approximate power, if $H_1: P = P_1 < P_0$, is

$$K(P_1) \doteq \Phi\left\{-z_{1-\alpha}\sqrt{\frac{P_0(1 - P_0)}{P_1(1 - P_1)}} - \delta\right\}. \tag{13.65}$$

For two-sided alternatives, the approximate critical region is $|\hat{P} - P_0| > P_0 + z_{1-\frac{1}{2}\alpha}\sqrt{P_0(1 - P_0)/n}$, and the approximate power with respect to a specific alternative, $H_1: P = P_1$, is

$$K(P_1) \doteq \Phi\left\{-z_{1-\frac{1}{2}\alpha}\sqrt{\frac{P_0(1 - P_0)}{P_1(1 - P_1)}} + |\delta|\right\}$$

$$+ \Phi\left\{-z_{1-\frac{1}{2}\alpha}\sqrt{\frac{P_0(1 - P_0)}{P_1(1 - P_1)}} - |\delta|\right\}. \tag{13.66}$$

In practice, the second term is often negligible.

Example 13.7.   Suppose that in Example 13.1 the simple alternative we have in mind is $P = \frac{13}{16} > \frac{3}{4}$. Then, in constructing a significance test, we should use a right-hand, one-sided test. Let us take $\alpha = 0.05$. We have $P_c = P_0 + z_{0.95}\sqrt{P_0(1 - P_0)/n} = 0.75 + 1.645 \times 0.0306 = 0.8003$. Since $0.77 < 0.8003$, we do not reject H_0. We now calculate the approximate power of this test if $H_1: P = \frac{13}{16}$ is valid. We have

$$z_{0.95} = 1.645; \quad \sqrt{\frac{P_0(1 - P_0)}{P_1(1 - P_1)}} = 1.1094; \quad \delta = 2.2645.$$

Hence,

$$-z_{0.95}\sqrt{\frac{P_0(1 - P_0)}{P_1(1 - P_1)}} + \delta = -1.645 \times 1.1094 + 2.2645 = 0.4395,$$

and $1 - \beta = \Pr\{Z < 0.4395\} = \Phi(0.4395) = 0.67$ (or $\beta = 0.33$). This is not a big power; therefore, in fact, we cannot discriminate very well between $H_0: P = \frac{3}{4}$ and $H_1: P = \frac{13}{16}$. In order to get larger power we have to increase the sample size.

13.15. SAMPLE SIZE WHEN α AND β ARE SPECIFIED

In Chapter 11 we asked this question: how large should the sample size be to ensure a given precision of estimate with a specified confidence coefficient, $1 - \alpha$? We now formulate a slightly different question concerned with testing hypotheses.

How large should the sample size, n, be to detect a difference $\mu_1 - \mu_0 = d$, with α and β having specified values?

It can be shown [see, e.g., Johnson and Leone (1964), Vol. I, Chapter 7] that, when μ_1 and μ_0 are expected values of normal distributions, each having the same (known) σ^2, n can be calculated from the formula

$$n = (z_{1-\alpha} + z_{1-\beta})^2 \left(\frac{\sigma}{d}\right)^2. \tag{13.67}$$

For *two-sided* alternatives (with $|\mu_1 - \mu_0| = d$) the formula is

$$n \doteq (z_{1-\frac{1}{2}\alpha} + z_{1-\beta})^2 \left(\frac{\sigma}{d}\right)^2. \tag{13.68}$$

When normal approximation is used for the distribution of binomial proportion, the required sample size for the *one-sided* alternatives is

$$n \doteq \left[\frac{z_{1-\alpha}\sqrt{P_0(1 - P_0)} + z_{1-\beta}\sqrt{P_1(1 - P_1)}}{P_1 - P_0}\right]^2. \tag{13.69}$$

If the second term in (13.66) is negligible, the approximate sample size for two-sided alternatives can be calculated, replacing $z_{1-\alpha}$ by $z_{1-\frac{1}{2}\alpha}$ in (13.69).

Example 13.8. What should be the sample size, n, in order to discriminate between $H_0: P = \frac{3}{4}$ and $H_1: P = \frac{13}{16} > \frac{3}{4}$, if $\alpha = 0.05$ and $1 - \beta = 0.90$ (i.e., $\beta = 0.10$)?

We apply formula (13.69). We have $z_{1-\alpha} = z_{0.95} = 1.645$ and $z_{1-\beta} = z_{0.90} = 1.28$; $P_1 - P_0 = \frac{1}{16} > 0$. We find $n \doteq 376$.

▼ 13.16. SEQUENTIAL TESTS

In all our problems of testing hypotheses, the sample size, n, was supposed to be fixed. In some cases, it would be much more economical if we could devise a procedure such that only one observation at a time is taken and, after it is obtained, we decide whether to accept H_0 or H_1 on the evidence so far available or, that a decision cannot yet be made, we continue the sampling. Methods which provide such techniques are called *sequential procedures*. Tests based on these procedures are termed *sequential tests*.

We do not attempt to discuss sequential methods in detail. The interested reader is referred to the book by Wald (1947), who originally initiated sequential analysis. The only sequential procedure presented here is the sequential probability ratio test of Wald.

13.16.1. Sequential Probability Ratio Test

Suppose that we consider two *simple* hypotheses about a parameter θ, that is, $H_0:\theta = \theta_0$ and $H_1:\theta = \theta_1$. The sequential probability ratio test (briefly, the S.P.R.T.) is constructed as follows. Let

$$\Pr\{\text{Accept } H_1 \mid H_0 \text{ true}\} = \alpha \quad \text{and} \quad \Pr\{\text{Accept } H_0 \mid H_1 \text{ true}\} = \beta$$

$$(13.70)$$

be the specified probabilities of errors of the first and second kinds, respectively. Let n be the number of observations at a certain stage. Let $p(x; \theta_0)$ be the discrete distribution of x when H_0 is true, and $p(x; \theta_1)$ when H_1 is true. Let $x_1, x_2, \ldots, x_n$ be a simple random sample of size n. The corresponding likelihood functions are:

$$L_n(\theta_0) = \prod_{i=1}^{n} p(x_i; \theta_0) \quad \text{when } H_0 \text{ is true};$$

$$(13.71)$$

$$L_n(\theta_1) = \prod_{i=1}^{n} p(x_i; \theta_1) \quad \text{when } H_1 \text{ is true}.$$

(We now use the subscripts n to indicate the sample size.)

The *probability ratio* is defined as

$$\lambda_n = \prod_{i=1}^{n} \frac{p(x_i; \theta_1)}{p(x_i; \theta_0)} = \frac{L_n(\theta_1)}{L_n(\theta_0)}. \tag{13.72}$$

Taking logarithms of both sides of (13.72), we have

$$\log \lambda_n = \sum_{i=1}^{n} \log \frac{p(x_i; \theta_1)}{p(x_i; \theta_0)}. \tag{13.73}$$

Let

$$y = \log \frac{p(x; \theta_1)}{p(x; \theta_0)} \tag{13.74}$$

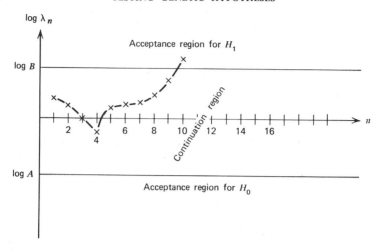

Figure 13.2.

and

$$y_i = \log \frac{p(x_i; \theta_1)}{p(x_i; \theta_0)}, \qquad i = 1, 2, \ldots, n. \qquad (13.74a)$$

Substituting (13.74a) into (13.73), we obtain

$$\log \lambda_n = \sum_{i=1}^{n} y_i. \qquad (13.75)$$

Put

$$A = \frac{\beta}{1 - \alpha}, \quad B = \frac{1 - \beta}{\alpha}. \qquad (13.76)$$

The rule of the S.P.R.T. is as follows:

(i) if $\log \lambda_n \geqslant \log B$, accept H_1;
(ii) if $\log \lambda_n \leqslant \log A$, accept H_0;   (13.77)
(iii) if $\log A < \log \lambda_n < \log B$, continue sampling.

Figure 13.2 gives a graphical representation of a typical sequential procedure in a general case. In the example given in the figure the decision "accept H_1" is reached after $n = 10$ observations.

13.16.2. Sequential Probability Ratio Test for Proportions

We wish to discriminate between two hypotheses, $H_0 : P = P_0$ and $H_1 : P = P_1$, $(P_1 > P_0)$, where P is the probability of occurrence of an event E

in a population. Let

$$x_i = \begin{cases} 1 & \text{if } E \text{ occurs at the } i\text{th trial,} \\ 0 & \text{if } E \text{ does not occur at the } i\text{th trial.} \end{cases}$$

The likelihood functions, $L_n(P_0)$ and $L_n(P_1)$, are

$$L_n(P_0) = \prod_{i=1}^{n} P_0^{x_i}(1 - P_0)^{1-x_i}, \tag{13.78a}$$

$$L_n(P_1) = \prod_{i=1}^{n} P_1^{x_i}(1 - P_1)^{1-x_i}; \tag{13.78b}$$

and the likelihood ratio, λ_n, is

$$\lambda_n = \frac{L_n(P_1)}{L_n(P_0)} = \prod_{i=1}^{n} \left(\frac{P_1}{P_0}\right)^{x_i}\left(\frac{1 - P_1}{1 - P_0}\right)^{1-x_i} = \left(\frac{P_1}{P_0}\right)^{r_n}\left(\frac{1 - P_1}{1 - P_0}\right)^{n-r_n}, \tag{13.79}$$

where

$$r_n = \sum_{i=1}^{n} x_i$$

is the number of times E occurs in the n trials.

From (13.77) we obtain the sequential procedure for discriminating between P_0 and P_1:
  (i) if

$$r_n \log \frac{P_1(1 - P_0)}{P_0(1 - P_1)} \geqslant \log \frac{1 - \beta}{\alpha} + n \log \frac{1 - P_0}{1 - P_1},$$

accept $H_1 : P = P_1$;
  (ii) if

$$r_n \log \frac{P_1(1 - P_0)}{P_0(1 - P_1)} \leqslant \log \frac{\beta}{1 - \alpha} + n \log \frac{1 - P_0}{1 - P_1},$$

accept $H_0 : P = P_0$;
  (iii) if

$$\log \frac{\beta}{1 - \alpha} < r_n \log \frac{P_1(1 - P_0)}{P_0(1 - P_1)} - n \log \frac{1 - P_0}{1 - P_1} < \log \frac{1 - \beta}{\alpha},$$

$$\left.\begin{array}{c}\\\\\\\\\\\\\\\\\\\\\\\\\\\\\\\end{array}\right\} \text{(13.80)}$$

continue the sampling.
  For brevity, let us denote

$$\frac{P_1}{P_0} = v_0, \quad \frac{1 - P_1}{1 - P_0} = v_1. \tag{13.81}$$

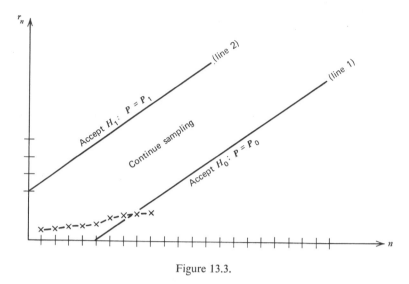

Figure 13.3.

Taking into account (13.76) and (13.81), we can write the sequential procedure (13.80) as follows:

(i) if $r_n (\log v_0 - \log v_1) \geqslant \log B - n \log v_1$, accept $H_1 : P = P_1$;

(ii) if $r_n (\log v_0 - \log v_1) \leqslant \log A - n \log v_1$, accept $H_0 : P = P_0$;

(iii) if $\log A < r_n (\log v_0 - \log v_1) + n \log v_1 < \log B$, continue sampling.
$$(13.82)$$

The sequential plan can be represented as in Fig. 13.3, where the number of trials, n, is plotted as abscissa, and the number of "successes," r_n, as ordinate. The two parallel lines,

and
$$
\left.
\begin{aligned}
r_n &= \frac{1}{\log v_0 - \log v_1} (\log A - n \log v_1) \quad \text{(line 1)} \\[2mm]
r_n &= \frac{1}{\log v_0 - \log v_1} (\log B - n \log v_1) \quad \text{(line 2)},
\end{aligned}
\right\}
\quad (13.83)
$$

determine the three regions for the sequential test (see Fig. 13.3). Note that these two lines have the common slope, s, say,

$$
s = \frac{-\log v_1}{\log v_0 - \log v_1} = \left(\log \frac{1 - P_0}{1 - P_1} \right) \Big/ \left(\log \frac{P_1}{P_0} - \log \frac{1 - P_1}{1 - P_0} \right). \quad (13.84)
$$

13.16.3. The Average (Expected) Sample Number of a Sequential Test for Proportions

The number of observations, n, required to reach a decision by using the S.P.R.T. is a random variable. We might be interested in knowing the average (expected) number of trials, $E(n)$ (called, briefly, ASN—"average sample number"), which we need to reach the decision (accept or reject H_0). The value of $E(n)$ depends on the true value of P. It can be shown [Wald (1949), Chapter 5] that, if the true proportion is P_0, the average sample number, $E(n \mid P_0)$, is approximately

$$E(n \mid P_0) \doteq \frac{(1 - \alpha)\log \dfrac{\beta}{1 - \alpha} + \alpha \log \dfrac{1 - \beta}{\alpha}}{P_0 \log \dfrac{P_1}{P_0} + (1 - P_0)\log\dfrac{1 - P_1}{1 - P_0}}$$

$$= \frac{(1 - \alpha)\log A + \alpha \log B}{P_0 \log v_0 + (1 - P_0)\log v_1}. \qquad (13.85)$$

If the true proportion is P_1, the average sample number, $E(n \mid P_1)$, is approximately

$$E(n \mid P_1) = \frac{\beta \log \dfrac{\beta}{1 - \alpha} + (1 - \beta)\log \dfrac{1 - \beta}{\alpha}}{P_1 \log \dfrac{P_1}{P_0} + (1 - P_1)\log\dfrac{1 - P_1}{1 - P_0}}$$

$$= \frac{\beta \log A + (1 - \beta)\log B}{P_1 \log v_0 + (1 - P_1)\log v_1}, \qquad (13.86)$$

provided the denominator is not zero.

Example 13.9. In Example 13.8 we wanted to discriminate between the hypotheses $P_0 = \frac{3}{4}$ and $P_1 = \frac{13}{16}$. Using $\alpha = 0.05$ and $\beta = 0.10$, we found that the required *fixed* sample size should be at least 376. Now we wish to know what should be the ASN for sequential procedure. We have

$$\log A = \log \frac{\beta}{1 - \alpha} = \log \frac{0.10}{0.95} = \log 0.1053 = -0.97757,$$

$$\log B = \log \frac{1 - \beta}{\alpha} = \log \frac{0.90}{0.05} = \log 18.00 = 1.25527,$$

$$\log v_0 = \log \frac{P_1}{P_0} = \log \frac{13}{12} = \log 1.0833 = 0.03475,$$

$$\log v_1 = \log \frac{1 - P_1}{1 - P_0} = \log \frac{3}{4} = \log 0.7500 = -0.12494.$$

(The logarithms used here are the common (decimal) logarithms; for the formulae applied here it would make no difference if the decimal or natural logarithms were used.)

We calculate from (13.85)

$$E(n \mid P_0) = \frac{0.95 \times (-0.97757) + 0.05 \times 1.25527}{0.75 \times 0.03475 + 0.25 \times (-0.12494)} \doteq 168.$$

If $H_0 : P = \frac{3}{4}$ is true, the ASN is about 168, which is less than half of the fixed sample size. We calculate from (13.86)

$$E(n \mid P_1) = \frac{0.10(-0.97757) + 0.90 \times (1.25527)}{0.8125 \times 0.03475 + 0.1875 \times (-0.12494)} \doteq 215.$$

Thus, if H_1 is true, the ASN is about 215.

As we can see, the practical advantage of sequential methods is that we usually reduce the sample size considerably, as compared with tests based on fixed sample sizes.

SUMMARY

In this chapter, we have presented methods by which genetic hypotheses regarding mode of inheritance can be rejected or accepted, with some probabilities of commiting errors of two kinds: rejecting a true hypothesis (error of the first kind) and accepting a false hypothesis (error of the second kind). The methods of constructing such tests, in the general case of parametric hypotheses, have been briefly outlined. Since the X^2-statistic is used in many genetic problems as the test criterion, different kinds of X^2-tests have been discussed in detail. We have presented significance tests about parameters, heterogeneity tests for data from different populations and/or from different experimental designs, and tests for association. We have distinguished between cases in which the null hypothesis is specified and those in which the values of the parameters are not given. We have shown, in a few examples, how these tests can be applied to data from experimental and natural populations and how at least some hypotheses which do not fit the data can be eliminated. It has also been shown that association tests are sometimes identical with heterogeneity tests, although the null hypotheses in the two cases are different. In all cases, the X^2-tests are only approximate. The approximation is fairly good, however, provided that the expected number in each multinomial class is not less than 3 (some authorities have suggested that it should be at least 5). Finally, we have introduced the basic ideas of sequential tests. In sequential procedures the sample size is not fixed *a priori*, and the decision whether to accept the null hypothesis or its alternative or whether to continue the sampling can be made each time that an observation has been obtained. A sequential test very often results in reduction of average sample size as compared with fixed sample sizes.

REFERENCES

Bodmer, W. F., and Payne, Rose, Theoretical consideration of leucocyte grouping using multispecific sera. In: *Histocompatibility Testing*, Munksgaard, Copenhagen, 1965, pp. 141–149.

Brownlee, K. A., *Statistical Theory and Methodology in Science and Engineering*, John Wiley & Sons, New York, 1960.

David, H. T., Goodness of fit. In: *International Encyclopedia of the Social Sciences*, The Macmillan Company and Free Press, New York, 1968, Vol. 6, pp. 199–207.

Dixon, W. J., and Massey, F. J., *Introduction to Statistical Analysis*, McGraw-Hill Book Company, New York, 1969 (third edition).

Grizzle, J. E., Continuity correction in the χ^2 test for 2 × 2 tables, *Amer. Statistician* **21** (1967), 28–32.

Johnson, N. L., and Leone, F. C., *Statistics and Experimental Design*, Vols, I and II, John Wiley & Sons, New York, 1964.

Lehmann, E. L., *Testing Hypotheses*, John Wiley & Sons, New York, 1959.

Vos, G. H., Kirk, R. L., and Steinberg, A. G., The distribution of the gamma globulin types Gm(a), Gm(b), Gm(x) and Gm-like in South and Southeast Asia and Australia, *Amer. J. Hum. Genet.* **15** (1963), 44–52.

Wald, A., *Sequential Analysis*, John Wiley & Sons, New York, 1947.

PROBLEMS

13.1. (*a*) Prove that the right-hand side of formula (13.51) can be written in the form

$$X^2_{\text{Total}} = \sum_{l=1}^{k} \frac{1}{P_{l0}} \sum_{g=1}^{m} \frac{r_{lg}^2}{n_g} - N. \tag{i}$$

(*b*) Prove that the right-hand side of (13.52) can be written in the form

$$X^2_{\text{Comb}} = \sum_{l=1}^{k} \frac{1}{P_{l0}} \frac{R_l^2}{N} - N. \tag{ii}$$

[*Hint:* See the proof of formula (12.73).]

13.2. Using results (i) and (ii) in Problem 13.1, show that the right-hand side of (13.53) for X^2_{Diff} is correct.

13.3. Using the results of Example 12.4, test whether the observed data fit the hypothesis that the "hybrid" population mates at random and is in equilibrium with respect to the *ABO* blood group system.

13.4. Let a random variable X be normally distributed as $N(\mu, \sigma^2)$. Construct a significance test for testing $H_0: \mu = \mu_0$ against $H: \mu \neq \mu_0$.

(*a*) Assume σ^2 is known.

(*b*) Assume σ^2 is unknown. (*Hint:* Use the *t*-distribution.)

13.5. Solve Problem 13.4 when the alternative is one-sided, $H: \mu > \mu_0$.

13.6. The confidence interval and the region of acceptance for the mean μ_0 of normal distribution with known σ are defined by the same probability condition, that is,

$$\Pr\left\{|\bar{x} - \mu_0| \leqslant z_{1-\frac{1}{2}\alpha} \frac{\sigma}{\sqrt{n}}\right\} = 1 - \alpha.$$

However, the confidence interval is

$$\bar{x} - z_{1-\frac{1}{2}\alpha} \frac{\sigma}{\sqrt{n}} \leqslant \mu_0 \leqslant \bar{x} + z_{1-\frac{1}{2}\alpha} \frac{\sigma}{\sqrt{n}}, \tag{i}$$

while the region of acceptance is

$$\mu_0 - z_{1-\frac{1}{2}\alpha} \frac{\sigma}{\sqrt{n}} \leqslant \bar{x} \leqslant \mu_0 + z_{1-\frac{1}{2}\alpha} \frac{\sigma}{\sqrt{n}}. \tag{ii}$$

Explain the similarities and differences in interpretation of the two statements, (i) and (ii). [*Hint:* Read two articles in *International Encyclopedia of the Social Sciences*, The Macmillan Company and Free Press, New York, 1963—(*a*) "Confidence intervals and regions" by J. Pfanzagl, Vol. 5, pp. 150–157, and (*b*) "Hypothesis testing" by E. L. Lehmann, Vol. 7, pp. 40–47.]

13.7. A certain trait (in plants) appears in two phenotypic forms, T and t, say. The following data from $n = 50$ intercrosses were obtained: $33T$ and $17t$.

(*a*) Test the hypothesis (H_0) that the trait T is a simple dominant (i.e., $T:t = 3:1$). Use $\alpha = 0.05$.

(*b*) Test the hypothesis (H_1) that the traits T and t are results of the complementary epistasis of two independent loci (i.e., $T:t = 9:7$). Use $\alpha = 0.05$.

As you can see, both of these hypotheses fit the data.

(*c*) Find the power of the test of H_0 with respect to H_1 (note that the test of H_0 should be the left-sided test). You will find that the power is not very large. To discriminate between H_0 and H_1, we have to increase the sample size.

(*d*) What should the sample size, n, be to effect discrimination between H_1 and H_2 with $\alpha = 0.05$ and $\beta = 0.05$?

13.8. Three sets of data on the genotype frequencies of the MN blood group system in the Puglie region, located at the tip of the "heel" of Italy, have been obtained by different investigators. The frequencies r_{lg} in the accompanying table have been recalculated from Table 5, given by Modiano et al. [*Ann. Hum. Genet. Lond.* 29 (1965), 25].

Reference	Sample size, n_g	Observed frequencies of genotypes, r_{lg}		
		MM	*MN*	*NN*
Introna (1953)	350	135	146	69
Liaci and Scakano (1959)	1000	417	424	159

(a) Test the hypothesis, $H_0^{(a)}$, that each of these two populations is in Hardy-Weinberg equilibrium.

(b) As you will see, this hypothesis is rejected. Test the hypothesis, H_0, that these populations are homogeneous with respect to another hypothesis, $H_0^{(x)}$.

(c) Let $H_0^{(b)}$ be a new hypothesis that these populations exhibit heterozygote disadvantage with relative fitness ϕ, that is, $NM:MN:NN = 1:\phi:1$ ($0 < \phi < 1$). Write the general form of the genotypic distribution under $H_0^{(b)}$.

(d) From (b) you will learn that the hypothesis H_0 about homogeneity is not rejected. Assuming that H_0 is true (i.e., we accept H_0), fit the hypothesis $H_0^{(b)}$ described in (c).

(e) Let $H_0^{(c)}$ be another hypothesis that the relative fitnesses of genotypes are $MM:MN:NN = 1:\phi_1:\phi_2$ ($0 > \phi_1 < 1, 0 < \phi_2 < 1$). Assuming that H_0, stated in (b), is true, test the hypothesis $H_0^{(c)}$.

13.9. Show that the maximum likelihood estimators for the marginal probabilities $P_{i.}$, $P_{.j}$ and the cell probabilities P_{ij}, under the assumption of independence, in an $r \times c$ contingency table as presented in Table 13.1, are

$$P_{i.} = \frac{n_{i.}}{N}, \quad P_{.j} = \frac{n_{.j}}{N}, \quad \text{and} \quad P_{ij} = \frac{n_{i.}}{N} \cdot \frac{n_{.j}}{N},$$

respectively.

(*Hint:* Note that when two classifications are independent, the likelihood function for an $r \times c$ contingency table is

$$L = \prod_{i=1}^{r} \prod_{j=1}^{c} P_{ij}^{n_{ij}} = \prod_{i=1}^{r} P_{.j}^{n_{i.}} \cdot \prod_{j=1}^{c} P_{.j}^{n_{.j}}.$$

13.10. The following dominant-recessive pairs of four traits in mice have been recorded by Little and Phillips [*Amer. Nat.* **47** (1913), 760]: A—agouti, a—non-agouti; B—black, b—nonblack (brown); D—density, d—diluteness; P—dark eye, p—pink eye.

The 16 possible phenotypes occurred with the following frequencies:

Phenotype:	ABDP[‡]		AbDP		ABDp		AbdP	
		aBDP		ABdP		abDP		aBdP
No. of mice:	436	127	103	130	103	40	31	37

Phenotype:	aBDp		ABdp		abDp		Abdp	
		AbDp		abdP		aBdp		abdp
No. of mice:	35	38	38	11	12	17	15	7

(a) Test the hypothesis that these loci are independent.

(b) Test the hypothesis that, for each locus separately, the phenotypic ratio is 3:1.

‡ Note that ABDP is, in fact, A—B—D—P—, etc

13.11. In sections 13.12.2 and 13.13 we considered a situation in which the multinomial parameters are functions of parameters $\boldsymbol{\theta}' = (\theta_1, \theta_2, \ldots, \theta_s)$ and experiments are of the same type, that is, for the gth experiment we have $P_{lg}(\boldsymbol{\theta}) = P_l(\boldsymbol{\theta})$, $l = 1, 2, \ldots, k$; $g = 1, 2, \ldots, m$. We may also consider a situation in which the experiments are of different types. Formulae (13.52a), (13.54), and (13.56) will still be valid, but it will be necessary to replace $P_l(\boldsymbol{\theta})$ by $P_{lg}(\boldsymbol{\theta})$, that is, $P_l(\boldsymbol{\theta}_0)$ by $P_{lg}(\boldsymbol{\theta}_0)$ in (13.52a), $P_l(\hat{\boldsymbol{\theta}}_g)$ by $P_{lg}(\hat{\boldsymbol{\theta}}_g)$ in (13.54), and $P_l(\hat{\hat{\boldsymbol{\theta}}})$ by $P_{lg}(\hat{\hat{\boldsymbol{\theta}}})$ in (13.56), respectively.

Apply the X^2-tests (13.54) and (13.55), using these adjustments, to the data of Example 13.5.

▼13.12. Derive formula (13.66), that is, for the approximate power when testing $H_0: P = P_0$ against $H: P \neq P_0$.

CHAPTER 14

Human Blood Groups

14.1. INTRODUCTION

The term "blood groups" is specifically applied to antigens which are present on surfaces of red cells and not to other substances also present in red cells (see section 14.2). The most important blood group systems, such as *ABO*, *MNSs*, and *Rh*, occur in several populations with appreciable frequencies and represent useful traits for analyzing the genetic structure of a particular population.

In this chapter, we shall briefly outline the principles of immunohematology, discuss the genetic complexity of different blood group systems, and present some statistical methods employed in estimating gene frequencies. The list of references given at the end of the chapter was compiled with the aim of helping readers who want to study more extensively the serology of blood groups and their use in the genetic analysis of different populations as well as the derivation of statistical methods especially useful in this field. A detailed discussion of the recent status of serology and the inheritance of different blood group systems is given in the book by Race and Sanger (1968).

14.2. SEROLOGY OF BLOOD GROUPS

Antigens are, by definition, structures able to evoke the production of substances called *antibodies* which, in turn, react with these antigens. An antigen represents a macromolecule with several sites of activity. An individual antigenic site on the antigen molecule which combines with a specific antibody is called an *antigenic determinant* or an *antigenic specificity*.

Among the components of blood are *red cells* (called also erythrocytes) suspended in the plasma. The plasma fluid (i.e., plasma without fibrinogen) is called *serum*. The red cells carry on their envelopes different blood substances. Some of these substances, composed mostly of sugars and proteins,

391

are blood antigens (termed also *agglutinogens*). Antibodies which react with agglutinogens are sometimes called *agglutinins*. If U is an antigen, then the antibody which reacts with U is called anti-U, and a serum which contains antibodies reacting with U is called *antiserum* against U or *anti*-U *serum*. A serum which reacts only with a specific antigenic determinant is a *unispecific* serum. Several sera are *multispecific*; they contain antibodies which can react with different antigenic specificities. Antigen is the substance itself, while the antigenic determinants, called also "blood factors," are the serological properties by which antigens are recognized [Wiener (1961)].

The basic rule of immunology is that the serum of a given individual does not possess, and cannot produce, antibodies against its own antigens. The "naturally" occurring human antibodies, anti-A and anti-B, have been found in sera of individuals of blood groups B and A, respectively. Also the serum of individuals of group O possesses both anti-A and anti-B natural antibodies.

Antibodies which do not occur naturally but are evoked in a serum by injecting or mixing specific antigen with an appropriate serum are called "immune" antibodies. For instance, A and B can be injected into rabbits, and the rabbits' sera will then produce anti-A and anti-B, respectively.

To determine an individual's blood group, a battery of diagnostic reagents (unispecific antisera) is required. It is not always possible to obtain such sera; but there are a few methods, well known to specialists in blood group typing, by which the amounts of different types of antibodies can be separated and estimated even though a serum is not unispecific.

14.3.   INHERITANCE OF BLOOD ANTIGENS

We recall the definition of a gene (or cistron) as a segment of DNA which is responsible for a single primary function. In biochemical genetics, this is equivalent to the segment of DNA which controls the synthesis of a single polypeptide chain (see section 1.2). Since most of the blood group antigens are composed of sugars and proteins, it follows that these antigens are not direct gene products, but may result from a sequence of biochemical reactions to which several genes from different loci may contribute. But it will still be useful to be able to use the terms "alleles" to distinguish among different forms associated with "hereditary units" which, from a biochemical point of view, may correspond to certain sequences of biochemical reactions.

We now recall the term "system" defined as a DNA region determining a sequence of genetic information which cannot be separated into subunits and which is inherited in the same way as if it corresponded to a series of alleles. An "antigen" will be now a product of such an "allele" and, for simplicity, we will consider such an "allele" as a form of a "gene," that is, as an "inheritance unit" (see section 1.2). Later in the text, we shall often omit quotation marks and use the terms allele or gene as defined above.

14.4. GENETICS OF THE A_1A_2BO-BLOOD GROUP SYSTEM AND ITS ASSOCIATION WITH THE H AND Se LOCI

14.4.1.   The A_1A_2BO and Hh Systems

The main blood groups, A, B, and O, were discovered by Landsteiner (1901). A year later, the AB group was found by two of his students. These four groups are determined by two antigenic substances, A and B, present on the surfaces of red cells, and by two naturally occurring antibodies, anti-A and anti-B. Anti-A occurs in the serum of B, anti-B in the serum of A, and both anti-A and anti-B in the serum of O blood group individuals. Using family data, Bernstein (1925) demonstrated that these data are consistent with the hypothesis that the ABO groups are inherited as a simple Mendelian trait with three alleles, A, B, O. The alleles A and B are co-dominants, whereas O is an "inactive" recessive. If a gene has no serologically detectable product and hence is antigenically inactive, it is called an *amorph*. Thus O is considered as an amorph. Here we denote the alleles (and the antigens associated with them) by the same letters, A, B, O; it is also customary to denote the alleles by L^A, L^B and L^O, respectively (in honor of Landsteiner).

Later discoveries have divided group A into two subgroups, A_1 and A_2. It has been shown that the serum previously called anti-A has two components: anti-A and anti-A_1 antibodies. Anti-A agglutinates all A cells (i.e., both A_1 and A_2), whereas anti-A_1 agglutinates only A_1 cells. It is assumed that A_1 and A_2 are alleles; A_2 is recessive to A_1, but both A_1 and A_2 are dominant over O. Therefore, we shall distinguish ten genotypes but only six phenotypes as follows: blood group O (genotype OO); A_1 (A_1A_1, A_1A_2, A_1O); A_2 (A_2A_2, A_2O); B (BB, BO); A_1B (A_1B); and A_2B (A_2B).

Another antibody, called anti-H, has been found in the saliva of A_1 or A_1B individuals and also in the extracts of certain seeds, such as those of *Ulex europaeus*. It has been observed that cells of the O group react strongly with anti-H. Cells of the A_2 and A_2B groups also react quite strongly with anti-H, whereas the reactions of the A_1 and B cells are rather weak. At first, it was supposed that the H substance was an antigenic determinant associated with the allele O. It was shown later, however, that this is not so, since the A_2B genotypes react quite strongly with anti-H. It is now the general opinion that there exists another locus H, with dominant allele H and its amorph h, and that this locus is independent of the ABO locus. Since the frequency of h is extremely low, almost all individuals are HH or Hh.

One biochemical hypothesis about the interaction of the Hh and ABO loci is that an antigen H (present in HH and Hh subjects) is a product of the gene H. The gene A or B interacting with the gene H converts, in large part, the antigen H into the antigen A or B, respectively, so that only a small amount of H is present in the saliva of A_1 or A_1B individuals. The allele O is an amorph with no conversion, so that only antigen H is detected in the saliva of O individuals. If an individual is hh, (which is a very rare case), no conversion occurs and neither A, B, nor O will be detected [see Race and Sanger (1968), p. 62]. Although this model has not been proved to be true, it does explain, at least, such phenomena as the "Bombay O_h" group, a case in which an offspring of an A_1 father and an O mother had the A_1B group [Race and Sanger (1968), pp. 23–26].

14.4.2.　Interaction of the ABO, Hh, and $Sese$ (Secretion) Loci

Further studies of the ABO and H blood group antigens have shown that they are not confined to the red cells but occur in almost all body tissues. It appears that there are two distinct forms of these antigens: (*a*) an alcohol-soluble form present in the red cells and all the tissues (except brain) but not present in secretions; and (*b*) a water-soluble form not present in the red cells but present in almost all body fluids and secretions of about 75–78% of White people. These persons are called *secretors*. Secretion of the A, B, and H antigens is controlled by another locus, Se, independent of the ABO and Hh loci, with two alleles, Se and se. The ability to secrete the A, B, and H substances is controlled by the gene Se, the function of its allele, se, being unknown. Thus $SeSe$ and $Sese$ are secretors, whereas $sese$ are non-secretors. Here we have an example of epistatic interaction of genes from different (independent) loci.

14.4.3.　Other Alleles at the ABO Locus

Other alleles, such as A_3 and some weak forms, called A_x or A_m types, have been reported. These are usually rare, however, and we will not consider them in our statistical calculations.

14.5. ESTIMATION OF GENE FREQUENCIES IN BLOOD GROUP SYSTEMS. GENERAL REMARKS

In deriving formulae for estimating gene frequencies, we assume that selection and mutation (*if* they operate) have only minor effects and can be neglected.

Let us use as an example the *ABO* blood group system. If we restrict ourselves to the early model, when only anti-A and anti-B sera were known, only three alleles in the system will be recognized, that is, *A*, *B*, and *O*. Hence we shall distinguish six genotypes but only four phenotypes (i.e., blood groups) as follows: group O (genotype *OO*); A (*AA*, *AO*); B (*BB*, *BO*), and AB (*AB*).

(i) Suppose that a *simple random sample* of n individuals from one generation (e.g., parents) is tested, and n_A, n_B, n_{AB} and n_O, with $n_A + n_B + n_{AB} + n_O = n$, are the observed frequencies of phenotypes A, B, AB, and O, respectively. Thus, the random variables n_A, n_B, n_{AB}, and n_O will have a joint multinomial distribution with parameters P_A, P_B, P_{AB}, and P_O, with $P_A + P_B + P_{AB} + P_O = 1$, as defined in Table 14.1.

We see that the multinomial parameters are functions of the gene frequencies, p, q, and r. In order to estimate p, q, and r, the likelihood function can be constructed, bearing in mind that $p + q + r = 1$, so that only two independent parameters need be estimated. In fact, before the maximum likelihood method was known, Bernstein (1925) introduced, for the *ABO* system, rather simple estimators which can be almost as efficient as the ML estimators in some situations. We shall discuss in outline, three methods of estimating gene frequencies, relevant to blood group systems:

(*a*) Bernstein's method.
(*b*) Maximum likelihood method.
(*c*) Gene counting method.

TABLE 14.1 Phenotype distribution in the *ABO* blood system

Phenotype	Genotype	Expected proportion	Observed frequency
A	*AA*, *AO*	$P_A = p^2 + 2pr$	n_A
B	*BB*, *BO*	$P_B = q^2 + 2qr$	n_B
AB	*AB*	$P_{AB} = 2pq$	n_{AB}
O	*OO*	$P_O = r^2$	n_O
Total		1	n

These methods, as applied to the *ABO* blood system, will be presented in sections 14.6, 14.7, and 14.8, respectively.

(ii) Simple random samples from *two generations* can also be used, bearing in mind that the sample from the offspring should be selected independently of the sample from the parents. For instance, the observed proportions of O offspring from matings A × O or matings B × O can be calculated, and methods similar to those presented in section 12.7 or section 12.8 can be applied.

(iii) Sampling from *related* pairs can also be used, utilizing the joint phenotype distributions of related pairs such as parent-child or sib-sib when more than two alleles are present at the locus. The appropriate joint phenotype distributions for parent-child can be obtained from formulae (7.25), and for sib-sib from formulae (7.72). Care should be taken about the mode of drawing the samples (compare section 12.9).

(iv) It is also possible to estimate blood gene frequencies by using *family data* with known sizes of offspring. The appropriate methods will be discussed in Chapter 18.

We shall now present methods (i)–(iii), based on a simple random sample from one generation.

14.6. BERNSTEIN'S METHOD OF ESTIMATING THE *ABO* GENE FREQUENCIES

In this method, we equate the expected proportions of phenotypes A, B, and O to the corresponding observed proportions. We have to solve the following equations:

$$p^2 + 2pr = \frac{n_A}{n}, \quad q^2 + 2pq = \frac{n_B}{n}, \quad r^2 = \frac{n_O}{n}, \tag{14.1}$$

with

$$n_A + n_B + n_{AB} + n_O = n. \tag{14.2}$$

From (14.1) we obtain

$$r = \sqrt{\frac{n_O}{n}}. \tag{14.3a}$$

Substituting (14.3a) into the two remaining equations and taking into account that p and q are nonnegative, we obtain

$$p = \sqrt{\frac{n_0 + n_A}{n}} - \sqrt{\frac{n_0}{n}}, \qquad (14.3b)$$

$$q = \sqrt{\frac{n_0 + n_B}{n}} - \sqrt{\frac{n_0}{n}}. \qquad (14.3c)$$

[Solutions (14.3) are known as *Wiener's estimators*.] But also from (14.1) we have

$$p = 1 - r - q = 1 - \sqrt{\frac{n_0 + n_B}{n}}$$

$$\text{and} \quad q = 1 - r - p = 1 - \sqrt{\frac{n_0 + n_A}{n}}.$$

These are known as *Bernstein's estimators*. If we denote Bernstein's estimators of r, p, and q by $\tilde{r}$, $\tilde{p}$, and $\tilde{q}$, respectively, we have

$$\tilde{r} = \sqrt{\frac{n_0}{n}}, \quad \tilde{p} = 1 - \sqrt{\frac{n_0 + n_B}{n}}, \quad \tilde{q} = 1 - \sqrt{\frac{n_0 + n_A}{n}}. \qquad (14.4)$$

To calculate the variances, we notice that $\tilde{p}$, $\tilde{q}$, and $\tilde{r}$ can be considered as functions of binomial proportions. Thus, for example, $y = (n_0/n)$ is binomially distributed with $P = r^2$, and $\text{Var}(y) = \text{Var}(n_0/n) = [r^2(1 - r^2)/n]$. Using the approximate formula (8.93), we obtain

$$\text{Var}(\tilde{r}) \doteq \left(\frac{d\tilde{r}}{dy}\right)^2_{y=r^2} \text{Var}(y) = \frac{1}{4r^2} \cdot \frac{r^2(1 - r^2)}{n} = \frac{1 - r^2}{4n} \qquad (14.5a)$$

(compare also Example 8.10).

The sample proportion $z = (n_B + n_0)/n$ can also be considered as a binomial variable with parameter $P_1 = q^2 + 2qr + r^2 = (q + r)^2$ so that

$$\text{Var}\left(\frac{n_B + n_0}{n}\right) = \frac{(q + r)^2[1 - (q + r)^2]}{n}.$$

Using again the approximate formula (8.93), we obtain

$$\text{Var}(\tilde{p}) \doteq \frac{1 - (q + r)^2}{4n}, \qquad (14.5b)$$

and in a similar way

$$\text{Var}(\tilde{q}) \doteq \frac{1 - (p + r)^2}{4n}. \qquad (14.5c)$$

It should be noticed that, although $p + q + r = 1$, the sum $\tilde{p} + \tilde{q} + \tilde{r}$ is not always, in practice, exactly equal to 1, though it is usually close to this value. Put

$$d = 1 - (\tilde{p} + \tilde{q} + \tilde{r}). \tag{14.6}$$

The *adjusted* Bernstein's estimators are calculated as

$$\tilde{p}' = \tilde{p}(1 + \tfrac{1}{2}d), \quad \tilde{q}' = \tilde{q}(1 + \tfrac{1}{2}d), \quad \tilde{r}' = (\tilde{r} + \tfrac{1}{2}d)(1 + \tfrac{1}{2}d). \tag{14.7}$$

Now $\tilde{p}' + \tilde{q}' + \tilde{r}' = 1 - \tfrac{1}{4}d^2$, which, as compared with (14.6), is closer to 1.

We also notice that Bernstein's formulae do not utilize the observations on the AB group. This can be done by applying the method of maximum likelihood.

14.7. THE MAXIMUM LIKELIHOOD ESTIMATORS OF THE *ABO* GENE FREQUENCIES

The likelihood function, L, for the sample of ABO phenotypes given in Table 14.1, is proportional to

$$(p^2 + 2pr)^{n_A}(q^2 + 2qr)^{n_B}(2pq)^{n_{AB}}(r^2)^{n_O}, \tag{14.8}$$

and its logarithm is

$$n_A \log (p^2 + 2pr) + n_B \log (q^2 + 2qr) + n_{AB} \log 2pq + 2n_O \log r. \tag{14.9}$$

Since $p + q + r = 1$, only two parameters, p and q, say, are functionally independent. To find the ML estimators, p and q, we may substitute $r = 1 - p - q$ into (14.9), so that $\log L$ becomes entirely a function of p and q.

On the other hand, it may be convenient (at least for simplifying the expressions) to retain r in the formulae, bearing in mind that, in fact, $r = 1 - p - q$. Thus, considering $\log L$ as a function $F[p, q, r(p, q)]$, we have

$$\frac{\partial \log L}{\partial p} = \frac{\partial F}{\partial p} + \frac{\partial F}{\partial r} \cdot \frac{\partial r}{\partial p},$$

$$\frac{\partial \log L}{\partial q} = \frac{\partial F}{\partial q} + \frac{\partial F}{\partial r} \cdot \frac{\partial r}{\partial q}. \tag{14.10}$$

In our case, $(\partial r/\partial p) = (\partial r/\partial q) = -1$. For instance, the partial derivative with respect to p of $\log (p^2 + 2pr)$ is

$$\frac{2p + 2r}{p^2 + 2pr} + \frac{2p}{p^2 + 2pr}(-1) = \frac{2r}{p^2 + 2pr},$$

and so on.

Thus, applying (14.10) to the function (14.9), we obtain the likelihood equations

$$
\left.
\begin{aligned}
U_p &= \frac{\partial \log L}{\partial p} = \frac{2r}{p^2 + 2pr} n_A - \frac{2}{q + 2r} n_B + \frac{1}{p} n_{AB} - \frac{2}{r} n_O = 0, \\
U_q &= \frac{\partial \log L}{\partial q} = \frac{2r}{q^2 + 2qr} n_B - \frac{2}{p + 2r} n_A + \frac{1}{q} n_{AB} - \frac{2}{r} n_O = 0, \\
r &= 1 - p - q.
\end{aligned}
\right\}
\tag{14.11}
$$

Since there are no straightforward algebraic solutions of (14.11), the numerical methods described in section 11.8 [formula (11.58)] should be used. For this purpose we have to evaluate the expected information matrix, **I**, that is, find the expected values of the second derivatives of function (14.9). By straightforward differentiation or by using the techniques described in section 12.12, we obtain the elements of the information matrix, **I**:

$$
\left.
\begin{aligned}
I_{pp} &= -E\left(\frac{\partial^2 \log L}{\partial p^2}\right) = \frac{2n(4r - 2q + 6pq + 2q^2 - 3pq^2)}{p(p + 2r)(q + 2r)}, \\
I_{qq} &= -E\left(\frac{\partial^2 \log L}{\partial q^2}\right) = \frac{2n(4r - 2p + 6pq + 2p^2 - 3pq^2)}{p(p + 2r)(q + 2r)}, \\
I_{pq} &= -E\left(\frac{\partial^2 \log L}{\partial p \, \partial q}\right) = \frac{2n(4r + 3pq)}{(p + 2r)(q + 2r)}.
\end{aligned}
\right\}
\tag{14.12}
$$

[See Stevens (1950).]

The inverse $\mathbf{I}^{-1}$ gives the (approximate) variance-covariance matrix, **V**, of the ML estimators, $\hat{p}$, $\hat{q}$, and $\hat{r}$:

$$
\left.
\begin{aligned}
I^{pp} &\doteq \mathrm{Var}(\hat{p}) = \frac{p}{4n}\left[(2 - p) - \frac{p^2 q}{2(r + pq)}\right], \\
I^{qq} &\doteq \mathrm{Var}(\hat{q}) = \frac{q}{4n}\left[(2 - q) - \frac{pq^2}{2(r + pq)}\right], \\
I^{pq} &\doteq \mathrm{Cov}(\hat{p}, \hat{q}) = -\frac{pq}{8n}\left(4 - \frac{pq}{r + pq}\right).
\end{aligned}
\right\}
\tag{14.13}
$$

Hence an approximate formula for the variance of dependent variable r is

$$
\begin{aligned}
\mathrm{Var}(\hat{r}) &\doteq \mathrm{Var}(\hat{p}) + \mathrm{Var}(\hat{q}) + 2\,\mathrm{Cov}(\hat{p}, \hat{q}) \\
&= \frac{1}{4n}\left[(1 - r^2) - \frac{pq(1 + r)^2}{2(r + pq)}\right].
\end{aligned}
\tag{14.14}
$$

[See Yasuda and Kimura (1968).]

A numerical method described in section 11.8 is usually applied, using a computer. Several statistical laboratories already possess standard programs for calculating the ML estimators of parameters which are functions of multinomial proportions.

14.8. GENE COUNTING ESTIMATORS OF THE *ABO* GENE FREQUENCIES

Gene counting methods for the estimation of gene frequencies, which we outlined briefly in Example 11.1 for a case of two alleles, have been discussed more generally by Ceppellini et al. (1955) and, as applied to the *ABO* and *ABO*-like systems, by Yasuda and Kimura (1968). We shall see later that the estimators obtained by gene counting are, in fact, identical with the ML estimators.

From Table 14.1, we see that the A phenotypes can be AA or AO in the proportions

$$h_{AA} = \frac{p^2}{p^2 + 2pr} = \frac{p}{p + 2r} \; ; \; h_{AO} = \frac{2pr}{p^2 + 2pr} = \frac{2r}{p + 2r}, \quad (14.15)$$

respectively. The allele A appears also in the AB phenotype. Among $2n$ genes present in a sample of n individuals, the proportion (relative frequency) of the A gene should be

$$p = \frac{1}{2n}\left(\frac{2p^2}{p^2 + 2pr}\, n_A + \frac{2pr}{p^2 + 2pr}\, n_A + n_{AB}\right)$$

$$= \frac{1}{2n}\left(\frac{p^2 + 2pr}{p^2 + 2pr}\, n_A + n_{AB} + \frac{p^2}{p^2 + 2pr}\, n_A\right)$$

$$= \frac{n_A + n_{AB}}{2n} + \frac{n_A}{2n}\, h_{AA}. \quad (14.16a)$$

In a similar way we find

$$q = \frac{n_B + n_{AB}}{2n} + \frac{n_B}{2n}\, h_{BB}, \quad (14.16b)$$

and

$$r = 1 - p - q = \frac{n_A + n_B + n_O}{2n} - \frac{n_A}{2n}\, h_{AA} - \frac{n_B}{2n}\, h_{BB}, \quad (14.16c)$$

where

$$h_{BB} = \frac{q^2}{q^2 + 2qr} = \frac{q}{q + 2r} \quad (14.17)$$

is the proportion of homozygotes BB among the B phenotypes.

Multiplying each of the equations (14.16) by $2n$ and subtracting (14.16c) from (14.16a) and from (14.16b), respectively, we obtain, after a very little algebra, exactly the likelihood equations (14.11). Thus:

System (14.16) *of gene counting method is equivalent with the maximum likelihood equations* (14.11).

Of course, the solutions of (14.16) are also not obvious, since h_{AA} and h_{BB} include p, q, and r as well. In practice, we start with some trial values, p_0, q_0, r_0. Substituting these into h_{AA} and h_{BB}, we obtain $h_{AA;0}$, $h_{BB;0}$, say. Using these, we calculate the new values, p_1, q_1, and r_1, say, from (14.16). Again substituting these into h_{AA} and h_{BB}, we calculate new values $h_{AA;1}$ and $h_{BB;1}$, say, to obtain the next approximations, and so on. The limiting values will be the same as those obtained by the ML method.

The difference between the gene-counting and the maximum likelihood methods is, in fact, only in the iteration procedures. The gene-counting method needs more iterations than the scoring method, but the iterations are simple and do not require matrix inversion.

It should be pointed out that the gene-counting method applies to any other system, as well as to the ABO blood groups. The appropriate equations can easily be constructed.

Example 14.1. We present an example given by Yasuda and Kimura (1968) and involving the northeastern Brazilian population, and apply the gene counting method in the estimation of gene frequencies. The observed frequencies of phenotypes are as follows: $n = 2128$, $n_A = 725$, $n_B = 258$, $n_{AB} = 72$, $n_O = 1073$. Hence,

$$\frac{n_A + n_{AB}}{2n} = 0.187265; \quad \frac{n_B + n_{AB}}{2n} = 0.077538;$$

$$\frac{n_A}{2n} = 0.170348; \quad \frac{n_B}{2n} = 0.060620.$$

As initial trial values, p_0, q_0, r_0, we take Bernstein's unadjusted estimators. These are

$$\tilde{r} = r_0 = \sqrt{\frac{1073}{2128}} = 0.710091; \quad \tilde{p} = p_0 = 1 - \sqrt{\frac{1331}{2128}} = 0.209133;$$

$$\tilde{q} = q_0 = 1 - \sqrt{\frac{1798}{2128}} = 0.080802.$$

Hence $h_{AA;0} = 0.128356$ and $h_{BB;0} = 0.053833$. Substituting these into (14.16), we obtain the first approximations, $p_1 = 0.209130$, $q_1 = 0.080801$, $r_1 = 0.710069$. Repeating this procedure, we obtain $p_2 = 0.209131$, $q_2 = 0.080801$, $r_2 = 0.710068$, which are also the maximum likelihood estimators, $\hat{p}$, $\hat{q}$, $\hat{r}$, obtained by the scoring method at the first iteration.

Substituting p_2, q_2, and r_2 into the first two equations of (14.13) and into (14.14), we find that the estimated standard deviations of the estimators are $s_{\hat{p}} = 0.0066$, $s_{\hat{q}} = 0.0043$, and $s_{\hat{r}} = 0.0074$.

14.9. EFFICIENCY OF BERNSTEIN'S ESTIMATORS

The variance of Bernstein's estimator, $\tilde{p}$, given in (14.5b), can be written as

$$\text{Var}(\tilde{p}) \doteq \frac{1 - (q + r)^2}{4n} = \frac{1 - (1 - p)^2}{4n} = \frac{p(2 - p)}{4n}. \qquad (14.18)$$

The relative efficiency of $\tilde{p}$ is

$$\text{Rel eff}(\tilde{p}) = \frac{\text{Var}(\hat{p})}{\text{Var}(\tilde{p})} \doteq \frac{p}{4n}\left[(2 - p) - \frac{p^2 q}{2(r + pq)}\right] : \frac{p(2 - p)}{4n}$$

$$= 1 - \frac{p^2 q}{2(r + pq)(2 - p)} < 1. \qquad (14.19)$$

If r is small and p close to 1, then

$$\text{Rel eff}(\tilde{p}) \doteq 1 - \frac{p}{2(2 - p)} \geqslant \frac{1}{2}.$$

If p is small, Rel. eff$(\tilde{p}) \doteq 1$. Thus the relative efficiency of Bernstein's estimator, $\tilde{p}$, lies between $\frac{1}{2}$ and 1. The same applies to Rel eff$(\tilde{q})$. Thus, if both p and q are small and r is rather large, Bernstein's estimators are almost as efficient as the ML estimators.

14.10. ESTIMATION OF THE GENE FREQUENCIES IN THE $A_1 A_2 BO$ BLOOD SYSTEM

When serological methods allow us to distinguish among four alleles, A_1, A_2, B, and O, we can estimate their frequencies, p_1, p_2, q, and r, respectively.

Using obvious notations, we can apply the following (inefficient) estimators due to Wellisch and Thomsen (1930):

$$
\left.\begin{aligned}
\tilde{p}_1 &= \sqrt{\frac{n_O + n_{A_1} + n_{A_2}}{n}} - \sqrt{\frac{n_O + n_{A_2}}{n}}, \\
\tilde{p}_2 &= \sqrt{\frac{n_O + n_{A_2}}{n}} - \sqrt{\frac{n_O}{n}}, \\
\tilde{q} &= \sqrt{\frac{n_O + n_B}{n}} - \sqrt{\frac{n_O}{n}}, \quad \tilde{r} = \sqrt{\frac{n_O}{n}}.
\end{aligned}\right\} \tag{14.20}
$$

The likelihood function can be constructed and the ML estimators obtained in the usual manner (see Problem 14.6).

The ML estimators can also be evaluated by the gene-counting method from the following equations, using iteration techniques as in section 14.8:

$$
\left.\begin{aligned}
p_1 &= \frac{n_{A_1} + n_{A_1 B}}{2n} + \frac{n_{A_1}}{2n} h_{A_1 A_1}, \quad p_2 = \frac{n_{A_2} + n_{A_2 B}}{2n} + \frac{n_{A_2}}{2n} h_{A_2 A_2}, \\
q &= \frac{n_B + n_{A_1 B} + n_{A_2 B}}{2n} + \frac{n_{A_2}}{2n} h_{BB}, \quad r = 1 - p_1 - p_2 - q,
\end{aligned}\right\} \tag{14.21}
$$

where the h's are the expected proportions of the $A_1 A_1$, $A_2 A_2$, and BB homozygotes among the phenotypes A_1, A_2, and B, respectively.

Example 14.2. The Brazilian sample from Example 14.1 was also subtyped with respect to the antigens A_1 and A_2. The following results were observed: $n_{A_1} = 563$, $n_{A_2} = 162$, $n_B = 258$, $n_{A_1 B} = 51$, $n_{A_2 B} = 21$, $n_O = 1073$, $n = 2128$. Using the iteration procedure determined by (14.21), Yasuda and Kimura (1968) obtained the following values of the ML estimates: $\hat{p}_1 = 0.1566$, $\hat{p}_2 = 0.0526$, $\hat{q} = 0.0808$, $\hat{r} = 0.7100$. Estimated standard deviations calculated by the ML-scoring method are $s_{\hat{p}_1} = 0.0058$, $s_{\hat{p}_2} = 0.0038$, $s_{\hat{q}} = 0.0043$, $s_{\hat{r}} = 0.0074$ (Yasuda and Kimura (1968)).

14.11. THE LEWIS BLOOD GROUP SYSTEM

In this system, there are two alleles, *Le* and *le*. The function of the *le* allele is not known, and it is considered to be an amorph of *Le*. Two Lewis antigens, Le^a and Le^b, are detected in the saliva of secretors. The *Le* gene is associated with the production of the Le^a antigen; the Le^b antigen does not have a corresponding gene and is a product of interaction of genes *Le* and *H*. Since H antigen appears only in secretors, Le^b has also been observed only in secretors. The antigen Le^b is then the product of interaction (epistasis) of three independent loci: *Le*, *H*, and *Se*. Table 14.2 represents the epistatic

TABLE 14.2[a] Genotype-phenotype relationships in the $(AB)H$, Lewis, and secretor systems

Genotype			Phenotype							
			Secretions			Red cells				
Hh	Lewis	Secretor	ABH	Le^a	Le^b	ABH	Le^a	Le^b	Lewis Designation	Frequency
H/H or H/h	Le/Le or Le/le	Se/Se or Se/se	+++	+	++	+++	− or +	++	Le(a − b+)	0.735
H/H or H/h	Le/Le or Le/le	se/se	−	+++	−	+++	+++	−	Le(a + b−)	0.231
H/H or H/h	le/le	Se/Se or Se/se	+++	−	−	+++	−	−	Le(a − b−)	0.028
H/H or H/h	le/le	se/se	−	−	−	+++	−	−	Le(a − b−)	0.006
h/h	Le/Le or Le/le	Se/Se or Se/se	−	+++	−	−	+++	−	Le(a + b−)	Very rare
h/h	Le/Le or Le/le	se/se	−	+++	−	−	+++	−	Le(a + b−)	Very rare
h/h	le/le	Se/Se or Se/se	−	−	−	−	−	−	Le(a − b−)	Very rare
h/h	le/le	se/se	−	−	−	−	−	−	Le(a − b−)	Very rare

Gene frequencies: Western European (Race and Sanger, 1962)

H nearly 1.000   Le 0.816   Se 0.523
h very rare   le 0.184   se 0.477

+ + + strong specific activity;
+ weak specific activity;
− no activity.

[a] Taken with permission from E. R. Giblett, *Genetic Markers in Human Blood*, F. A. Davis Co., Philadelphia, 1969, p. 283; adapted from W. Watkins, *Science* **152** (1966), 174.

404

model for these loci and gives the genotype and gene frequencies for Western Europeans. This is another interesting example of epistasis in human genetics.

14.12. THE *P* BLOOD GROUP SYSTEM

In this system, three alleles, P_1, P_2, and p, with appreciable frequencies and two rare alleles, P_1^k, P_2^k are recognized. It appears that the antigen P_1 (associated with the allele P_1) reacts with two antisera, anti-P and anti-P_1, whereas P_2 reacts only with anti-P. No function of the allele p has been detected. Thus, at the present stage, it is assumed that P_1 is dominant over P_2 and p, while P_2 is dominant over p. (Note the resemblance to alleles A_1, A_2, and O in the A_1A_2BO system.) For practical purposes, P_1 and P_2 are sometimes combined into one allele, P, and the individuals PP and Pp are denoted as $P+$, while pp are $P-$ (compare alleles A and O in the ABO system).

Although the P system seems to be simple, the antigens do not appear to be direct products of the alleles. An interesting tentative biochemical model of interacting loci in a series of reactions is discussed by Giblett (1969), p. 299.

14.13. THE *MNSs* BLOOD GROUP SYSTEM

The MN blood groups were discovered by Landsteiner and Levine (1927), using anti-M and anti-N sera produced by immunizing rabbits with human blood. It was suggested and demonstrated, from family data, that the mode of inheritance of antigens M and N is consistent with a single locus having two codominant alleles, since the MN genotypes are distinguishable. Using monkey antisera, Wiener (1952) demonstrated that the antigens M and N react with the spectrum of antibodies so that several blood factors must be present on the M and N molecules.

The genetics of the MN blood group system seemed to be simple until a new serum, called anti-S (S for "Sydney"; Walsh, personal communication), and closely related to the MN system was discovered by Walsh and Montgomery (1947) and studied further by Sanger and Race (1947). By analogy with Fisher's theory of closely linked loci in the *Rhesus* system (see section 14.15), it has been suggested that there might be another locus, S, with alleles S and s, closely linked with the locus MN. The allele s was thought to be an "inactive" recessive until Levine et al. (1951) discovered anti-s serum. Although both S and s have corresponding antisera, the nomenclature has not been changed.

Another approach has been suggested by Wiener [e.g., Wiener (1952)]. He postulates four alleles, L^{MS}, L^{Ms}, L^{NS}, and L^{Ns}, and calls M, N, S, and s blood factors (i.e., antigenic determinants).

By virtue of what has been said about a "system" and the alleles associated with it, both models are acceptable. In fact, both are in present use.

Several other antigens closely associated with the $MNSs$ locus were discovered later. Detailed discussion of the $MNSs$ locus, which appears to be rather complex, can be found in Chapter 3 of the book by Race and Sanger (1968).

To illustrate techniques for estimating allele frequencies, we restrict ourselves to four "alleles" in the system and use, for convenience, Wiener's nomenclature. The number of recognizable phenotypes depends on which antisera are used in blood typing. With anti-M, anti-N, and anti-S only six phenotypes can be recognized (as in Table 14.3). Using additionally anti-s, we can distinguish nine phenotypes (only $L^{MS}L^{Ns}$ and $L^{Ms}L^{NS}$ are indistinguishable).

14.14. ESTIMATION OF GENE FREQUENCIES IN THE $MNSs$ BLOOD SYSTEM, USING ANTI-M, ANTI-N, AND ANTI-S REAGENTS

We will denote the (relative) frequencies of the four alleles, L^{MS}, L^{Ms}, L^{NS} and L^{Ns}, by m_S, m_s, n_S, and n_s, respectively. We have

$$m_S + m_s + n_S + n_s = 1. \tag{14.22}$$

We will now outline the methods of maximum likelihood and of gene counting for estimating the gene frequencies, m_S, m_s, n_S, and n_s, for the case when only anti-M, anti-N, and anti-S antisera are used, that is, when only six phenotypes are distinguished. The general techniques when anti-s is also used are pretty much the same, except that the algebraic manipulations are longer.

The phenotypes will be denoted by the symbols of the antigens present, which can be detected by one or the other of all of anti-M, anti-N, or anti-S; the symbol s will be used if no S is present in a phenotype. Thus, for instance, the phenotype MS denotes $L^{MS}L^{MS}$ and $L^{MS}L^{Ms}$ genotypes, while Ms corresponds to $L^{Ms}L^{Ms}$ (see Table 14.3).

To simplify the notation, we will also denote the *observed* (absolute) frequencies of the phenotypes by the same symbols (but in parentheses) as the phenotypes themselves. Thus MS denotes the phenotype MS, and (MS) the *observed number* of phenotypes MS in a sample of size T. (Note that the symbols M, N and m, n have been used as symbols of antigens and their

TABLE 14.3 Phenotype distribution in the *MNSs* blood system (using anti-M, anti-N, and anti-S)

No.	Phenotype	Genotypes	Observed phenotype frequency	Expected proportion of phenotypes
1	MS	$L^{MS}L^{MS}$, $L^{MS}L^{Ms}$	(MS)	$m_S^2 + 2m_S m_s = P_{MS}$
2	Ms	$L^{Ms}L^{Ms}$	(Ms)	$m_s^2 = P_{Ms}$
3	MNS	$L^{MS}L^{NS}$, $L^{MS}L^{Ns}$, $L^{Ms}L^{Ns}$	(MNS)	$2m_S n_S + 2m_S n_s + 2m_s n_S = P_{MNS}$
4	MNs	$L^{Ms}L^{Ns}$	(MNs)	$2m_s n_s = P_{MNs}$
5	NS	$L^{NS}L^{NS}$, $L^{NS}L^{Ns}$	(NS)	$n_S^2 + 2n_S n_s = P_{NS}$
6	Ns	$L^{Ns}L^{Ns}$	(Ns)	$n_s^2 = P_{Ns}$
	Total		T	1

relative frequencies, respectively, so the sample of all individuals tested will be denoted by T.)

Table 14.3 shows the observed frequencies of the six phenotypes and also their expected proportions, which play the roles of multinomial parameters.

14.14.1. The Maximum Likelihood Estimators

The phenotype frequencies follow a multinomial distribution with parameters, $P_{MS}, P_{Ms}, \ldots$, etc., given in the last column of Table 14.4, which are functions of gene frequencies.

The logarithm of the likelihood function, L, for this multinomial distribution (apart from the additive constant), is

$$\log L \sim (MS) \log (m_S^2 + 2m_S m_s) + (Ms) \log m_s^2$$
$$+ (MNS) \log 2(m_S n_S + m_s n_S + m_S n_s) + (MNs) \log 2m_s n_s$$
$$+ (NS) \log (n_S^2 + 2n_S n_s) + (NS) \log n_s^2. \tag{14.23}$$

It is evident that the likelihood equations will not be simple, and there are no explicit solutions for gene frequency estimators. But we can set up these equations and also the elements of the expected information matrix and then use a computer program.

It should be noticed that, although there are four parameters, m_S, m_s, n_S, and n_s to be estimated, only three are functionally independent because of relation (14.22). We may then substitute

$$n_s = 1 - m_S - m_s - n_S \tag{14.24}$$

into (14.23) and evaluate the partial derivatives with respect to m_S, m_s, and n_S. On the other hand, it is convenient to retain all parameters as they are, bearing in mind what was said in section 14.7 on a similar occasion in regard to the likelihood function for the *ABO* blood system.

We now apply the scoring techniques described in section 12.12. Table 14.4 gives all terms necessary for setting up the maximum likelihood equations and the expected information matrix.

According to formula (12.47), the likelihood equations will be obtained by multiplying column (2) by the reciprocals of column (3) and by the column of the appropriate derivative, and then adding the resulting products over all rows (phenotypes) and equating them to zero. For instance,

$$\frac{\partial \log L}{\partial m_S} = \frac{2(MS)(m_S + m_s)}{m_S^2 + 2m_S m_s} + \frac{(MNS)(n_S + n_s - m_S)}{m_S n_S + m_s n_S + m_S n_s}$$
$$- \frac{(MNs)}{n_s} - \frac{2(NS)n_s}{n_S^2 + 2n_S n_s} - \frac{2(Ns)}{n_s} = 0.$$

TABLE 14.4 Components of the ML scores in the $MNSs$ blood system

(1) No. of phenotype	(2) Observed phenotype frequency	(3) P_l	(4) $\dfrac{\partial P_l}{\partial m_S}$	(5) $\dfrac{\partial P_l}{\partial m_s}$	(6) $\dfrac{\partial P_l}{\partial n_S}$
1	(MS)	$m_S^2 + 2m_S m_s$	$2(m_S + m_s)$	$2m_s$	0
2	(Ms)	m_s^2	0	$2m_s$	0
3	(MNS)	$2(m_S n_S + m_s n_S + m_S n_s)$	$2(n_S + n_s - m_S)$	$2(n_S - m_s)$	$2m_s$
4	(MNs)	$2m_s n_s$	$-2n_s$	$2(n_s - m_s)$	$-2m_s$
5	(NS)	$n_S^2 + 2n_S n_s$	$-2n_S$	$-2n_S$	$-2n_S$
6	(Ns)	n_s^2	$-2n_S$	$-2n_S$	$-2n_S$

In a similar way we obtain $(\partial \log L/\partial m_s)$ and $(\partial \log L/\partial n_S)$ and equate each to zero.

The elements of the information matrix, $\mathbf{I}$, can be evaluated from Table 14.4, applying formulae (12.52) and (12.53). Thus, for example, the element $I_{\text{MS},\text{Ms}}$ will be [from (12.52)]

$$I_{\text{MS},\text{Ms}} = T\left[\frac{4(m_S + m_s)m_s}{m_S^2 + 2m_S m_s} + \frac{2(n_S + n_s - m_S)(n_S - m_S)}{m_S n_S + m_s n_S + m_S n_s}\right.$$
$$\left. - \frac{2(n_s - m_s)}{n_s} + \frac{4n_s}{n_S + 2n_s} + 4\right],$$

and so on.

The inverse $\mathbf{I}^{-1}$ can be evaluated at some initial values, and an iteration procedure, as given by (11.58), can be applied until final values of the required precision are obtained. Obviously this can be very laborious with a desk calculator, but it is relatively simple if an electronic computer is available.

14.14.2. The Gene Counting Method

As with the *ABO* system, we can set up the gene counting equations. The estimators will be identical with the ML estimators, but, of course, different iteration procedures will be used.

For instance, the frequency, m_S, of the gene L^{MS} can be evaluated by counting this gene in all genotypes, that is,

$$m_S = \frac{1}{2T}\left[\frac{2m_S^2}{m_S^2 + 2m_S m_s}(MS) + \frac{2m_S m_s}{m_S^2 + 2m_S m_s}(MS)\right.$$
$$\left. + \frac{2(m_S n_S + m_S n_s)}{2(m_S n_S + m_s n_S + m_S n_s)}(MNS)\right]$$
$$= \frac{1}{2T}\left[\frac{m_S^2 + 2m_S m_s}{m_S^2 + 2m_S m_s}(MS) + \frac{m_S^2}{m_S^2 + 2m_S m_s}(MS)\right.$$
$$\left. + \frac{m_S(n_S + n_s)}{m_S n_S + m_s n_S + m_S n_s}(MNS)\right]$$
$$= \frac{(MS)}{2T}(1 + h_{m_S m_s}) + \frac{(MNS)}{2T}h_{m_S n}, \tag{14.25}$$

where

$$h_{m_S m_s} = \frac{m_S^2}{m_S^2 + 2m_S m_s} \tag{14.26}$$

is the expected proportion of homozygotes $L^{MS}L^{MS}$ among the individuals of blood group MS, and

$$h_{m_S n} = \frac{m_S(n_S + n_s)}{m_S n_S + m_s n_S + m_S n_s},$$ (14.27)

with $n_S + n_s = n$, is the expected proportion of heterozygotes with one allele being L^{MS} among individuals with the MNS phenotype.

The remaining equations can be set up in a similar way. It can be seen that the iteration procedures are not so simple here as in the case of the *ABO* system.

Another method, which leads to simpler equations and gives estimators with fairly high efficiency, has been suggested by DeGroot and Li (1960). The interested reader is referred to their paper.

Example 14.3. The following data represent the observed frequencies of the MNSs blood groups tested on $T = 1419$ English people [Race and Sanger (1968)]:

Blood group:	MS	Ms	MNS	MNs	NS	Ns
Observed frequency:	295	107	379	322	102	214.

R. A. Fisher (see Race and Sanger (1968), p. 91) obtained the following maximum likelihood estimates of the gene frequencies: $m_S = 0.2472$, $m_s = 0.2831$, $n_S = 0.0802$, $n_s = 0.3895$.

If we do not take into account the factor S, and recognize only the MN level, the frequency of the "allele" M will be $m = m_S + m_s = 0.5303$, and of the "allele" N will be $n = n_S + n_s = 0.4697$.

14.15. THE RHESUS BLOOD GROUP SYSTEM

14.15.1. Rh-Positives and Rh-Negatives

The Rhesus (or, briefly, the *Rh*) system is the most complicated and is next in importance to the *ABO* system. It was discovered by Landsteiner and Wiener (1940), who injected blood of the Rhesus monkey (*Macacus rhesus*) into rabbits and guinea pigs. Sera containing the resulting antibodies agglutinated the blood cells of not only the Rhesus monkey but also about 85% of a sample of 448 New York Whites tested with this antiserum. The substance reacting with such sera became known as Rhesus factor; persons possessing it were termed Rh+ (Rhesus positive), while those lacking it were designated Rh− (Rhesus negative).

By testing parents and their offspring, it was soon demonstrated that Rh is genetically controlled. On the basis of reaction with Landsteiner and

Wiener's anti-Rh serum, it was assumed that an "allele" *Rh* (dominant) is responsible for the presence of this factor and an "allele" *rh* (reccessive) for its absence. Thus *RhRh* and *Rhrh* genotypes are Rh+, and *rhrh* are Rh−. The anti-Rh serum was obtained not only from rabbits injected with monkey blood but also from human beings sensitized with Rh. If an Rh− person receives a transfusion from an Rh+ donor, the serum of the Rh− recipient is able to produce antibodies against Rh.

14.15.2. The Rhesus Complex

Soon after the discovery of the anti-Rh serum, more sera giving distinct but regular reactions were obtained and knowledge of the *Rh* system has become complex, as also have the notations to describe it. As more antisera have been detected, the notation has developed, but no simple and easily memorized system of symbols has been attained as yet. There are two theories, using different symbols, in regard to the genetics of the Rh complex: Fisher's theory of closely linked loci, and Wiener's theory of a single locus with multiple alleles.

Fisher's theory is as follows: there are *three closely linked loci*, each with two alleles, denoted by *D, d* (former *Rh, rh*), *C, c,* and *E, e.* (The order of loci is *D, C, E.*) The antigens, denoted by D, C, E, correspond to the genes *D, C, E* and can be detected by antisera anti-D (former anti-Rh), anti-C, and anti-E. The last two were additionally found antisera. Two other sera, anti-c and anti-e, have been later detected. Although Fisher deduced that there should also be anti-d, this serum has not, as yet, been found. A set of three genes (a "gene complex"), located in one chromosome, is considered as a "hereditary unit," which can be transmitted to the offspring. Thus $2^3 = 8$ different type of "gametes" and 27 genotypes with respect to this system can be recognized.

Wiener's theory assumes a single locus with eight alleles, each controlling one antigen possessing several Rh blood factors. The factors can be detected serologically and correspond to Fisher's genes. The "alleles" in Wiener's model correspond to "three-gene complexes" in Fisher's model.

The symbols introduced by Wiener, as mentioned above, are different from Fisher's. The correspondence between the two systems is shown below.

Fisher's genes:	D	(d)	C	c	E	e
Wiener's blood factors:	Rh_0	(Hr_0)	rh'	hr'	rh''	hr''.

Table 14.5 summarizes the notations used by Fisher and Wiener. The last column gives the frequencies of some "gene complexes" (Wiener's "alleles") (from Race and Sanger (1968), p. 178) in England.

TABLE 14.5 Comparison of notation in the *Rh* system
according to Fisher's and to Wiener's hypotheses

No.	Fisher's model of closely linked loci		Wiener's model of multiple alleles			Frequency of gene complexes in England
	Gene complex	Antigens	Allele	Antigen	Antigenic factors	
1	*dce*	(d), c, e	r	rh	hr$'$, hr$''$	0.3886
2	*dCe*	(d), C, e	r'	rh$'$	rh$'$, hr$''$	0.0098
3	*dcE*	(d), c, E	r''	rh$''$	hr$'$, rh$''$	0.0119
4	*dCE*	(d), C, E	r^y	rh$_y$	rh$'$, rh$''$	. . .
5	*Dce*	D, c, e	R^0	Rh$_0$	Rh$_0$, hr$'$, hr$''$	0.0257
6	*DCe*	D, C, e	R^1	Rh$_1$	Rh$_0$, rh$'$, hr$''$	0.4076
7	*DcE*	D, c, E	R^2	Rh$_2$	Rh$_0$, hr$'$, rh$''$	0.1411
8	*DCE*	D, C, E	R^z	Rh$_z$	Rh$_0$, rh$'$, rh$''$	0.0024

The most frequent are *DCe* and *dce*, and the rarest is *DCE*.

With the passing of time, more and more "allelic antigens" have been found. For instance, at locus *C*, antigens C^u, C^w, C^x, and c^v are recognized, among which only C^w is not rare. Its properties are similar to those of C. In England, the frequency of gene complex $C^w De$ (in Wiener's notation, allele R^{1w}) is 0.0129. (Note that the frequencies in Table 14.5 do not add up to 1.) To locus *E*, rare allelic antigens E^u, E^w, and e^x have been assigned; to locus *D*, D^u. At the present time, at least 27 "gene complexes" are recognized. Of course, although the old terminology (Fisher's "*DCE* loci" or Wiener's "*Rh* alleles") is still used, scientists in this field are trying to determine what, in fact, these "entities" represent in terms of cistrons. It seems that *C* and *E* may be in one cistron, but the situation of *D* is uncertain. For a detailed discussion of recent studies of the Rh complex, see Race and Sanger (1968), Chapter 5.

The maximum likelihood method for calculating relative frequencies of "genes" and "gene complexes" was first given by Fisher (1946, 1947). The techniques for setting up likelihood functions are analogous to those discussed for the *MNSs* system.

14.16. THE LUTHERAN BLOOD GROUP SYSTEM AND ITS LINKAGE WITH THE SECRETOR LOCUS

Two alleles, Lu^a and Lu^b, are assigned to the secretor locus, and antigens associated with this locus can be detected by anti-Lu^a and anti-Lu^b sera.

In the English population, estimated gene frequencies are $Lu^a = 0.0390$, $Lu^b = 0.9610$ [Race and Sanger (1968), p. 250]. An interesting feature of this system is that it is linked with the secretor locus Se. The recombination fraction has been estimated to be about 0.15.

14.17. THE X-LINKED BLOOD GROUP SYSTEM

The blood groups discussed in preceding sections are associated with autosomal loci. The exciting discovery of an X-linked blood group has been reported by Mann et al. (1962). The locus is called Xg. So far an anti-Xg^a serum has been detected, and the allele, called Xg^a has been assigned to the antigen Xg^a. The "silent" allelic partner of Xg^a is called Xg; no serum has been found as yet. The distribution of Xg^a, Xg alleles is pretty stable in populations of Whites and has been estimated as $Xg^a = 0.659$, $Xg = 0.341$ [Race and Sanger (1968), p. 525]. It is expected that the Xg-locus will be very helpful as a genetic marker for analyzing several X-linked traits, since both genes exhibit appreciable frequencies.

14.18. OTHER BLOOD GROUP SYSTEMS

Several other autosomal blood group systems have been discovered. Their names derived mostly from those of the persons in whom they were first observed. Thus there are Kell, Duffy, Kidd, Diego, and Dombrock blood group systems. The inheritance and serology of these groups are presented in the book by Race and Sanger (1968).

14.19. BLOOD GROUPS AND DISEASES

14.19.1. Hemolytic Disease of the Newborn

If the mother is Rh− (in previous notation, lacking the antigen D) and the father is Rh+ (possessing the antigen D), the fetus may be Rh+. The Rh+ fetus may pass the cells through the placenta to the mother, whose cells may produce antibodies against Rh+ (anti-D) and become sensitized. The circulating antibodies remain in the mother. During her next pregnancy, the existing antibodies react with the cells of the fetus and damage its red cells, causing the so-called *hemolytic disease*. This can occur also from incompatibility with respect to c, E, or C, but it is not so severe.

Hemolytic disease can also be due to incompatibility at the ABO blood group when the mother is of the O and the fetus of the A group. It is usually of milder form, however, than that due to Rh incompatibility. An especially unfavorable situation occurs when the mother is O and Rh−, and the fetus is A and Rh+.

14.19.2. The ABO Blood Groups and Susceptibility to Ulcers and Carcinoma

The association between susceptibility to certain diseases and the ABO blood groups has been studied by several workers, utilizing data from blood-typing surveys, from hospitals, and from pedigree analysis. A large number of such data, with discussion, have been collected and analyzed by Clarke (1961); a brief summary of the recent status of this research is given by Race and Sanger (1966, Chapter 20).

It appears that a higher proportion of people with group O suffers from duodenal and gastric ulcer than of those with group A, B, or AB. Duodenal ulcer is more common in nonsecretors than in secretors, and its incidence in group O nonsecretors seems markedly higher than in secretors of the remaining groups, and even in group O secretors.

It also appears that white people of group A are more susceptible to stomach carcinoma than those of group O or B.

However, the role of the *ABO* system in susceptibility to these diseases is unknown, and some research workers in the field are inclined to believe that it is not important. They feel that association with infectious diseases is a more significant factor. A brief review of the literature on this topic is given by Giblett (1969), p. 314.

14.20. SELECTION AT THE *ABO* AND THE *MNSs* LOCI

The question of possible selective forces acting on the ABO systems has been reviewed by Chung and Morton (1961). They suggest that ABO incompatibility [i.e., a mother of group O may be immunized by her A (or also, perhaps, B) fetus] is the main force of selection. Heterozygote advantage is also possible, but as a rather weak selective force.

It has been conjectured in the literature that the intercrosses $MN \times MN$ may produce an excess of MN children, and this may indicate some selection in the MNSs blood groups. It seems more likely, however, that this apparent phenomenon was due rather to technical errors, and selection at the *MN* locus has not been proved as yet [see Race and Sanger (1968), Chapter 3].

SUMMARY

The modes of inheritance and the serology of three main blood group systems, A_1A_2BO, $MNSs$, and Rh, and also several relatively new systems, such as the Lewis, Lutheran, P, and X-linked Xg, have been discussed. The interactions of different nonallelic genes (epistasis) in the production of antigenic substances have been emphasized (e.g., the interaction of ABO and H loci or of Le, H, and Se loci). Two controversial hypotheses, Fisher's of closely linked loci and Wiener's of a single locus with multiple alleles, as well as their nomenclature for the $MNSs$ and Rhesus systems, have been presented, and their relationship to the modern definition of "gene" has been briefly analyzed. The maximum likelihood method and its equivalent, the gene-counting method of estimating gene frequencies, have been derived in detail for the ABO and the $MNSs$ systems.

REFERENCES

Anderson, J., The ABO, secretors and Lewis systems: some aspects of the action of the secretor gene. In: *Modern View on the ABO Blood Groups and Secretor Status*, Series Haemotologica, Munsksgaard, Copenhagen, Vol. II, 1, 1969, pp. 34–82.

Archer, G. T., Haemolitic disease of the newborn. In: *Practical Human Genetics for General Practitioners*, Abbott Lab., Pty. Australia, 1968, pp. 53–55.

Bernstein, F., Zusammenfassende Betrachtungen über die erblichen Blutenstructuren des Menschen, *Z. Abstamm. Vererbgsl.* 37 (1925), 237–270.

Ceppellini, R., Siniscalco, M., and Smith, C. A. B., The estimation of gene frequencies in random mating populations, *Ann. Hum. Genet.* 20 (1955), 97–115.

Chung, C. S., and Morton, N. E., Selection at the ABO locus, *Amer. J. Hum. Genet.* 13 (1961), 9–27.

Clarke, C. A., Blood group and disease. In: *Progress in Medical Genetics*, Vol. 1, ed. by A. G. Steinberg, Grune & Stratton, New York, 1961, pp. 81–119.

DeGroot, M. H., Efficiency of gene frequency estimates for the ABO systems, *Amer. J. Hum. Genet.* 8 (1956a), 39–43.

DeGroot, M. H., The covariance structure of maximum likelihood gene frequency estimates for the MNS system, *Amer. J. Hum. Genet.* 8 (1956b), 229–235.

DeGroot, M. H. and Li, C. C., Simplified method of estimating the MNS gene frequencies, *Ann. Hum. Genet. Lond.* 24 (1960), 109–115.

Fisher, R. A., The fittings of gene frequencies to data on Rhesus reactions, *Ann. Eugen. Lond.* 13 (1946), 150–155.

Fisher, R. A., Note on the calculation of rhesus allelomorphs, *Ann. Eugen. Lond.* 13 (1947), 223–224.

Fisher, R. A., Standard calculations for evaluating a blood group system, *Heredity* 5 (1951), 95–102.

Giblett, Eloise, R., *Genetic Markers in Human Blood*, F. A. Davis Co., Philadelphia, 1969, Chapter 9.

Heiken, A., A genetic study of the $MNSs$ blood group system, *Hereditas* 53 (1965), 187–211.

Landsteiner, K., Über Agglutinationserscheinungen normalen menschlichen Blutes, *Wien. Klin. Wschr.* **14** (1901), 1132–1134.

Landsteiner, K., and Levine, P., A new agglutinable factor differentiating individual human bloods, *Proc. Soc. Exp. Biol.* **24** (1927), 600–602.

Landsteiner, K., and Wiener, A. S., An agglutinable factor in human blood recognized by immune sera for Rhesus blood, *Proc. Soc. Exp. Biol.* **43** (1940), 223.

Levine, P., Kuhmichel, A. B., Wigod, M., and Koch, E., A new blood factor's, allelic to S, *Proc. Soc. Exp. Biol.* **78** (1951), 218–220.

Li, C. C., The components of sampling variance of *ABO* gene frequency estimates, *Amer. J. Hum. Genet.* **8** (1956), 133–137.

Mann, J. D., Cahan, A., Gelb, A. G., Fisher, N., Hamper, J., Tippett, P., Sanger, R., and Race, R. R., A sex-linked blood group, *Lancet* **i** (1962), 8–10.

Mi, M. P., and Morton, N. E., Blood factor association, *Vox Sang.*, **11** (1966), 434–449.

Modiano, G., Benerecetti-Santachiara, A. S., Gonono, F., and Zei, G., An analysis of ABO, MN, Hp, Tf, and G-6-PD types in a sample from the human population of the Lecce Province, *Ann. Hum. Genet. Lond.* **29** (1965), 19–31.

Mourant, A. E., The *Distribution of the Human Blood Groups*, Blackwell and Mott, Oxford, 1954.

Nei, M., and Imaizumi, Y., Genetic structure of human populations, I. Local differentiation of blood groups gene frequencies in Japan, II. Differentiation of *ABO* blood group gene frequencies in small areas in Japan, *Heredity* **21** (1966), 9–35, 183–190, 461–472.

Race, R. R., and Sanger, Ruth, *Blood Groups in Man*, Blackwell Scientific Publications, Oxford, 1966 (fourth edition), and F. A. Davis Co., Philadelphia, 1968 (fifth edition).

Reed, T. E., Tests of models representing selections in mother-child data on ABO blood groups, *Amer. J. Hum. Genet.* **8** (1956), 257–268.

Roberts, F., *An Introduction to Human Blood Groups*, W. Heinemann, London, 1960.

Sanger, Ruth, and Race, R. R., Subdivision of the MN blood groups in man, *Nature* **160** (1947), 505.

Stevens, W. L., Estimation of blood group gene frequencies, *Ann. Eugen. Lond.* **8** (1938), 362–375.

Stevens, W. L., Statistical analysis of the A-B-O-blood groups, *Hum. Biol.* **22** (1950), 191–217.

Walsh, R. J., and Montgomery, C., A new human isoaglutinin subdividing the MN blood groups, *Nature* **160** (1947), 504.

Wellisch, S., and Thomsen, O., Über die vier-Gene Hypothese Thomsens, *Hereditas*, **14** (1930), 50–52.

Wiener, A. S , Heredity of the M-N-S blood types. Theoretic-statistical consideration, *Amer. J. Hum. Genet.* **4** (1952), 37–53.

Wiener, A. S., *An Rh-Hr Syllabus*, Grune & Stratton, New York, 1954.

Wiener, A. S., Principles of blood group serology and nomenclature, *Transfusion* **1** (1961), 308–320.

Wiener, A. S., and Wexler, I. B., *Heredity of the Blood Groups*, Grune & Stratton, New York, 1958.

Yasuda, N., and Kimura, M., A gene counting method of maximum likelihood in *ABO* and *ABO*-like systems, *Ann. Hum. Genet.* **31** (1968), 409–420.

PROBLEMS

14.1. Let n_1, n_2, n_3, and n_4, with $n_1 + n_2 + n_3 + n_4 = n$, be multinomial random variables with parameters P_1, P_2, P_3, and P_4, with $P_1 + P_2 + P_3 + P_4 = 1$, respectively. Let $p_i = n_i/n$ $(i = 1, 2, \ldots, 4)$ be the proportions in a sample of size n $(n_1 + n_2 + n_3 + n_4 = n)$. Also let $u = p_1 + p_3$ and $v = p_2 + p_3$.

Find the following covariances: $\text{Cov}(p_3, u)$, $\text{Cov}(p_3, v)$, and $\text{Cov}(u, v)$.

14.2. Using the results of Problem 14.1 and applying formula (8.95), show that the variance of $d = 1 - (\tilde{p} + \tilde{q} + \tilde{r})$, where $\tilde{p}$, $\tilde{q}$, and $\tilde{r}$ are given by formula (14.4) and are Bernstein's gene frequency estimators of A, B, and O alleles, respectively, is approximately

$$\frac{pq}{2n(1 - p)(1 - q)}.$$

(*Hint:* Notice that $P_1 = p^2 + 2pr$, $P_2 = q^2 + 2qr$, and $P_3 = r^2$.)

14.3. (*a*) Derive the remaining gene-counting equations for the *MNSs* blood system as described in section 14.14.2. (*b*) Show that these are equivalent to the maximum likelihood equations for this system.

14.4. Using the data given in Example 14.3 on the MNSs blood groups, calculate the expected frequencies of different phenotypes in this sample of $T = 1419$ individuals, and test the hypothesis that the data fit the model. (*Hint:* Use the X^2-test of goodness of fit.)

14.5. (*a*) Using the data of Example 14.3, combine the corresponding genotypes so that only MM, MN, and NN phenotypes will be distinguished. (*b*) Using the method of maximum likelihood, estimate the frequencies, m and n, of "alleles" M and N, respectively. Compare your values with those given in Example 14.3.

14.6. (*a*) Using techniques similar to those outlined in section 14.14 for the *MNSs* blood system, set up the maximum likelihood equations and the elements of the expected information matrix for the $A_1 A_2 BO$ system.

(*b*) Show that the gene counting equations, given by formula (14.21), are equivalent to the ML equations.

14.7. The data below represent the frequencies of the ABO blood group phenotypes in different parts of the London region [Stevens, *Human Biology* **22** (1950), 191–217.] Test the hypothesis of homogeneity among these groups. (*Hint:* Use the X^2-tests for heterogeneity discussed in section 13.10.)

No.	Place	Sample size	Phenotype			
			A	B	AB	O
1	Hammersmith	1432	600	120	45	667
2	London, Central (English)	3536	1575	303	113	1545
3	London, Central (Scottish)	160	59	24	6	71
4	Slough	4032	1585	334	104	2009

14.8. On the assumption of two alleles, P (dominant) and p (recessive), in the P blood system, the data in Tables I and II represent samples from two generations: Table I, from Denmark (Copenhagen), and Table II, from England (London).

TABLE I The P blood groups of 303 Danish families
and their 802 children

Matings (parents)		Children		
Type	No. of families	Total	P+	P−
P+ × P+	194	524	471	53
P+ × P−	93	240	169	71
P− × P−	16	38	0	38
Total	303	802	640	162

Table II. The P blood groups of 306 English families and
their 700 children

Matings (parents)		Children		
Type	No. of families	Total	P+	P−
P+ × P+	190	440	384	56
P+ × P−	107	243	163	80
P− × P−	9	17	0	17
Total	306	700	447	153

(These data are taken from a book by Race and Sanger, *Blood Groups in Man*, Blackwell Scientific Publications, Oxford, 1950, pp. 91–93. The original sources of the data are given in this book.) In the tables P+ denotes dominants (i.e., *PP* or *Pp*), and P− indicates recessive (i.e., *pp*). For each set of data solve the following problems.

(*a*) Using the observations on the parents (which represent a simple random sample from the parent generation), estimate the frequencies of genes *P* and *p*.

To test the conformity of the distributions of P blood groups in two generations, some people would calculate Snyder s ratios by substituting estimates obtained from the data on the parents, as shown in section 12.8. Thus, $\hat{S}'_{10}$ given in (12.29) will play the role of binomial parameters in the sample of n children. One may also apply the X^2-test of goodness of fit, as discussed in section 13.7.

(*b*) Using significance level $\alpha = 0.05$, show that for the Danish sample X^2 is insignificant in both types of matings, but for the English sample it is significant for P+ × P+ matings. What is your explanation?

(Note that the children do not represent simple random samples from the second generation; these samples consist of all the children of the given sample of parents.

In view of what has been said in section 12.7, the calculation of Snyder's ratios is not appropriate. On the other hand, the samples of parents in both examples are fairly large, so that using Snyder's ratios might be approximately valid.)

14.9. The data in the accompanying table represent the sample distributions of the ABO blood groups among controls, and among stomal and duodenal ulcer patients, from London populations [data from Clarke (1961)].

Blood group Sample	O	A	B	AB	Total
Control	4578	4219	890	313	10,000
Stomal ulcer	181	96	18	5	300
Duodenal ulcer	298	214	39	13	564

(*a*) Is there any evidence that stomal ulcer is associated with the O blood group? [*Hint:* Construct the 2 × 2 contingency table with columns "O" and "non-O" (i.e., A + B + AB), and rows "Control" and "Stomal ulcer" and apply the X^2-test for association.]

(*b*) Is there any evidence that duodenal ulcer is associated with the O blood group?

(*c*) Is there any evidence that susceptibility to stomal ulcer is lower in individuals of the A blood group than in those of the B or AB blood group?

14.10. (*a*) Derive the gene counting equations for the frequencies of Gm^a, Gm^{ax}, and Gm^b genes at the Gm locus described in Problem 12.1. (*b*) Using the data of Problem 12.2, show that the gene-counting estimates are equivalent to the estimates obtained by the maximum likelihood method

CHAPTER 15

Autosomal Linkage in Experimental
Populations

15.1. INTRODUCTION

Concepts of genetic linkage among autosomal loci, crossing over, and the recombination fraction were introduced in Chapter 1. We have also shown that at equilibrium linked and independent loci cannot be distinguished, so that linkage cannot be detected in large, randomly mating populations from frequency data alone. In experimental animals and plants, linkage can be estimated from intercrosses or backcrosses, and in human beings from family or pedigree data, using methods and tests especially derived for this purpose. The recombination fraction, which measures the degree of linkage between two loci, can be used to locate the genes along chromosomes, that is, in preparing chromosome maps. The problem can be complicated by partial manifestation or differential viability of gametes or zygotes. The topic of linkage is broad, and the mathematical and statistical methods applied to it have been extensively treated in specialized books, such as the one by Bailey (1961), and in a series of methodological papers, some of which are cited in the references.

Since the scope of this book does not permit discussion of all the details of this subject, we shall present only a few basic methods useful in the detection and estimation of linkage in experimental animals and plants when experiments such as backcrosses or intercrosses can be carried out. Linkage in human beings is a special topic and will only be mentioned in Chapter 18.

15.2. ORTHOGONAL FUNCTIONS OF RANDOM VARIABLES

We first introduce a concept of orthogonal functions, called also orthogonal contrasts, which play an important role in experimental statistics and,

421

in particular, will be useful in the detection of linkage. Let us first consider a simple example.

Suppose that a certain character is controlled by a single locus with two alleles, A and a. Let T_1, T_2, and T_3 be some values of this trait associated with the genotypes AA, Aa, and aa, respectively. If we wish to compare the average value of the character of both homozygotes with that of the hetero-zygote, we may take the difference

$$\frac{T_1 + T_3}{2} - T_2 \tag{15.1}$$

or, equivalently,

$$C = (T_1 + T_3) - 2T_2. \tag{15.2}$$

If this is zero, the average value of the character in both homozygotes is equal to that in the heterozygote; otherwise, this is not so. Of course, if T_1, T_2, T_3 are random variables, we would like to know how large $|C|$ can be in order to raise doubts that it is due to random sampling. A function of type (15.1) or (15.2) is called a *contrast*. We compare $\frac{1}{2}(T_1 + T_3)$ *versus* T_2. The problem can be generalized.

Let $T_1, T_2, \ldots, T_k$ be k variables, and $w_1, w_2, \ldots, w_k$ be certain nonzero weights assigned to these variables. Let

$$C_1 = \alpha_{11}T_1 + \alpha_{12}T_2 + \cdots + \alpha_{1k}T_k = \sum_{i=1}^{k} \alpha_{1i}T_i, \tag{15.3}$$

and suppose that the coefficients α_{1i} ($i = 1, 2, \ldots, k$) satisfy the condition

$$w_1\alpha_{11} + w_2\alpha_{12} + \cdots + w_k\alpha_{1k} = \sum_{i=1}^{k} w_i\alpha_{1i} = 0. \tag{15.4}$$

The function C_1 is called a *contrast* (relative to the weights, w_i's). If two functions,

$$C_1 = \sum_{i=1}^{k} \alpha_{1i}T_i \quad \text{and} \quad C_2 = \sum_{i=1}^{k} \alpha_{2i}T_i,$$

satisfy the conditions

$$\sum_{i=1}^{k} w_i\alpha_{1i} = 0, \quad \sum_{i=1}^{k} w_i\alpha_{2i} = 0, \tag{15.5}$$

and

$$\sum_{i=1}^{k} w_i\alpha_{1i}\alpha_{2i} = 0, \tag{15.6}$$

then these functions are called *orthogonal functions* or *orthogonal contrasts* (relative to the weights, w_i's).

For k variables it is possible to construct a set of $k - 1$ (but no more than $k - 1$) functionally independent orthogonal contrasts such that

$$\sum_{i=1}^{k} w_i \alpha_{ti} = 0, \qquad t = 1, 2, \ldots, k - 1; \tag{15.7a}$$

$$\sum_{i=1}^{k} w_i \alpha_{ti} \alpha_{ui} = 0, \qquad t, u = 1, 2, \ldots, k - 1 \quad (t \neq u). \tag{15.7b}$$

There exists more than one set of $k - 1$ independent contrasts associated with the same k variables T's and weights w's. Of particular interest is the situation in which the T's are multinomial variates with $T_i = r_i$, $\sum r_i = n$, and with $w_i = P_i$, $\sum P_i = 1$, where the P_i are the corresponding multinomial parameters. Thus

$$C_t = \sum_{i=1}^{k} \alpha_{ti} r_i, \qquad t = 1, 2, \ldots, k - 1 \tag{15.8}$$

are functions of multinomial variates. It can be shown (see also Problem 15.1) that, if conditions (15.7a) are satisfied, we have

$$E(C_t) = 0, \qquad t = 1, 2, \ldots, k - 1, \tag{15.9a}$$

and

$$\text{Var}(C_t) = n \sum_{i=1}^{k} \alpha_{ti}^2 P_i = V_t, \qquad t = 1, 2, \ldots, k - 1. \tag{15.9b}$$

If additionally (15.7b) is satisfied, we have

$$\text{Cov}(C_t, C_u) = 0, \qquad t, u = 1, 2, \ldots, k - 1; \ (t \neq u). \tag{15.9c}$$

It can also be shown that the quantity

$$X^2 = \sum_{i=1}^{k} \frac{(r_i - nP_i)^2}{nP_i} \tag{15.10}$$

can be partitioned into $k - 1$ statistically independent X^2's, of the form

$$X^2(C_t) = \frac{C_t^2}{V_t}, \qquad t = 1, 2, \ldots, k - 1. \tag{15.11}$$

When n is large, these are approximately distributed as $k - 1$ independent χ^2, each with 1 d.f. We now apply the method of orthogonal contrasts to detect linkage from backcross and intercross experiments.

15.3. DETECTION OF LINKAGE FROM DOUBLE BACKCROSSES

We will consider two loci, A and B, each with dominance A over a and B over b, respectively. Suppose that we observe the frequencies of phenotypes from backcrosses $ABab \times abab$. (i.e., in the offspring of these backcrosses). If there is no linkage, the four phenotypes will have the expected ratios $1:1:1:1$. Let r_1, r_2, r_3, r_4, with $\sum r_i = n$, be the observed frequencies. Thus X^2 calculated from (15.10), with $P_i = \frac{1}{4}$ $(i = 1, 2, 3, 4)$, provides a test of independent segregation of loci A and B. If X^2 is significant, there can be several reasons for this. For instance, we may suspect that the segregation ratios at one locus or the other or at both loci are disturbed by differential viabilities or reduced manifestation, or that some interaction between these loci occurs, so that some gametes or zygotes, considered with respect to both loci jointly, have different manifestation or survival rates. Instead, it might happen that the loci are linked.

To test whether the segregation ratios are not disturbed, that is, are $1:1$ at each locus, we can use the contrast "A versus a" whatever the "level" of B is, and similarly B versus b. But if there is linkage, we should expect an excess of $ABab$ and $abab$ if F_1 were in coupling (AB/ab) or an excess of $Abab$ or $aBab$ if F_1 were in repulsion (Ab/aB). Thus the contrast ($ABab + abab$) versus ($Abab + aBab$) will be appropriate for the detection of linkage.

Table 15.1 lists the set of coefficients α_{it} $(i = 1, 2, 3; t = 1, 2, 3, 4)$ for three mutually independent orthogonal contrasts. Here all weights are the same, equal to $\frac{1}{4}$. Hence

$$X_A^2 = \frac{[(r_1 + r_2) - (r_3 + r_4)]^2}{n} \tag{15.12}$$

and

$$X_B^2 = \frac{[(r_1 + r_3) - (r_2 + r_4)]^2}{n}, \tag{15.13}$$

each with 1 d.f., are test criteria for the segregation ratios at locus A and B,

TABLE 15.1   Orthogonal contrasts for detection of linkage from backcrosses $AaBb \times aabb$

Phenotype:	$ABab$	$Abab$	$aBab$	$abab$
Obs. freq., r_i:	r_1	r_2	r_3	r_4
Weight, P_i:	$\frac{1}{4}$	$\frac{1}{4}$	$\frac{1}{4}$	$\frac{1}{4}$
Contrasts:				
A vs. a	$+1$	$+1$	-1	-1
B vs. b	$+1$	-1	$+1$	-1
Linkage	$+1$	-1	-1	$+1$

respectively, while

$$X_L{}^2 = \frac{[(r_1 + r_4) - (r_2 + r_3)]^2}{n} \qquad (15.14)$$

is a test criterion for the detection of linkage.

Example 15.1. The data below represent the results from backcrosses in Upland cotton [Stephens (1955)]. The loci (with dominance) considered here are $R - r$ [petal spot (R) and normal (r)] and $Y - y$ [green (Y) and yellow-green (y) color of plants], respectively. The backcross was $Rr\,Yy \times rryy$.

No.:	1	2	3	4	
Phenotype:	$R-Y-$	$R-yy$	$rr\,Y-$	$rryy$	Total
Observed frequency, r_i:	77	14	24	75	190
Expected frequency, nP_i:	47.5	47.5	47.5	47.5	190 .

Under the null hypothesis that the expected proportion of each phenotype is $P_i = \frac{1}{4}$, we obtain $nP_i = 47.5$, $i = 1, \ldots, 4$. The calculated X^2 with 3 d.f. is 69.495, and it is significant at the significance level $\alpha < 0.0001$. We split it up into three independent X^2's each with 1 d.f., using formulae (15.12)–(15.14). The results are shown in the table.

Contrast	D.f.	X^2	Significance level, α
R vs. r	1	0.337	$0.50 < \alpha < 0.60$
Y vs. y	1	0.758	$0.30 < \alpha < 0.40$
Linkage	1	68.400	$\alpha < 0.0001$
Total	3	69.495	$\alpha < 0.0001$

The $X_L{}^2 = 68.400$ is highly significant, whereas "segregation X^2's" are insignificant. Since $(r_1 + r_4) > (r_2 + r_3)$, we assume that the F_1 individuals (with genotypes $Rr\,Yy$) were in coupling, that is, RY/ry.

15.4. DETECTION OF LINKAGE FROM INTERCROSSES

Linkage can also be detected by using the intercrosses $AaBb \times AaBb$. Under the null hypothesis of independent segregation of two loci the expected numbers of phenotypes in F_2 are as given in the third row of Table 15.2. The total X^2 calculated from (15.10) is $X^2 = 35.877$. To construct a contrast C_1, "A versus a," we have to find α's satisfying the condition

TABLE 15.2   Orthogonal contrasts for detection of linkage from intercrosses $AaBb \times AaBb$

Phenotype:		$A-B-$	$A-bb$	$aaB-$	$aabb$
Obs. freq., r_i:		r_1	r_2	r_3	r_4
Weight, P_i:		$\frac{9}{16}$	$\frac{3}{16}$	$\frac{3}{16}$	$\frac{1}{16}$
Contrasts:					
A vs. a	(C_1)	$+1$	$+1$	-3	-3
B vs. b	(C_2)	$+1$	-3	$+1$	-3
Linkage	(C_3)	$+1$	-3	-3	$+9$

$\frac{1}{16}(9\alpha_{11} + 3\alpha_{12} + 3\alpha_{13} + \alpha_{14}) = 0$. We notice that $\alpha_{11} = \alpha_{12} = 1$ and $\alpha_{13} = \alpha_{14} = -3$ satisfy this condition. The contrast C_2, "B versus b," not only should satisfy the condition $\frac{1}{16}(9\alpha_{21} + 3\alpha_{22} + 3\alpha_{23} + \alpha_{24}) = 0$, but also must be orthogonal to C_1, that is, we must have $9\alpha_{11}\alpha_{21} + 3\alpha_{12}\alpha_{22} + 3\alpha_{13}\alpha_{23} + \alpha_{14}\alpha_{24} = 0$. It is easy to see that the coefficients α_{2i} given in Table 15.2 in the fifth row ("B vs. b") satisfy this condition. The contrast C_3, "linkage," not only must satisfy the condition $\frac{1}{16}(9\alpha_{31} + 3\alpha_{32} + 3\alpha_{33} + \alpha_{34}) = 0$, but also must be orthogonal to both C_1 and C_2. It is easy to see from Table 15.2 that C_1, C_2, and C_3 are a complete set of mutually orthogonal contrasts.

It should be noticed that the coefficients of the contrast for linkage are products of the coefficients of the contrasts for segregation ratios. (See Tables 15.1 and 15.2.) It can be shown that this property holds quite generally.

We now evaluate the variances. For instance, the variance of the contrast "A versus a" is $V_A = \frac{1}{16}n(1 \times 9 + 1 \times 3 + 9 \times 3 + 9 \times 1)$ $= (48n/16) = 3n$. Thus, using formula (15.11), we obtain

$$X_A^2 = \frac{[(r_1 + r_2) - 3(r_3 + r_4)]^2}{3n}. \tag{15.15}$$

In a similar manner, we obtain

$$X_B^2 = \frac{[(r_1 + r_3) - 3(r_2 + r_4)]^2}{3n} \tag{15.16}$$

and

$$X_L^2 = \frac{[(r_1 + 9r_4) - 3(r_2 + r_3)]^2}{9n}, \tag{15.17}$$

each X^2 having 1 d.f.

Example 15.2. The data are taken from the source as those of Example 15.1 and represent the frequencies of phenotypes in F_2 with respect to loci

R and Y in Upland cotton.

Phenotype:	$R-Y-$	$R-yy$	$rrY-$	$rryy$	Total
r_i:	68	11	13	21	113
nP_i:	63.56	21.19	21.19	7.06	113 .

Using formulae (15.15)–(15.17), we obtain the X^2-analysis shown in the table.

Contrast	D.f.	X^2	Significance level, α
R vs. r	1	1.560	$0.20 < \alpha < 0.30$
Y vs. y	1	0.664	$0.40 < \alpha < 0.50$
Linkage	1	33.653	$\alpha < 0.0001$
Total	3	35.877	$\alpha < 0.0001$

The data from intercrosses confirm the hypothesis that these two loci are linked.

It should be noticed that, if there is no dominance, the same method of linkage detection can be used; however, a different set of orthogonal contrasts, utilizing the information on observed genotypes, must be set up (compare Problem 15.2).

15.5. DETECTION OF LINKAGE FROM TRIPLE BACKCROSSES

Using techniques of orthogonal contrasts, we can derive X^2-statistics for detecting linkage among several loci. Table 15.3 represents an example of

TABLE 15.3 Orthogonal contrasts for the offspring of triple backcross $AaBbCc \times aabbcc$

Phenotype:[a]	ABC	ABc	AbC	Abc	aBC	aBc	abC	abc
Obs. freq. r_i:	r_1	r_2	r_3	r_4	r_5	r_6	r_7	r_8
Contrasts:								
1. A vs. a	+1	+1	+1	+1	−1	−1	−1	−1
2. B vs. b	+1	+1	−1	−1	+1	+1	−1	−1
3. Linkage of A and B	+1	+1	−1	−1	−1	−1	+1	+1
4. C vs. c	+1	−1	+1	−1	+1	−1	+1	−1
5. Linkage of A and C	+1	−1	+1	−1	−1	+1	−1	+1
6. Linkage of B and C	+1	−1	−1	+1	+1	−1	−1	+1
7.	+1	−1	−1	+1	−1	+1	+1	−1

[a] For abbreviation, we omit the symbols abc in each phenotype (e.g., ABC is, in fact, $ABCabc$ and so on).

detecting linkage among three loci, A, B, and C, using backcrosses $AaBbCc \times aabbcc$. It is easy to see that the contrasts in Table 15.3 represent a complete set of seven mutually orthogonal functions. The weights for each term are the same ($\frac{1}{8}$), so we omit them from the table. Also the seventh contrast has no genetic meaning and was used merely to complete the set.

15.6. ESTIMATION OF THE STRENGTH OF LINKAGE

So far we have only discussed methods for the detection of linkage. We wish now to estimate the *strength* of the linkage, expressed in terms of the recombination fraction, λ. We recall the definition of λ.

DEFINITION 15.1. *The expected proportion* (*probability*) *of recombinants* (*i.e., genotypes different from those of the parents*) *in the offspring due to crossing over is called the recombination fraction,* λ.

The range for λ is from 0 when loci are so close that they are indistinguishable, to $\frac{1}{2}$ when linkage cannot be distinguished from independent segregation.

In a sample of size n the phenotype frequencies have a multinomial distribution with parameters equal to the expected proportions of the corresponding phenotypes. Thus setting up the likelihood function and applying the standard procedure to find the maximum likelihood estimator of λ seems to be straightforward for any experimental design such as a back- or intercross. It should be pointed out that we need to distinguish whether F_1 is in coupling (AB/ab) or in repulsion (Ab/aB). Here we only outline the ML techniques for the offspring of intercrosses in coupling (i.e., when matings are $AB/ab \times AB/ab$).

On the assumption that the recombination fractions in males and females are the same, the phenotype distribution in F_2 will be as given in Table 2.4c. The individual scores and expected amounts of information for a sample of size n are given in Table 15.4.

The total score, U_λ, is

$$U_\lambda = 2(1 - \lambda) \left[-\frac{r_1}{2 + (1 - \lambda)^2} + \frac{r_2 + r_3}{1 - (1 - \lambda)^2} - \frac{r_4}{(1 - \lambda)^2} \right]. \quad (15.18)$$

By putting $(1 - \lambda)^2 = \theta$, the equation $U_\lambda = 0$ will take the form

$$n\theta^2 - (r_1 - 2r_2 - 2r_3 - r_4)\theta - 2r_4 = 0. \quad (15.19)$$

This quadratic equation with respect to θ can be solved directly, giving $\hat{\theta}$ and hence $\hat{\lambda} = 1 - \sqrt{\hat{\theta}}$ which should be calculated for the value of $\hat{\theta}$ yielding $\hat{\lambda} \geqslant 0$. The expected average amount of information is given by the term in the last row and the last column of Table 15.4. Thus the variance of $\hat{\lambda}$ is

TABLE 15.4 Estimation of linkage from intercrosses in coupling. The ML method

Class (Phenotype in F_2)	Class number, l	Class frequency, r_l	$P_l(\lambda)$	$\dfrac{\partial P_l}{\partial \lambda}$	$\dfrac{1}{P_l}\dfrac{\partial P_l}{\partial \lambda}$	$\dfrac{1}{P_l}\left(\dfrac{\partial P_l}{\partial \lambda}\right)^2$
$A-B-$	1	r_1	$\frac{1}{4}[2+(1-\lambda)^2]$	$-\frac{1}{2}(1-\lambda)$	$-2\dfrac{1-\lambda}{2+(1-\lambda)^2}$	$\dfrac{(1-\lambda)^2}{2+(1-\lambda)^2}$
$A-bb$	2	r_2	$\frac{1}{4}[1-(1-\lambda)^2]$	$\frac{1}{2}(1-\lambda)$	$2\dfrac{1-\lambda}{1-(1-\lambda)^2}$	$\dfrac{(1-\lambda)^2}{1-(1-\lambda)^2}$
$aaB-$	3	r_3	$\frac{1}{4}[1-(1-\lambda)^2]$	$\frac{1}{2}(1-\lambda)$	$2\dfrac{1-\lambda}{1-(1-\lambda)^2}$	$\dfrac{(1-\lambda)^2}{1-(1-\lambda)^2}$
$aabb$	4	r_4	$\frac{1}{4}(1-\lambda)^2$	$-\frac{1}{2}(1-\lambda)$	$-2\dfrac{1}{1-\lambda}$	1
Total		$n \quad 1$				$i_\lambda = \dfrac{2[1+2(1-\lambda)^2]}{[1-(1-\lambda)^2][2+(1-\lambda)^2]}$

approximately

$$\text{Var}(\hat{\lambda}) \doteq \frac{[1 - (1 - \lambda)^2][2 + (1 - \lambda)^2]}{2n[1 + 2(1 - \lambda)^2]}. \tag{15.20}$$

Example 15.3. We use the data from Example 15.2 to estimate λ between R and Y in Upland cotton. From (15.19) we have $113\theta^2 + \theta - 42 = 0$. We obtain $\hat{\theta} = (1 - \hat{\lambda})^2 \doteq 0.6053$. Hence $\hat{\lambda} = 1 - \sqrt{0.6053} \doteq 0.222$. The estimated variance of $\hat{\lambda}$ is approximately

$$\text{Var}(\hat{\lambda}) \doteq \frac{(1 - 0.6053)(2 + 0.6053)}{2 \cdot 113(1 + 2 \cdot 0.6053)} = 0.0021.$$

It can be shown (see Problem 15.3) that for the offspring of backcrosses $AB/ab \times abab$ (coupling) the likelihood equation is

$$U_\lambda = \frac{r_2 + r_3}{\lambda} - \frac{r_1 + r_4}{1 - \lambda} = 0, \tag{15.21}$$

where r_1 through r_4 are the observed phenotype frequencies as shown in Table 15.1. Solving (15.21) with respect to λ, we obtain the ML estimator of λ, that is,

$$\hat{\lambda} = \frac{r_2 + r_3}{n}. \tag{15.22}$$

Its variance is

$$\text{Var}(\hat{\lambda}) = \frac{\lambda(1 - \lambda)}{n}. \tag{15.23}$$

If the backcross is $Ab/aB \times abab$, we obtain

$$\hat{\lambda} = \frac{r_1 + r_4}{n}, \tag{15.24}$$

and $\text{Var}(\hat{\lambda})$ will be the same as in (15.23).

The test of independence (or the test for linkage) will now be equivalent o the test of the hypothesis $H_0: \lambda = \frac{1}{2}$. Under H_0, the statistic

$$Z^2 = \frac{(\hat{\lambda} - \frac{1}{2})^2}{\text{Var}(\hat{\lambda})} \tag{15.25}$$

will be approximately distributed as χ^2 with 1 d.f. [compare formula (13.19)]. [Note that $\text{Var}(\hat{\lambda})$ should be evaluated from the appropriate formula by substituting $\lambda = \frac{1}{2}$.] It is easy to show (Problem 15.7) that (15.25) evaluated for the offspring of backcrosses and intercrosses will be identical with the X^2 given by (15.14) and (15.17), respectively.

If data from different experiments are available, the combined likelihood function can be used (compare section 12.11). Also these data can be tested for conformity with the linkage hypothesis, by the heterogeneity tests described in Chapter 13.

If the recombination fractions in males and females are different, two parameters, λ_1 and λ_2, say, must be estimated. The maximum likelihood method can again be applied in the usual manner. These cases have been discussed in detail by Bailey (1961).

15.7. DISTURBANCE FACTORS IN LINKAGE

In all preceding sections we have assumed that deviations from the segregation ratios expected under the hypothesis that the loci are independent are due only to linkage. But this might not be the case. It will be shown in Chapter 16 that, if incomplete penetrance of certain genes or differential viability of gametes or zygotes occurs, linkage may be confounded with, or indistinguishable from, other models of inheritance, if only backcrosses or intercrosses are considered. Thus, the tests for detection of linkage discussed in sections 15.3–15.5 or the ML estimator of the recombination fraction may, in fact, detect or estimate relative viability or a combination of linkage with relative viability (compare section 16.4, for instance). To discriminate among different hypotheses, especially designed experiments are required. Some such experiments will be outlined in Chapter 16. The deviations of estimators in cases of incomplete penetrance and differential viability are considered in detail by Bailey (1961).

15.8. DOUBLE CROSSING OVER

Suppose that we consider three loci, A, B, C, arranged as in Fig. 15.1, on a chromosome. Let $\lambda_{AB} = \lambda_1$ be the recombination fraction between loci A and B, $\lambda_{BC} = \lambda_2$ be the recombination fraction between B and C, and $\lambda_{AC} = \lambda_{1+2}$ be the recombination fraction between A and C. The fractions λ_1, λ_2, and λ_{1+2} could be estimated independently by methods described for two loci. On the other hand, we may be able to derive λ_{1+2} from λ_1 and λ_2.

Suppose that crossing over occurs randomly in segments AB and BC. If these segments are rather small in length, the occurrence of crossing over in AB may exclude the occurrence of crossing over in BC, so that crossovers in these two segments will represent mutually exclusive events. In this case,

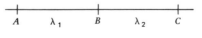

Figure 15.1.

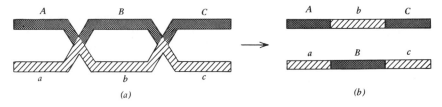

Figure 15.2.

the probability λ_{1+2} will be simply

$$\lambda_{1+2} = \lambda_1 + \lambda_2. \tag{15.26}$$

But if A and C are far apart, the events may be not exclusive and the situation called *double crossing over* can occur as shown in Fig. 15.2. Under these circumstances, we have

$$\lambda_{1+2} = \lambda_1(1 - \lambda_2) + \lambda_2(1 - \lambda_1) = \lambda_1 + \lambda_2 - 2\lambda_1\lambda_2. \tag{15.27}$$

The term $2\lambda_1\lambda_2$ is due to double crossovers.

15.9.   INTERFERENCE AND THE COEFFICIENT OF COINCIDENCE

It is now well established (from cytological analysis and from statistical analysis of recombination) that double crossing over does not occur as often as expected. It appears that the occurrence of crossing over in a segment reduces the chance of its occurrence in the adjacent segment. The failure of crossing over to occur randomly is called *interference*.

Let λ_{12} be the probability of occurrence of double crossing over (i.e., that crossing over occurs in segments AB and BC simultaneously). If there is no interference, we have

$$\lambda_{12} = \lambda_1\lambda_2. \tag{15.28}$$

If $\lambda_{12} < \lambda_1\lambda_2$ we have *positive interference*; if $\lambda_{12} > \lambda_1\lambda_2$, *negative interference*.

DEFINITION 15.2.   *The ratio of the expected proportion (probability) of double crossovers to the product of expected proportions (probabilities) of single crossovers in two adjacent segments, that is,*

$$\gamma = \frac{\lambda_{12}}{\lambda_1\lambda_2}, \tag{15.29}$$

is called the coefficient of coincidence.

It is an index which expresses the strength of interference. Thus, if interference occurs, we have

$$\lambda_{1+2} = \lambda_1 + \lambda_2 - 2\lambda_{12} = \lambda_1 + \lambda_2 - 2\gamma\lambda_1\lambda_2, \tag{15.30}$$

where now the term $2\lambda_{12} = 2\gamma\lambda_1\lambda_2$ is due to double crossovers. We notice that in most cases $\lambda_{12} < \lambda_1\lambda_2$, so that $\gamma < 1$.

15.10. ARRANGEMENT OF LOCI ON A CHROMOSOME

One way to establish the order of three loci, A, B, and C, is to estimate the recombination fractions, λ_{AB}, λ_{BC}, and λ_{AC}. Even if (15.26) does not hold precisely, we usually have $\lambda_{1+2} > \lambda_1$ and $\lambda_{1+2} > \lambda_2$. Therefore, the largest value of λ indicates that the corresponding loci (shown in the subscript of λ) are furthest apart.

The order of loci can also be predicted without actually calculating the three recombination fractions. The most convenient experiment for this purpose is a triple backcross. Experiments of this type are sometimes called *three-point testcrosses*.

First we notice that the most frequent types of offspring will be parental phenotypes. Thus, for example, if the testcross is ABC × abc (more precisely, *ABC/abc × abc/abc*), the most frequent phenotypes will be ABC and abc; if the testcross is ABc × abC, the most frequent will be ABc and abC; and so on. Thus, by observing the two kinds of phenotypes with the total largest frequency, n_0, say, in the offspring, we are able to establish the parental types.

From Fig. 15.2b we see that double recombinants are those with the middle gene exchanged. Let n_{12} be the observed number of double recombinants. Usually n_{12} is the lowest frequency in the whole sample. Thus from double recombinants we can estimate the order of loci on the chromosome. The phenotypes in which the external genes exchanged are single crossovers in either the first or the second segment. We denote the observed frequencies of such phenotypes by n_1 and n_2, respectively.

15.11. ESTIMATION OF THE COEFFICIENT OF COINCIDENCE FROM THREE-POINT TESTCROSS

Let us consider the offspring of the backcross $ABC/abc \times abc/abc$. In a sample of size n, the distribution of noncrossovers, single crossovers, and double crossovers is multinomial, as given in Table 15.5.

Since there are four multinomial categories and three parameters, λ_1, λ_2, and λ_{12}, to be estimated, we shall calculate the maximum likelihood estimators by equating the observed proportions to their expectations. We obtain

$$\hat{\lambda}_1 = \frac{n_1 + n_{12}}{n}, \quad \hat{\lambda}_2 = \frac{n_2 + n_{12}}{n}, \quad \hat{\lambda}_{12} = \frac{n_{12}}{n}. \tag{15.31}$$

TABLE 15.5   The distribution of recombinants from three-point backcross test $ABc/abc \times abc/abc$ with interference

Phenotype:	ABC, abc	aBC, Abc	ABc, abc	AbC, aBc	
		Single recombinants		Double	
Type of recombination:	Nonrecombinants	First segment	Second segment	recombinants	Total
Observed frequency:	n_0	n_1	n_2	n_{12}	n
Multinomial parameters, P_i:	$P_1 = 1 - \lambda_1 - \lambda_2 + \lambda_{12}$	$P_2 = \lambda_1 - \lambda_{12}$	$P_3 = \lambda_2 - \lambda_{12}$	$P_4 = \lambda_{12}$	1

Hence, the ML estimator of the coefficient of coincidence is

$$\hat{\gamma} = \frac{\hat{\lambda}_{12}}{\hat{\lambda}_1 \hat{\lambda}_2} = \frac{n \cdot n_{12}}{(n_1 + n_{12})(n_2 + n_{12})}. \tag{15.32}$$

Applying formula (12.55), we find that the variance of $\hat{\gamma}$ is approximately

$$\mathrm{Var}(\hat{\gamma}) \doteq \frac{\gamma}{n} \cdot \frac{1 - \gamma(\lambda_1 \lambda_2 + \lambda_1 + \lambda_2) + 2\gamma^2 \lambda_1 \lambda_2}{\lambda_1 \lambda_2}. \tag{15.33}$$

[See also Stevens (1936).]

The method can be extended to more than three loci—four, five, etc., point testcrosses can be constructed.

Example 15.4.   The data in the table represent the results from the three-point testcross of three recessive mutants in corn [from Srb et al. (1965), Chapter 9]. These alleles are v = virescent seedling, gl = glossy seedling, va = variable sterile. The corresponding normal (dominant) alleles are denoted by $+ + +$. The testcross is $+ + +/v\ gl\ va \times v\ gl\ va/v\ gl\ va$ or, briefly, $+ + + \times v\ gl\ va$.

Phenotype	Observed frequency		Type of recombination
$+ \quad + \quad +$	235	505	Parental types
$v \quad gl \quad va$	270		
$v \quad + \quad +$	60	122	Single recombinants in segment I
$+ \quad gl \quad va$	62		
$+ \quad + \quad va$	40	88	Single recombinants in segment II
$v \quad gl \quad +$	48		
$+ \quad gl \quad +$	7	11	Double recombinants
$v \quad + \quad va$	4		
Total	726		

(a) Estimate the order of loci on the chromosome, the corresponding recombination fractions, and the coefficient of coincidence.

The parental phenotypes are $+ + + \times v\ lg\ va$. From double recombinants we find that the order of loci is v, gl, va. We have $n_0 = 505, n_1 = 122, n_2 = 88, n_{12} = 11$.

The estimated recombination fractions [from (15.31)] are as follows: between v and gl, $\hat{\lambda}_{v,gl} = \hat{\lambda}_1 = (122 + 11)/726 = 133/726 = 0.1832$; between gl and va, $\hat{\lambda}_{gl,va} = \hat{\lambda}_2 = (88 + 11)/726 = 99/726 = 0.1364$; and $\hat{\lambda}_{12} = 11/726 = 0.152$.

The estimated coefficient of coincidence (from (15.32)) is

$$\hat{\gamma} = \frac{726 \times 11}{133 \times 99} = \frac{7986}{13167} = 0.6065.$$

(b) Perform a test of significance for interference. The hypothesis of no interference is $H_0 : \lambda_{12} = \lambda_1 \lambda_2$ or, equivalently, $H_0 : \gamma = 1$. Under H_0, the variance of $\hat{\gamma}$ is [from (15.33), substituting $\gamma = 1$]

$$\mathrm{Var}(\hat{\gamma}) \doteq \frac{1}{n} \frac{1 - \lambda_1 - \lambda_2 + \lambda_1 \lambda_2}{\lambda_1 \lambda_2}.$$

Substituting the estimates of λ_1 and λ_2, we obtain $\mathrm{Var}(\hat{\gamma} \mid H_0) = 0.03888$. As a test criterion for H_0 we use the statistic given in (13.19), that is,

$$Z^2 \doteq \frac{(\hat{\gamma} - 1)^2}{\mathrm{Var}(\hat{\gamma} \mid H_0)},$$

which is approximately distributed as X^2 with 1 d.f. For our data we obtain $(-0.3935)^2/0.0388 = 3.98$. This is significant at $\alpha = 0.05$ ($X^2_{0.95} = 3.84$), so the hypothesis that there is no interference is rejected.

15.12. CHROMOSOME MAPPING

Genes are arranged linearly on a chromosome. It may be useful to know how far apart from each other they are located, that is, to construct a *chromosome map*. It is difficult to measure the "distance" between the loci in any standard units of length, but the percentage of recombinants is used as a relative measure. One per cent of recombinants is called 1 *centimorgan* (CM) and serves as a unit by which map distance on the chromosome is measured.

(i) If two loci, A and C, say, are relatively close to each other so that double crossing over can be neglected, their "distance" can be evaluated from the three-point testcross, using formula (15.26). But when they are not very close, the "distance" should be evaluated from (15.27) if no interference occurs, or from (15.30) otherwise. It will be convenient, then, to introduce some other measure which has the property of being additive.

▼ (ii) Suppose first that there is *no interference*, that is, crossing over is a random process ($\gamma = 1$). Let d be the distance between two loci, A and B. Then the number of crossovers, x, say, has the Poisson distribution with parameter d, that is,

$$p(x; d) = \frac{d^x}{x!} e^{-d}, \qquad x = 0, 1, \dots . \tag{15.34}$$

Of course, the recombination is observable only if x is *odd*. Thus the expected proportion of recombinants, λ, is

$$\lambda = \sum_{x=0}^{\infty} p[(2x + 1); d] = e^{-d} \sum_{x=0}^{\infty} \left[\frac{d^{2x+1}}{(2x + 1)!} \right] = \tfrac{1}{2}(1 - e^{-2d}) \tag{15.35}$$

[Haldane (1919)]. From (15.35) we obtain

$$d = d(\lambda) = -\tfrac{1}{2} \ln (1 - 2\lambda). \tag{15.36}$$

The quantity d is called the *map distance*; it possesses the additive property. Hence, if three loci are in the order A, B, C, and the map distance between A and B is d_1 and that between B and C is d_2, the distance between A and C is the sum $d_{1+2} = d_1 + d_2$.

(iii) If there is *interference*, another map distance, called the *Kosambi coefficient*, should be used. Kosambi's theory on map distance is beyond the scope of this book. The reader is referred to the original papers by Kosambi (1944) and Owen (1950, 1951) and to the book by Bailey (1961).

SUMMARY

The subject matter of this chapter has been the statistical methods of detecting and estimating linkage in plants and animals, using double back-crosses and intercrosses. The X^2-analysis, by setting up orthogonal contrasts, has been applied in detection of linkage. The scores and expected amounts of information for the maximum likelihood method have been derived for estimation of the recombination fraction. The coefficient of coincidence has been defined as the ratio of probability of double crossing over to the product of probabilities of single crossovers in two adjacent segments, and its ML estimator, using the three-point testcross, has been derived. Methods of assigning loci on chromosomes (chromosome mapping) have been briefly outlined, and "map distance" has been defined.

REFERENCES

Bailey, N. T. J., *Introduction to the Mathematical Theory of Genetic Linkage*, Clarendon Press, Oxford, 1961.

Haldane, J. B. S., The combination of linkage values, and the calculation of distance between the loci of linked factors, *J. Genet.* **8** (1919), 299–309.

Jennings, H. S., The numerical relations in the crossing-over of the genes, with a critical examination of the theory that the genes are arranged in a linear series, *Genetics* **8** (1923), 393–457.

Kosambi, D. D., The estimation of map distance from recombination values, *Ann. Eugen. Lond.* **12** (1944), 172–175.

Mather, K., *The Measurement of Linkage in Heredity*, John Wiley & Sons, New York, 1957.

Muller, H. J., The mechanism of crossing-over, *Amer. Nat.* **50** (1916), 193–434.

Owen, A. R. G., The theory of genetic recombination, *Advances in Genetics*, **3** (1950), 117–157.

Owen, A. R. G., An extension of Kosambi's formula, *Nature* **168** (1951), 208.

Owen, A. R. G., The analysis of multiple linkage data, *Heredity* **7** (1953), 247–264.

Shult, E. E., and Lindegren, C. C., A general theory of crossing-over, *J. Genet.* **54** (1956), 343–355.

Srb, A. M., Owen, R. D., and Edgar, R. S., *General Genetics*, W. H. Freeman & Co., San Francisco, 1965.

Stephens, S. G., Linkage in Upland cotton, *Genetics* **40** (1955), 903–917.

Stevens, W. L., The analysis of interference, *J. Genet.* **32** (1936), 51–64.

PROBLEMS

15.1. Let r_i $(i = 1, 2, \ldots, k)$ be multinomial variates, with $\sum r_i = n$, and P_i $(i = 1, 2, \ldots, k)$ be multinomial parameters, with $\sum P_i = 1$. Let

$$C_t = \sum_{i=1}^{k} \alpha_{ti} r_i, \qquad t = 1, 2, \ldots, \tag{i}$$

be a linear function of r's such that its coefficient α's satisfy the condition

$$\sum_{i=1}^{k} \alpha_{ti} P_i = 0. \tag{ii}$$

(a) Show that

$$E(C_t) = 0, \quad V_t = \mathrm{Var}(C_t) = n \sum_{i=1}^{k} \alpha_{ti}^2 P_i.$$

[*Hint:* Notice that $E(r_i) = nP_i$, $\mathrm{Var}(r_i) = nP_i(1 - P_i)$, and $\mathrm{Cov}(r_i, r_j) = -nP_i P_j$.]

(b) Let C_t and C_u be two functions defined as in (i), each satisfying condition (ii). Show that, if additionally to (ii)

$$\sum_{i=1}^{k} \alpha_{ti} \alpha_{ui} P_i = 0, \tag{iii}$$

then

$$\text{Cov}(C_t, C_u) = 0.$$

[*Hint:* Apply formula (12.57).]

(c) Explain why the statistic C_t^2/V_t is approximately distributed as χ^2 with 1 d.f. (provided $nP_i > 3$).

15.2. Let A_1, A_2 be two codominant alleles at locus A, and B_1, B_2 be two codominant alleles at locus B. Consider the population of intercrosses $A_1A_2B_1B_2 \times A_1A_2B_1B_2$.

(a) Write down the genotype (so also the phenotype) distribution under the hypothesis that the loci segregate independently (notice that there will be nine distinct genotypes).

(b) Write down the contrast C_A for testing the hypothesis that the segregation ratios at locus A are 1:2:1. (*Hint:* The expected number of homozygotes A_1A_1 and A_2A_2 should be the same as that of heterozygotes, A_1A_2.) Set up the X^2-test for testing this hypothesis.

(c) Find, in a similar way, orthogonal contrast C_B and the independent X^2-statistic for testing the hypothesis that the segregation ratios at locus B are 1:2:1.

(d) Set up the contrast, C_L, for linkage orthogonal with C_A and C_B, and derive X_L^2. (*Hint:* Apply the rule of products of coefficients mentioned in section 15.4.)

15.3. Set up the likelihood function for estimation of the recombination fraction, λ, from the offspring of the following backcrosses:

(a) $AB/ab \times abab$ (coupling),

(b) $Ab/aB \times abab$ (repulsion),

and find the ML estimators of λ and their variances.

15.4. Solve Problem 15.3 assuming that the recombination fractions in males and females are λ_1 and λ_2, respectively.

▼ **15.5.** Show that for the offspring of intercrosses $Ab/aB \times Ab/aB$ (repulsion) the expected average amount of information is approximately

$$i(\lambda) \doteqdot \frac{2(1 + 2\lambda^2)}{(1 - \lambda^2)(2 + \lambda^2)}.$$

15.6. Calculate the relative efficiencies of the estimators of λ obtained from the following experiments:

(a) $AB/ab \times AB/ab$,

(b) $Ab/aB \times Ab/aB$,

as compared with the $\text{Var}(\hat{\lambda})$ obtained from the backcross experiment.

15.7. Show that the linkage X_L^2 given by formula (15.4) for backcross experiment and by (15.17) for intercross experiment is identical with the X_L^2 for association, given by (15.25), for each case. [*Hint:* For each case calculate $\text{Var}(\hat{\lambda})$ by substituting $\lambda = \frac{1}{2}$ in the appropriate formula for the variance of $\hat{\lambda}$.]

CHAPTER 16

Statistically Equivalent Models
of Inheritance

16.1. IDENTICAL PHENOTYPE DISTRIBUTIONS UNDER DIFFERENT MODES OF INHERITANCE

We saw in Chapter 13 that the same body of data might fit different genotype or phenotype distributions derived under different hypotheses regarding the mode of inheritance. For instance, the data of Example 13.1 fitted the hypothesis of a single locus ($P = \frac{3}{4}$) as well as that of mutual epistasis ($P = \frac{13}{16}$). This was due to the sampling error; increasing the sample size or using different sampling techniques may discriminate between these two hypotheses.

The problems which we shall discuss here are different. There are some situations in which, for a given experimental design (such as intercross or backcross, for instance), the resulting phenotypic distributions appear to be identical for different models of inheritance. This usually occurs when Mendelian ratios are disturbed by such parameters as incomplete penetrance or different viabilities of gametes or zygotes, but these are not the only cases. As we will see later, this happens because we are estimating not a single parameter, but a certain function of a number of parameters which cannot be "statistically" separated.

DEFINITION 16.1. *If for a certain design the distributions of phenotypes in the resulting population corresponding to two or more different models of inheritance are indistinguishable, we will call these models statistically equivalent with respect to this design.*

DEFINITION 16.2. *If for two or more models of inheritance there exists an experimental design which leads to different phenotype distributions for each model, we will call it a discriminatory design with respect to these models.*

"Models of inheritance" are associated here with different hypotheses about the transmission of heredity units. We wish to show here how important it is to find an appropriate discriminatory design so that, by the methods described in Chapter 13, we can select the right hypothesis and reject the others.

In this chapter, we shall discuss a few kinds of statistically equivalent models and some discriminatory designs for each group of such models.

16.2.   INCOMPLETE PENETRANCE AND EXPRESSIVITY

16.2.1.   Penetrance

If a certain genotype occasionally fails to produce its usual phenotypic effect, we say that it shows only *partial manifestation* or that the genotype is *incompletely penetrant.*

For a dominant character at a single locus, incomplete penetrance usually occurs in heterozygotes Aa which appear as recessives aa (we write this briefly as $Aa \rightarrow aa$), or in homozygotes aa which appear as heterozygous dominants (i.e., $aa \rightarrow Aa$). But it is also possible that homozygotes AA may appear as aa (i.e., $AA \rightarrow aa$). A few remarks on what is understood by "incomplete penetrance" may be useful.

1. We note that some authors speak about "penetrance of a gene" instead of a genotype. For instance, if some $Aa \rightarrow aa$ they say that the gene A is "incompletely penetrant" in heterozygotes Aa. But in some situations it may be misleading to speak about gene penetrance. What, for example, does it mean to say that in the case of complete dominance the heterozygote Aa does not exhibit the effect of gene a? This means that a is never "penetrant" in Aa; but this phenomenon has been defined as recessiveness of the gene a. For this reason and similar ones we will speak about *penetrance of a genotype* rather than of a gene although the phrase "penetrance of a gene" is permissible.

2. Another situation which may introduce some ambiguity in the understanding of "penetrance of a genotype" is the problem of genes called suppressors or inhibitors. In Example 3.12 we considered a locus C with genes C (for colored onion bulbs) and c (for white bulbs). The phenotype distribution, under the hypothesis of complete dominance, should be $3:1$. However, another dominant gene, I, at another independent locus, when present in the genotype, inhibits the appearance of the color so that only phenotypes $iiC-$ will be colored. The phenotypic ratios are as follows: White:Colored $= 13:3$. Do we say that the genotype CC or Cc in the presence of II or Ii, fails to exhibit color? Yes, but the expected proportion

of such individuals can be deduced from Mendelian segregation ratios and the phenomenon has been defined as mutual epistasis and thus cannot be called incomplete penetrance.

3. At the beginning of this section we used the expression "occasionally fails"; we mean that such situations occur only from time to time and we cannot say how often, without knowing the mechanism which stops a genotype being penetrant. In fact, very little is known about the causes of reduced penetrance. It is usually assumed that genetic background, that is, a set of other genes different from those which control the trait under consideration, and environmental conditions are the main causes why a genotype does not show its usual phenotypic effect. This may be true, but only within certain limits.

Fuller et al. (1950) studied susceptibility to audiogenic seizure in mice. Mice of the C57BL strain never reacted to the stimulus, whereas almost all mice of the DBA strain convulsed. The hybrids of F_1 were variable. Moreover, their susceptibilities to audiogenic seizure were not constant for the same animal from day to day. One mouse could be resistant on some days and fail on others; another mouse could convulse on the first day, recover, and then convulse again. If the parental strains are assumed to be homogeneous, the individuals in F_1 should have the same genetic background. We also can assume that the conditions in a given experiment were uniform. Thus these mice should react in a similar way — all should show either susceptibility or resistance. How then can this variability in response be explained?

One explanation is to assume that certain genic combinations represent some "unbalanced" states which can be upset in prenatal life by different kinds of physiological factors and then, in interaction with uncontrolled ("random"), even small external causes, may fail to produce the phenotypic effect. Only a certain proportion of genotypes can be so affected; the rest will be penetrant.

We now introduce a definition of penetrance.

DEFINITION 16.3. *By the penetrance parameter (or, briefly, penetrance) of a given genotype we will understand the expected proportion of individuals showing the usual phenotypic effect, associated with the mode of inheritance for this genotype.*

16.2.2. Expressivity

The term *expressivity* is used to describe the kind of phenotypic effect produced by a genotype when it is penetrant. Usually the qualitative traits with full penetrance show clear-cut phenotypic effects, but traits with a possibility of reduced penetrance are not uniform in their expressions.

For instance, polydactyly in man (excess number of fingers or toes) can occur on the right or the left hand or foot (or both). Susceptibility to audiogenic seizure in mice, mentioned above, ranges in intensity from so severe as to cause the death of some mice to sufficiently mild that other mice recover. Guinea pigs normally have three toes on the hind feet. There are some strains, however, which develop an atavistic fourth little toe. These fourth toes show different expressivity from perfect to very poor.

Variable expressivity, in general, is not necessarily associated with reduced penetrance and can be explained in many ways. For instance:

(i) There may exist special genes called *modifiers* which change only slightly the effects of a major gene. The roles of such genes are not well known, and it is difficult to establish their segregation ratios.

(ii) Environmental conditions such as temperature or food, combined with genetic background, may change a phenotypic effect within certain limits.

(iii) Finally, the character may be controlled, not by a single major gene, but by genes from several loci with cumulative effects. This could then resemble, in some ways, quantitative traits. In such cases the genetics can be worked out (see, for instance, Example 5.9, which deals with the color of wheat kernels).

16.2.3. Estimation of Penetrance

By virtue of what has been said above, the penetrance parameter, β, say, can be estimated provided that: (*a*) the mode of inheritance of the trait under consideration is known, and (*b*) the relative frequency of cases in which the genotype fails to exhibit the character approaches a limit as the sample size becomes large.

On the other hand, by introducing an unknown parameter β, confounded with segregation ratios, we may fit several hypotheses to the same data.

In the next section, we will consider specific models for reduced penetrances of different genotypes and show that these models are "statistically equivalent" with respect to backcrosses and intercrosses.

16.3. SEVERAL LOCI WITH ADDITIVE PHENOTYPIC EFFECTS VERSUS SINGLE LOCUS. EQUIVALENT MODELS

We will now consider a character as an "all or none" phenomenon, without regard to the possible different expressivities of this character. We discuss first the situation in which the character is controlled by *two loci* and is a dominant (double or single). For convenience we will call the individuals that exhibit the character *responders* and denote them by R; the individuals that show recessive conditions (i.e., *aabb*) we will call non-responders and denote them by $\bar{R}$. We shall construct a model based on the following assumptions:

1. The double recessive does not exhibit the character. Given two recessive genes at locus B (i.e., given *bb*), the genotypes have the following penetrances: if AA is present in the genotype, only a proportion α_1 exhibits the character; if Aa is present in the genotype, only a proportion β_1 exhibits the character. Similarly, given the recessive condition *aa*, proportions α_2 and β_2 exhibit the character if BB or Bb, respectively, is present in the genotype.

2. The phenotypic effects are *additive*. For instance, the proportion of responders with genotype $AABb$ is $(\alpha_1 + \beta_2)$. In particular, the proportion of responding double heterozygotes $AaBb$ is $\beta_1 + \beta_2 = \beta$, say.

3. Additionally, we assume $\alpha_1 + \alpha_2 = 1$, that is, the double homozygote $AABB$ is fully penetrant.

Let $AABB$ and $aabb$ be the genotypes of the parental strains Pt_1 and Pt_2, respectively. For three basic populations, Pt_1, Pt_2, and F_1 resulting from crosses $Pt_1 \times Pt_2$, the expected proportions of responders are as follows:

$$
\begin{array}{llll}
\text{Population:} & Pt_1 & Pt_2 & F_1 \\
\text{Genotype:} & ABAB & abab & AB/ab \\
\Pr\{R\}: & 1 & 0 & \beta_1 + \beta_2 = \beta.
\end{array}
\tag{16.1}
$$

We will now consider the backcrosses to the parental strain Pt_2 (i.e., to the double recessive *aabb*), and intercrosses, $F_1 \times F_1$, and derive the corresponding phenotypic distributions for the offspring. For backcrosses to the parental strain Pt_1 (i.e., $ABAB$) we only outline the problem, since these experiments are not appropriate for studying dominance when penetrance is complete. To consider more general cases we assume that loci A and B are *linked* with recombination fraction λ.

16.3.1. Repeated Backcrosses $AB/ab \times abab$

The F_1 generation of genotypes AB/ab produces the following

$$
\begin{array}{lllll}
\text{Gametes:} & AB & Ab & aB & ab \\
\text{With probabilities:} & \frac{1}{2}(1 - \lambda) & \frac{1}{2}\lambda & \frac{1}{2}\lambda & \frac{1}{2}(1 - \lambda).
\end{array}
\tag{16.2}
$$

The genotype distribution in the offspring of $F_1 \times Pt_2$ (i.e., backcrosses $AB/ab \times abab$) is

$$
\begin{array}{lllll}
\text{Genotype:} & AB/ab & Abab & aBab & abab \\
\text{Distribution:} & \frac{1}{2}(1 - \lambda) & \frac{1}{2}\lambda & \frac{1}{2}\lambda & \frac{1}{2}(1 - \lambda).
\end{array}
\tag{16.3}
$$

Let $\Pr\{R \mid F_1 \times Pt_2\}$ be the expected proportion of responders in the offspring of $F_1 \times Pt_2$. We have

$$\Pr\{R \mid F_1 \times Pt_2\} = \tfrac{1}{2}(1 - \lambda)(\beta_1 + \beta_2) + \tfrac{1}{2}\lambda\beta_1 + \tfrac{1}{2}\lambda\beta_2$$
$$= \tfrac{1}{2}(1 - \lambda)\beta + \tfrac{1}{2}\lambda\beta = \tfrac{1}{2}\beta. \tag{16.4}$$

We now find the phenotype distribution in the offspring of double backcrosses $F_1 \times Pt_2 \times Pt_2$. The first generation of $F_1 \times Pt_2$ produces gametes with the probabilities

$$\Pr\{AB \mid F_1 \times Pt_2\} = \tfrac{1}{4}(1 - \lambda)^2,$$
$$\left.\begin{array}{l}\Pr\{Ab \mid F_1 \times Pt_2\} = \Pr\{aB \mid F_1 \times Pt_2\} = \tfrac{1}{2}[\tfrac{1}{2}(1 - \lambda)\lambda + \tfrac{1}{2}\lambda] = \tfrac{1}{4}\lambda(2 - \lambda), \\ \Pr\{ab \mid F_1 \times Pt_2\} = \tfrac{1}{2}[\tfrac{1}{2}(1 - \lambda)^2 + \tfrac{1}{2}\lambda + \tfrac{1}{2}\lambda + \tfrac{1}{2}(1 - \lambda)] = \tfrac{1}{4}[2 + (1 - \lambda)^2].\end{array}\right\}$$
$$\tag{16.5}$$

Probabilities (16.5) also represent the genotype distribution in the second generation of the backcrosses, that is, in the offspring of double backcrosses $F_1 \times Pt_2 \times Pt_2$. We have now

Genotype:	AB/ab	$Abab$	$aBab$	$abab$
Distribution:	$\tfrac{1}{4}(1 - \lambda)^2$	$\tfrac{1}{4}\lambda(2 - \lambda)$	$\tfrac{1}{4}\lambda(2 - \lambda)$	$\tfrac{1}{4}[2 + (1 - \lambda)^2]$.

$$\tag{16.6}$$

The proportion of responders in the population given by (16.6) is

$$\Pr\{R \mid F_1 \times Pt_2 \times Pt_2\} = \tfrac{1}{4}[(1 - \lambda)^2(\beta_1 + \beta_2) + \lambda(2 - \lambda)(\beta_1 + \beta_2)]$$
$$= \tfrac{1}{4}\beta = (\tfrac{1}{2})^2\beta. \tag{16.7}$$

The result can be generalized for k-tuple backcrosses. It can be shown that the proportion of responders in the offspring of k-tuple backcross

$$\underbrace{F_1 \times Pt_2 \times \cdots \times Pt_2}_{k \text{ times}}$$

is

$$\Pr\{R \mid \underbrace{F_1 \times Pt_2 \times \cdots \times Pt_2}_{k \text{ times}}\} = (\tfrac{1}{2})^k\beta. \tag{16.8}$$

16.3.2. Intercrosses $AB/ab \times AB/ab$ and Their Subsequent Generations

The genotype distribution in the offspring of the intercrosses $F_1 \times F_1$ ($AB/ab \times AB/ab$) is given in Table 2.4a. Using this table, we find the expected proportion of responders in the offspring of $F_1 \times F_1$ (i.e., in F_2).

We have

$\Pr\{R \mid F_1 \times F_1\}$

$= \frac{1}{4}[(1 - \lambda)^2(1 + \beta_1 + \beta_2 + \beta_1 + \beta_2)$

$\quad + \lambda(1 - \lambda)(\alpha_1 + \beta_2 + \beta_1 + \alpha_2 + \alpha_1 + \beta_2 + \beta_1 + \beta_1$

$\quad + \alpha_2 + \beta_2 + \beta_1 + \beta_2) + \lambda^2(\alpha_1 + \beta_1 + \beta_2 + \alpha_2)]$

$= \frac{1}{4}[(1 - \lambda)^2(1 + 2\beta) + 2\lambda(1 - \lambda)(1 + 2\beta) + \lambda^2(1 + 2\beta)] = \dfrac{1 + 2\beta}{4}.$

$$(16.9)$$

We now find the distribution in the offspring of $F_2 \times F_1$. The vector of gametes produced by the F_2 generation can be calculated from formulae (4.35).

We now have from (16.2)

$$g_1 = g_4 = \tfrac{1}{2}(1 - \lambda); \quad g_2 = g_3 = \tfrac{1}{2}\lambda. \qquad (16.10)$$

Hence, from (4.33),

$$\Delta_0 = g_1 g_4 - g_2 g_3 = \tfrac{1}{4}[(1 - \lambda)^2 - \lambda^2] = \tfrac{1}{4}(1 - 2\lambda). \qquad (16.11)$$

And then from (4.35)

$\Pr\{AB \mid F_1 \times F_1\} = g_1^{(1)} = g_1 - \lambda\Delta_0$

$\qquad = \tfrac{1}{2}(1 - \lambda) - \tfrac{1}{4}\lambda(1 - 2\lambda) = \tfrac{1}{4}[2 - \lambda(3 - 2\lambda)], \quad (16.12a)$

and also

$$\Pr\{ab \mid F_1 \times F_1\} = g_4 - \lambda\Delta_0 = \tfrac{1}{2}[2 - \lambda(3 - 2\lambda)]. \qquad (16.12b)$$

Similarly,

$$\Pr\{Ab \mid F_1 \times F_1\} = g_2 + \lambda\Delta_0 = \tfrac{1}{2}\lambda + \tfrac{1}{4}\lambda(1 - 2\lambda) = \tfrac{1}{4}\lambda(3 - 2\lambda), \qquad (16.12c)$$

and also

$$\Pr\{aB \mid F_1 \times F_1\} = g_3 + \lambda\Delta_0 = \tfrac{1}{4}\lambda(3 - 2\lambda). \qquad (16.12d)$$

Let us now consider randomly mating populations $F_2 \times F_1$. The genotype distribution in the offspring of $F_2 \times F_1$ is given in the form of a 4×4 matrix in Table 16.1. From this table we calculate the proportion of responders in the offspring of $F_2 \times F_1$, that is,

$\Pr\{R \mid F_2 \times F_1\} = \frac{1}{8}\{(1 - \lambda)[2 - \lambda(3 - 2\lambda)(1 + \beta + \beta)]$

$\qquad + [(1 - \lambda)\lambda(3 - 2\lambda)(\alpha_1 + \beta_2 + \beta_1 + \alpha_2 + \beta_1 + \beta_2)]$

$\qquad + \lambda[2 - \lambda(3 - 2\lambda)](\alpha_1 + \beta_2 + \beta_1 + \alpha_2 + \beta_1 + \beta_2)$

$\qquad + \lambda^2(3 - 2\lambda)(\alpha_1 + \beta + \beta + \alpha_2)\}$

$\qquad = \frac{1}{8}(1 + 2\beta)[2(1 - \lambda) + 2\lambda] = \dfrac{1 + 2\beta}{4}. \qquad (16.13)$

TABLE 16.1 Genotype distribution in the offspring of $F_2 \times F_1$ crosses

F_2 \\ F_1	AB $\frac{1}{2}(1-\lambda)$	Ab $\frac{1}{2}\lambda$	aB $\frac{1}{2}\lambda$	ab $\frac{1}{2}(1-\lambda)$
AB $\frac{1}{4}[2-\lambda(3-2\lambda)]$	$ABAB$ $\frac{1}{8}(1-\lambda)[2-\lambda(3-2\lambda)]$	$ABAb$ $\frac{1}{8}\lambda[2-\lambda(3-2\lambda)]$	$ABaB$ $\frac{1}{8}\lambda[2-\lambda(3-2\lambda)]$	AB/ab $\frac{1}{8}(1-\lambda)[2-\lambda(3-2\lambda)]$
Ab $\frac{1}{4}\lambda(3-2\lambda)$	$ABAb$ $\frac{1}{8}(1-\lambda)\lambda(3-2\lambda)$	$AbAb$ $\frac{1}{8}\lambda^2(3-2\lambda)$	Ab/aB $\frac{1}{8}\lambda^2(3-2\lambda)$	$Abab$ $\frac{1}{8}(1-\lambda)\lambda(3-2\lambda)$
aB $\frac{1}{4}\lambda(3-2\lambda)$	$ABaB$ $\frac{1}{8}(1-\lambda)\lambda(3-2\lambda)$	Ab/aB $\frac{1}{8}\lambda^2(3-2\lambda)$	$aBaB$ $\frac{1}{8}\lambda^2(3-2\lambda)$	$aBab$ $\frac{1}{8}(1-\lambda)\lambda(3-2\lambda)$
ab $\frac{1}{4}[2-\lambda(3-2\lambda)]$	AB/ab $\frac{1}{8}(1-\lambda)[2-\lambda(3-2\lambda)]$	$Abab$ $\frac{1}{8}\lambda[2-\lambda(3-2\lambda)]$	$aBab$ $\frac{1}{8}\lambda[2-\lambda(3-2\lambda)]$	$abab$ $\frac{1}{8}(1-\lambda)[2-\lambda(3-2\lambda)]$

This is identical with (16.9), that is, with the result for the offspring of $F_1 \times F_1$ mating. It can be shown that the same result will be obtained also for the offspring of $F_2 \times F_2$ (see Problem 16.1) and for further generations.

16.3.3. Repeated Backcrosses $AB/ab \times ABAB$

The procedure is analogous to that for backcrosses $AB/ab \times abab$. It is left to the reader to show that

$$\Pr\{R \mid F_1 \times Pt_1\} = \frac{1 + \beta}{2}, \tag{16.14}$$

$$\Pr\{R \mid (F_1 \times Pt_1) \times Pt_1\} = \frac{3 + \beta}{4}, \tag{16.15}$$

and finally, for the offspring of k-tuple backcrosses,

$$\Pr\{R \mid F_1 \times \underbrace{Pt_1 \times \cdots \times Pt_1}_{k \text{ times}}\} = 1 - (\tfrac{1}{2})^k(1 - \beta) = \frac{(2^k - 1) + \beta}{2^k}. \tag{16.16}$$

16.3.4. Equivalent Models

We shall now take note of a few consequences of the results obtained in the preceding sections.

(i) The phenotypic distributions of responders and nonresponders do not depend on the recombination fraction, λ, and so they also are the same when loci are *independent*.

(ii) Putting $\alpha_1 = 1$, $\beta_1 = \beta$, and $\alpha_2 = \beta_2 = 0$, we have the situation of a single locus with incomplete penetrance, β, in heterozygotes.

(iii) Of particular interest is the case in which not only $\alpha_1 + \alpha_2 = 1$ but also $\beta_1 + \beta_2 = \beta = 1$. For a single locus, this means dominance with complete penetrance. For two loci, this means that not only the double homozygotes $ABAB$ but also the double heterozygotes AB/ab are fully penetrant and exhibit the dominant character. If, additionally, $\alpha_1 = \alpha_2 = \beta_1 = \beta_2 = \tfrac{1}{2}$, that is, the contributions of all loci to phenotypic expressions are equal, then also the single heterozygotes $AaBB$ and $AABb$ are fully penetrant and exhibit the dominant character. Therefore, if a hypothesis of one locus with complete dominance (or, equivalently, with recessiveness) fits some data obtained from intercrosses or backcrosses, it does not mean that this is a "proof" of single-locus inheritance (see also section 16.6). In particular, care should be taken when there are several ways in which the character can express itself; in such cases multiloci hypotheses are quite likely to be relevant. Of course, no decisive statement can be reached until appropriate discriminatory experiments have been designed and performed.

(iv) It is not difficult to see that the results obtained for two loci will apply to n loci, each with two alleles, incomplete penetrance and additive effects such that

$$\sum_{i=1}^{n} \alpha_i = 1, \quad \text{and} \quad \sum_{i=1}^{n} \beta_i = \beta.$$

In conclusion, the following hypotheses on the mode of inheritance lead to the same phenotypic distribution if the breeding populations are backcrosses or intercrosses.

The character is inherited as dominant in the following:

H_1: Two or more *linked* loci with penetrance being an *additive* function of the number of genes from each single locus;

H_2: Two or more *independent* loci with penetrance as determined for H_1;

H_3: *Single* locus with incomplete (or complete) penetrance in heterozygotes.

The distributions of phenotypes (i.e., expected proportions of responders and nonresponders), identical for each of these three hypotheses, for backcrosses and intercrosses, are summarized in Table 16.2.

TABLE 16.2 Phenotype distributions under H_1, H_2, and H_3, in the offspring of backcrosses and intercrosses

Matings	$\Pr\{R\}$	$\Pr\{\bar{R}\} = 1 - \Pr\{R\}$
Pt_1	1	0
Pt_2	0	1
F_1	β	$1 - \beta$
$F_1 \times Pt_2$	$\frac{1}{2}\beta$	$\frac{1}{2}(2 - \beta)$
$F_1 \times \underbrace{Pt_2 \times \cdots \times Pt_2}_{k \text{ times}}$	$\beta/2^k$	$(2^k - \beta)/2^k$
$\left. \begin{array}{l} F_1 \times F_1 \\ F_1 \times \underbrace{Pt_2 \times \cdots \times Pt_2}_{k \text{ times}} \end{array} \right\}$	$\frac{1}{4}(1 + 2\beta)$	$\frac{1}{4}(3 - 2\beta)$
$F_1 \times Pt_1$	$\frac{1}{2}(1 + \beta)$	$\frac{1}{2}(1 - \beta)$
$F_1 \times \underbrace{Pt_1 \times \cdots \times Pt_1}_{k \text{ times}}$	$[(2^k - 1) + \beta]/2^k$	$(1 - \beta)/2^k$

16.4. A DISCRIMINATORY DESIGN FOR MODELS WITH ADDITIVE PHENOTYPIC EFFECTS

In some cases, one way of discriminating among hypotheses regarding a single locus versus multiple loci is to perform an *assortative mating*. For instance, one can mate the responders from a segregating population (e.g., from the offspring of intercrosses or backcrosses) with responders or with nonresponders from the same segregating population or from a parental population.

We will here present an experiment in which the responders, R, from backcross $F_1 \times Pt_2$ are crossed back again to Pt_2. We denote this briefly by $R(F_1 \times Pt_2) \times Pt_2$.

16.4.1. Truncated Genotype Distribution in the Offspring of Matings $R(F_1 \times Pt_2) \times Pt_2$. Two Linked Loci

The distribution of the genotypes in the offspring of backcrosses ($F_1 \times Pt_2$) for two linked loci is given in (16.3). The proportion of responders is $\frac{1}{2}\lambda$. Let $Pr\{Genotype \mid R(F_1 \times Pt_2)\}$ be the probability distribution of the genotypes in the offspring of ($F_1 \times Pt_2$), given that they are responders. This is the conditional or *truncated* distribution:

$$
\begin{array}{llll}
\text{Genotype in } R(F_1 \times Pt_2): & AB/ab & Abab & aBab \\
Pr\{\text{Genotype} \mid R(F_1 \times Pt_2)\}: & 1 - \lambda & \lambda\beta_1/\beta & \lambda\beta_2/\beta.
\end{array}
\tag{16.17}
$$

The proportions of gametes produced by these genotypes are as follows:

$$
Pr\{AB \mid R(F_1 \times Pt_2)\} = \tfrac{1}{2}(1 - \lambda)(1 - \lambda) = \tfrac{1}{2}(1 - \lambda)^2, \quad (16.18a)
$$

$$
Pr\{Ab \mid R(F_1 \times Pt_2)\} = \frac{1}{2}\frac{\lambda(1 - \lambda)\beta}{\beta} + \frac{1}{2}\frac{\lambda\beta_1}{\beta} = \frac{1}{2}\frac{\lambda}{\beta}[\beta_1 + (1 - \lambda)\beta]. \quad (16.18b)
$$

Similarly,

$$
Pr\{aB \mid R(F_1 \times Pt_2)\} = \frac{1}{2}\frac{\lambda}{\beta}[\beta_2 + (1 - \lambda)\beta], \quad (16.18c)
$$

and

$$
Pr\{ab \mid R(F_1 \times Pt_2)\} = \tfrac{1}{2}(1 - \lambda)^2 + \frac{1}{2}\frac{\lambda}{\beta}(\beta_1 + \beta_2) = \tfrac{1}{2}(1 - \lambda - \lambda^2). \quad (16.18d)
$$

The probabilities given in (16.18a–d) are also the expected proportions of genotypes AB/ab, $Abab$, $aBab$, and $abab$, respectively, in the offspring of the backcross $R(F_1 \times Pt_2) \times Pt_2$.

Thus, the expected proportion of responders in the offspring of $R(F_1 \times Pt_2) \times Pt_2$ backcrosses is

$\Pr\{R \mid [R(F_1 \times Pt_2) \times Pt_2]\}$

$$= (\beta_1 + \beta_2) \times \frac{1}{2} \frac{(1 - \lambda)^2 \beta}{\beta}$$

$$+ \beta_1 \times \frac{1}{2} \frac{\lambda}{\beta} [\beta_1 + (1 - \lambda)\beta] + \beta_2 \times \frac{1}{2} \frac{\lambda}{\beta} [\beta_2(1 - \lambda)\beta]$$

$$= \frac{1}{2\beta} [(1 - \lambda)^2 \beta^2 + \lambda(1 - \lambda)\beta(\beta_1 + \beta_2) + \lambda(\beta_1^2 + \beta_2^2)]$$

$$= \frac{1}{2\beta} [(1 - \lambda)\beta^2(1 - \lambda + \lambda) + \lambda(\beta_1^2 + \beta_2^2)]$$

$$= \frac{1}{2\beta} [(1 - \lambda)\beta^2 + \lambda(\beta_1^2 + \beta_2^2)]. \tag{16.19}$$

In particular, when $\lambda = \frac{1}{2}$, that is, when the loci are independent, we obtain

$$\Pr\{R \mid [R(F_1 \times Pt_2) \times Pt_2]\} = \frac{1}{4\beta} [\beta^2 + (\beta_1^2 + \beta_2^2)]. \tag{16.20}$$

When, additionally, $\beta_1 = \beta$ and $\beta_2 = 0$, that is, when there is a single locus with penetrance β in the heterozygote, we obtain

$$\Pr\{R \mid [R(F_1 \times Pt_2) \times Pt_2]\} = \frac{1}{2}\beta. \tag{16.21}$$

Table 16.3 summarizes the results.

TABLE 16.3 Expected proportions of responders in the offspring of "truncated" backcross $[R(F_1 \times Pt_2) \times Pt_2]$

Genetic hypothesis	$\Pr\{R \mid [R(F_1 \times Pt_2) \times Pt_2]\}$	
		$\beta_1 = \beta_2 = \frac{1}{2}$ $(\beta = 1)$
Two linked loci with additive effects and incomplete penetrance	$\dfrac{1}{2\beta} [(1 - \lambda)\beta^2 + \lambda(\beta_1^2 + \beta_2^2)]$	$\frac{1}{4}(2 - \lambda)$
Two independent loci with additive effects and incomplete penetrance	$\dfrac{1}{4\beta} [\beta^2 + (\beta_1^2 + \beta_2^2)]$	$\frac{3}{8}$
Single locus	$\frac{1}{2}\beta$	$\frac{1}{2}$

It will be of particular interest to compare two cases: (i) $\lambda = \frac{1}{2}, \beta_1 = \beta_2 = \frac{1}{2}$, that is, two independent loci with equal additive effects such that the heterozygote AB/ab is fully penetrant (since $\beta = \beta_1 + \beta_2 = 1$); and (ii) a single locus with complete penetrance ($\beta = 1$). Formulae (16.20) and (16.21) yield $\frac{3}{8}$ and $\frac{1}{2}$, respectively. This means that, if the sample is large enough, this breeding test can distinguish between the hypotheses of (i) two loci with additive effects and (ii) a single locus with complete penetrance.

16.4.2. Family Study

Assortative matings cannot be experimentally performed in human populations. But we can select families of certain mating types (e.g., responders × responders or responders × nonresponders) and analyze the segregation ratios of offspring in such families. Family data of this type represent the best material when the mode of inheritance of a certain trait is the subject of the investigation. Family study can be used for both experimental and nonexperimental populations. Chapters 17 and 18, in particular, are devoted to the problems of statistical analysis of family data (segregation analysis). Such data are very important in studying human genetic traits.

16.5. DIFFERENTIAL VIABILITY

In the preceding sections we considered different genetical hypotheses which generated the same phenotype distribution. One way to distinguish among them was to use an experiment involving assortative mating.

Another class of equivalent models is that in which *not only the phenotype but also the genotype distributions* under different genetic hypotheses concerning mode of inheritance are indistinguishable. In this section, we will consider some examples of such models.

By *differential viability* we will understand unequal survival rates of different genes or their combinations, so that some genotypes are more likely to survive than others and the ordinary Mendelian ratios are disturbed. We might recognize differential viability of gametes before fertilization occurs, or differential viability of zygotes which die *in utero*, or differential viability of individuals in early or even adult life. In our models, however, we will confine ourselves to problems in which differential viability occurs in *gametes*.

We restrict ourselves to two loci, each with two alleles (dominant and recessive), and will discuss the following two hypotheses:

H_4: Loci are *independent* and different gametes have different viabilities;
H_5: Loci are *linked* and different gametes have different viabilities.

Suppose that, under H_4, the *relative* viabilities of the gametes AB, Ab, aB with respect to the gamete ab are

$$\begin{array}{lcccc} \text{Gamete:} & AB & Ab & aB & ab \\ \text{Rel. viability:} & \phi_1 & \phi_2 & \phi_3 & 1, \end{array} \tag{16.22}$$

with $0 \leqslant \phi_i \leqslant 1$, $i = 1, 2, 3$. Therefore, the distribution of gametes produced by F_1 (i.e., by individuals $AaBb$) is

$$\begin{array}{lcccc} \text{Gamete:} & AB & Ab & aB & ab \\ \text{Distribution:} & \phi_1/D & \phi_2/D & \phi_3/D & 1/D, \end{array} \tag{16.23}$$

where

$$D = \phi_1 + \phi_2 + \phi_3 + 1. \tag{16.24}$$

Suppose that under H_5 the recombination fraction is λ and the relative viabilities of gametes AB, Ab, aB, and ab are ϕ'_1, ϕ'_2, ϕ'_3, and 1, respectively, with $0 \leqslant \phi'_i \leqslant 1$ $i = 1, 2, 3$. The distribution of gametes produced by F_1 (i.e., by AB/ab) is

$$\begin{array}{lcccc} \text{Gamete:} & AB & Ab & aB & ab \\ \text{Distribution:} & (1 - \lambda)\phi'_1/D' & \lambda\phi'_2/D' & \lambda\phi'_3/D' & (1 - \lambda)/D', \end{array} \tag{16.25}$$

where

$$D' = (1 - \lambda)(1 + \phi'_1) + \lambda(\phi'_2 + \phi'_3). \tag{16.26}$$

Putting

$$\frac{D'}{1 - \lambda} = D \tag{16.27}$$

and denoting

$$\phi'_1 = \phi_1; \quad \frac{\lambda\phi'_2}{1 - \lambda} = \phi_2; \quad \frac{\lambda\phi'_3}{1 - \lambda} = \phi_3, \tag{16.28}$$

we obtain the gamete distribution (16.25) in the form (16.23). The parameter λ is "confounded" with ϕ'_i ($i = 1, 2, 3$), so that (16.23) and (16.25) are indistinguishable unless, of course, the theoretical values of either λ or ϕ's or both are known.

Therefore, if only matings $F_1 \times Pt_1$, $F_1 \times Pt_2$, or $F_1 \times F_1$ are performed, we cannot distinguish between hypotheses H_4 and H_5 (i.e., independent or linked loci with differential gamete viabilities). But it is easy to see that the mechanisms of the gamete production under the two hypotheses are different. Thus, the proportions of gametes produced by backcrosses or intercrosses will also be different for H_4 and H_5. Therefore, further breeding tests such as $F_1 \times Pt_1 \times Pt_1$ or $F_1 \times Pt_2 \times Pt_2$ or $F_2 \times F_1$ can be used to distinguish between linked and independent loci with differential viabilities of gametes.

16.5.1. Differential Viability of Gametes Ab and aB

Of particular interest is the case in which $\phi_1 = 1$, $0 < \phi_2 < 1$, and $0 < \phi_3 < 1$ — that is, in which the gametes Ab and aB are less viable than the gametes AB and ab.

Suppose that additionally we have for linked loci $\phi_1' = \phi_2' = \phi_3' = 1$. Then from (16.26) we obtain $D' = 2$, and from (16.28) we obtain

$$\phi_1' = \phi_1 = 1, \quad \frac{\lambda\phi_i'}{1 - \lambda} = \phi_i = \phi, \quad \text{for} \quad i = 2, 3. \qquad (16.29)$$

This means that, when only breeding tests $F_1 \times Pt_1$, $F_1 \times Pt_2$, or $F_1 \times F_1$ are performed, it is impossible to distinguish between the following hypotheses:

H_6: Loci are independent, but gametes Ab and aB have relative viabilities ϕ $(0 < \phi < 1)$ as compared to AB or ab;

H_7: Loci are linked and all the gametes are fully viable.

Further breedings are necessary.

16.6. SIMPLE VERSUS DOUBLE (OR MULTIPLE) RECESSIVE. INDISTINGUISHABLE SNYDER'S RATIOS

In Chapter 7 we derived formulae for the expected proportions of recessive offspring when the parental matings are Dom × Rec or Dom × Dom, and when a single locus is involved. These proportions are known as Snyder's ratios, S_{10} and S_{11}, respectively. In sections 12.7 and 12.8 we discussed how to estimate q, that is, the frequency of the gene a in a population, by using Snyder's ratios.

We will now show that, in fact, Snyder's ratios do not distinguish between single and double (in general, multiple) recessives and that the estimated q can be the estimated product q_1q_2 of the frequencies of genes a and b at loci A and B, respectively. We will present the analysis in full detail only for parental matings Dom × Rec (8 genotype matings). The matings Dom × Dom include 36 genotype matings, and the derivation of Snyder's ratio is left to the reader as an exercise (see Problem 16.5).

Let p_1, q_1 be the frequencies of genes A and a at locus A, and p_2, q_2 be the frequencies of genes B, b at locus B. Suppose that only two phenotypes are observed: $aabb$ (i.e., double recessive) and non-$aabb$ (dominant). We assume that the population is in equilibrium with respect to these loci. Table 16.4 represents the relative frequencies of parental matings Dom × Rec

TABLE 16.4 Probabilities for parental matings Dom × Rec
and their offspring. Double recessive

Genotype matings	Probability of mating	Expected proportion of recessives in offspring
$AABB \times aabb$	$2p_1^2p_2^2q_1^2q_2^2$	0
$AaBB \times aabb$	$4p_1p_2^2q_1^3q_2^2$	0
$AABb \times aabb$	$4p_1^2p_2q_1^2q_2^3$	0
$AAbb \times aabb$	$2p_1^2q_1^2q_2^4$	0
$aaBB \times aabb$	$2p_2^2q_1^4q_2^2$	0
$Aabb \times aabb$	$4p_1q_1^3q_2^4$	$\frac{1}{2}$
$aaBb \times aabb$	$4p_2q_1^4q_2^3$	$\frac{1}{2}$
$AaBb \times aabb$	$8p_1p_2q_1^3q_2^3$	$\frac{1}{4}$
Total	$2q_1^2q_2^2(1 - q_1^2q_2^2)$	

and the expected proportions of recessive offspring for each given parental mating. The expected proportion of recessive offspring in the population of matings Dom × Rec is

Pr{Rec offspring | Dom × Rec}

$$= S_{10} = \frac{\frac{1}{2} \cdot 4q_1^3q_2^3(p_1q_2 + p_2q_1) + \frac{1}{4} \cdot 8p_1p_2q_1^3q_2^3}{2q_1^2q_2^2(1 - q_1^2q_2^2)}$$

$$= \frac{q_1q_2(1 - q_1q_2)}{(1 + q_1q_2)(1 - q_1q_2)} = \frac{q_1q_2}{1 + q_1q_2}. \tag{16.30}$$

If we denote q_1q_2 by q, we obtain $S_{10} = q/(1 + q)$, which is indistinguishable from Snyder's ratio S_{10} for a single recessive.

It can be shown that for matings Dom × Dom we will have

$$S_{11} = S_{10}^2 = \frac{q_1^2q_2^2}{(1 + q_1q_2)^2}. \tag{16.31}$$

The results can be generalized for k-tuple recessives. If q_i is the frequency of the recessive allele at the ith locus, the Snyder's ratios are

$$S_{10} = \frac{\prod_{i=1}^{k} q_i}{1 + \prod_{i=1}^{k} q_i} \quad \text{and} \quad S_{11} = S_{10}^2. \tag{16.32}$$

Putting $\prod q_i = q$, these cannot be distinguished from Snyder's ratios for a single recessive.

To discriminate between single and double (in general, multiple) recessives, methods of segregation analysis, using family data, should be applied (compare section 18.7).

SUMMARY

When testing hypotheses on the mode of inheritance, one should take care to use a properly designed experiment. It is customary in plant and animal breeding to use data from backcrosses and/or intercrosses for testing such hypotheses. It has been shown in this chapter that in several cases, backcrosses and intercrosses are not sufficient to distinguish among certain types of mode of inheritance. The following kinds of statistically equivalent hypotheses have been discussed:

(i) The genotype distributions have different forms and different parameters, but the phenotype distributions are indistinguishable. These models usually include such assumptions as incomplete penetrance and the additive effects of genes. Also duplicate epistasis belongs to this group. By using experiments of assortative matings in plant and animal breeding, or family study for human populations, one can discriminate among different hypotheses.

(ii) The genotype distributions are indistinguishable because genetic parameters are confounded. These models usually introduce differential viability of gametes. Further breeding, such as double backcrosses or intercrosses, or, again, family analysis is necessary to distinguish among various hypotheses.

Of course, in several situations, biological tests (biochemical, cytogenetical, etc.) can be much more effective in discriminating among some hypotheses than the use of statistical analysis.

REFERENCES

Bailey, N. T. J., *Introduction to the Mathematical Theory of Genetic Linkage*, Clarendon Press, Oxford, 1961.

Fuller, J. L., Easler ,C., and Smith, M. E., Inheritance of audiogenic seizure susceptibility in the mouse, *Genetics* **35** (1950), 622–632.

Li, C. C., Some general properties of recessive inheritance, *J. Amer. Hum. Genet.* **5** (1953), 269–279.

Stern, C., *Principles of Human Genetics*, W. H. Freeman & Co., San Francisco, 1960.

Wright, S., The results of crosses between inbred strains of guinea pigs, differing in number of digits, *Genetics* **19** (1934), 534–551.

PROBLEMS

16.1. (*a*) Find the genotype distribution in the offspring of $F_2 \times F_2$, assuming two linked loci with recombination fraction λ.

(*b*) Show that, under the assumptions concerning additive effects stated in section 16.3, the expected proportion of responders in the offspring of $F_2 \times F_2$ is $\Pr\{R \mid F_2 \times F_2\} = (1 + 2\beta)/4$.

16.2. (*a*) Find the proportion of responders in the offspring of the "truncated" backcross $[R(F_1 \times Pt_1) \times Pt_1]$, assuming two independent loci with α_1, α_2, β_1, β_2 defined as in section 16.3.

(*b*) Show that for $\alpha_1 = \alpha_2 = \beta_1 = \beta_2 = \frac{1}{2}$ this mating test cannot distinguish between hypotheses of a single locus with complete penetrance and of two loci with additive effects.

16.3. Find the genotype distributions in the offspring of the $F_1 \times Pt_2 \times Pt_2$, assuming that two loci, each with two alleles and different viabilities of gametes, are involved, when:

(*a*) the loci are independent;

(*b*) the loci are linked.

(For the notation and discussion see section 16.5.)

16.4. Construct the model for the designs in Problem 16.3, assuming complete penetrance and equal viabilities of gametes. Consider cases (*a*) and (*b*).

16.5. Suppose that a certain character is due to a double recessive, *aabb*. Let q_1 and q_2 be the frequencies of genes *a* and *b*, respectively.

(*a*) Write down all kinds (36) of genotype matings for Dom × Dom and evaluate their frequencies, assuming that the population is in equilibrium with respect to these loci.

(*b*) Show that the expected proportion of matings Dom × Dom is $(1 - q_1^2 q_2^2)^2$.

(*c*) Show that the expected proportion of recessives in the offspring of Dom × Dom is

$$S_{11} = \frac{q_1^2 q_2^2}{(1 + q_1 q_2)^2}.$$

(*d*) Find the expected proportions of segregating and non-segregating families, each of size *s* (compare section 8.8).

16.6. Consider the following genotype distribution:

$$\begin{array}{ccc} AA & Aa & aa \\ p^2 & 2pq & q^2 . \end{array}$$

Let W_{11}, W_{12}, and W_{22} be the fitnesses of the genotypes AA, Aa, and aa, respectively. After selection, the gene frequencies, p' and q', are, respectively,

$$p' = \frac{W_{11}p + W_{12}q}{\overline{W}}\, p, \quad q' = \frac{Wp_{12} + W_{22}q}{\overline{W}}\, q,$$

with

$$\overline{W} = W_{11}p^2 + 2W_{12}pq + W_{22}q^2.$$

Suppose that all three genotypes are distinct, and that a sample with n_{11} of AA, n_{12} of Aa, and n_{22} of aa ($n_{11} + n_{12} + n_{22} = n$) has been observed in the next generation of a randomly mating population.

Show that the X^2-test of the hypothesis H_0: $W_{11} = W_{12} = W_{22}$ is indistinguishable from the X^2-test of the hypothesis H_0': $W_{11} W_{22} = W_{12}^2$ [Lewontin and Cockerham, *Evolution* **13** (1959), 561–564].

16.7. (*a*) Show that the two hypotheses, H_1: the trait is controlled by a single locus with two alleles, dominant and recessive, and H_2: the trait is the result of the complementary epistatis of two closely linked loci (with recombination fraction $\lambda = 0$), cannot be distinguished from phenotypic ratios in the offspring of intercrosses $AB/ab \times AB/ab$.

(*b*) A discriminatory experiment can be: $AB/ab \times Ab/aB$. (*Hint:* recall Problem 3.6.)

CHAPTER 17

Segregation Ratios in Families.
Simple Modes of Inheritance

17.1. RANDOM SAMPLING AND FAMILY RECORDS

In studies of mode of inheritance and the gene frequencies of genetic traits in natural (e.g., human) populations, statistical data can be obtained in four main ways:

(i) By selecting a random sample of individuals from the whole population.

(ii) By selecting a random sample of pairs from a population consisting of related pairs.

(iii) By selecting two independent random samples from two generations.

(iv) By tracing the trait in sibships of several families; such studies can be extended to more than one generation.

Applications of simple random sampling techniques (i)–(iii) were discussed in previous chapters, especially in Chapters 11–13. We pointed out some conditions under which it would be difficult to obtain such samples. There are also some situations in which these samples are insufficient for testing the mode of inheritance (e.g., linkage cannot be detected, or discrimination between single or multiple autosomal recessives is impossible).

More genetic information is obtained from studies of all the members in a sibship, which we will also call a *family*. The observations in such samples are, of course, not independent, and their analysis requires special statistical methods called *segregation analysis*.

Weinberg (1912*a*, 1912*b*, 1927), Lenz (1929), and Hogben (1931) were the first to point out that in handling family data adjustments for family size and for the way in which a family has been brought into the record should be taken into account.

The second step in this analysis, utilizing the maximum likelihood principle, was taken by Fisher (1934) and Haldane (1932, 1938, 1949) and extended by Bailey (1951a, 1951b) and Smith (1956, 1959). Only simple modes of inheritance, associated with a single locus (simple recessive and simple rare dominant), could be tested by their methods. These will also be the topic of this chapter. We shall introduce here the basic definitions and review the statistical methods which have been used in estimation and in testing hypotheses.

Advances in segregation analysis will be the topic of Chapter 18.

17.2. RARE AND COMMON GENETIC TRAITS

Methods of segregation analysis are chosen depending on whether the trait is rare or common. The only exception is a simple recessive — the same formulae are used whether the recessive trait is rare or common. It is then important to realize what is meant by rare and common traits.

By a *rare genetic trait* we will understand a trait for which the gene frequency(ies) associated with this trait is very low, less than 0.01 (i.e., 1 %), say. A genetic trait for which the gene frequency(ies) is greater than 0.01 will be called a *common genetic trait*. The value 0.01 was arbitrarily chosen, and it may be argued that 0.001 or 0.0001 should be used instead. But the important point of our definitions is that they are expressed in terms of *gene frequencies* rather than in terms of prevalence or incidence. For instance, if the onset of a trait depends on the age, or penetrance parameter, the incidence or prevalence rates might be quite low, while the gene frequencies can be fairly high.

There are a number of deleterious traits in human populations which are genetically controlled and which are called *hereditary diseases*. Fortunately most of them are rare. An individual with such a trait is called *affected;* without the trait he is called *normal*. In our definitions, by "trait" we shall mean an "abnormal trait."

DEFINITION 17.1. *If a trait is genetically controlled, then for given genotypes of the parents which (potentially) could produce affected children there is a constant nonzero probability, θ, called the segregation probability (or the segregation parameter) that a child born in such a family will be affected.*

In fact, θ is the expected proportion of affected children in a very large (infinite) number of families with given parental genotypes. The segregation parameter, θ, can be used as one of the criteria in classifying the mode of inheritance.

Let D and d be the dominant and recessive genes, respectively, at the locus under consideration. If D is the normal gene, then DD and Dd are phenotypically normal whereas dd is affected. If d is the normal gene, then dd is phenotypically normal whereas Dd and DD are affected. In the case where D is normal, normal parents may have affected children. It seems to be useful to summarize the criteria by which simple dominant and recessive traits (rare and common) can be recognized.

17.2.1. Autosomal Dominant

(1) The trait is transmitted from generation to generation without skipping. (2) Every affected child has at least one affected parent (except for sporadic cases). (3) Both sexes are affected equally. (4) When the trait is *rare*, the matings are usually Normal × Affected; and since the genotypes $dd \times DD$ are then unlikely, the genotypes of such parents are usually $dd \times Dd$. The segregation probability is, in this case, $\theta = \frac{1}{2}$. When the trait is *common*, matings $dd \times DD$ with $\theta = 1$ are also likely to occur. Affected × Affected matings give rise to more possibilities.

17.2.2. Autosomal Recessive

(1) All children with both parents affected are also affected. (2) Both sexes are affected equally. (3) The risk of having an affected child is greater in consanguineous marriages than in marriages between unrelated individuals. (4) If the trait is *rare*, the parental matings in which the trait can appear are Normal × Normal with genotypes $Dd \times Dd$ and $\theta = \frac{1}{4}$. If the trait is *common*, the Normal × Normal parents who can produce affected children will still be $Dd \times Dd$ with $\theta = \frac{1}{4}$, but also the families Normal × Affected with genotypes $Dd \times dd$ can occur more often and produce affected children.

17.2.3. X-Linked Dominant

(1) Affected males transmit the trait to all their daughters but not to their sons. The X-linked dominant can be distinguished from the autosomal dominant only through affected fathers, not through affected mothers. (2) Every affected child has at least one affected parent (except for sporadic cases). (3) If the trait is *rare*, the affected females are usually Dd. They transmit the trait with $\theta = \frac{1}{2}$ to their children of either sex. (4) If p is the frequency of gene A in the population, the proportion of affected males is p and of affected females is $p^2 + 2pq = p(1 + q)$. If p is very small (and so q is near to 1), then $p(1 + q) \doteq 2p$. This means that for *rare* traits there will

be almost twice as many affected females as affected males. (5) If the X-linked dominant is *common*, females can also be DD, so that marriages $♀$Affected $× ♂$Normal can be $♀DD × ♂d$ with $θ = 1$, and $♀Dd × ♂d$ with $θ = \frac{1}{2}$.

17.2.4. X-Linked Recessive

(1) The usual Normal × Normal matings in which the trait can appear will be $♀Dd × ♂D$, so that all the daughters will be normal but one half of them are expected to be carriers of the gene d, while one half of the sons will be affected. (2) The matings $♀$Normal $× ♂$Affected, in which the trait appears, are usually $♀Dd × ♂d$. In such families all sons will be affected, while one half of the daughters will be affected and one half will be normal carriers. (3) For *rare* X-linked recessive, we deduce from (1) and (2), that the proportion of affected females will be approximately the square of the proportion of affected males. (4) If the trait is *common* the matings $♀dd × ♂D$ and $♀dd × ♂d$ are also likely to occur.

17.3. SEGREGATING AND NONSEGREGATING FAMILIES. ASCERTAINMENT

In this and in the next section we introduce some terminology and definitions which are necessary as a "communication language" among people dealing with family data. Unfortunately, the terminology is not well established and is often confusing. According to the suggestion in the Report of the Chicago Conference (1966), we shall use chiefly the terminology introduced by Morton (1959, 1962) with, however, a few exceptions which will be pointed out in the text.

DEFINITION 17.2. *Families with exactly one affected child are called simplex families; those with more than one affected child, multiplex families.*

DEFINITION 17.3. *An affected child in a simplex family is called an isolated case; an affected child in a multiplex family, a familial case.*

Isolated cases can be classified into two classes, depending on the sources which cause the trait. *Chance isolated* cases are due to segregation at a given locus controlling the trait; these are of the same origin as familial cases. *Sporadic* cases are different from familial and chance isolated cases and can be due to mutation, phenocopy, diagnostic errors, etc.

The familial and chance isolated cases have a defined segregation probability, $θ > 0$, whereas the sporadic cases do not.

DEFINITION 17.4.   *Families for which the segregation probability θ (θ > 0) is defined and is not small are called high-risk type families. Families for which the probability of occurrence of an affected child is so small that the risk of having more than one affected child is negligible, are called low-risk type families.*

Practically, low-risk type families are those in which sporadic cases occur.

In section 8.8 we introduced the concept of segregating and nonsegregating families. We now define them more precisely.

DEFINITION 17.5.   *Excluding sporadic cases, families with at least one affected child will be called segregating families, and those with no affected children will be designated as nonsegregating families.*

From this definition, the multiplex and simplex (with chance isolated cases) families are in the category of segregating families. On the other hand, nonsegregating families are either those in which the parents have genotypes such that they are not able to produce affected children (e.g., matings $DD \times Dd$ cannot produce offspring dd; for these matings $\theta = 0$), or those which are potentially able to produce affected children but have θ less than 1 so that affected children may not occur (e.g., $Dd \times Dd$ can produce dd only with $\theta = \frac{1}{4}$).

Methods of estimation of the segregation parameter, θ, depend on how the families have been brought into the record or, in other words, how the families have been *ascertained*.

DEFINITION 17.6.   *By a method of ascertainment of families we understand a sampling procedure by which data about traits in families are collected.*

Broadly speaking, two main ways of obtaining information about a certain trait can be recognized:

(i) through the parents;
(ii) through the affected children.

DEFINITION 17.7.   *A random ascertainment through the parents is a simple random sample of families from a population without consideration of the phenotypes of the offspring.*

Morton (1959) called this *complete selection*. He used the phrase in the sense that all kinds of families, segregating and nonsegregating, are selected in the sample. However, it is not "complete" in the sense that not all families with affected children will be selected. For this reason, we shall simply call the method "ascertainment through the parents." It is of practical use only when the trait is common rather than rare, since for rare traits and reasonable sample sizes of families we might not find any familial cases at all. This method will be briefly discussed in section 17.17.

17.4. ASCERTAINMENT THROUGH THE AFFECTED CHILDREN. TERMINOLOGY AND DEFINITIONS

We shall first define a few terms necessary for further discussion.

DEFINITION 17.8. *A proband is an affected individual who has been detected independently of the other members of the family and through whom the family can be ascertained.*

If there is more than one affected child in a family and these children become known through the first detected individual, they are not probands. However, if some (or all) of them have been detected independently (for instance, from different sources), they will also be probands. A family with r affected children can have at most r *different* probands. Moreover, the same affected individual can be recorded more than once from different independent sources, so that it is possible to have more than r probands in a family with r affected children.

DEFINITION 17.9. *The first proband recorded in a family is called the index case.*

Definitions 17.8 and 17.9 were established by Morton (1959). Another term, "propositus," is often used instead of "proband" in the English literature [Bailey (1951b)]. In fact, there is also some ambiguity between "propositus," "proband," and "index case." Stern (1960, p. 134), and Li (1961, p. 61) use these terms as synonyms. We shall distinguish them, however, as defined above.

DEFINITION 17.10. *The probability, π, that an affected individual, I, is a proband,*

$$\Pr\{I \text{ is a proband} \mid I \text{ affected}\} = \pi, \tag{17.1}$$

is called the ascertainment probability.

(More precisely, π is the *probability of ascertainment of a proband*.)

The probability that a family with r affected children will have c of them as probands is

$$\Pr\{c\} = \binom{r}{c}\pi^c(1 - \pi)^{r-c}, \qquad c = 0, 1, \ldots, r, \tag{17.2}$$

and the probability that there will be at *least one proband* in a family with r affected children is

$$\Pr\{c > 0\} = 1 - (1 - \pi)^r = \alpha_r. \tag{17.3}$$

The value of α_r is, in fact, the probability of *ascertainment of a family*, since it is necessary and sufficient for a family to have at least one proband to be ascertained.

DEFINITION 17.11.   *Selection of families having at least one affected child is called ascertainment through affected children.*

Morton (1959) called this "incomplete selection," in the sense that only segregating families (and some with sporadic cases) are selected. Since we reserve the terms "complete" and "incomplete" for attributes associated with the values of ascertainment probability, π, rather than with the method of ascertainment, we shall retain the phrase "ascertainment through affected children" for the method. In the later part of this chapter, we shall consider only problems in which ascertainment is through the children; therefore we shall omit this phrase unless ambiguity arises.

DEFINITION 17.12.   *If for a certain trait observed in a given population the ascertainment probability, π, is equal to 1 ($\pi = 1$), the ascertainment will be called complete, while for $0 < \pi < 1$ the ascertainment will be called incomplete.*

By definition, $\pi = 1$ means that all affected individuals become probands, and since also, for $\pi = 1$, $\alpha_r = 1 - (1 - \pi)^r = 1$ for $r = 1, 2, \ldots$, each family independently of its size will be ascertained.

"Complete" ascertainment, introduced in Definition 17.12, is called by Morton "truncate" ascertainment. We avoid this term since it is not commonly used by other authors.

We now consider the opposite extreme, in which only a very small proportion of all affected children can actually be brought into the record. This is the case when π is small. Using the binomial expansion of $(1 - \pi)^r$, we have

$$(1 - \pi)^r = 1 - r\pi + \frac{r(r - 1)}{2} \pi - \cdots .$$

When π is small, we have approximately

$$(1 - \pi)^r \doteq 1 - r\pi$$

or

$$\alpha_r = 1 - (1 - \pi)^r \doteq r\pi, \tag{17.4}$$

both of which express the fact that, if the probability of ascertainment is very small ($\pi \to 0$), then the probability, α_r, that a family with r affected children will be recorded is nearly proportional to the number of affected children. Since r is not very large, there is practically no chance that there will be more than one proband in families which have been brought into the record. We summarize this in the following definition.

DEFINITION 17.13. *If the ascertainment probability, π, is so small that there is practically no chance for a family to be brought into the record by more than one proband, we speak of a single ascertainment.*

(Note that a single ascertainment is an extreme case of incomplete ascertainment.)

If π is not very small but is less than 1 ($0 < \pi < 1$), then also $0 < \alpha_r < 1$. This means that some families may be brought into the records by more than one proband, but the chance that a family will be recorded, α_r, is still less than 1.

DEFINITION 17.14. *If $0 < \pi < 1$, then also $0 < \alpha_r < 1$ and the ascertainment is called incomplete multiple ascertainment.*

[Note that: (*a*) the condition $0 < \pi < 1$, with π not too small, allows that the same proband can be ascertained more than once from several independent sources; and (*b*) for the limiting case $\pi = 1$, we have complete ascertainment or, more precisely, "complete multiple ascertainment."]

Summarizing the results, we have distinguished three kinds of *ascertainment of families* through the children, depending on the probability of ascertainment (of a proband), π:

(i) Complete ascertainment, when $\pi = 1$ and so $\alpha_r = 1$ for $r = 1, 2, \ldots, s$; the ascertainment of a family is independent of the number of affected children in this family. All affected children are probands.

(ii) Incomplete multiple ascertainment, when $0 < \pi < 1$ and so $0 < \alpha_r < 1$, $r = 1, 2, \ldots, s$; usually there is more than one proband per family, although by definition there can be a single proband who is ascertained more than once.

(iii) Incomplete single ascertainment, when π is small and so $\alpha_r \doteq \pi r$, $r = 1, 2, \ldots, s$; here α_r is proportional to the number of affected children in the family. Practically, there will be no more than one proband per ascertained family.

To ensure a valid segregation analysis, it is important that the probands and ascertainments should be correctly identified. We must realize that π is the probability that an affected individual will become a proband. Therefore, if all affected children in a family are observed, but only some are probands and the others are detected through the probands, π is still less than 1 and should be estimated. Some methods of estimating π will be discussed in section 17.15. In collecting the data, it is, then, important to count not only the number of affected children but also the number of probands and the number of independent sources from which probands were recorded.

If all affected children are probands, it is usually assumed that $\pi = 1$ and the methods described for complete ascertainment are applied. However, the assumption $\pi = 1$ is true only if all affected children *in a population* are ascertained. In fact, π is usually less than 1 and θ is usually overestimated.

If there is only one proband per family, some authors suggest using single ascertainment. It seems from experience, unfortunately, to be practically never true that $\pi \to 0$, and the methods of single ascertainment almost always underestimate θ. However, if π is unknown, the best that we can do is to carry out the analysis for these two limiting cases to obtain some idea about the estimated θ. For this reason we will present in detail the analysis for two limiting situations of complete and single ascertainment.

17.5. THE DISTRIBUTION OF AFFECTED CHILDREN IN HIGH-RISK SEGREGATING FAMILIES

(*a*) The theory derived in the remaining sections applies only to simple autosomal traits: one-locus recessive and one-locus rare dominant. The parental phenotypic matings considered here will be Normal × Normal and Normal × Affected. By virtue of what was said in section 17.2 about rare dominant and recessive, each of these phenotypic mating types will, in fact, correspond to a certain genotype mating type, such as $Dd \times Dd$ or $Dd \times dd$. For such matings there will be defined segregation probability θ.

The derivation of segregation distribution and of likelihood function is associated with one of the parental mating types (Normal × Normal or Normal × Affected).

(*b*) We restrict our discussion to high-risk segregating families ascertained through the children.

(*c*) We assume that the ascertainment probability π ($0 < \pi < 1$) is constant for a given population and trait.

Let s denote the family size, r the number of affected children, and $\alpha_r = 1 - (1 - \pi)^r$ the ascertainment probability for a family with r affected children (i.e., the probability that such a family is, at least once, ascertained). The probability that in a family of size s there are r affected children is

$$\Pr\{r\} = \binom{s}{r}\theta^r(1 - \theta)^{s-r}, \qquad r = 0, 1, 2, \ldots, s. \qquad (17.5)$$

Probability function (17.5) represents the distribution of affected children in a family of size s. It is sometimes called the *segregation distribution*.

The probability that in a family of size s there are exactly r affected children, and that such a family will be ascertained, is the joint probability

$$\Pr\{r\}\alpha_r = \binom{s}{r}\theta^r(1 - \theta)^{s-r}[1 - (1 - \pi)^r], \qquad r = 1, 2, \ldots, s. \qquad (17.6)$$

The probability that in a family of size s there is at least one affected child *and* that such a family will be ascertained is

$$\sum_{r=1}^{s} \Pr\{r\}\alpha_r = \sum_{r=1}^{s} \binom{s}{r} \theta^r (1 - \theta)^{s-r}[1 - (1 - \pi)^r]$$

$$= \sum_{r=1}^{s} \binom{s}{r} \theta^r (1 - \theta)^{s-r} - \sum_{r=1}^{s} \binom{s}{r} [(1 - \pi)\theta]^r (1 - \theta)^{r-s}$$

$$= [1 - (1 - \theta)^s] - \{[(1 - \pi)\theta + (1 - \theta)]^s - (1 - \theta)^s\}$$

$$= 1 - (1 - \pi\theta)^s. \tag{17.7}$$

Thus the distribution of affected children in a family of size s, *given that the family has been ascertained*, that is, the *segregation distribution in an ascertained family*, is

$$\frac{\Pr\{r\}\alpha_r}{\sum_{j=1}^{s} \Pr\{j\}\alpha_j} = \frac{\binom{s}{r} \theta^r (1 - \theta)^{s-r}[1 - (1 - \pi)^r]}{1 - (1 - \pi\theta)^s}, \qquad r = 1, 2, \ldots, s. \tag{17.8}$$

17.5.1. Complete Ascertainment

In a particular case, when $\pi = 1$, the segregation distribution (17.8) takes the form

$$\frac{\Pr\{r\}\alpha_r}{\sum_{j=1}^{s} \Pr\{j\}\alpha_j} = \binom{s}{r} \frac{\theta^r (1 - \theta)^{s-r}}{1 - (1 - \theta)^s} = \Pr\{r \mid r > 0\}, \quad r = 1, 2, \ldots, s. \tag{17.9}$$

As we can see, this is also the distribution of affected children in a family of size s with at least one affected child, that is, a *truncated binomial distribution*.

17.5.2. Single Ascertainment

When π is very small, $\alpha_r \doteq r\pi$, so that (17.6) takes the form

$$\Pr\{r\}\alpha_r \doteq \binom{s}{r} \theta^r (1 - \theta)^{s-r} r\pi, \qquad r = 1, 2, \ldots, s. \tag{17.10}$$

Hence,

$$\sum_{r=1}^{s} \Pr\{r\}\alpha_r \doteq \pi \sum_{r=1}^{s} r \binom{s}{r} \theta^r (1 - \theta)^{s-r} = \pi s \theta, \tag{17.11}$$

and finally (17.8) takes the form

$$\frac{\Pr\{r\}\alpha_r}{\sum_{j=1}^{s}\Pr\{j\}\alpha_j} \doteq \frac{\binom{s}{r}\theta^r(1-\theta)^{s-r}\pi r}{\pi s\theta} = \frac{(s-1)!}{(r-1)!(s-r)!}\theta^{r-1}(1-\theta)^{s-r}$$

$$= \binom{s-1}{r-1}\theta^{r-1}(1-\theta)^{s-r}, \qquad r = 1, 2, \ldots, s. \qquad (17.12)$$

This is the binomial distribution in a $(s-1)$-size sibship. Therefore the distribution of affected children in a family recorded by a single ascertainment is just the same as the (binomial) distribution of affected sibs among the $s-1$ sibs of the proband.

17.6. MAXIMUM LIKELIHOOD ESTIMATOR OF THE SEGREGATION PARAMETER θ UNDER MULTIPLE ASCERTAINMENT $(0 < \pi \leqslant 1)$, WITH KNOWN π. FIXED NUMBERS OF ASCERTAINED FAMILIES

We introduce the following notation:

s = the family size (i.e., the size of the sibship);

S = the maximal family size in the sample;

r = the number of affected children in the sibship of size s $(r = 1, 2, \ldots, s)$;

a_{rs} = the recorded number of families of size s, each with r affected children;

$n_s = \sum_{r=1}^{s} a_{rs}$ = the total number of recorded families of size s. We first assume that n_s is *fixed*. Therefore, all distributions, moments, estimators, etc., derived in this section and the next three will be, in fact, conditional, assuming that n_s is given. Since this applies to all quantities, we drop this condition in written formulae, so that $L_s(\theta)$ is in fact $L_s(\theta \mid n_s)$, etc.;

$N = \sum_{s=1}^{S} n_s = \sum_{s=1}^{S}\sum_{r=1}^{s} a_{rs}$ = the total number of recorded families of all sizes, that is, the sample size which is fixed, together with fixed S;

$r_s = \sum_{r=1}^{s} r a_{rs}$ = the number of affected children in all n_s families;

$R = \sum\limits_{s=1}^{S} r_s = \sum\limits_{s=1}^{S} \sum\limits_{r=1}^{s} ra_{rs} =$ the total number of affected children in the whole sample;

$t_s = sn_s =$ the number of recorded children in all n_s families;

$T = \sum\limits_{s=1}^{S} t_s = \sum\limits_{s=1}^{S} sn_s =$ the total number of recorded children in the whole sample.

Let $P_{rs}(\theta)$ be the probability of the occurrence of r affected children in an ascertained family. It is, in fact, the probability given by (17.8), that is,

$$P_{rs}(\theta) = \frac{\binom{s}{r} \theta^r (1 - \theta)^{s-r} [1 - (1 - \pi)^r]}{1 - (1 - \pi\theta)^s}. \qquad (17.8)$$

The probability that in n_s families (each of size s) there will be a_{1s} families with 1 affected child, a_{2s} families with 2 affected children, $\ldots$, a_{rs} families with r affected children, $\ldots$, and finally, a_{ss} families with s affected children, where

$$\sum\limits_{r=1}^{s} a_{rs} = n_s,$$

will be a multinomial distribution given by

$$\Pr\{a_{1s}, a_{2s}, \ldots, a_{ss}; \theta\} = \frac{n_s!}{a_{1s}! a_{2s}! \cdots a_{ss}!} P_{1s}^{a_{1s}}(\theta) P_{2s}^{a_{2s}}(\theta) \cdots P_{ss}^{a_{ss}}(\theta)$$

$$= n_s! \prod\limits_{r=1}^{s} \frac{P_{rs}^{a_{rs}}(\theta)}{a_{rs}!}. \qquad (17.13)$$

The right-hand side of (17.13), considered as a function of θ, represents the likelihood, $L_s(\theta)$, for the sample of n_s ascertained families, each of size s.

Substituting $P_{rs}(\theta)$, given by (17.8), into (17.13), we obtain

$$L_s(\theta) = \frac{n_s!}{\prod\limits_{r=1}^{s} a_{rs}!} \prod\limits_{r=1}^{s} \left\{ [1 - (1 - \pi)^r] \binom{s}{r} \frac{\theta^r (1 - \theta)^{s-r}}{1 - (1 - \pi\theta)^s} \right\}^{a_{rs}}, \qquad (17.14)$$

and the logarithm of $L_s(\theta)$ is

$$\log L_s(\theta) = \text{const} + \sum\limits_{r=1}^{s} \{ r \log \theta + (s - r)\log(1 - \theta)$$

$$- \log[1 - (1 - \pi\theta)^s] \} a_{rs}. \qquad (17.14a)$$

Denoting now the score of $L_s(\theta)$ by $U_s(\theta)$, we obtain

$$\frac{\partial \log L_s(\theta)}{\partial \theta} = U_s(\theta) = \sum_{r=1}^{s} \left(\frac{r}{\theta} - \frac{(s-r)}{1-\theta} - \frac{\pi s(1-\pi\theta)^{s-1}}{1-(1-\pi\theta)^s} \right) a_{rs}$$

$$= \sum_{r=1}^{s} \left[\frac{r}{\theta(1-\theta)} - \frac{s}{1-\theta} - \frac{\pi s(1-\pi\theta)^{s-1}}{1-(1-\pi\theta)^s} \right] a_{rs}$$

$$= \frac{1}{\theta(1-\theta)} \sum_{r=1}^{s} r a_{rs} - \frac{s}{1-\theta} \sum_{r=1}^{s} a_{rs} - \frac{\pi s(1-\pi\theta)^{s-1}}{1-(1-\pi\theta)^s} \sum_{r=1}^{s} a_{rs}$$

$$= \frac{r_s}{\theta(1-\theta)} - \frac{sn_s}{1-\theta} - \frac{\pi s n_s(1-\pi\theta)^{s-1}}{1-(1-\pi\theta)^s}$$

$$= \frac{r_s}{\theta(1-\theta)} - \frac{sn_s \left[1-(1-\pi\theta)^s - \pi(1-\theta)(1-\pi\theta)^{s-1}\right]}{(1-\theta)[1-(1-\pi\theta)^s]}$$

$$= \frac{1}{1-\theta} \left[\frac{r_s}{\theta} - \frac{1-(1-\pi)(1-\pi\theta)^{s-1}}{1-(1-\pi\theta)^s} t_s \right]. \tag{17.15}$$

The overall score, $U(\theta)$, for the whole sample is,

$$U(\theta) = \sum_{s=1}^{S} U_s(\theta) = \frac{1}{1-\theta} \left[\frac{R}{\theta} - \sum_{s=1}^{S} \frac{1-(1-\pi)(1-\pi\theta)^{s-1}}{1-(1-\pi\theta)^s} t_s \right]. \tag{17.16}$$

To obtain the ML estimator of θ, we have to solve the equation $U(\theta) = 0$ with respect to θ, that is,

$$\frac{R}{\theta} = \sum_{s=1}^{S} \frac{1-(1-\pi)(1-\pi\theta)^{s-1}}{1-(1-\pi\theta)^s} t_s. \tag{17.17}$$

We can see that θ in (17.17) is not a simple function of R, but it can be calculated by iteration.

17.6.1. Expected Values of a_{rs} and r_s

In order to calculate $\mathrm{Var}(\hat{\theta})$ we shall find the expected amount of information, $I(\theta)$, and therefore we must find the expected value of a_{rs}. For fixed n_s, a_{rs} is a multinomial variable, so that $E(a_{rs}) = n_s P_{rs}(\theta)$, $r = 1, 2, \ldots, s$. Substituting (17.8) for $P_{rs}(\theta)$, we obtain

$$E(a_{rs}) = n_s \frac{\binom{s}{r} \theta^r (1-\theta)^{s-r} [1-(1-\pi)^r]}{1-(1-\pi\theta)^s}, \qquad r = 1, 2, \ldots, s. \tag{17.18}$$

The expected value of r_s, $E(r_s)$, is

$$E(r_s) = \sum_{r=1}^{s} r E(a_{rs})$$

$$= \frac{n_s}{1 - (1 - \pi\theta)^s} \left\{ \sum_{r=1}^{s} r \binom{s}{r} \theta^r (1 - \theta)^{s-r} - \sum_{r=1}^{s} r \binom{s}{r} [(1 - \pi)\theta]^r (1 - \theta)^{s-r} \right\}.$$

$$(17.19)$$

The first term in the braces (curly brackets) of (17.19) is the mean of a binomial distribution, that is,

$$\sum_{r=1}^{s} r \binom{s}{r} \theta^r (1 - \theta)^{s-r} = s\theta. \qquad (17.20a)$$

We now evaluate the second term. We shall use the fact that

$$r \binom{s}{r} = s \binom{s-1}{r-1}.$$

We have

$$\sum_{r=1}^{s} r \binom{s}{r} [(1 - \pi)\theta]^r (1 - \theta)^{s-r}$$

$$= s(1 - \pi)\theta \sum_{r=1}^{s} \binom{s-1}{r-1} [(1 - \pi)\theta]^{r-1} (1 - \theta)^{s-r}$$

$$= s(1 - \pi)\theta [(1 - \pi)\theta + (1 - \theta)]^{s-1}$$

$$= s\theta(1 - \pi)(1 - \pi\theta)^{s-1}. \qquad (17.20b)$$

Substituting (17.20a) and (17.20b), respectively, into (17.19), we obtain

$$E(r_s) = \frac{s\theta n_s}{1 - (1 - \pi\theta)^s} [1 - (1 - \pi)(1 - \pi\theta)^{s-1}]. \qquad (17.21)$$

17.6.2. The Expected Amount of Information, $I(\theta)$, and the Variance, $\mathrm{Var}(\hat{\theta})$

Let us take $U_s(\theta)$ in the form

$$\frac{\partial \log L_s(\theta)}{\partial \theta} = U_s(\theta) = \frac{r_s}{\theta(1 - \theta)} - \frac{sn_s}{1 - \theta} - \frac{\pi sn_s(1 - \pi\theta)^{s-1}}{1 - (1 - \pi\theta)^s}. \qquad (17.22)$$

Therefore,

$$-\frac{\partial^2 \log L_s(\theta)}{\partial \theta^2} = -\frac{\partial U_s(\theta)}{\partial \theta}$$

$$= -\frac{r_s(2\theta - 1)}{\theta^2(1 - \theta)^2} + \frac{sn_s}{(1 - \theta)^2}$$

$$-\frac{\pi^2 s n_s(s - 1)(1 - \pi\theta)^{s-2}[1 - (1 - \pi\theta)^s]}{[1 - (1 - \pi\theta)^s]^2}$$

$$-\frac{\pi^2 s^2 n_s(1 - \pi\theta)^{s-1}(1 - \pi\theta)^{s-1}}{[1 - (1 - \pi\theta)^s]^2}$$

$$= -\frac{r_s(2\theta - 1)}{\theta^2(1 - \theta)^2} + \frac{sn_s}{(1 - \theta)^2}$$

$$-\frac{\pi^2 s n_s(1 - \pi\theta)^{s-2}}{[1 - (1 - \pi\theta)^s]^2}\{s - [1 - (1 - \pi\theta)^s]\} \quad (17.23)$$

The expected amount of information, $I(\theta)$, will be obtained by substituting $E(r_s)$ into (17.23) for r_s. We have

$$I_s(\theta) = -\frac{\theta s n_s[1 - (1 - \pi)(1 - \pi\theta)^{s-1}](2\theta - 1)}{[1 - (1 - \pi\theta)^s]\theta^2(1 - \theta)^2} + \frac{sn_s}{(1 - \theta)^2}$$

$$-\frac{\pi^2 s n_s(1 - \pi\theta)^{s-2}}{[1 - (1 - \pi\theta)^s]^2}\{s - [1 - (1 - \pi\theta)^s]\}, \quad (17.24)$$

and the total expected amount of information, $I(\theta)$, is

$$I(\theta) = \sum_{s=1}^{S} I_s(\theta). \quad (17.25)$$

Thus the variance of $\hat{\theta}$, $\mathrm{Var}(\hat{\theta})$, is approximately

$$\mathrm{Var}(\hat{\theta}) \doteq \frac{1}{I(\theta)}. \quad (17.26)$$

17.7. MAXIMUM LIKELIHOOD ESTIMATOR OF θ UNDER COMPLETE ASCERTAINMENT $(\pi = 1)$

We shall now evaluate the maximum likelihood estimate of θ, assuming that the ascertainment is complete, that is, $\pi = 1$. All formulae of section 17.6 should now hold for $\pi = 1$. Therefore,

$$U_s(\theta) = \frac{1}{1 - \theta}\left[\frac{r_s}{\theta} - \frac{1}{1 - (1 - \theta)^s}sn_s\right]; \quad (17.27)$$

$$U(\theta) = \sum_{s=1}^{S} U_s(\theta) = \frac{1}{1 - \theta}\left[\frac{R}{\theta} - \sum_{s=1}^{S}\frac{sn_s}{1 - (1 - \theta)^s}\right]; \quad (17.28)$$

and the estimator $\hat{\theta}$ will be obtained by solving (by iteration) the equation

$$\frac{R}{\theta} = \sum_{s=1}^{S} \frac{sn_s}{1 - (1 - \theta)^s} . \tag{17.29}$$

Furthermore,

$$E(a_{rs}) = n_s \frac{\binom{s}{r} \theta^r (1 - \theta)^{s-r}}{1 - (1 - \theta)^s} , \qquad r = 1, 2, \ldots, s, \tag{17.30}$$

and

$$E(r_s) = \frac{s\theta n_s}{1 - (1 - \theta)^s} . \tag{17.31}$$

We now evaluate

$$-\frac{\partial^2 \log L_s(\theta)}{\partial \theta^2} = -\frac{\partial U_s(\theta)}{\partial \theta} = -\frac{r_s(2\theta - 1)}{\theta^2 (1 - \theta)^2} + \frac{sn_s[1 - (1 - \theta)^s - s(1 - \theta)^s]}{(1 - \theta)^2[1 - (1 - \theta)^s]^2}$$

$$= \frac{-1}{(1 - \theta)^2} \left\{ \frac{2\theta - 1}{\theta^2} r_s - \frac{[1 - (1 - \theta)^s] - s(1 - \theta)^s}{[1 - (1 - \theta)^s]^2} sn_s \right\}. \tag{17.32}$$

Substituting $E(r_s)$ from (17.31) for r_s in (17.32), we obtain the expected amount of information, $I_s(\theta)$, as

$$I_s(\theta) = \frac{-1}{(1 - \theta)^2} \left\{ \frac{2\theta - 1}{\theta^2} \cdot \frac{\theta sn_s}{1 - (1 - \theta)^s} - \frac{1 - (1 - \theta)^s - s(1 - \theta)^s}{[1 - (1 - \theta)^s]^2} sn_s \right\}$$

$$= \frac{-sn_s \{(2\theta - 1)[1 - (1 - \theta)^s] - \theta[1 - (1 - \theta)^s + s(1 - \theta)^s]\}}{\theta(1 - \theta)^2[1 - (1 - \theta)^s]^2}$$

$$= \frac{sn_s}{\theta(1 - \theta)^2[1 - (1 - \theta)^s]^2} \{-(1 - \theta)[1 - (1 - \theta)^s] - \theta s(1 - \theta)^s\}$$

$$= \frac{1}{\theta(1 - \theta)} \cdot \frac{1 - (1 - \theta)^s - s\theta(1 - \theta)^{s-1}}{[1 - (1 - \theta)^s]^2} \cdot sn_s. \tag{17.33}$$

The overall expected amount of information, $I(\theta)$, is

$$I(\theta) = \sum_{s=1}^{S} I_s(\theta) = \frac{1}{\theta(1 - \theta)} \sum_{s=1}^{S} \frac{1 - (1 - \theta)^s - s\theta(1 - \theta)^{s-1}}{[1 - (1 - \theta)^s]^2} \cdot sn_s, \tag{17.34}$$

and

$$\mathrm{Var}(\hat{\theta}) \doteq \frac{1}{I(\theta)} . \tag{17.35}$$

17.8. EVALUATION OF SCORES, AMOUNTS OF INFORMATION, AND HETEROGENEITY TESTS

17.8.1. Estimation of θ

The formulae for $U(\theta)$ and $I(\theta)$, even in the simple case for $\pi = 1$, are rather complicated, especially if S is not small. The computation can be facilitated by preparing appropriate tables of coefficients. To prepare such tables we use the following approach.

The distribution of affected children in ascertained families of size s is given by (17.9), that is,

$$\Pr\{r \mid r > 0\} = \binom{s}{r} \frac{\theta^r (1 - \theta)^{s-r}}{1 - (1 - \theta)^s}, \qquad r = 1, 2, \ldots, s. \qquad (17.9)$$

This is a truncated binomial distribution. It can be shown (the proof is left to the reader) that, for a truncated binomial variate, the expected value of r for given sample size s, which we denote by A_s, is

$$E(r \mid r > 0) = A_s = \frac{s\theta}{1 - (1 - \theta)^s}, \qquad (17.36)$$

and the variance of r, B_s, say, is

$$\mathrm{Var}(r \mid r > 0) = B_s = \frac{s\theta(1 - \theta)}{1 - (1 - \theta)^s} - \frac{s^2 \theta^2 (1 - \theta)^s}{[1 - (1 - \theta)^s]^2}. \qquad (17.37)$$

Hence

$$E(r_s) = A_s n_s \qquad (17.38)$$

[see 17.31)], and

$$\mathrm{Var}(r_s) = B_s n_s. \qquad (17.39)$$

The score (17.27) can now be written as

$$U_s(\theta) = \frac{1}{\theta(1 - \theta)}(r_s - n_s A_s), \qquad (17.40)$$

and

$$U(\theta) = \sum_{s=1}^{S} U_s(\theta) = \frac{1}{\theta(1 - \theta)} \left(\sum_{s=1}^{S} r_s - \sum_{s=1}^{S} n_s A_s \right) = \frac{1}{\theta(1 - \theta)} \left(R - \sum_{s=1}^{S} n_s A_s \right). \qquad (17.41)$$

We shall now express $I_s(\theta)$, given by (17.33), in terms of B_s.

The right-hand side of (17.33) can be written as

$$I_s(\theta) = \frac{1}{\theta^2(1-\theta)^2}\left[\frac{s\theta(1-\theta)}{1-(1-\theta)^s} - \frac{s^2\theta^2(1-\theta)^s}{[1-(1-\theta)^s]^2}\right]n_s$$

$$= \frac{1}{\theta^2(1-\theta)^2}\,n_s B_s. \tag{17.42}$$

Thus,

$$I(\theta) = \sum_{s=1}^{S} I_s(\theta) = \frac{1}{\theta^2(1-\theta)^2}\sum_{s=1}^{S} n_s B_s, \tag{17.43}$$

and

$$\mathrm{Var}(\hat{\theta}) \doteq \frac{\theta^2(1-\theta)^2}{\displaystyle\sum_{s=1}^{S} n_s B_s}. \tag{17.44}$$

When a trait is simple *recessive*, the parental matings are usually $Dd \times Dd$ (both normal heterozygotes) and $\theta = \frac{1}{4}$. When a trait is a rare simple *dominant*, the parental matings are usually $Dd \times dd$ (Affected $\times$ Normal), with $\theta = \frac{1}{2}$.

Values of $A_s = A_s(\theta)$ and $B_s = B_s(\theta)$, for $\theta = \frac{1}{4}$ and $\theta = \frac{1}{2}$, and several s have been evaluated by Smith (1956). These are given in Table 17.1.

TABLE 17.1[a]   Values A_s, B_s, and $[1 - (1-\theta)^s]$ for $\theta = \frac{1}{4}$ and $\theta = \frac{1}{2}$

s	Simple autosomal recessive, $\theta = \frac{1}{4}$			Simple autosomal dominant, $\theta = \frac{1}{2}$		
	A_s	B_s	$1 - (\frac{3}{4})^s$	A_s	B_s	$1 - (\frac{1}{2})^s$
1	1.000	0.000	0.250	1.000	0.000	0.500
2	1.143	0.122	0.438	1.333	0.222	0.750
3	1.297	0.263	0.578	1.714	0.490	0.875
4	1.463	0.420	0.684	2.133	0.782	0.938
5	1.639	0.592	0.763	2.581	1.082	0.969
6	1.825	0.776	0.822	3.048	1.379	0.984
7	2.020	0.970	0.867	3.528	1.667	0.992
8	2.223	1.172	0.900	4.016	1.945	0.996
9	2.433	1.380	0.925	4.509	2.215	0.998
10	2.649	1.592	0.944	5.005	2.478	0.999
11	2.871	1.805	0.958	5.503	2.737	1.000
12	3.098	2.020	0.968	6.001	2.992	1.000
13	3.329	2.234	0.976	6.501	3.245	1.000
14	3.563	2.446	0.982	7.000	3.497	1.000
15	3.801	2.658	0.987	7.500	3.748	1.000
16	4.040	2.867	0.990	8.000	3.999	1.000

[a] Taken with permission from C.A.B. Smith, *Ann. Hum. Genet. Lond.* **20** (1956), p. 265.

The following result may be of some historical interest. We can easily show that formula (17.41) can be written as

$$U(\theta) = \frac{1}{\theta(1 - \theta)} [R - E(R)], \tag{17.45}$$

where

$$E(R) = \sum_{s=1}^{S} n_s A_s = \sum_{s=1}^{S} \frac{s\theta n_s}{1 - (1 - \theta)^s}. \tag{17.46}$$

The equation $U(\theta) = 0$ is equivalent to $R = E(R)$, and it gives exactly the likelihood equation (17.29). In fact, Weinberg (1912a) obtained the estimator of θ by solving the equation $R = E(R)$, that is, by setting R equal to its expected value, and called this method of estimating θ the *direct method*. The maximum likelihood method was not known at that time; but, as we have seen, Weinberg's estimator is, in this case, identical with the ML estimator.

17.8.2. Testing Hypotheses

For a given parental mating we can specify the parameter $\theta = \theta_0$ under the null hypothesis. The three kinds of Z^2-tests discussed in section 13.11 now take the following forms:

$$Z_{\text{Total}}^2 = \sum_{s=1}^{S} \frac{[U_s(\theta_0)]^2}{I_s(\theta_0)} = \sum_{s=2}^{S} \frac{(r_s - n_s A_s)^2}{n_s B_s}, \tag{17.47}$$

with the number of degrees of freedom equal to the number of different family sizes with nonzero frequencies and excluding families with $s = 1$, and A_s, B_s evaluated for $\theta = \theta_0$. We notice that families with $s = r = 1$ affected children contribute nothing to the total Z^2, since all children are affected; therefore we assume that $U_1(\theta)/I_1(\theta_0) = 0/0$ is equal to zero. For this reason we lose 1 d.f.

$$Z_{\text{Comb}}^2 = \frac{[U(\theta_0)]^2}{I(\theta_0)} = \frac{\left(R - \sum\limits_{s=1}^{S} n_s A_s\right)^2}{\sum\limits_{s=1}^{S} n_s B_s} \tag{17.48}$$

with 1 d.f. And, finally,

$$Z_{\text{Diff}}^2 = Z_{\text{Total}}^2 - Z_{\text{Comb}}^2 \tag{17.49}$$

If we go through all the numerical procedures, estimate θ from (17.29), and calculate $I(\theta_0)$ from (17.34), the test statistic for $H_0: \theta = \theta_0$ can also be

$$Z^2 = (\hat{\theta} - \theta_0)^2 I(\theta_0), \tag{17.50}$$

with 1 d.f. It is, in fact, identical with Z^2_{Comb} given by (17.48) [see formula (13.23)].

Example 17.1. Microcephaly is characterized by a distinctly small skull and brain, an underdeveloped body, and some neurophathic symptoms. The inheritance of this trait is not established. It has been suggested that the partial deletion of chromosome 18 may cause microcephaly [Porter (1968), p. 50]. It has been also found that the exposure of the pregnant mother to X-radiation can cause microcephaly in the fetus. Komai et al. (1955) collected data from 77 families, each with two normal parents, in Japan, and suggested that microcephaly may be inherited as a *simple autosomal recessive.*

Although we are aware that this hypothesis is not the right one, we use their data to demonstrate the calculating techniques. We will see that the data fit the hypothesis $H_0: \theta = \frac{1}{4}$. Therefore this is a cautionary example of a situation where statistically the data fit the hypothesis, but we must check the biological background before we actually accept the hypothesis.

We first illustrate the calculations, assuming $\pi = 1$. The left-hand side of Table 17.2 represents the data, and the right-hand side the necessary calculations, using the values of A_s and B_s from Table 17.1, for $\theta = \frac{1}{4}$.

Assuming $H_0: \theta = \frac{1}{4}$ is true, we calculate the expected amount of information from (17.43), that is,

$$I(\theta_0) = (\tfrac{4}{1} \cdot \tfrac{4}{3})^2 \times 51.566 = 1466.766.$$

The approximate variance of $\hat{\theta}$ is

$$\sigma_\theta^2 = \text{Var}(\hat{\theta}) \doteq 1/I(\theta_0) = 1/1466.766 = 0.000682,$$

and $\sigma_\theta \doteq 0.0262$. From formula (17.41) we also calculate $U(\theta_0)$, that is

$$U(\theta_0) = \tfrac{4}{1} \cdot \tfrac{4}{3}(142 - 131.990) = 53.387.$$

If one is interested in the estimate of θ, he can take as a first "guess" $\theta = \theta_0 = \frac{1}{4}$ and calculate the first approximation θ_1, say, from the formula $\theta_1 \doteq \theta_0 + U(\theta_0)/I(\theta_0)$ (compare section 11.7.2). In our example, we have

$$\theta_1 \doteq 0.2500 + 53.387/1466.766 = 0.2500 + 0.0364 = 0.2864.$$

The second approximation, θ_2, say, can be calculated from the formula $\theta_2 \doteq \theta_1 + U(\theta_1)/I(\theta_1)$. Of course, now Table 17.1 is no longer valid. Repeating the iteration procedure, we find $\hat{\theta} = 0.2855$.

The evaluation of the three Z^2 is straightforward. Let us take the significance level $\alpha = 0.05$. We have $\chi^2_{0.95;1} = 3.841$. Since $Z^2_{Comb} = 1.943 < 3.841$ one may suggest that the data fit the hypothesis of simple recessive. To test about homogeneity, we find $\chi^2_{0.95;8} = 15.51$. Since $Z^2_{Diff} = 16.484 > 15.51$, we reject the hypothesis that these family data are homogeneous. As has been

TABLE 17.2. Data on microcephaly in Japan. Calculations for testing $H_0: \theta = \frac{1}{4}$ (Both parents normal)

Family size, s	No. of families, n_s	No. of all children, sn_s	No. of affected children, r — No. of families with r affected children, a_{rs}					Affected children — Obs. r_s	Expected $n_s A_s$	$n_s B_s$	Z^2 with 1 d.f.	$\sum_{r=1}^{s} r^2 a_{rs}$
			1	2	3	4	5					
2	6	12	6					6	6.858	0.732	1.006	6
3	10	30	5	1	4			19	12.970	2.630	13.825	45
4	14	56	8	4	2			22	20.482	5.880	0.392	42
5	14	70	6	5	3			25	22.946	8.288	0.509	53
6	13	78	5	5	1	1	1	27	23.725	10.088	1.063	75
7	10	70	4	3	2		1	21	20.200	9.700	0.066	59
8	5	40	3		2			9	11.115	5.860	0.763	21
10	4	40	2	1			1	9	10.596	6.368	0.400	31
12	1	12				1		4	3.098	2.020	0.403	16
Total	77(N)	408(T)	39(A_1)	19	14	2	3	142(R)	131.990	51.078	18.427 Z^2_{Total} with 9 d.f.	348

Z^2_{Comb} with 1 d.f. → 1.943

Z^2_{Diff} with 8 d.f. → 16.484

mentioned before, the data have been "forced" to the hypothesis of a simple recessive and there is no sufficient biological background to assume that this hypothesis is true.

17.9. APPROXIMATE WEINBERG'S SIB METHOD

Weinberg (1912a) suggested a method of estimating θ which he called the "*simple sib method.*" As Fisher (1934) clearly pointed out, this method is valid only when ascertainment is *complete* (so that each affected child is a proband). This method consists of counting the siblings of each independently ascertained child and taking the ratio of total counts of affected to the total counts of all (affected and normal) siblings of each proband. Suppose that in a family of size s there are r affected children and each child has $(r-1)$ affected siblings. If the ascertainment is complete, each child is a proband, so the number of counts of siblings of each proband is $r(r-1)$. If there are a_{rs} such families then the number of such counts is $(r-1)ra_{rs}$. Counting all values, for $r = 1, 2, \ldots, s$, the total number of counts of affected siblings of each proband is equal to

$$\sum_{r=1}^{s} (r-1)ra_{rs}.$$

Similarly, the number of counts of normal siblings is

$$\sum_{r=1}^{s} (s-r)ra_{rs}.$$

Thus, the total number of counts of all siblings of probands is

$$\sum_{r=1}^{s} (r-1)ra_{rs} + \sum_{r=1}^{s} (s-r)_{rs}ra_{rs} = (s-1)\sum_{r=1}^{s} ra_{rs}.$$

Let $\breve{\theta}_s$ denote Weinberg's estimator of θ, based on the data from n_s families, each of size s. It is calculated as

$$\breve{\theta}_s = \frac{\sum_{r=1}^{s} (r-1)ra_{rs}}{(s-1)\sum_{r=1}^{s} ra_{rs}} = \frac{\sum_{r=1}^{s} r^2 a_{rs} - r_s}{(s-1)r_s}. \tag{17.51}$$

To find the approximate variance of $\breve{\theta}_s$, we notice that $\breve{\theta}_s$ is a function of random variables a_{rs} with expectations (for fixed s) given by (17.30). Using the multivariable Taylor expansion about the set of values $\{a_{rs}\} = \{E(a_{rs})\}$ and applying formula (8.95), we can evaluate the approximate variance of $\breve{\theta}_s$.

On the other hand, we notice that (17.51) can be written in the form of a ratio

$$\breve{\theta}_s = \frac{u_s}{v_s},$$

with

$$u_s = \sum_{r=1}^{s} (r-1)ra_{rs}, \quad v_s = (s-1)\sum_{r=1}^{s} ra_{rs}.$$

A formula for approximate variance of the ratio of two random variables, derived from the general formula (8.95), is given in Problem 8.8. The calculations are rather tedious, but finally the approximate variance of $\breve{\theta}_s$ can be obtained in the form

$$\text{Var}(\breve{\theta}_s) \doteq \theta(1-\theta) \cdot \frac{[2 + (s-3)\theta][1 - (1-\theta)^s]}{\theta(s-1)sn_s}. \tag{17.52}$$

[For details of the derivation see Bailey (1961), section 14.32.]

By analogy with the ML-estimation method, let us call the reciprocal of $\text{Var}(\breve{\theta}_s)$ the "amount of information" and denote it, for Weinberg's estimator $\breve{\theta}_s$, by $I_s'(\theta)$. We then have

$$I_s'(\theta) = \frac{1}{\text{Var}(\breve{\theta}_s)} \doteq \frac{1}{\theta(1-\theta)} \cdot \frac{\theta(s-1)sn_s}{[2 + (s-3)\theta][1 - (1-\theta)^s]}. \tag{17.53}$$

If families of all sizes ($s = 2, 3, \ldots, S$) are taken into account, the "pooled" Weinberg's estimator, $\breve{\theta}$, is

$$\breve{\theta} = \frac{\displaystyle\sum_{s=2}^{S} \sum_{r=1}^{s} (r-1)ra_{rs}}{\displaystyle\sum_{s=2}^{S} (s-1)\sum_{r=1}^{s} ra_{rs}} = \frac{\displaystyle\sum_{s=2}^{S} \sum_{r=1}^{s} r^2 a_{rs} - R}{\displaystyle\sum_{s=2}^{S} sr_s - R}. \tag{17.54}$$

We now have

$$\breve{\theta} = \frac{\displaystyle\sum_{s=2}^{S} U_s}{\displaystyle\sum_{s=2}^{S} V_s} = \frac{U}{V},$$

and to find the variance of $\breve{\theta}$ we can again apply the approximate formula (8.95).

However, we may find another approximation by using the following heuristic approach. By analogy with the ML method, let us call the sum

$$I'(\theta) = \sum_{s=2}^{S} I_s'(\theta) = \sum_{s=2}^{S} \frac{1}{\text{Var}(\breve{\theta}_s)}$$

$$\doteq \sum_{s=2}^{S} \frac{1}{\theta(1-\theta)} \cdot \frac{\theta(s-1)sn_s}{[2 + (s-3)\theta][1 - (1-\theta)^s]} \tag{17.55}$$

the "total amount of information" about θ obtained through Weinberg's estimator, $\breve{\theta}$. Since each $\breve{\theta}_s$ is not fully efficient, each $I'_s(\theta)$ is smaller than the corresponding $I_s(\theta)$ (for the ML estimator, $\hat{\theta}$) and hence the value obtained for $\mathrm{Var}(\breve{\theta}_s)$ will be greater than $\mathrm{Var}(\hat{\theta}_s) \doteq (1/I_s(\theta))$, where $I(\theta_s)$ is given in (17.33), for the same n_s. However, if the efficiencies of the $\breve{\theta}_s$'s are not too low, $I'(\theta)$ will give a reasonable (though underestimated) approximation to $I(\theta)$, and we may calculate the variance of the "pooled" Weinberg's estimator, $\breve{\theta}$, from the approximate formula

$$\mathrm{Var}(\breve{\theta}) \doteq \frac{1}{I'(\theta)}. \tag{17.56}$$

The variance will again be larger than $\mathrm{Var}(\hat{\theta})$ [obtained from (17.35), when substituting (17.34), and where $\hat{\theta}$ is the "pooled" ML estimator of θ].

17.10. AN APPROXIMATE METHOD OF LI AND MANTEL FOR ESTIMATING θ

Another method for estimating θ when there is complete ascertainment has been described by Li and Mantel (1968). For families with fixed sizes s $(s = 1, 2, \ldots, S)$, they proposed an estimator of θ, which we now denote by $\tilde{\theta}_s$, of the form

$$\tilde{\theta}_s = \frac{r_s - a_{1s}}{t_s - a_{1s}}. \tag{17.57}$$

When the families of all available sizes are taken, the "pooled" estimator, $\tilde{\theta}$, of Li and Mantel is

$$\tilde{\theta} = \frac{R - A_1}{T - A_1}, \tag{17.58}$$

where

$$A_1 = \sum_{s=1}^{S} a_{1s}.$$

[For justification for using estimator (17.58), see the original paper by Li and Mantel (1968).]

These authors have also shown that

$$\mathrm{Var}(\tilde{\theta}_s) \doteq \theta(1 - \theta) \frac{[1 - (1 - \theta)^s][1 - (1 - \theta)^s + (s - 2)\theta(1 - \theta)^{s-1}]}{[1 - (1 - \theta)^{s-1}]^2 s n_s}. \tag{17.59}$$

Similarly as for Weinberg's estimator, let us here denote the "amount of information" by $I_s''(\theta)$, that is,

$$I_s''(\theta) = \frac{1}{\text{Var}(\tilde{\theta}_s)}$$

$$\doteq \frac{1}{\theta(1-\theta)} \cdot \frac{[1-(1-\theta)^{s-1}]^2 s n_s}{[1-(1-\theta)^s][1-(1-\theta)^s+(s-2)\theta(1-\theta)^{s-1}]} \cdot$$

(17.60)

Again using the same heuristic approach as for $I'(\theta)$ [for details see also Li and Mantel (1968)], we calculate the total "amount of information" as

$$I''(\theta) = \sum_{s=1}^{S} I_s''(\theta) = \sum_{s=1}^{s} \frac{1}{\text{Var}(\tilde{\theta}_s)}, \tag{17.61}$$

and the approximate variance as

$$\text{Var}(\tilde{\theta}) \doteq \frac{1}{I''(\theta)}. \tag{17.62}$$

17.11. EFFICIENCIES OF THE APPROXIMATE ESTIMATORS $\breve{\theta}_s$ AND $\tilde{\theta}_s$

We find, from formula (17.33), that $i_s(\theta)$, the average expected amount of information *per sibship of size s* (using the maximum likelihood method) is,

$$i_s(\theta) = \frac{I_s(\theta)}{n_s} \doteq \frac{s}{\theta(1-\theta)} \cdot \frac{1-(1-\theta)^s - s\theta(1-\theta)^{s-1}}{[1-(1-\theta)^s]^2}. \tag{17.63}$$

The average "amounts of information" for the approximate methods are as follows.

For Weinberg's method [from (17.54)]:

$$i_s'(\theta) = \frac{I_s'(\theta)}{n_s} \doteq \frac{s}{\theta(1-\theta)} \cdot \frac{(s-1)\theta}{[2+(s-3)\theta][1-(1-\theta)^s]}; \tag{17.64}$$

For Li's and Mantel's method [from (17.60)]:

$$i_s''(\theta) = \frac{I_s''(\theta)}{n_s}$$

$$\doteq \frac{s}{\theta(1-\theta)} \cdot \frac{[1-(1-\theta)^{s-1}]^2}{[1-(1-\theta)^s][1-(1-\theta)^s+(s-2)\theta(1-\theta)^{s-1}]}. \tag{17.65}$$

Thus, the (asymptotic) efficiencies of $\breve{\theta}_s$ and $\tilde{\theta}_s$ with respect to the ML estimator, $\hat{\theta}$, are, respectively,

$$\text{Rel eff}(\breve{\theta}_s) \doteq \frac{i'_s(\theta)}{i_s(\theta)} \quad \text{and} \quad \text{Rel eff}(\tilde{\theta}_s) \doteq \frac{i''_s(\theta)}{i_s(\theta)}. \qquad (17.66)$$

Table 17.3 gives some numerical values of average "amounts of information" per sibship of size s and the efficiencies of estimators $\breve{\theta}_s$ and $\tilde{\theta}_s$ for

TABLE 17.3 Average "amounts of information" per sibship of size s and their efficiencies for Weinberg's, and Li and Mantel's estimators for various s and $\theta = \frac{1}{4}$

s	Max. likelihood $i_s(\theta)$	Weinberg $i'_s(\theta)$	Li and Mantel $i''_s(\theta)$	$\dfrac{i'_s(\theta)}{i_s(\theta)}$	$\dfrac{i''_s(\theta)}{i_s(\theta)}$
2	3.48	3.48	3.48	1.000	1.000
3	7.48	6.92	7.37	0.925	0.985
4	11.95	10.40	11.66	0.870	0.976
5	16.84	13.98	16.34	0.830	0.970
6	22.07	17.70	21.38	0.802	0.969
7	27.59	21.53	26.73	0.780	0.969
8	33.34	25.53	32.36	0.766	0.971
9	39.25	29.65	38.20	0.755	0.973
10	45.28	33.90	44.20	0.749	0.976
11	51.34	38.27	50.30	0.745	0.980
12	57.46	42.78	56.46	0.745	0.983
13	63.54	47.36	62.63	0.745	0.986
14	69.58	52.02	68.78	0.748	0.989
15	75.60	56.74	74.88	0.751	0.990
16	81.55	61.57	80.93	0.755	0.992

various s, when the true value of θ is $\theta = \frac{1}{4}$ (simple autosomal recessive). More extensive tables of approximate efficiencies of $\breve{\theta}_s$ and $\tilde{\theta}_s$ are given by Gart (1968). As we can see from Table 17.3, the approximate method of Li and Mantel is almost fully efficient.

Example 17.2. We demonstrate the application of these two approximate methods to the data given in Example 17.1.

(a) Weinberg's Estimate of θ. We calculate $\breve{\theta}$ from formula (17.54). The last column in Table 17.2 gives

$$\sum_{s=2}^{S} \sum_{r=1}^{s} r^2 a_{rs} = 348.$$

We calculate

$$\sum_{s=2}^{S} sr_s = 801.$$

Hence

$$\theta = \frac{348 - 142}{801 - 142} = \frac{206}{659} = 0.3126.$$

The variance of $\breve{\theta}$ is now calculated from the approximate formula (17.56), which can be written in the form

$$\text{Var}(\breve{\theta}) \doteq \frac{1}{\sum_{s=1}^{S} n_s i_s'(\theta)}. \tag{17.67}$$

It yields

$$\text{Var}(\breve{\theta}) \doteq \frac{1}{1182.83} = 0.000845; \quad \sigma_{\breve{\theta}} \doteq 0.0291,$$

(b) Li and Mantel's Estimate $\tilde{\theta}$. We calculate $\tilde{\theta}$ from formula (16.58):

$$\tilde{\theta} = \frac{142 - 39}{408 - 39} = \frac{103}{369} = 0.2791.$$

The variance of $\tilde{\theta}$ will now be calculated from the approximate formula (17.62), which can also be written in the form

$$\text{Var}(\tilde{\theta}) \doteq \frac{1}{\sum_{s=1}^{S} n_s i_s''(\theta)}. \tag{17.68}$$

It yields

$$\text{Var}(\tilde{\theta}) \doteq \frac{1}{1426.88} = 0.000701; \quad \sigma_{\tilde{\theta}} = 0.0265.$$

We notice that Li and Mantel's estimate of θ (and its variance) is very close to that obtained by the ML method.

17.12. MAXIMUM LIKELIHOOD ESTIMATOR OF θ UNDER SINGLE ASCERTAINMENT ($\pi \to 0$)

In this situation only a small proportion of affected individuals is recorded, and each family is usually ascertained by one proband. We have shown that the probability $P_{rs}(\theta)$, that is, the probability of occurrence of r affected children in an ascertained family, is given by (17.12):

$$P_{rs}(\theta) = \binom{s-1}{r-1} \theta^{r-1}(1 - \theta)^{s-r}, \quad \begin{matrix} s = 2, 3, \ldots, S; \\ r = 1, 2, \ldots, s. \end{matrix} \tag{17.12}$$

Following the notation and arguments given in section 17.6, we obtain the likelihood function, $L_s(\theta)$, for the single ascertainment. It is analogous to that in (17.14) and is

$$L_s(\theta) = \frac{n_s!}{\prod\limits_{r=1}^{s} a_{rs}!} \prod_{r=1}^{s} \left[\binom{s-1}{r-1} \theta^{r-1}(1-\theta)^{s-r} \right]^{a_{rs}}, \quad s = 2, 3, \ldots, S, \quad (17.69)$$

and so

$$\log L_s(\theta) = \text{const} + \sum_{r=1}^{s} [(r-1)\log\theta + (s-r)\log(1-\theta)]a_{rs}. \quad (17.69a)$$

Differentiating with respect to θ, we obtain

$$\frac{\partial \log L_s(\theta)}{\partial \theta} = U_s(\theta)$$

$$= \sum_{r=1}^{s} \left(\frac{r-1}{\theta} - \frac{s-r}{1-\theta} \right) a_{rs}$$

$$= \frac{1}{\theta(1-\theta)} \left[\left(\sum_{r=1}^{s} ra_{rs} - \sum_{r=1}^{s} a_{rs} \right) - \theta \left(s \sum_{r=1}^{s} a_{rs} - \sum_{r=1}^{s} a_{rs} \right) \right]$$

$$= \frac{1}{\theta(1-\theta)} [(r_s - n_s) - \theta(sn_s - n_s)]$$

$$= \frac{1}{\theta(1-\theta)} [(r_s - n_s) - \theta(t_s - n_s)]. \quad (17.70)$$

Hence the total score for families of all sizes, $U(\theta)$, is

$$U(\theta) = \sum_{s=2}^{S} U_s(\theta) = (R - N) - \theta(T - N). \quad (17.71)$$

The ML estimator is the root of the equation $U(\theta) = 0$ and yields

$$\hat{\theta} = \frac{R - N}{T - N}. \quad (17.72)$$

It is left to the reader to show that

$$I_s(\theta) = \frac{(s-1)n_s}{\theta(1-\theta)} = \frac{t_s - n_s}{\theta(1-\theta)} = \frac{n_s(s-1)}{\theta(1-\theta)}. \quad (17.73)$$

We now have

$$I(\theta) = \sum_{s=2}^{S} I_s(\theta) = \frac{\sum\limits_{s=2}^{S} sn_s - \sum\limits_{s=2}^{S} n_s}{\theta(1-\theta)} = \frac{T - N}{\theta(1-\theta)}, \quad (17.74)$$

and hence

$$\text{Var}(\hat{\theta}) = \frac{1}{I(\theta)} = \frac{\theta(1-\theta)}{T - N}. \quad (17.75)$$

17.13. WEINBERG'S "PROBAND METHOD"

Applying Weinberg's formula (17.51) when there is only *one* proband in a family, we obtain

$$\breve{\theta}_s = \frac{\sum\limits_{r=1}^{s}(r-1)\cdot 1 \cdot a_{rs}}{(s-1)\sum\limits_{r=1}^{s} a_{rs}} = \frac{r_s - n_s}{t_s - n_s}, \tag{17.76}$$

and the "pooled" Weinberg's estimator over families of all sizes is

$$\breve{\theta} = \frac{R - N}{T - N}. \tag{17.77}$$

Weinberg (1912*b*) called this method of estimation the *proband method*. In fact, it is a particular case of his sib method. We notice that in this case the estimator $\breve{\theta}$ is identical with the ML estimator $\hat{\theta}$ given by (17.72).

17.14. TEST OF GOODNESS OF FIT UNDER SINGLE ASCERTAINMENT

Since for an ascertained family the probability of the affected $r - 1$ sibs among the $s - 1$ sibs of the proband is given by (17.12), then

$$E(a_{rs}) = n_s \left[\binom{s-1}{r-1} \theta^{r-1}(1-\theta)^{s-r} \right]. \tag{17.78}$$

Hence

$$E(r_s - n_s) = E\left[\sum_{r=1}^{s}(r-1)a_{rs} \right] = \sum_{r=1}^{s}(r-1)E(a_{rs}) = (s-1)\theta n_s$$

$$= \theta(sn_s - n_s) = \theta(t_s - n_s), \tag{17.79}$$

and

$$E(R - N) = \sum_{s=2}^{S} E(r_s - n_s) = \theta(T - N). \tag{17.80}$$

The heterogeneity Z_{Total}^2 criterion, if $H_0: \theta = \theta_0$ is specified, is

$$\begin{aligned} Z_{\text{Total}}^2 &= \sum_{s=2}^{S} \frac{U_s^{\,2}(\theta_0)}{I(\theta_0)} = \sum_{s=2}^{S} \left\{ \frac{[(r_s - n_s) - \theta_0(t_s - n_s)]^2}{\theta_0^{\,2}(1-\theta_0)^2} \cdot \frac{\theta_0(1-\theta_0)}{t_s - n_s} \right\} \\ &= \frac{1}{1-\theta_0} \sum_{s=2}^{S} \frac{[(r_s - n_s) - \theta_0(t_s - n_s)]^2}{\theta_0(t_s - n_s)} \\ &= \frac{1}{1-\theta_0} \sum_{s=2}^{S} \frac{[(r_s - n_s) - E(r_s - n_s)]^2}{E(r_s - n_s)}. \end{aligned} \tag{17.81}$$

If $\theta = \theta_0$ is specified, the number of degrees of freedom is equal to the number of individual Z^2's, each calculated for given n_s with 1 d.f. (See also section 17.8.2.)

When $\theta = \theta_0$ is specified by H_0, the test criterion Z^2_{Comb} is

$$Z^2_{\text{Comb}} = \frac{\left[\sum\limits_{s=2}^{S} U_s(\theta_0)\right]^2}{\sum\limits_{s=2}^{S} I_s(\theta_0)} = \frac{1}{1 - \theta_0} \frac{[(R - N) - \theta_0(T - N)]^2}{\theta_0(T - N)},$$

$$= \frac{1}{1 - \theta_0} \frac{[(R - N) - E(R - N)]^2}{E(R - N)} \qquad (17.82)$$

with 1 d.f., and

$$Z^2_{\text{Diff}} = Z^2_{\text{Total}} - Z^2_{\text{Comb}}. \qquad (17.83)$$

Example 17.3. To illustrate the method, let us apply it to the data of Example 17.1. We calculate $\hat{\theta}$ from (17.72):

$$\hat{\theta} = \frac{142 - 77}{408 - 77} = \frac{65}{331} = 0.1964.$$

To test H_0: $\theta = \frac{1}{4}$ we use Z^2_{Comb}, given by (17.82):

$$Z^2_{\text{Comb}} = \frac{4}{3} \cdot \frac{(65 - 82.75)^2}{82.75} = 5.08,$$

which is now significant at $\alpha = 0.05$ ($z_{0.95} = 3.84$).

17.15. THE ASCERTAINMENT PROBABILITY, π, AND ITS ESTIMATION

To ensure a valid segregation analysis, it is important that the probands be correctly identified. The first requirement for segregation analysis is that the probands be ascertained *independently*. It is indeed difficult to see how this condition can be satisfied. It is quite likely that, if one affected person from a family enters a particular hospital, an affected sibling will also enter the same hospital. Moreover, different sources of ascertainment are not likely to be independent. A blind person may be a member of certain associations for sightless people, be a patient of a given doctor, enter a hospital, etc.

However, there might be some occasions on which this condition of independence can, at least approximately, be satisfied. Under such circumstances it would be worthwhile to attempt to estimate the ascertainment probability, π. We will present here a few models, all based on the assumption that probands are ascertained *independently*, which could be applied in estimating π.

17.15.1. Estimation of π from Proband Distribution

Let us consider ascertained families, each with r affected children. If c is the number of probands in a family with r affected children, the distribution of the number of probands, c, in an ascertained family with r affected children (we will call it the *proband distribution*) is a truncated binomial with parameters r and π, that is,

$$\Pr\{c \mid c > 0, r > 0\} = \binom{r}{c} \frac{\pi^c (1 - \pi)^{r-c}}{1 - (1 - \pi)^r}, \qquad c = 1, 2, \ldots, r. \quad (17.84)$$

We can easily see that this distribution resembles distribution (17.8) by substituting π for θ, c for r, and r for s. Thus the methods of estimating π will be the same as those discussed in section 17.7 for estimating θ.

Sometimes, if a certain trait exhibits many sporadic cases, it is safer to restrict ourselves to families with two or more probands. In this case, the proband distribution is

$$\Pr\{c \mid c > 1; r > 1\} = \binom{r}{c} \frac{\pi^c (1 - \pi)^{r-c}}{1 - (1 - \pi)^r - r\pi(1 - \pi)^{r-1}},$$

$$c = 2, 3, \ldots, r. \quad (17.85)$$

Using standard procedure, we can also obtain the ML estimator of π for this case.

17.15.2. Estimation of θ and π Simultaneously, Using the Joint Likelihood Function

We only outline the method briefly, since its basic ideas are similar to those presented in section 17.6, except that now we also assume that π is unknown. Let $P_{crs}(\theta, \pi)$ denote the probability that a family of size s with r affected children is *independently* ascertained c times, and let a_{crs} be the number of such families in a sample of n_s families which have been ascertained. If θ is the segregation probability for families of a given mating type and π the ascertainment probability, it is easy to show that

$$P_{crs}(\theta, \pi) = \frac{1}{1 - (1 - \pi\theta)^s} \binom{s}{r} \theta^r (1 - \theta)^{s-r} \times \binom{r}{c} \pi^c (1 - \pi)^{r-c}. \quad (17.86)$$

Thus the likelihood function is

$$L_s(\theta, \pi) = n_s! \prod_{r=1}^{s} \prod_{c=1}^{r} \frac{[P_{crs}(\theta, \pi)]^{a_{crs}}}{a_{crs}!}. \quad (17.87)$$

Let

$$r_s = \sum_{r=1}^{s} \sum_{c=1}^{r} r a_{crs}; \quad t_s = sn_s = s \sum_{r=1}^{s} \sum_{c=1}^{r} a_{crs}; \quad c_s = \sum_{r=1}^{s} \sum_{c=1}^{r} c a_{crs}. \quad (17.88)$$

Taking logarithms of both sides of (17.87) and differentiating with respect to θ and π, we find the scores

$$
\left.
\begin{aligned}
\frac{\partial \log L_s(\theta, \pi)}{\partial \theta} = U_{\theta;s} &= \frac{s\pi(1 - \pi\theta)^{s-1}}{1 - (1 - \pi\theta)^s} + \frac{r_s}{\theta} - \frac{t_s - r_s}{1 - \theta}, \\
\frac{\partial \log L_s(\theta, \pi)}{\partial \pi} = U_{\pi;s} &= \frac{s\theta(1 - \pi\theta)^{s-1}}{1 - (1 - \pi\theta)^s} + \frac{c_s}{\pi} - \frac{r_s - c_s}{1 - \pi}.
\end{aligned}
\right\} \quad (17.89)
$$

The overall scores are, of course,

$$U_\theta = \sum_{s=1}^{S} U_{\theta;s}; \quad U_\pi = \sum_{s=1}^{S} U_{\pi;s}. \quad (17.90)$$

The ML estimators, $\hat{\theta}$ and $\hat{\pi}$, are the roots of the two simultaneous equations

$$U_\theta = 0 \quad \text{and} \quad U_\pi = 0. \quad (17.91)$$

These are not simple equations, and computer programs should be used to solve them.

It can be shown [Bailey (1961), section 14.52] that the elements of the expected information matrix, $I_s(\theta, \pi)$, are

$$
\left.
\begin{aligned}
I_{\theta\theta;s} &= \frac{sn_s}{\theta(1 - \theta)} \cdot \frac{1 - (1 - \pi)(1 - \pi\theta)^{s-2}}{1 - (1 - \pi\theta)^s}, \\
I_{\theta\pi;s} = I_{\pi\theta;s} &= sn_s \cdot \frac{(1 - \pi\theta)^{s-2}}{1 - (1 - \pi\theta)^s}, \\
I_{\pi\pi;s} &= \frac{sn_s}{\pi(1 - \pi)} \frac{1 - \pi(1 - \theta)(1 - \pi\theta)^{s-2}}{1 - (1 - \pi\theta)^s}.
\end{aligned}
\right\} \quad (17.92)
$$

The elements of the "pooled" information matrix, $I(\theta, \pi)$, are obtained in the usual manner, that is, by summation of the corresponding elements of the matrix $I_s(\theta, \pi)$ over all s.

17.15.3. Other Methods

(a) *Weinberg's General Proband Method.* For the case of c probands per family, ($c = 1, 2, \ldots, r$) Weinberg (1927) suggested an estimator of θ:

$$\overset{\vee}{\theta} = \frac{\displaystyle\sum_{s=2}^{S} \sum_{r=1}^{s} \sum_{c=1}^{r} c(r - 1)a_{crs}}{(s - 1)\displaystyle\sum_{r=1}^{s} \sum_{c=1}^{r} ca_{crs}}. \quad (17.93)$$

[We see the analogy with formula (17.51).] He called this method the *general proband method*.

R. A. Fisher (1934) suggested for the estimation of π the analogous formula

$$\check{\pi} = \frac{\sum\limits_{s=2}^{S}\sum\limits_{r=1}^{s}\sum\limits_{c=1}^{r} c(c-1)a_{crs}}{\sum\limits_{s=2}^{S}\sum\limits_{r=1}^{s}\sum\limits_{c=1}^{r} c(r-1)a_{crs}}. \tag{17.94}$$

These estimators are consistent but not efficient. Their variances have also been calculated by Fisher (1934).

(b) Morton's Truncated Poisson Distribution. Morton (1959) considered the situation in which a proband can be *independently* ascertained from more than one source and assumed that the number of ascertainments per proband follows a truncated Poisson distribution. He derived the formulae for the score and the amount of information. The interested reader is referred to his original paper.

The application of any of these methods requires that the ascertainments of the probands be independent. As mentioned previously, it is difficult to see how this condition can be satisfied. If it holds approximately, however, the most straightforward and fairly satisfactory method of estimating π seems to be the one based on the proband distribution and described in section 17.15.1.

17.16. THE NUMBER OF FAMILIES UNDER RISK

As was pointed out in section 17.6, all formulae were derived *conditionally* on a fixed value of n_s. Let N_s be the number of families, each of size s, and of a given mating type. Among these, n_s observed families have at least one affected child. Thus n_s may be considered as a random variable with the probability distribution

$$P(n_s) = \binom{N_s}{n_s}[1 - (1 - \pi\theta)^s]^{n_s}[(1 - \pi\theta)^s]^{N_s - n_s}, \tag{17.95}$$

where $n_s = 0, 1, \ldots, N_s$. It is a binomial distribution with parameters N_s and $P = 1 - (1 - \pi\theta)_s$. We have

$$E(n_s) = N_s[1 - (1 - \pi\theta)^s], \tag{17.96}$$

and so

$$N_s = \frac{E(n_s)}{1 - (1 - \pi\theta)^s}. \tag{17.96a}$$

Substituting n_s for its expected value $E(n_s)$, we can use

$$\hat{N}_s = \frac{n_s}{1 - (1 - \pi\theta)^s} \qquad (17.97)$$

as an estimator of N_s.

Now, the likelihood function, for the parameters θ, π, and N_s, $L(\theta, \pi, N_s \mid n_s)$, is of the form

$$L_s(\theta, \pi, N_s \mid n_s) = L_s(\theta, \pi \mid n_s)P(n_s). \qquad (17.98)$$

Here $L_s(\theta, \pi \mid n_s)$ is given in (17.87). [Note that $P(n_s)$ is a function of N_s.]

If we wish to find the estimators of π, θ, and N_s, we have to solve the system of equations

$$\frac{\partial \log L_s(\theta, \pi, N_s \mid n_s)}{\partial \theta} = 0, \quad \frac{\partial \log L_s(\theta, \pi, N_s \mid n_s)}{\partial \pi} = 0,$$

$$\frac{\partial \log L_s(\theta, \pi, N_s \mid n_s)}{\partial N_s} = 0. \qquad (17.99)$$

It can be shown [see Berger and Gold (1967)] that system (17.99) is equivalent to the system

$$\frac{\partial \log L_s(\theta, \pi \mid n_s)}{\partial \theta} = 0, \quad \frac{\partial \log L_s(\theta, \pi \mid n_s)}{\partial \pi} = 0, \quad \frac{\partial \log P(n_s)}{\partial N_s} = 0. \quad (17.100)$$

Therefore, to obtain the estimators $\hat{\theta}$, $\hat{\pi}$, and $\hat{N}_s$ from combined data on sibships of different sizes, we have to solve system (17.91) combined with (17.97), that is,

$$U_\theta = 0, \quad U_\pi = 0, \quad \text{and} \quad n_s = N_s[1 - (1 - \pi\theta)^s]. \qquad (17.101)$$

In particular cases, we will have the following:

(a) for complete ascertainment ($\pi = 1$)

$$n_s = \hat{N}_s[1 - (1 - \theta)^s]; \quad \text{hence} \quad \hat{N}_s = \frac{n_s}{1 - (1 - \theta)^s}; \qquad (17.102)$$

(b) for single ascertainment (π small)

$$n_s \doteq \frac{\hat{N}_s}{s\pi\theta}, \quad \text{hence} \quad \hat{N} \doteq \frac{n_s}{s\pi\theta}. \qquad (17.103)$$

It is now easy to see that all the formulae for expected amounts of information (under different ascertainments, when n_s is considered as a *random* variable) can be immediately "adjusted" by substituting in all of the formulae derived in sections 17.6–17.15 the expected value $E(n_s) = N_s[1 - (1 - \pi\theta)^s]$, for n_s.

It should also be noted that some formulae given by Fisher (1934) and Haldane (1938) were also expressed in terms of N_s instead of n_s. They differ, therefore, from those given in this chapter.

Example 17.4. We wish to estimate the number of mating types $Dd \times Dd$ (i.e., the number of families with two normal parents under the risk of having an affected child) for the data of Example 17.1 on microcephaly. Assuming that $H_0: \theta = \frac{1}{4}$ is true, we calculate $\hat{N}_s$ from formula (17.102), using the column "$1 - (\frac{3}{4})^s$" of Table 17.1. We have

s:	2	3	4	5	6	7	8	10	12	Total
n_s:	6	10	14	14	13	10	5	4	1	77
$\hat{N}_s$:	13.7	17.3	20.5	18.3	15.8	11.5	5.6	4.2	1.0	107.9.

17.17. ASCERTAINMENT THROUGH THE PARENTS. RECESSIVE TRAIT

If the trait is not too rare, sampling may be through the parents. If the affected individuals are dominants, the phenotypic mating type Dom $\times$ Dom will include more than one genotypic mating. However, we may then take our observations with respect to the recessive trait even if it is called "normal"; and, as pointed out before, segregation analyses for rare and common simple recessives are the same. If we restrict ourselves to families which have at least one recessive child (i.e., segregating families), the segregation distribution in such families will be

$$\Pr\{r \mid r > 0\} = \frac{\binom{s}{r} \theta^r (1 - \theta)^{s-r}}{1 - (1 - \theta)^s}, \qquad r = 1, 2, \ldots, s. \quad (17.104)$$

This is identical with the distribution for complete ascertainment through the children, given by (17.9). Hence

The segregation analysis for families with at least one recessive child ascertained through the parents is the same as that for complete ascertainment through the children.

17.18. DIFFICULTIES IN ANALYSIS OF FAMILY DATA

In this chapter, we have presented several statistical methods for testing hypotheses about simple modes of inheritance, using family data of one generation, when the phenotypes of both parents are known. These methods are valid only under certain assumptions, which have been stated in the text. However, there are many situations for which the assumptions are not valid. We have already mentioned that it is not always certain that the probands are ascertained from independent sources, so estimation of π is always difficult. Some characters differ in the age of onset. Selection of families through children would then be incomplete, and tracing such traits through pedigree appears to be more appropriate. We have seen, from Example 17.1, that a hypothesis can be "forced" and may fit some data even if it is not true. Careful examination of the biological assumptions underlying any hypothesis is the most important feature, and statistical tests can only aid in the evaluation of statistical data, not completely "prove" that a model is the right one. Eight papers presented at the Conference on Problems and Methods in Human Genetics (1954) discuss in greater detail several difficulties in segregation analysis.

SUMMARY

Methods of statistical analysis (estimation and testing hypotheses on the segregation ratios), using data obtained from sibships of different sizes, have been presented. We have been concerned only with problems in which a character is an autosomal recessive or a rare dominant, controlled by a single locus. Definitions of such terms as "proband," "ascertainment," and "segregation probability" have been introduced and explained. Maximum likelihood estimators for the segregation parameter, under different methods of ascertainment, have been derived. Also the "historical" estimators of Weinberg and a new estimator proposed by Li and Mantel have been presented, and their efficiencies calculated for a simple recessive.

REFERENCES

Bailey, N. T. J., The estimation of the frequencies of recessives with incomplete multiple selection, *Ann. Eugen. Lond.* **16** (1951a), 215–222.

Bailey, N. T. J., A classification of methods of ascertainment and analysis in estimating the frequencies of recessives in man, *Ann. Eugen. Lond.* **16** (1951b), 223–225.

Bailey, N. T. J., *Introduction to the Mathematical Theory of Genetic Linkage*, Clarendon Press, Oxford, 1961.

Berger, Agnes, and Gold, Ruth L., On estimating recessive frequencies from truncated samples, *Biometrics* **23** (1967), 356–360.

Conference on Problems and Methods in Human Genetics, Proceedings ed. by W. E. Heston. In: *Amer. J. Hum. Genet.* **6** (March 1954), 45–194 (8 papers with discussions).

Chung, C. S., Robinson, O. W., and Morton, N. E., A note on deaf mutism, *Ann. Hum. Genet. Lond.* **23** (1959), 357–366.

Fisher, R. A., The effect of methods of ascertainment upon the estimation of frequencies, *Ann. Eugen. Lond.* **6** (1934), 13–25.

Gart, J. J., A simple, nearly efficient alternative to the simple sib method in the complete ascertainment case, *Ann. Hum. Genet. Lond.* **31** (1968), 283–291.

Haldane, J. B. S., A method for investigating recessive characters in man, *J. Genet.* **25** (1932), 251–256.

Haldane, J. B. S., The estimation of the frequencies of recessive conditions in man, *Ann. Eugen. Lond.* **8** (1938), 256–262.

Haldane, J. B. S., A test for homogeneity of records of familial abnormalities, *Ann. Eugen. Lond.* **14** (1949), 339–341.

Halperin, M., The use of χ^2 in testing effect of birth order, *Ann. Hum. Genet. Lond.* **18** (1953), 99–106.

Hogben, L., The genetic analysis of familial traits, I. Single gene substitution, *J. Genet.* **25** (1931), 97–112.

Komai, T., Kishimoto, K. C., and Ozbai, Y., Genetic study of microcephaly based on Japanese material, *Amer. J. Hum. Genet.* **7** (1955), 51–65.

Lenz, F., Methoden der menschlichen Erblichkeitsforschung, *Handbuch hyg. Untersuch.* Jena, 1929, Vol. III, p. 700.

Li, C. C., Some general properties of recessive inheritance, *Amer. J. Hum. Genet.* **5** (1953), 269–279.

Li, C. C., *Human Genetics*, McGraw-Hill Book Company, New York, 1961.

Li, C. C., and Mantel, N., A simple method of estimating the segregation ratio under complete ascertainment, *Amer. J. Hum. Genet.* **20** (1968), 61–81.

Morton, N. E., Genetic tests under incomplete ascertainment, *Amer. J. Hum. Genet.* **11** (1959), 1–16.

Morton, N. E., Segregation and linkage. In: *Methodology in Human Genetics*, ed. by J. Burdette, Holden-Day, San Francisco, 1962, pp. 17–52.

Neel, J., Problems in the estimation of the frequency of uncommon inherited traits, *Amer. J. Hum. Genet.* **6** (1954), 51–60.

Porter, I H., *Heredity and Disease*, McGraw-Hill Book Company, New York, 1968.

Report on the Chicago Conference, *National Foundation*, Vol. II, No. 3, December 1966.

Smith, C. A B., A test for segregation ratios in family data, *Ann. Hum. Genet. Lond.* **20** (1956), 257–265.

Smith, C. A. B., A note on the effects of method of ascertainment on segregation ratios, *Ann. Hum. Genet.* **23** (1959), 311–323.

Stern, C., *Principles of Human Genetics*, W. H. Freeman & Co., San Francisco, 1960.

Weinberg, W., Methode und Fellerquellen der Untersuchung auf Mendelsche Zahlen beim Menschen, *Arch. Rass.-u. Ges. Biol.* **6** (1912a), 165–174.

Weinberg, W., Zur Vererbung der Anlage der Blutenkrankheit mit methodol. Ergänzugen meiner Geschwistermethode, *Arch. Rass.-u. Ges. Biol.* **6** (1912b), 694–709.

Weinberg, W., Mathematische Grundlagen der Probandenmethode, *I. Indukt. Abstamm. u. Vererb. Lehre* **48** (1927), 179–228.

PROBLEMS

17.1. Hypersensitivity to hay fever is a common trait, and it has been suggested that the cause is a *single-locus autosomal recessive*. Its gene frequency for a certain population has been estimated to be $\hat{q} = 0.4020$ [Tips, *Amer. J. Hum. Genet.* **6** (1954), 323–343]. The accompanying table represents the affected offspring in families where both parents were normal. Although the ascertainment was through a single proband in each family, carry out your calculations for two limiting cases of ascertainment (complete and single).

Family size, s	No. of families, n_s	No. of affected children, r_s	No. of all children, $sn_s = t_s$
1	4	4	4
2	15	16	30
3	18	26	54
4	8	12	32
5	8	14	40
6	5	12	30
7	2	4	14
8	4	13	32
9	1	4	9
10	1	2	10
Total	66(N)	107(R)	255(T)

(a) Estimate the segregation parameter, θ. Find the estimated Var($\hat{\theta}$).

(b) Test the hypothesis H_0: $\theta = \frac{1}{4}$. Calculate Z^2_{Total}, Z^2_{Comb}, and Z^2_{Diff}.

(c) Assuming that H_0: $\theta = \frac{1}{4}$ is true, find the estimate of the total number of families in which both parents are heterozygotes (i.e., $Dd \times Dd$), assuming complete ascertainment.

(d) Estimate the total number of matings $Dd \times Dd$, using the estimate of gene frequency $\hat{q} = 0.4020$ and assuming that the population is in equilibrium. [*Remark:* The number obtained in (d) includes families of sizes greater than S and also families with no children, so it should be rather larger than that in (c).]

(e) Calculate the proportion of segregating families among matings $Dd \times Dd$, using $\hat{q} = 0.4020$. Compare with the results in (c).

17.2. Spherocytosis is a blood disorder characterized clinically by anemia, jaundice, and enlargement of the spleen. It has been suggested that the disease, which is *rare*, is inherited as a *simple autosomal dominant*. The accompanying table represents the data obtained from nine families of parental matings Normal × Affected [Abrams and Battle, *Amer. J. Hum. Genet.* **4** (1952), 350–355]. Ascertainment was complete ($\pi = 1$).

Family size, s	No. of families, n_s	No. of affected children, r_s	No. of all children, $sn_s = t_s$
1	2	2	2
2	2	3	4
3	2	4	6
5	1	4	5
10	1	3	10
11	1	5	11
Total	9(N)	21(R)	38(T)

(a) Estimate the segregation parameter, θ.

(b) Test the hypothesis H_0: $\theta = \frac{1}{2}$.

17.3. Congenital aniridia is bilateral absence of the iris. The accompanying table represents data from 59 ascertained families in which one parent was affected [Shaw et al., *Amer. J. Hum. Genet.* **12** (1960), 389–415].

Family size, s	No. of families, n_s	No. of affected children, r_s	No. of all children, $sn_s = t_s$
1	8	6	8
2	15	9	30
3	9	10	27
4	6	8	24
5	9	16	45
6	3	6	18
7	6	22	42
10	1	4	10
11	1	8	11
12	1	6	12
Total	59	95	227

Supposing that this is a single locus, rare trait:

(a) Estimate the segregation parameter, θ, and find the observed amounts of information, assuming that the ascertainment is (i) complete, (ii) single, (iii) incomplete multiple with $\pi = 0.64$.

(b) Test the hypothesis that the trait is due to a *simple autosomal recessive* (i.e., H_0: $\theta = \frac{1}{2}$) for each of the cases (i), (ii), (iii).

17.4. Calculate the efficiencies of Weinberg's sib method and Li and Mantel's method when the hypothesis is that a *rare* trait is due to a *simple autosomal dominant*, the parents are Normal × Affected, and ascertainment is complete through the children (i.e., calculate a table similar to Table 17.3, but for $\theta = \frac{1}{2}$).

CHAPTER 18

▼ Complex Segregation Analysis

18.1. MORE ADVANCED MODELS FOR ANALYZING FAMILY DATA. ADVANTAGES AND DIFFICULTIES

In Chapter 17 we discussed models in which a given phenotypic mating usually corresponds to one genotypic mating, that is, we discussed traits which are inherited as simple recessives or simple rare dominants. The segregation distribution has included only one segregation parameter, θ, which had to be estimated. The only complication encountered in the previous chapter occurred when the ascertainment probability, π, as well as θ, had to be estimated from the same set of family data. The problems which we wish to discuss in this chapter are generalizations of those considered in Chapter 17.

We define *complex segregation analysis* as the statistical methods used in handling family data when two or more distinct, and functionally independent, segregation parameters are involved in the analysis.

Complex segregation analysis will include the following situations:

(*a*) There are two phenotypic forms of a certain genetic character; neither form is rare, and both are controlled by one (or more) loci.

(*b*) There are more than two phenotypic forms controlled by a certain genetic system (e.g., multiallelic inheritance or multiloci inheritance with or without epistasis and/or with or without linkage).

(*c*) Some of the models discussed in (*a*) or (*b*) can be complicated by incomplete penetrance or different fitnesses of genotypes.

A general model for such situations consists of sets of multinomial distributions, each set associated with families of given size, s. The multinomial parameters in each set, derived for a given parental phenotypic mating type, are weighted sums of other multinomial parameters associated with genotypic matings. By comparison with Chapter 17, statistical analysis is now more difficult not only mathematically but also technically, and more extensive computer facilities are necessary.

This chapter is designated as "difficult" since it requires the reader to be familiar with differential calculus, simple numerical methods, and the use of computer programs. One may doubt whether it should be included in this book at all. My feeling is that a human geneticist, analyzing family data, should be at least acquainted with models other than those of a single locus with two alleles, discussed in Chapter 17. Use of such models greatly increases the possibilities of utilizing family data in estimating simultaneously several parameters, including gene frequencies. If sections 11.8, 12.11, 12.12, 13.11, and 13.12 and Chapter 17 have been understood, there should be no difficulty in comprehending the present chapter. Following the instructions included in the subsequent sections, one can set up his model (hypothesis) and use standard computer programs, obtainable in several statistical laboratories, to analyze family data.

In this chapter, we shall be concerned mostly with general techniques for constructing models for complex segregation analysis, rather than giving numerical examples. We restrict ourselves here to *autosomal* traits and will first consider a relatively simple situation in which the classification is *dichotomous:* an individual exhibits a trait (in which case we will call him "affected") or he does not exhibit a trait (he will be called "normal").

18.2. A GENERAL MODEL FOR COMPLEX SEGREGATION ANALYSIS. RANDOM SAMPLING OF PARENTS

In this section, we will consider models for use when the experimental designs are associated with a *random sample of families*. We will distinguish situations in which the information on the phenotypes of parents is (1) not available and (2) available.

18.2.1. No Information Available about the Parental Phenotypes

Let H_0 be a certain hypothesis about the mode of inheritance of a particular trait. We first consider families of *fixed size s;* therefore the phrase "families of size s" will often be omitted in the text. In constructing a probability model for H_0 the following steps are useful.

(i) Let f be the number of possible genotypes associated with the (autosomal) mode of inheritance. It is sometimes convenient to distinguish between the genotype of the mother and that of the father. In such cases the maximum number of possible matings is f^2.

Let ψ_{uv} be the expected proportion of genotypic mating $u \times v$. For the whole population, we have

$$\sum_{u=1}^{f} \sum_{v=1}^{f} \psi_{uv} = 1. \tag{18.1}$$

Assuming that the female and male genotype frequencies are equal, and the genotype l has the frequency ψ_l, with $\sum \psi_l = 1$, for *random* mating we have

$$\psi_{uv} = \psi_u \psi_v, \quad u, v = 1, 2, \ldots, f. \tag{18.2}$$

(ii) Let θ_{uv} be the probability that a child born in a family with parental genotypes $u \times v$ will be affected. Thus the (conditional) probability, $P_{rs(uv)}$, that in such a family (of size s) there will be r affected children is

$$P_{rs(uv)} = \binom{s}{r} \theta_{uv}^r (1 - \theta_{uv})^{s-r}, \tag{18.3}$$

for $r = 0, 1, \ldots, s$; $u, v = 1, 2, \ldots, f$. Note that $P_{rs(uv)}$, considered as a function of r, represents segregation distribution in a family with parental genotypes u and v. (When $\theta_{uv} = 0$ or $\theta_{uv} = 1$, we define $0^0 = 1$.)

(iii) Let P_{rs} be the probability that a family of size s, selected at random from the whole population, will have r affected children. Thus

$$P_{rs} = \sum_{u=1}^{f} \sum_{v=1}^{f} \psi_{uv} P_{rs(uv)}, \quad r = 0, 1, 2, \ldots, s. \tag{18.4}$$

The probability P_{rs} is, in fact, the expected proportion of s-size families with r affected children.

The probabilities ψ_{uv} can be functions of some genetic parameters such as gene frequencies, inbreeding coefficient, or fitnesses of genotypes. Some of these parameters may be functionally dependent (e.g., gene frequencies add up to 1). Similarly, the parameters θ_{uv} can be functions of other parameters such as penetrance, fitnesses, or linkage and again some of these genetic parameters may be functionally dependent. However, it is not difficult to find the total number, M, of parameters which are distinct and *functionally independent*. Let us denote these parameters by $\gamma' = (\gamma_1, \gamma_2, \ldots, \gamma_M)$. Thus P_{rs} can be a function of M functionally independent parameter γ_i's, that is,

$$P_{rs} = P_{rs}(\gamma_1, \gamma_2, \ldots, \gamma_M) = P_{rs}(\gamma). \tag{18.5}$$

(iv) Suppose that we observe n_s families, each of size s. Let a_{rs} be the number of these families which have r affected children. We have

$$\sum_{r=0}^{s} a_{rs} = n_s. \tag{18.6}$$

For a given sample size, n_s, the quantities $a_{0s}, a_{1s}, \ldots, a_{ss}$ satisfying (18.6) can be considered as multinomial random variables with multinomial parameters, $P_{0s}, P_{1s}, \ldots, P_{ss}$, with

$$\sum_{r=0}^{s} P_{rs} = 1.$$

that is,

$$\Pr\{a_{0s}, a_{1s}, \ldots, a_{ss}; \boldsymbol{\gamma} \mid n_s\} = n_s! \prod_{r=0}^{s} \left(\frac{P_{rs}^{a_{rs}}}{a_{rs}!}\right). \tag{18.7}$$

We notice that

$$\begin{aligned} E(a_{rs}) &= n_s P_{rs}; \quad \mathrm{Var}(a_{rs}) = n_s P_{rs}(1 - P_{rs}); \\ \mathrm{Cov}(P_{rs}, P_{r's}) &= -n_s P_{rs} P_{r's}. \end{aligned} \tag{18.8}$$

For given $a_{0s}, \alpha_{1s}, \ldots, a_{ss}$, the probability in (18.7), considered as a function of the γ's, is the likelihood function, $L_s(\boldsymbol{\gamma})$, and will be used in estimating $\gamma_1, \gamma_2, \ldots, \gamma_M$.

(v) In evaluating the probabilities from (18.4) we have allowed, in the general case, for f^2 different segregation distributions, $P_{rs(uv)}$. In fact, there might be not so many. In the first place, the matings $u \times v$ and $v \times u$ might be considered as the same, so that there will be only $f(f + 1)/2$ different possible matings. Second, the same segregation parameters can be associated with different matings. For instance, if an affected child is supposed to be a double recessive, then for two matings, *Aabb* × *aabb* and *aaBb* × *aabb*, we have $\theta = \frac{1}{2}$. Similarly for four matings, *AaBb* × *aabb*, *Aabb* × *Aabb*, *aaBb* × *Aabb*, and *aaBb* × *aaBb*, we have $\theta = \frac{1}{4}$ (see Table 18.2). We say in such cases that these genotypic matings have the same *segregation pattern*, that is, they have the same segregation distribution.

Suppose that for a given model (determined by H_0) there are m distinct segregation patterns. Let ϕ_t be the expected proportion of families with the tth segregation pattern, in the population. We have

$$\sum_{t=1}^{m} \phi_t = 1. \tag{18.9}$$

Let θ_t be the segregation probability for the tth pattern, and

$$P_{rst} = \binom{s}{r} \theta_t^r (1 - \theta_t)^{s-r}, \qquad r = 0, 1, \ldots, s \tag{18.10}$$

be the segregation distribution for the tth pattern. Thus P_{rs}, given by (18.4), takes the form

$$P_{rs} = \sum_{t=1}^{m} \phi_t P_{rst}, \qquad r = 0, 1, \ldots, s. \tag{18.11}$$

The likelihood function, $L_s(\boldsymbol{\gamma})$, defined in (18.7) will be of the same form, with expression (18.11) substituted for P_{rs}.

18.2.2. Parental Phenotypes Are Known

When selecting families, we usually observe the phenotypes of the parents. As in Chapter 17, we shall distinguish phenotypic matings: (a) Affected × Affected, (b) Normal × Affected, (c) Normal × Normal. For each type of phenotypic matings we can repeat what has been said in section 18.2.1, with the exception that now ψ_{uv} will be the conditional probabilities for genotype mating (uv) in the reduced sample space determined by a given phenotypic mating. For instance, in a population under equilibrium with respect to a single locus with alleles A and a, the mating $Aa \times aa$ has the probability $4pq^3$; but if we restrict ourselves to the population of phenotypic matings Normal × Affected, we have the conditional probability $\Pr\{Aa \times aa \mid \text{Norm} \times \text{Aff}\} = 2q/(1 + q)$ (see Table 18.1). Similar remarks apply to the probabilities ϕ_t.

18.3. MODELS FOR COMPLEX SEGREGATION ANALYSIS. SELECTION THROUGH THE CHILDREN

We will confine ourselves to situations with m distinct segregation patterns within a given type of parental phenotypic mating, so that formula (18.11) for P_{rs} will be of general use. As we have seen from section 18.2, there is no difficulty in switching to formula (18.4), applied to whole or truncated populations.

Let π be the ascertainment probability. Assuming that π is constant and the same for each genotype mating, the probability P_{rst} that a family of size s, within the tth segregation pattern, will have r affected children and will be ascertained is

$$P_{rst} = P_{rst}(\gamma) = [1 - (1 - \pi)^r]\binom{s}{r}\theta_t^r(1 - \theta_t)^{s-r}, \qquad (18.12)$$

for $t = 1, 2, \ldots, m; r = 1, 2, \ldots, s$. [Compare (17.6).]

Let Q_{st} be the probability that a randomly chosen family of size s, within the tth segregation pattern, has been ascertained. From formula (17.7) we have

$$Q_{st} = \sum_{r=1}^{s} P_{rst} = 1 - (1 - \pi\theta_t)^s. \qquad (18.13)$$

The probability that any randomly selected family of size s has been ascertained is

$$1 - \sum_{t=1}^{m} \phi_t(1 - \pi\theta_t)^s = \sum_{t=1}^{m} \phi_t - \sum_{t=1}^{m} \phi_t(1 - \pi\theta_t)^s$$

$$= \sum_{t=1}^{m} \phi_t[1 - (1 - \pi\theta_t)^s] = \sum_{t=1}^{m} \phi_t Q_{st} \qquad (18.14)$$

(note that $\sum \phi_t = 1$).

Let P_{rs} now denote the probability that an ascertained family of size s has r affected children:

$$P_{rs} = P_{rs}(\gamma) = \frac{\sum\limits_{t=1}^{m} \phi_t P_{rst}}{\sum\limits_{t=1}^{m} \phi_t Q_{st}}, \qquad r = 1, 2, \ldots, s. \qquad (18.15)$$

In particular, when $m = 1$, P_{rs} takes the form (17.8). For an observed sample of n_s ascertained families, the likelihood function, $L_s(\gamma)$, takes the form (18.7), with P_{rs} defined by (18.15).

For *complete* ascertainment ($\pi = 1$) all formulae are valid after substituting $\pi = 1$.

For *single* ascertainment, we use approximation (17.11) so that

$$Q_{st} \doteq s\pi\theta_t. \qquad (18.16)$$

Substituting (18.16) into (18.15) and denoting

$$P'_{rst} = \binom{s-1}{r-1} \theta_t^r (1 - \theta_t)^{s-r}, \qquad (18.17)$$

we can easily show that (18.15) takes the form

$$P_{rs} = P'_{rs} = \frac{\sum\limits_{t=1}^{m} \phi_t P'_{rst}}{\sum\limits_{t=1}^{m} \phi_t \theta_t}. \qquad (18.18)$$

When $m = 1$, P'_{rs} takes the form (17.12).

18.4. LIKELIHOOD EQUATIONS AND INFORMATION MATRIX

We now wish to derive the ML estimators for $\gamma' = (\gamma_1, \gamma_2, \ldots, \gamma_m)$. We confine ourselves to models in which $P_{rs}(\gamma)$ is given by (18.15). As we shall see later, there is no difficulty in giving the likelihood equations when P_{rs} is defined by (18.11) or (18.4).

We first express (18.15) in more convenient form. Let

$$A = A_{rs}(\gamma) = \sum_{t=1}^{m} \phi_t P_{rst} \quad \text{and} \quad B = B_s(\gamma) = \sum_{t=1}^{m} \phi_t Q_{st}. \quad (18.19)$$

Then (18.15) can be written in the form

$$P_{rs} = P_{rs}(\gamma) = \frac{A_{rs}(\gamma)}{B_s(\gamma)} = \frac{A}{B}. \quad (18.20)$$

We now apply the results of section 12.12 in order to derive the maximum likelihood estimators of M functionally independent parameters, $\gamma' = (\gamma_1, \gamma_2, \ldots, \gamma_M)$. For a given sample of n_s ascertained families, each of size s, the likelihood function, $L_s(\gamma)$, is given by (18.7) and can be written as

$$L_s = L_s(\gamma) = \frac{n_s!}{\prod\limits_{r=1}^{s} a_{rs}!} \prod_{r=1}^{s} P_{rs}^{a_{rs}} = \frac{n_s!}{\prod\limits_{r=1}^{s} a_{rs}!} \prod_{r=1}^{s} \left(\frac{A}{B}\right)^{a_{rs}}. \quad (18.21)$$

Hence

$$\log L_s = \text{const} + \sum_{r=1}^{s} a_{rs} P_{rs} = \text{const} + \sum_{r=1}^{s} (\log A - \log B) a_{rs}. \quad (18.22)$$

If we take into account families with sizes $s = 1, 2, \ldots, S$, then the logarithm of the overall likelihood function, $L = L(\gamma)$, is

$$\log L = \sum_{s=1}^{S} \log L_s. \quad (18.23)$$

Let $u_{i;s}(r) = u_{i;s}(\gamma \mid r)$ denote the individual score with respect to the parameter γ_i, for given s, in the rth multinomial class. We have

$$u_{i;s}(r) = u_{i;s}(\gamma \mid r) = \frac{1}{P_{rs}} \frac{\partial P_{rs}}{\partial \gamma_i} = \frac{1}{A} \frac{\partial A}{\partial \gamma_i} - \frac{1}{B} \frac{\partial B}{\partial \gamma_i}, \quad (18.24)$$

for $i = 1, 2, \ldots, M; r = 1, 2, \ldots, s$.

It should be noted that in the general case the derivative $(\partial A / \partial \gamma_i)$ is of the form

$$\frac{\partial A}{\partial \gamma_i} = \sum_{t=1}^{m} \left(\frac{\partial A}{\partial \phi_t} \frac{\partial \phi_t}{\partial \gamma_i} + \frac{\partial A}{\partial P_{trs}} \frac{\partial P_{trs}}{\partial \gamma_i}\right), \quad i = 1, 2, \ldots, M. \quad (18.25)$$

The derivative $(\partial B / \partial \gamma_i)$ has a similar form. The total score for the sample of n_s families, each of size s, $U_{i;s}$, say, is

$$U_{i;s} = \sum_{r=1}^{s} a_{rs} u_{i;s}(r), \quad i = 1, 2, \ldots, M; \quad (18.26)$$

and the overall score with respect to the parameter γ_i, U_i, say, is

$$U_i = \sum_{s=1}^{S} U_{i;s}. \quad (18.27)$$

To obtain the maximum likelihood estimates we have to solve M (linearly independent) equations of the form

$$U_i = \sum_{s=1}^{S} \sum_{r=1}^{s} \left(\frac{1}{A} \frac{\partial A}{\partial \gamma_i} - \frac{1}{B} \frac{\partial B}{\partial \gamma_i} \right) a_{rs} = 0, \qquad i = 1, 2, \ldots, M. \quad (18.28)$$

A unique set of solutions exists, provided

$$M \leqslant \sum_{s=2}^{S} (s - 1) = \tfrac{1}{2} S(S - 1),$$

as is usually the case.

The elements of the expected information matrix, $\mathbf{I}_s$, are

$$\left. \begin{aligned} I_{ii;s} &= \sum_{r=1}^{s} E(a_{rs}) u_{i;s}^{2}(r), \quad i = 1, 2, \ldots, M; \\[2mm] I_{ij;s} &= \sum_{r=1}^{s} E(a_{rs}) u_{i;s}(r) u_{j;s}(r), \qquad i, j = 1, 2, \ldots, M \quad (i \neq j). \end{aligned} \right\} \quad (18.29)$$

The elements of the overall expected information matrix, $\mathbf{I}$, are

$$I_{ii} = \sum_{s=1}^{S} I_{ii;s}; \quad I_{ij} = \sum_{s=1}^{S} I_{ij;s}, \qquad i, j = 1, 2, \ldots, s \qquad (i \neq j), \quad (18.30)$$

so that

$$\mathbf{I} = \sum_{s=1}^{S} \mathbf{I}_s. \quad (18.31)$$

18.5. TESTING HYPOTHESES

In testing hypotheses about parameters $\mathbf{\gamma}' = (\gamma_1, \gamma_2, \ldots, \gamma_M)$ we apply the theory presented in section 13.11.

(i) Suppose that all M parameters are specified by the null hypothesis $H_0 : \mathbf{\gamma} = \mathbf{\gamma}_0$. Let

$$\mathbf{U}_s'(\mathbf{\gamma}_0) = (U_{1;s}(\mathbf{\gamma}_0), U_{2;s}(\mathbf{\gamma}_0), \ldots, U_{s;s}(\mathbf{\gamma}_0)) \quad (18.32)$$

be the vector of scores $U_{i;s}$ evaluated at the point $\mathbf{\gamma} = \mathbf{\gamma}_0$, and let $\mathbf{I}_s(\mathbf{\gamma}_0)$ be the expected information matrix. Following formula (13.44), we calculate the statistic

$$Z_{\text{Total}}^2 = \sum_{s=2}^{S} \mathbf{U}_s'(\mathbf{\gamma}_0) \mathbf{I}_s^{-1}(\mathbf{\gamma}_0) \mathbf{U}_s(\mathbf{\gamma}_0). \quad (18.33)$$

If H_0 is true and the family data are homogeneous with respect to a model defined by H_0, then Z_{Total}^2 is approximately distributed as χ^2 with

$\frac{1}{2}M(2S - M - 1)$ degrees of freedom.* We notice that the summation over s begins with $s = 2$, since the families with $s = 1$ children contribute nothing to Z^2_{Total}. Let

$$U(\gamma_0) = \sum_{s=1}^{S} U_s(\gamma_0) \quad \text{and} \quad I(\gamma_0) = \sum_{s=1}^{S} I_s(\gamma_0) \tag{18.34}$$

be, respectively, the overall vector of scores and the expected information matrix of combined data obtained from families of all sizes. If H_0 is true "on the average," then the statistic

$$Z^2_{\text{Comb}} = U'(\gamma_0)I^{-1}(\gamma_0)U(\gamma_0) \tag{18.35}$$

is approximately distributed as χ^2 with M d.f. For testing homogeneity with respect to any model determined by a hypothesis $H_0^{(x)}$, we may calculate

$$Z^2_{\text{Diff}} = Z^2_{\text{Total}} - Z^2_{\text{Comb}} \tag{18.36}$$

[with $\frac{1}{2}M(2S - M - 1) - M = \frac{1}{2}M(2S - M - 3)$ d.f.].

(ii) If *all* M parameters are *not specified*, we substitute the pooled estimates $\hat{\hat{\gamma}}' = (\hat{\hat{\gamma}}_1, \hat{\hat{\gamma}}_2, \ldots, \hat{\hat{\gamma}}_M)$, obtained by solving system (18.28), into (18.33). If H_0 is true and the family data are homogeneous, the statistic

$$Z^2_{\text{Total}} = Z^2_{\text{Diff}} = \sum_{s=2}^{S} U'_s(\hat{\hat{\gamma}})I_s^{-1}(\hat{\hat{\gamma}})U_s(\hat{\hat{\gamma}}) \tag{18.37}$$

is approximately distributed as χ^2 with $\frac{1}{2}M(2S - M - 3)$ d.f. (Note that $Z^2_{\text{Comb}} = 0$.)

(iii) If among M parameters *only K are specified*, then the statistics (18.33) and (18.35) can be calculated for K of the γ_0's and $M - K$ of the $\hat{\hat{\gamma}}$'s. The degrees of freedom have to be appropriately decreased by subtracting $(M - K)$, that is, the number of estimated parameters, from the degrees of freedom given in (i).

(iv) Taking terms P_{rst} in $A_{rs}(\gamma)$ from $r = 0$ (instead of $r = 1$) and putting $B_s(\gamma) = 1$, defined by (18.19), we are back in the situation when the P_{rst} are defined by (18.10) and hence P_{rs} by (18.11). All the results of section 18.4 (including terms for $r = 0$) and of section 18.5 hold for family data obtained by random selection of families through the parents.

---

* We notice that when $s \leqslant M$ the rank of the matrix I_s is equal to $s - 1$, so that for $s \leqslant M$ each term in (18.33) contributes only $s - 1$ d.f. The remaining $S - M$ terms, for which $s > M$, each contributes M d.f. Hence the total number of degrees of freedom is

$$1 + 2 + 3 \cdots + (M - 1) + (S - M)M = \frac{1}{2}(M(M - 1) + M(S - M)) = \frac{1}{2}M(2S - M - 1).$$

Of course, if some sizes s are missing, the formula for the number of degrees of freedom should be appropriately modified.

It should be mentioned that, if information on the parental phenotypic mating is available, and the whole population is sampled, we can obtain more efficient estimators by using the method of "pooled" estimators from combined data (see section 12.11) than by treating the data as a single sample from the whole population (compare sections 11.2.3 and 11.7.1).

To illustrate the theory developed in sections 18.2–18.5 a few examples will be useful.

18.6. ESTIMATION OF GENE FREQUENCIES FROM FAMILY DATA. SEPARATION OF NONSEGREGATING AND SEGREGATING FAMILIES

When a trait is common and usually not harmful (e.g., blood groups), selection is often *through the parents*. Such data may serve at least three purposes: (a) testing hypotheses on mode of inheritance; (b) estimating gene frequencies; (c) both (a) and (b).

In sections 12.5–12.9 we discussed methods of estimating gene frequencies by using random samples obtained from different model populations. It has been emphasized that the usual custom of pooling all the data from families and calling them a "simple random sample" may be incorrect and can lead to biased estimates. Here we want to show how to utilize family data in estimating gene frequencies without pooling them into one sample. We restrict ourselves to a single locus with alleles A (dominant) and a (recessive). The term "affected" will here be replaced by the term "recessive" (i.e., aa genotypes).

Let p and q, with $p + q = 1$, be the frequencies of genes A and a, respectively, in a population. We assume that the population is in equilibrium with respect to locus A. We wish to test the hypothesis, H_0, that the trait under consideration is inherited as a simple recessive and, using the same family data, to estimate the gene frequencies p and q.

We may apply the theory presented in section 18.2 to the model given in Table 8.1. However, since some families will not be able to have recessive children and others will not, in fact, have them (even though this is potentially possible), a more appropriate (and more efficient) method is to divide the families into two classes: (a) those with no recessive children (nonsegregating families) and (b) those with at least one recessive child (segregating families). Table 18.1 shows this division and gives the segregation distributions when H_0 is true.

We now set up the likelihood functions for the general case in which both gene frequency, q, and the proportion of *recessive* children in segregating families, θ, have to be estimated from the same experiment.

TABLE 18.1 Separation of segregating and nonsegregating families under H_0 (ascertainment through the parents)

Matings	Expected proportions of matings		Segregation parameter	Nonsegregating families		Segregating families	
	In population	In a given mating type		Probability of no recessive children	Proportions of families with no recessive children in a given mating type	Segregation distribution	Proportion of families with at least one recessive child
Dominant × Recessive							
$AA \times aa$	$2p^2q^2$	$\dfrac{1-q}{1+q} = 1-h$	0	1	$(1-h) \times 1$		0
$Aa \times aa$	$4pq^3$	$\dfrac{2q}{1+q} = h$	$\tfrac{1}{2}$	$(1-\tfrac{1}{2})^8$	$h \times (1-\tfrac{1}{2})^8$	$\dbinom{s}{r}(\tfrac{1}{2})^r(1-\tfrac{1}{2})^{s-r}$	$h[1 - (1-\tfrac{1}{2})^8]$
Total	$2q^2(1-q^2)$	1			$1 - h[1 - (1-\tfrac{1}{2})^8]$		$h[1 - (1-\tfrac{1}{2})^8]$
Dominant × Dominant							
$AA \times AA$	p^4	$\left(\dfrac{1-q}{1+q}\right)^2 = (1-h)^2$	0	1	$(1-h)^2 \times 1$		0
$AA \times Aa$	$4p^3q$	$\dfrac{4q(1-q)}{(1+q)^2} = 2h(1-h)$	0	1	$2h(1-h) \times 1$		0
$Aa \times Aa$	$4p^2q^2$	$\left(\dfrac{2q}{1+q}\right)^2 = h^2$	$\tfrac{1}{4}$	$(1-\tfrac{1}{4})^8$	$h^2(1-\tfrac{1}{4})^8$	$\dbinom{s}{r}(\tfrac{1}{4})^r(1-\tfrac{1}{4})^{s-r}$	$h^2[1 - (1-\tfrac{1}{4})^8]$
Total	$(1-q^2)^2$	1			$1 - h^2[1 - (1-\tfrac{1}{4})^8]$		$h^2[1 - (1-\tfrac{1}{4})^8]$

507

We will consider only two values for r:

$$r = \begin{cases} 0 & \text{if no recessive child (nonsegregating family);} \\ 1 & \text{otherwise, that is, at least one recessive child (segregating family).} \end{cases}$$

Thus we shall calculate P_{0s} and P_{1s}, respectively. We want to take into account the parental phenotypic mating types. Thus for (1) Dominant $\times$ Recessive we calculate $P_{0s}^{(1)}$ and $P_{1s}^{(1)}$ and for (2) Dominant $\times$ Dominant we calculate $P_{0s}^{(2)}$ and $P_{1s}^{(2)}$, respectively. Similarly we shall distinguish $L_s^{(1)}$ and $L_s^{(2)}$.

18.6.1. Parental Mating Dominant $\times$ Recessive

Let β be the (unknown) segregation probability for $Aa \times aa$ mating. Since our classification consists of nonsegregating and segregating families, we have, for mating Dom $\times$ Rec:

$$\phi_1^{(1)} = \frac{1-q}{1+q} = 1 - h, \quad \phi_2^{(1)} = \frac{q}{1+q} = h;$$

$$\theta_1^{(1)} = 1, \quad \theta_2^{(1)} = (1 - \beta)^s.$$

Thus

$$P_{0s1}^{(1)} = 1; \quad P_{1s1}^{(1)} = 0; \quad P_{0s2}^{(1)} = (1 - \beta)^s; \quad P_{1s2}^{(1)} = 1 - (1 - \beta)^s.$$

Hence

$$P_{0s}^{(1)} = \phi_1^{(1)} P_{0s1}^{(1)} + \phi_2^{(1)} P_{0s2}^{(1)}$$

$$= (1 - h) \cdot 1 + h(1 - \beta)^s = 1 - h[1 - (1 - \beta)^s], \quad (18.38a)$$

and

$$P_{1s}^{(1)} = \phi_1^{(1)} P_{1s1}^{(1)} + \phi_2^{(1)} P_{1s2}^{(1)} = 1 - P_{0s}^{(1)} = h[1 - (1 - \beta)^s]. \quad (18.38b)$$

Let $n_s^{(1)}$ be the number of families, each of size s, selected from matings Dom $\times$ Rec, and $a_{0s}^{(1)}$ the number of nonsegregating families. Hence the number of segregating families, $a_{1s}^{(1)}$, equals $n_s^{(1)} - a_{0s}^{(1)}$.

For observed sample values, the likelihood function, $L_s^{(1)}(q, \beta)$, is

$$L_s^{(1)}(q, \beta) = \begin{bmatrix} n_s^{(1)} \\ a_{0s}^{(1)} \end{bmatrix} [P_{0s}^{(1)}]^{a_{0s}^{(1)}} [1 - P_{0s}^{(1)}]^{n_s^{(1)} - a_{0s}^{(1)}} \quad (18.39)$$

If we take into account families of $s = 1, 2, \ldots, S$ sizes, then the overall likelihood function, $L^{(1)}(q, \beta)$, is

$$L^{(1)}(q, \beta) = \prod_{s=1}^{S} L_s^{(1)}(q, \beta). \quad (18.40)$$

If, however, we know that the trait under consideration is recessive, we substitute, in (18.38), $\beta = \frac{1}{2}$ and estimate only q, using (18.40).

18.6.2. Parental Mating Dominant × Dominant

Using arguments similar to those in section 18.6.1, we obtain

$$P_{0s}^{(2)} = 1 - h^2[1 - (1 - \tfrac{1}{2}\beta)^s], \qquad (18.41a)$$

$$P_{1s}^{(2)} = 1 - P_{0s}^{(2)} = h^2[1 - (1 - \tfrac{1}{2}\beta)^s]. \qquad (18.41b)$$

The likelihood functions, $L_s^{(2)}(q, \beta)$ and $L^{(2)}(q, \beta)$, will be similar in form to (18.39) and (18.40), respectively, changing the superscript (1) to (2). If H_0 is true, we substitute, in (18.41), $\tfrac{1}{2}\beta = \tfrac{1}{4}$ and estimate q.

We may also use the combined likelihood function

$$L(q, \beta) = \prod_{g=1}^{2} \prod_{s=1}^{S} L_s^{(g)}(q, \beta) \qquad (18.42)$$

in estimating q and β (see section 12.11). The general methods developed in sections 18.4 and 18.5 should be applied, using computer facilities.

18.7. INTERACTION OF LOCI. DOUBLE AUTOSOMAL RECESSIVE

We will now give an example of complex segregation analysis with ascertainment *through the children*. The hypothesis H_0 will be that any affected individual is a double recessive, *aabb*.

If it is suspected that a trait is a *rare* double recessive, the parental mating Normal × Affected is likely to be *AaBb* × *aabb* with segregation parameter $\theta = \tfrac{1}{4}$, and the parental mating Normal × Normal is likely to be *AaBb* × *AaBb* with $\theta = \tfrac{1}{16}$. In such cases, the segregation analysis presented in section 17.6 is applicable.

However, if a trait is *common*, some other genotypic matings occur. Table 18.2 shows all genotype matings which are able to produce affected children for parental matings Norm × Aff and Norm × Norm. The matings Aff × Aff produce only affected children. Under the null hypothesis, H_0, segregation distributions are given in the fourth and fifth columns, respectively.

18.7.1. Matings Normal × Affected

Let θ be the segregation parameter for single heterozygote × double recessive matings (*Aabb* × *aabb* or *aaBb* × *aabb*). As we can see from Table 18.2, we have $\theta_1 = \theta$, $\theta_2 = \tfrac{1}{2}\theta$, and $\phi_1 = 1 - h$, $\phi_2 = h$.

TABLE 18.2 Segregation model for (common) double recessive (segregating families)[a]

Matings	Expected proportions of matings which can produce affected children		Segregation parameter, θ_i (under H_0)	Segregation distribution (under H_0)
	In population	In a given phenotypic mating type		
Norm × Aff:		One parent unaffected		
$Aabb \times aabb$	$4p_1q_1^3q_2^4$	$\dfrac{p_1q_2}{p_1+p_2}$ $\Big\}\,1-h$	$\tfrac{1}{2}$	$\binom{s}{r}(\tfrac{1}{2})^r(\tfrac{1}{2})^{s-r}$
$aaBb \times aabb$	$4q_1^4p_2q_2^3$	$\dfrac{q_1p_2}{p_1+p_2}$	$\tfrac{1}{2}$	,,
$AaBb \times aabb$	$8p_1q_1^3p_2q_2^3$	$\dfrac{2p_1p_2}{p_1+p_2}=h$	$\tfrac{1}{4}$	$\binom{s}{r}(\tfrac{1}{4})^r(\tfrac{3}{4})^{s-r}$
Total	$4q_1^3q_2^3(p_1+p_2)$	1		

Both parents unaffected

Norm × Norm:				
$Aabb \times Aabb$	$4p_1^2q_1^2q_2^4$	$\dfrac{p_1^2q_2^2}{(p_1+p_2)^2}$	$\frac{1}{4}$	$\dbinom{s}{r}\left(\frac{1}{4}\right)^r\left(\frac{3}{4}\right)^{s-r}$
$aaBb \times Aabb$	$8p_1q_1^3p_2q_2^3$	$\dfrac{2p_1q_1p_2q_2}{(p_1+p_2)^2}$ $\Big\}(1-h)^2$	$\frac{1}{4}$	,,
$aaBb \times aaBb$	$4q_1^4p_2^2q_2^2$	$\dfrac{q_1^2p_2^2}{(p_1+p_2)^2}$	$\frac{1}{4}$	,,
$AaBb \times Aabb$	$16p_1^2q_1^2p_2q_2^3$	$\dfrac{4p_1^2p_2q_2}{(p_1+p_2)^2}$ $\Big\}2h(1-h)$	$\frac{1}{8}$	$\dbinom{s}{r}\left(\frac{1}{8}\right)^r\left(\frac{7}{8}\right)^{s-r}$
$AaBb \times aaBb$	$16p_1q_1^3p_2^2q_2^2$	$\dfrac{4p_1q_1p_2^2}{(p_1+p_2)^2}$	$\frac{1}{8}$	,,
$AaBb \times AaBb$	$16p_1^2q_1^2p_2^2q_2^2$	$\dfrac{4p_1^2p_2^2}{(p_1+p_2)^2}=h^2$	$\frac{1}{16}$	$\dbinom{s}{r}\left(\frac{1}{16}\right)^r\left(\frac{15}{16}\right)^{s-r}$
Total	$4q_1^2q_2^2(p_1+p_2)^2$	1		

[a] Taken from Elandt-Johnson, *Amer. J. Hum. Genet.* 22 (1970), p. 138.

511

18.7.2. Matings Normal × Normal

Let θ be the segregation parameter for single heterozygote × single heterozygote matings. From Table 18.2, we have $\theta_1 = \theta$, $\theta_2 = \frac{1}{2}\theta$, $\theta_3 = \frac{1}{4}\theta$, $\phi_1 = (1 - h)^2$, $\phi_2 = 2h(1 - h)$, $\phi_3 = h^2$.

Suppose that we wish to test $H_0:\theta = \theta_0$ (i.e., $\theta_0 = \frac{1}{2}$ for Norm × Aff and $\theta_0 = \frac{1}{4}$ for Norm × Norm). For each mating type, we construct the likelihood function $L(\theta, h)$, find the pooled estimators, $\hat{\hat{\theta}}$ and $\hat{\hat{h}}$, and evaluate Z^2_{Total}, Z^2_{Comb}, and Z^2_{Diff} with $\theta = \theta_0$ specified and $h = \hat{\hat{h}}$ unspecified (see sections 18.4 and 18.5).

The results from both types of parental matings can be combined in the usual way.

Example 18.1. Psoriasis is a chronic inflammatory disease of the skin characterized by rounded, erythematous, dry, scaling patches of various sizes covered by grayish white scales. It has been suggested [Steinberg et al. (1951)] that the disease may be due to an autosomal double recessive. Table 18.3 represents some data for the condition that both parents were unaffected. [Steinberg et al. (1951)]. The ascertainment probability, π, is unknown, and we perform the analysis for both limiting cases, $\pi = 1$ and π very small.

It is left to the reader to show that, under the assumption that the trait is *rare*, the hypothesis on double recessive (i.e., $H_0:\theta = \frac{1}{16}$) does not fit these data (see also Problem 18.2.)

When the trait is assumed to be *common*, we follow the instruction given in section 18.7.2, using the lower part of Table 18.2 (both parents unaffected). We test $H_0:\theta = \theta_0 = \frac{1}{4}$. The results are as follows.

Complete Ascertainment ($\pi = 1$). We obtained, using the computer program, $\hat{\hat{\theta}} = 0.0991$, $\hat{\hat{h}} = 0.7617$. The variance-covariance matrix, evaluated at $\theta = \theta_0 = \frac{1}{4} = 0.2500$ and $h = \hat{\hat{h}} = 0.7617$, is

$$\mathbf{I}^{-1}(\theta_0, \hat{\hat{h}}) \doteq \mathbf{V}(\hat{\hat{\theta}}, \hat{\hat{h}}) = \begin{bmatrix} 0.6709 & 0.3175 \\ 0.3175 & 0.1508 \end{bmatrix}.$$

In this problem, we have $S = 13$, $M = 2$, $K = 1$. We now test the hypothesis $H_0:\theta = \theta_0 = 0.2500$. We obtain $Z^2_{\text{Total}} = 46.32$ with $\nu = 22$ d.f., $Z^2_{\text{Comb}} = 36.61$ with $\nu = 1$ d.f., $Z^2_{\text{Diff}} = 9.71$ with $\nu = 21$ d.f.

Incomplete Ascertainment (π *very small*). Here $\hat{\hat{\theta}} = 0.0549$, $\hat{\hat{h}} = 0.7958$;

$$\mathbf{I}^{-1}(\theta_0, \hat{\hat{h}}) \doteq \mathbf{V}(\hat{\hat{\theta}}, \hat{\hat{h}}) = \begin{bmatrix} 0.2710 & 0.1393 \\ 0.1393 & 0.0720 \end{bmatrix}.$$

TABLE 18.3   Psoriasis in 409 families. Both parents unaffected

Family size, s	No. of families, n_s	No. of all children, sn_s	No. of affected children, r				Total No. of affected children, r_s
			1	2	3	4	
			No. of families with r affected children, a_{rs}				
1	22	22	22				22
2	50	100	45	5			55
3	72	216	67	5			77
4	61	244	55	6			67
5	62	310	59	3			65
6	37	222	32	5			42
7	28	196	26	2			30
8	24	192	22	1	1		27
9	24	216	22	2			26
10	13	130	11	1	1		16
11	7	77	5	2			9
12	3	36	2	1			4
13	6	78	5			1	9
Total	409(N)	2039(T)	373(A_1)	33	2	1	449(R)

Then $Z^2_{\text{Total}} = 115.69$ with $\nu = 21$ d.f., $Z^2_{\text{Comb}} = 110.20$ with $\nu = 1$ d.f., $Z^2_{\text{Diff}} = 5.49$ with $\nu = 20$ d.f.

It appears from both analyses that the data are homogeneous (Z^2_{Diff} in both calculations is insignificant), but the hypothesis that psoriasis is a common double recessive trait does not fit the data. One reason for the lack of fit may be that the onset of this disease is age-dependent; thus the data in Table 18.2 might be incomplete.

18.8. COMPLEX SEGREGATION ANALYSIS FOR MULTIPLE CLASSIFICATION

The results of sections 18.2–18.5 can be extended to problems in which a certain genetically controlled character is classified into k ($k > 2$) mutually exclusive classes, $C_1, C_2, \ldots, C_k$. Examples are, blood groups O, A, B, AB at the three-allelic ABO locus, or four phenotypes in offspring of matings Double Dom × Double Rec. Let $\theta_{l(uv)}$ ($l = 1, 2, \ldots, k$) be the probability that a child born in a family with parental genotypes ($u \times v$) is classified in the C_l category. We have

$$\sum_{l=1}^{k} \theta_{l(uv)} = 1. \tag{18.43}$$

Let r_l ($l = 1, 2, \ldots, k$) be the observed number of children within the category C_l in a family of size s. We have

$$\sum_{l=1}^{k} r_l = s. \tag{18.44}$$

Thus for a family of size s, given that the parental genotypes are (uv), we have the following distribution of children among the k categories:

Category:	C_1	C_2	$\cdots$	C_k	Total
Probability:	$\theta_{1(uv)}$	$\theta_{2(uv)}$	$\cdots$	$\theta_{k(uv)}$	1
Number of children:	r_1	r_2	$\cdots$	r_k	$s.$

The quantities $r_1, r_2, \ldots, r_k$ considered as random variables have the (conditional) multinomial distribution (provided, we define $0^\circ = 1$)

$$P_{s(uv)}(r_1, r_2, \ldots, r_k) = \frac{s!}{r_1! \, r_2! \cdots r_k!} \theta_{1(uv)}^{r_1} \theta_{2(uv)}^{r_2} \cdots \theta_{k(uv)}^{r_k}. \tag{18.45}$$

We note that the quantity in (18.45) is analogous to $P_{rs(uv)}$, defined in (18.3) for $k = 2$. To avoid long subscripts we have used the notation $P_{s(uv)}(r_1, r_2, \ldots, r_k)$ instead of $P_{r_1, r_2, \ldots, r_k; s(uv)}$.

By analogy to (18.3) we will call (18.45) a (multinomial) *segregation distribution* and the parameters $\theta_{l(uv)}$ ($l = 1, 2, \ldots, k$) the *segregation probabilities* (or segregation parameters), given genotypic mating (uv).

Let $P_s(r_1, r_2, \ldots, r_k)$ (instead of $P_{r_1, r_2, \ldots, r_k; s}$) be the probability that a family of size s will have r_1 children in the category C_1, r_2 children in the category C_2, $\ldots$, and, finally, r_k children in the category C_k. Retaining the notation ψ_{uv} as defined in section 18.2, we have for selection through the parents

$$P_s(r_1, r_2, \ldots, r_k) = \sum_{u=1}^{f} \sum_{v=1}^{f} \psi_{uv} P_{s(uv)}(r_1, r_2, \ldots, r_k). \tag{18.46}$$

This is analogous to P_{rs}, given in (18.4). When selection is through the children, the probability in (18.45) has to be calculated *conditionally* according to which children (belonging to one or more C_l classes) are used as probands.

Similarly, when information about the phenotypic mating of the parents is available, we take ψ_{uv} as the conditional probabilities, given this mating. Modification, using ϕ_t instead of ψ_{uv}, is straightforward [see section 18.2 (v)].

Let n_s be the number of families of size s, and $r_{1\alpha}, r_{2\alpha}, \ldots, r_{k\alpha}$, with

$$\sum_{l=1}^{k} r_{l\alpha} = s,$$

be the *observed* numbers of children in classes $C_1, C_2, \ldots, C_k$, respectively, in the αth family ($\alpha = 1, 2, \ldots, n_s$). The likelihood function for the whole sample, $L_s(\gamma)$, can be written as

$$L_s(\gamma) = \prod_{\alpha=1}^{n_s} P_s(r_{1\alpha}, r_{2\alpha}, \ldots, r_{k\alpha}). \tag{18.47}$$

The further procedures for estimating the parameters γ's and testing hypotheses follow the general techniques described in sections 18.4 and 18.5.

18.9. ESTIMATION OF GENE FREQUENCIES AND RECOMBINATION FRACTION OF TWO LINKED LOCI, USING FAMILY DATA

The general method described in section 18.8 has several applications. We may use it in the estimation of gene frequencies at a multiallelic locus [see Elandt-Johnson (1971)]. It can also be used in detection of linkage and estimation of the recombination fraction. As we know, linkage in human beings cannot be detected in a sample randomly selected from an entire population, assumed to be in equilibrium. For its detection, family data are used. Several procedures have been discussed by Morton (1955) and Smith (1968). In most cases it is assumed that the parental genotypes are known, or that ascertainment is through the children so that the parental genotypes can be identified. [However, Morton (1955), for his sequential test, has considered briefly the situation in which the genotypes of the parents are unknown but the gene frequencies in the population are known.]

Let the locus for the "main" trait (i.e., the trait in which we are interested) be G, and let A denote a locus for a "test" trait (i.e., a trait with which we wish to find the linkage). Usually test traits are well-known genetic markers, such as blood group loci. Often we want to find whether any other trait (e.g., some disease) which is genetically controlled is linked with this genetic marker. As an example, we will consider the situation in which G has two alleles, G (dominant) and g (recessive), and similarly A and a are dominant and recessive, respectively, at locus A. We shall distinguish four phenotypic classes, GA, Ga, gA, and ga. (Note that Ga denotes double dominant, i.e., $G–A–$; ga denotes double recessive, $gaga$; while Ga ($G–aa$) and gA ($ggA–$) are single dominants.)

Let λ be the (unknown) recombination fraction; p_1, q_1, with $p_1 + q_1 = 1$, the frequencies of genes G and g; and p_2, q_2, with $p_2 + q_2 = 1$, the frequencies of genes A and a, respectively. As an example, we consider a parental mating Ga × ga [i.e., Double Dom × Double Rec ($G–A– \times ggaa$)] and derive the likelihood function in order to estimate the three functionally independent parameters, p_1, p_2, and λ, if subject to assumptions that the parental population is in equilibrium, the mating is random, there is no selection, penetrance is complete, and the viabilities of genotypes are the same. Table 18.4 represents the model for the complex segregation analysis. Note that for genotypic matings double heterozygote × double homozygote we have to consider the linkage phase, coupling (GA/ga) or repulsion (Ga/gA). We assume that the ratio of coupling to repulsion is $1:1$.

Table 18.4 is self-explanatory. We introduce the notation

$$\frac{2q_1}{1 + q_1} = h_1, \quad \frac{2q_2}{1 + q_2} = h_2. \tag{18.48}$$

Note that $\frac{1}{2}h_1$ and $\frac{1}{2}h_2$ are Snyder's ratios, defined in section 7.13. We have

$$\phi_1 = (1 - h_1)(1 - h_2), \quad \phi_2 = (1 - h_1)h_2,$$
$$\phi_3 = (1 - h_2)h_1, \quad \phi_4 = \phi_5 = \frac{1}{2}h_1h_2. \tag{18.49}$$

Let us now denote, for simplicity, the multinomial probability $P_{st}(r_1, r_2, \ldots, r_k)$ by P_{st}. Provided in each formula, that the appropriate r's are each not equal to zero, and all sum to s, we have

$$P_{s1} = (1)^s, \quad P_{s2} = \frac{s!}{r_1! \, r_2!}(\tfrac{1}{2})^{r_1+r_2}, \quad P_{s3} = \frac{s!}{r_1! \, r_3!}(\tfrac{1}{2})^{r_1+r_3},$$

$$P_{s4} = \frac{s!}{r_1! \, r_2! \, r_3! \, r_4!}(\tfrac{1}{2})^s \lambda^{r_2+r_3}(1 - \lambda)^{r_1+r_4}, \tag{18.50}$$

$$P_{s5} = \frac{s!}{r_1! \, r_2! \, r_3! \, r_4!}(\tfrac{1}{2})^s \lambda^{r_1+r_4}(1 - \lambda)^{r_2+r_3}.$$

Hence

$$P_s(r_1, r_2, r_3, r_4) = \sum_{t=1}^{5} \phi_t P_{st}. \tag{18.51}$$

Substituting observed values $r_{1\alpha}$, $r_{2\alpha}$, $r_{3\alpha}$, $r_{4\alpha}$ for the αth family into (18.51) and multiplying the results from n_s tested families, we obtain the likelihood function $L_s(h_1, h_2, \lambda)$, as defined in (18.47). For evaluating the estimates $\hat{h}_1$, $\hat{h}_2$, $\hat{\lambda}$ from data, a computer program should be used.

Example 18.2. We give now a simple example showing how to construct the likelihood function, $L_s(h_1, h_2, \lambda)$ from the observed data.

Suppose that we observe phenotypes of the offspring in three families, F_1, F_2, F_3, each with $s = 5$ children, within a phenotypic mating Ga × ga.

TABLE 18.4 A segregation model for detection and estimation of linkage and gene frequencies. Parental phenotypic mating, GA × ga

Segregation type, t	Genotype matings	Expected proportion of genotype matings:		Segregation parameter, θ_{lt}				No. of children			
		In population	In a given phenotypic mating, ϕ_t	GA	Ga	gA	ga				
1	$GGAA \times ggaa$	$2p_1^2q_1^2p_2^2q_2^2$	$\dfrac{(1-q_1)(1-q_2)}{(1+q_1)(1+q_2)}$ $= (1-h_1)(1-h_2)$	1	0	0	0				
2	$GGAa \times ggaa$	$4p_1^2q_1^2p_2q_2^3$	$\dfrac{2(1-q_1)q_2}{(1+q_1)(1+q_2)} = (1-h_1)h_2$	$\frac{1}{2}$	$\frac{1}{2}$	0	0				
3	$GgAA \times ggaa$	$4p_1q_1^3p_2^2q_2^2$	$\dfrac{2(1-q_2)q_1}{(1+q_1)(1+q_2)}$ $= (1-h_2)h_1$	$\frac{1}{2}$	0	$\frac{1}{2}$	0				
4	$GA/ga \times ggaa$	$4p_1q_1^3p_2q_2^3$	$\dfrac{2q_1q_2}{(1+q_1)(1+q_2)} = \frac{1}{2}h_1h_2$	$\frac{1}{2}(1-\lambda)$	$\frac{1}{2}\lambda$	$\frac{1}{2}\lambda$	$\frac{1}{2}(1-\lambda)$				
5	$Ga/gA \times ggaa$	$4p_1q_1^3p_2q_2^3$	$\dfrac{2q_1q_2}{(1+q_1)(1+q_2)} = \frac{1}{2}h_1h_2$	$\frac{1}{2}\lambda$	$\frac{1}{2}(1-\lambda)$	$\frac{1}{2}(1-\lambda)$	$\frac{1}{2}\lambda$				
Total		$2p_1p_2q_1^2q_2^2$ $\times [(1+q_1)(1+q_2)]$	1	r_1	r_2	r_3	r_4				

The results are as follows:

Family, F_α	Phenotype of offspring			
	GA	Ga	gA	ga
	r_1	r_2	r_3	r_4
F_1	2	3	0	0
F_2	1	2	2	0
F_3	5	0	0	0.

We simplify the notation slightly. Let F_α denote the αth family ($\alpha = 1, 2, \ldots, n_s$), and defining $0° = 1$, we write

$$C_s(\alpha) = \frac{s!}{r_{1\alpha}! \, r_{2\alpha}! \ldots r_{k\alpha}!}, \quad P_{st}(F_\alpha) = \theta_{1t}^{r_{1\alpha}} \theta_{2t}^{r_{2\alpha}} \cdots \theta_{kt}^{r_{k\alpha}}.$$

Hence

$$P_s(r_{1\alpha}, r_{2\alpha}, \ldots, r_{k\alpha}) = P_s(F_\alpha) = C_s(\alpha) \sum_{t=1}^{m} \phi_t P_{st}(F_\alpha).$$

We now calculate the probabilities $P_5(F_\alpha)$:

$$P_5(F_1) = \frac{5!}{2! \, 3!} [\phi_2(\tfrac{1}{2})^5 + \phi_4(\tfrac{1}{2})^5 \lambda^3(1 - \lambda)^2 + \phi_5(\tfrac{1}{2})^5 \lambda^2(1 - \lambda)^3].$$

Since here $\phi_4 = \phi_5$, we obtain $P_5(F_1) = 10(\tfrac{1}{2})^5 [\phi_2 + \phi_4 \lambda^2(1 - \lambda)^2]$. Substituting for ϕ_2 and ϕ_4 the expressions given in Table 18.4, we obtain $P_5(F_1)$ as a function of h_1, h_2, λ, that is, $P_5(F_1) = P_5(h_1, h_2, \lambda; F_1)$.

In a similar way we calculate

$$P_5(F_2) = \frac{5!}{1! \, 2! \, 2! \, 0!} (\tfrac{1}{2})^5 [\phi_4 \lambda^4(1 - \lambda) + \phi_5 \lambda(1 - \lambda)^4]$$

$$= 15(\tfrac{1}{2})^5 \phi_4 \lambda(1 - \lambda)[\lambda^3 + (1 - \lambda)^3].$$

Finally, for F_3 we have

$$P_5(F_3) = \phi_1 \cdot 1^5 + \phi_2(\tfrac{1}{2})^5 + \phi_3(\tfrac{1}{2})^5 + \phi_4(\tfrac{1}{2})^5(1 - \lambda)^5 + \phi_5(\tfrac{1}{2})^5 \lambda^5$$

$$= (\tfrac{1}{2})^5 \{2^5 \phi_1 + \phi_2 + \phi_3 + \phi_4[\lambda^5 + (1 - \lambda)^5]\}.$$

The likelihood function, $L_5(h_1, h_2, \lambda)$, for these three families is

$$L_5(h_1, h_2, \lambda) = P_5(F_1)P_5(F_2)P_5(F_3).$$

We can also analyze Table 18.4 in a different way. We may separate families into nonsegregating and segregating. Thus the genotypes of segregating families will be $GgAa \times ggaa$; they will now be known except for the phase. The segregating families will be more appropriate to use when the gene frequencies q_1 and q_2 are known and we wish to investigate linkage.

More examples can be found in most of the papers listed in the references at the end of the chapter.

SUMMARY

Models for segregation analysis when there is more than one distinct segregation parameter within a given type of parental phenotypic matings have been constructed. A general form of multinomial likelihood function has been derived, taking into account the contributions of families with different parental genotypic matings. Likelihood functions for common double recessives and for two linked loci have been constructed in order to estimate segregation parameters and the recombination fraction, respectively, and to test appropriate hypotheses. Problems of using family data in estimating gene frequencies have been discussed.

REFERENCES

Barrai, I., Mi, M. P., Morton, N. E., and Yasuda, N., Estimation of prevalence under incomplete selection, *Amer. J. Hum. Genet.* **17** (1965), 221–236.

Dewey, W. J., Barrai, I., Morton, N. E., and Mi, M. P., Recessive genes in severe mental defect, *Amer. J. Hum. Genet.* **17** (1965), 237–256.

Elandt-Johnson, Regina C., Segregation analysis for complex modes of inheritance, *Amer. J. Hum. Genet.* **22** (1970), 129–144.

Elandt-Johnson, Regina C., Complex segregation analysis, II. Multiple classification, *Amer. J. Hum. Genet.* **23** (1971), 17–32.

Hogben, L., The genetic analysis of familial traits, II. Double gene substitutions, with special reference to hereditary dwarfism, *J. Genet.* **25** (1932*a*), 211–240.

Hogben, L., The genetic analysis of familial traits, III. Matings involving one parent exhibiting a trait determined by a single gene substitution with special reference to sex-linked conditions, *J. Genet.* **25** (1932*b*), 293–314.

Mi, M. P., Segregation analysis, *Amer. J. Hum. Genet.* **19** (1967), 313–321.

Morton, N. E., Genetic tests under incomplete ascertainment, *Amer. J. Hum. Genet.* **11** (1959), 1–16.

Morton, N. E., Segregation and linkage. In: *Methodology in Human Genetics*, ed. by J. Burdette, Holden-Day, San Francisco, 1962, pp. 17–52.

Morton, N. E., Models and evidence in human population genetics. In: *Genetics Today*, Vol. 3 (Proceedings of the XI International Congress on Genetics, The Hague, The Netherlands, 1963), ed. by S. J. Geerts, Pergamon Press, Oxford, pp. 935–951.

Morton, N. E., Sequential tests for detection linkage, *Amer. J. Hum. Genet.* **7** (1955), 277–318.

Peritz, E., On some models for segregation analysis, *Ann. Hum. Genet. Lond.* **30** (1966), 183–192.

Smith, C. A. B., Linkage scores and corrections in simple two- and three-generation families, *Ann. Hum. Genet. Lond.* **32** (1968), 127–150.

Steinberg, A. G., Becker, S. W., and Fitzpatrick, T. B., A genetic and statistical study of psoriasis, *Amer. J. Hum. Genet.* **3** (1951), 267–281.

Steinberg, A. G., Becker, S. W., and Fitzpatrick, T. B., A further note on the genetics of psoriasis, *Amer. J. Hum. Genet.* **4** (1952), 373–375.

PROBLEMS

18.1. Construct the segregation model for a simple autosomal dominant when the trait is common.

18.2. Using the data of Example 18.1, show that the hypothesis that psoriasis is a double rare recessive does not fit these data. (*Hint:* Test H_0: $\theta = \frac{1}{16}$, using methods described in section 17.8.)

18.3. Suppose that we wish to estimate gene frequencies of the ABO blood groups from family data and that the selection is through the parents.

(*a*) How many phenotypic matings do you distinguish?

(*b*) Let us confine ourselves to matings A × A. Construct the model for segregation analysis.

(*c*) Divide the families into segregating and nonsegregating and derive the likelihood function for estimating gene frequencies. (Note that segregation parameters are known.)

18.4. Repeat (*b*) and (*c*) from Problem 18.3 for matings A × AB.

18.5. Derive the segregation model for detection and estimation of the recombination fraction, λ, and the gene frequencies of two loci, G and A, when the parental phenotypic mating is GA × GA. (*Hint:* Compare section 18.9 and assume that the matings, $GA/ga \times GA/ga$, $GA/ga \times GA/gA$, and $Ga/gA \times Ga/gA$, are in the ratios $1:2:1$.)

CHAPTER 19

Histocompatibility Testing

19.1. IMMUNOGENETIC MECHANISM OF HISTOCOMPATIBILITY

When an individual (recipient) needs a tissue or an organ transplant, the first problem is to find a suitable donor. This is an individual whose tissue will not be rejected when it is grafted to the recipient. We say that such a donor is *histocompatible* (i.e., compatible with respect to tissue) with the recipient. The methods of selecting a suitable (matching) donor are called *histocompatibility testing.*

When the donor and the recipient are two unrelated individuals, the graft is, in most cases, rejected. But if the donor and the recipient are two identical twins or are two animals from the same homozygous strain, the graft is usually accepted. On the other hand, if, after the rejection of a graft from a histoincompatible donor a second graft from the same donor is placed on the recipient, it will be rejected more rapidly than the first one — the recipient has become "immunized" against the foreign tissue.

These two phenomena — histocompatibility in genetically identical individuals and accelerated rejection of the second graft in others — indicate that the mechanism of tissue rejection is as follows:

(*a*) *Genetically controlled* — the fate of transplanted tissue depends on the genetic constitution of the donor-recipient pair.

(*b*) In the basic concepts *immunological* — the graft from the donor, transplanted to the recipient, "releases" specific substances which we will call *transplantation antigens* or *histocompatibility antigens.* If the antigen(s) present in the donor's tissue are absent in the recipient's tissue, certain cells of the recipient, located mostly in lymph nodes and spleen, produce specific *antibodies* against the donor's antigens. The first graft is rejected, and in addition the recipient is sensitized against the donor's antigen(s). Thus, the second graft from the same donor, or from any other donor who possesses at least one antigen against which the recipient has been immunized, is rejected in accelerated fashion.

521

The situation is analogous to that in blood group systems, where the fate of transfusion can be determined in terms of relationships among three factors: gene, antigen, antibody. Although the immunogenetic mechanism of tissue transplantation is also determined by the gene-antigen-antibody relationship, it does not seem, at present, to be so simple as in the case of blood groups. The main difficulty is that not all transplantation antigens and their roles in different tissues are precisely known. It appears that some red cell antigens which are blood antigens are also responsible for the fate of transplanted tissue. For example, the *H*-2 locus in mice, the *Ag-B* locus in rats, and the *B* locus in chickens, which are blood group loci, are also important histocompatibility loci. There is no doubt that ABO blood antigens in man are relevant to histocompatibility [Ceppellini et al. (1965, 1966) and Rogers (1957)]. But the most important histocompatibility locus in man seems to be connected with leucocyte antigens and is called the *HL-A* system (Human Leucocyte – locus *A*). On the other hand, there may be systems as yet unknown which are also relevant to transplantation.

19.2. HISTOCOMPATIBILITY MODELS. GENERAL ASSUMPTIONS

Histocompatibility testing is a new, but very important, field in medicine. A great deal of work has been done, and much more will be required, on the biochemical and physiological aspects of transplantation. In this chapter, we only sketch the basic ideas, concepts, and definitions and show how probabilistic models can be applied to transplantation. We will construct a few probability models and experimental designs which may be helpful in histocompatibility testing. As has been mentioned, the problem is not very easy at present, and we shall consider these models as only "first approximations." Our models will be based on certain general assumptions.

1. First, we recall the definition of an antigen. An antigen is usually defined as a substance which stimulates the production of an antibody. In this context, we shall understand by "antigen" the *antigenic product* associated with an allele. (The antigen controlled by an allele A_i will also be denoted by A_i.) A substance present in the total antigenic product of an allele, which is able to react with a unispecific antibody, will be called an *antigenic determinant* or an *antigenic specificity*. Thus, an antigen may consist of several antigenic specificities and may react with several antibodies. The sequence of reactions, which ultimately results in graft destruction, can be written in the simplified form

$$\text{allele} \rightarrow \text{antigen} \overset{\nearrow}{\underset{\searrow}{\rightarrow}} \text{antibodies.} \qquad (19.1)$$

If, for convenience, we call the total spectrum of antibodies which react with a given antigen, A_i, just "*antibody* against A_i" and denote it by $\overline{A}_i$, then our model (19.1) can be simply written as

$$\text{allele} \rightarrow \text{antigen} \rightarrow \text{antibody}. \tag{19.2}$$

2. We assume that the histocompatibility alleles are *codominant*. However, some antigens react so weakly that they may be assumed to be "antigenically inactive," and by analogy with the *O* allele in the *ABO* blood group system we will call them recessives. In fact, if the time after transplantation is long enough, the presence of such an allele in the donor and its absence in the recipient may lead ultimately to graft rejection.

3. It has been demonstrated in mice and rats that there exist some "factors" (the authors did not call them antigens) associated with the Y chromosome [Eichwald and Silmser (1955) and Billingham et al. (1962)] and with the X chromosome [Bailey (1963)]. The immune response to "X-factor" is rather weak, but the "Y-factor" appears to be of some importance in certain strains. In estimating the number of autosomal loci, the safe procedure will be to graft female-to-female and, perhaps, female-to-male.

4. In all models for randomly mating populations (e.g., human beings) we assume genetic equilibrium with respect to all histocompatibility loci.

Some other assumptions for different models will be specified later.

It should be mentioned that only tissue transplantation will be discussed here. In fact, experimental work in histocompatibility testing started with the transplantation of several kinds of tumors in mice. Since several loci in mice control the growth of tumors as well as the growth of tissue, the *histocompatibility antigens* are defined as products associated with histocompatibility genes. It is not yet precisely known whether the same antigens (or only parts of the same antigens) are responsible for tissue and tumor growth. Thus *transplantation antigen* is defined as the antigenic product associated with a histocompatibility gene which is particularly responsible for graft rejection. It might be identical with the corresponding histocompatibility antigen, might be part of it, or might be a separate structure.

The reader interested in the earliest experiments with tumor transplantation is referred to *Biology of the Laboratory Mouse*, ed. by G. D. Snell, Blakiston, Philadelphia, 1941.

19.3. SOME TERMINOLOGY IN TISSUE TRANSPLANTATION

It will be useful to introduce first a few definitions commonly used in the literature on tissue transplantation.

Autograft is a graft transferred from one place to another on the same individual.

Orthotopic graft is a graft which is placed on the host in anatomically the same position as that previously held in the donor.

Isogenic or *isogeneic* means genetically identical.

Allogenic or *allogeneic* indicates similarity of origin.

Isograft is a graft between genetically identical individuals (of the same species and the same genetic constitution). Grafts between members of an inbred line, between F_1 hybrids of two inbred lines, or between identical twins are isografts. Synonyms are *isogenic, isogeneic,* or *syngeneic graft.*

Allograft denotes a graft between genetically different individuals of the same species; for example, grafts between two strains of mice or between human beings are allografts. Synonym is *allogenic graft.* In the earlier literature (before 1962) the term *homograft* was commonly used.

Xenograft is a graft from a different species, for example, from a dog to a monkey. A term previously used was *heterograft.*

Most of the experimental work on the immunogenetics of transplantation has been done on mice. A few definitions, applied to strains obtained in a special way for such studies, will be useful.

Isogenic strains are strains which are genetically identical.

Coisogenic strains are strains genetically identical except for a difference at a single locus. Coisogenicity is not always attainable and for particular purposes may be unnecessary.

Congenic strains are strains which are identical with respect to certain groups of loci except for a difference at a single locus in such a group.

Strains which are identical with respect to all histocompatibility loci are called *isohistogenic*, and strains which are identical with respect to all histocompatibility loci except for a difference at a single such locus can be called *coisohistogenic*. According to the definition, coisohistogenic strains are a special case of congenic strains.

The term *congenic resistant* (CR) is used to denote a strain which has the same histocompatibility loci as the original strain except at the locus in which the allele of susceptibility to a certain tumor, present in the original strain, is replaced by the allele of resistance to this tumor. (More precisely, this is a "coisohistogenic resistant.")

The method of obtaining CR strains was originally devised by G. D. Snell (1948) and is described in the papers mentioned above and also in an article by Snell and Stimpfling (1966). If, for instance, strain A is susceptible to a certain tumor and strain B is resistant to this tumor, then the congenic resistant strain is denoted by A.B. It is derived from A and has the same background, except at the locus in which the allele of susceptibility, present in A, has been replaced by the allele of resistance present in B. Congenic resistant strains can also be obtained with respect to two or more loci.

Recently, several CR strains have been obtained in various laboratories and have been used in studies of the antigenic properties of *H* loci in mice.

19.4. IMMUNOLOGICAL UNRESPONSIVENESS AND TOLERANCE

It is generally assumed that graft rejection is a consequence of the formation of antibodies and their reaction with the foreign antigen(s). There are ways in which the immune response can be avoided, reduced, or suppressed, and sometimes, in certain circumstances, the graft may ultimately be accepted.

Reduction or suppression can be effected by using chemical agents (antimetabolites, alkalating agents, or antitumor agents). More than one hundred different immunosuppressive drugs have been used in order to delay graft rejection (e.g., in kidney transplantation). The best known are Azathioprine, Actinomycin C, and Imuran. The chemical agents act by blocking antibody formation and so delaying the destruction of cells [Schwartz (1965)].

Immunological unresponsiveness is usually observed when the drugs are applied. Graft survival is prolonged under immunosuppressive therapy, and in some cases renal grafts have survived even if the recipient has been withdrawn from the therapy [see, for discussion, Murray et al. (1964)].

McKhann (1964) has shown that a long-lasting unresponsiveness can be induced in certain strains of *adult* mice by injection of large doses of spleen cells; low doses elicit immunity. The term *immunological paralysis* is used when unresponsiveness is induced in adult life, using large doses of antigen(s).

Enhancement is another phenomenon in which a successful establishment of tumor or delayed rejection of allogenic tissue is observed as a consequence of the presence of a specific antiserum in the host. It can be evoked by pretreatment (of certain strains of mice) with either lyophylized tissue or fresh homogenates obtained from members of donor strains.

The term "immunological tolerance" refers, not to the unresponsiveness obtained by immunosuppressive therapy, to immunological paralysis, or to enhancement, but to reduction of immune response acquired as a result of special conditions in the very earliest stage of life [Owen (1959)]. "*Immunological tolerance* is a specific state of unresponsiveness to antigen(s) in adult life as a consequence of exposure to these antigens *in utero* during *neonatal* life. Especially notable in this connection is failure to give allograft response, so that allogeneic tissue graft may be retained for an indefinite period" [Humphrey and White (1963)].

Specific immunological tolerance may be induced by injection of large amounts of antigen(s) during embryonic, neonatal, or early postnatal life. Animals so treated fail to respond either for the whole life-span (*complete tolerance*), or at least for a certain period (*partial tolerance*), to the antigen(s) in question but respond normally to unrelated antigens [Humphrey and White (1963)].

19.5. TYPES OF DESIGNS FOR HISTOCOMPATIBILITY TESTING IN LABORATORY ANIMALS

The phenomenon of tissue rejection has been (and still is) extensively studied in highly inbred strains and their progeny, of laboratory animals such as mice or rats. We restrict our consideration to constructing genetic and immunogenetic models for tissue rejection. Two main problems are of interest to us:

(1) How many loci are involved in histocompatibility?
(2) How many loci may be responsible for early graft rejection?

Typical experiments for studing these problems are those based on skin grafting. Generally, two main types of experimental designs are used in skin grafting on laboratory animals:

(i) *Single-grafting designs*, in which only *one graft* from a donor is placed on a recipient. Typical experiments are the following: grafting from one or another parental strain to the F_2 progeny; grafting within the F_2, B_1, or B_2 populations;* grafting from an related donor (e.g., sib → sib); reciprocally exchanged grafts between two unrelated or related individuals.

(ii) *Double-grafting designs*, in which *two grafts*, each from a different donor, are placed simultaneously on the same recipient. Here again the triplets, of one recipient and two donors, can be unrelated (e.g., from both parental strains to the F_2 progeny) or related (e.g., three siblings).

19.6. ESTIMATION OF THE NUMBER OF HISTOCOMPATIBILITY LOCI IN LABORATORY ANIMALS. SINGLE, ONE-WAY $(D \rightarrow R)$ GRAFTING DESIGNS

In planning experiments from which we wish to estimate the number of histocompatibility loci, we should take into account the following aspects.

---

* By B_1 we denoted the offspring of $F_1 \times Pt_1$, and by B_2 the offspring of $F_1 \times Pt_2$.

(a) It has been observed that in nonimmunized animals no rejection occurs before the 9th postoperative day. However, the immune response depends on the donor-recipient genetic relationship. It has been observed in some congenic strains that rejection of grafts is completed between the 9th and 11th (in some strains, between the 9th and 13th) days after grafting, whereas in other strains 200–300 days are required [Graff et al. (1966a, 1966b)]. This is due to differences in immune responses: the first is "strong" whereas the other is "weak" (compare section 19.11). Therefore, if the experiment is completed after a certain time, in fact, the number of loci which are "active" up to this time can be estimated. It is now customary to wait 200 days before the experiment is completed, that is, before the rejected and non-rejected grafts are counted.

(b) Using inbred strains and their generations, it is only possible to estimate the number of loci with respect to which these two strains differ.

(c) Applying skin grafts, we estimate the number of loci controlling the growth of skin tissue. The "skin loci" may not be the same for other tissues.

We now describe an experimental design in which donor-recipient pair [briefly, (D, R)] is selected, and a graft from the donor is placed on the recipient. We will call this a *single, one-way grafting design* and use for it the notation $D \to R$.

Let us first consider a single histocompatibility locus with two alleles, A_1 and A_2. Let x be a random variable* determining the histoincompatibility of a donor (D) with a recipient (R) and taking the values:

$$x = \begin{cases} 0 & \text{if } D \text{ is a compatible donor with } R; \\ 1 & \text{if } D \text{ is an incompatible donor with } R. \end{cases} \tag{19.3}$$

Table 19.1 represents a sample space for histocompatibility in a single-grafting design, $D \to R$, and Table 19.2 the general case of joint genotype distribution of the pairs (R, D).

TABLE 19.1 Histocompatibility model for grafting $D \to R$

R \ D	A_1A_1	A_1A_2	A_2A_2
A_1A_1	0	1	1
A_1A_2	0	0	0
A_2A_2	1	1	0

TABLE 19.2 Joint genotype distribution of pairs (R, D)

R \ D	A_1A_1	A_1A_2	A_2A_2
A_1A_1	p_{11}	p_{12}	p_{13}
A_1A_2	p_{21}	p_{22}	p_{23}
A_2A_2	p_{31}	p_{32}	p_{33}

* Note that in this chapter we use lower case letters for random variables.

We find

$$\Pr\{x = 0\} = p_{11} + p_{21} + p_{22} + p_{23} + p_{33}, \qquad (19.4)$$

and

$$\Pr\{x = 1\} = 1 - \Pr\{x = 0\} = p_{12} + p_{13} + p_{31} + p_{32}. \qquad (19.5)$$

Let us consider populations of donors and recipients with respect to n *independent* and *autosomal* loci, each with codominant alleles. Let $\Pr\{+\}$ be the probability that D is compatible with R with respect to all loci, and $\Pr\{-\}$ be the probability that D is incompatible with R (we notice that D is incompatible with R if he is incompatible with respect to at least one locus). Thus

$$\Pr\{+\} = [\Pr\{x = 0\}]^n \qquad (19.6)$$

and

$$\Pr\{-\} = 1 - \Pr\{+\} = 1 - [\Pr\{x = 0\}]^n. \qquad (19.7)$$

Let us now consider some particular designs.

19.6.1. Grafting from Parental (Pt) Strains to F_2 Progeny $(Pt_1 \to F_2 \text{ or } Pt_2 \to F_2)$

These are well-known experiments in which a recipient is selected at random from F_2 and a donor is chosen from parental strain Pt_1 or Pt_2. In this case, the distributions of donors and recipients are independent. For instance, in grafting $Pt_1 \to F_2$, where Pt_1 is the parental strains with all individuals A_1A_1, we have $\Pr\{A_1A_1\} = 1$, while the genotype distribution of F_2 is $\Pr\{A_1A_1\} = \frac{1}{4}$, $\Pr\{A_1A_2\} = \frac{1}{2}$, and $\Pr\{A_2A_2\} = \frac{1}{4}$. Since selection of donors and recipients is independent, the joint distribution in Table 19.2 is the product of marginal distributions of donors and recipients. Thus, we have $p_{11} = \frac{1}{4}$, $p_{21} = \frac{2}{4}$, $p_{31} = \frac{1}{4}$, and the other probabilities in Table 19.2 are equal to zero.

The same values apply when grafting $Pt_2 \to F_2$, where Pt_2 individuals are A_2A_2. Hence,

$$\Pr\{x = 0 \mid Pt_i \to F_2\} = \tfrac{3}{4}, \quad \Pr\{x = 1 \mid Pt_i \to F_2\} = \tfrac{1}{4}, \qquad (19.8)$$

and

$$\Pr\{+ \mid Pt_i \to F_2\} = (\tfrac{3}{4})^n, \quad \Pr\{- \mid Pt_i \to F_2\} = 1 - (\tfrac{3}{4})^n, \qquad (19.9)$$

for $i = 1, 2$.

Suppose that N_1 (surgically correct) grafts from Pt_1 to F_2 were performed and that K_1 grafts were not rejected (were compatible). Thus, the random variable K_1 is binomially distributed with the binomial parameter $P = (\tfrac{3}{4})^n$ or, in other words,

$$\Pr\{K_1\} = \binom{N_1}{K_1} P^{K_1}(1 - P)^{N_1 - K_1} = \binom{N_1}{K_1}(\tfrac{3}{4})^{nK_1}[1 - (\tfrac{3}{4})^n]^{N_1 - K_1}. \qquad (19.10)$$

From formula (11.24) we obtain the ML estimator of P:

$$\hat{P} = \frac{K_1}{N_1} \quad \text{or} \quad (\tfrac{3}{4})^{\hat{n}} = \frac{K_1}{N_1}. \tag{19.11}$$

Hence

$$\hat{n} = \frac{\log (K_1/N_1)}{\log \tfrac{3}{4}}. \tag{19.12}$$

Similarly, if N_2 grafts from $Pt_2 \rightarrow F_2$ were performed and K_2 were not rejected, the ML estimator for the number of loci (n) is

$$\hat{n} = \frac{\log(K_2/N_2)}{\log \tfrac{3}{4}}. \tag{19.13}$$

Of course, we can use the joint likelihood function for the combined experiment and find the combined estimator which is, in this case,

$$\hat{n} = \frac{\log[(K_1 + K_2)/(N_1 + N_2)]}{\log \tfrac{3}{4}}. \tag{19.14}$$

(Compare section 12.11.)

Example 19.1. One of the first experiments in which parental grafts were applied to F_2 progeny was performed by Barnes and Krohn (1957). Strains A and CBA were used. A period of only 100 days was taken as experiment completion time, and it is not clear whether the effect of the "Y-factor" was taken into account. Hence the estimate of the number of loci, $\hat{n}$, will probably be too small, but we use these data for illustration. (See Table 19.3.)

The pooled estimate of P is $\hat{P} = \frac{3}{274} = 0.01095$; hence

$$\hat{n} = (\log 0.01095)/(\log 0.75) = 15.69 \doteq 16.$$

Using $\hat{P}$ as a "true" value of P, we calculate the "expected" frequencies; for example, $E(K_1) \doteq N_1\hat{P} = 120 \times 0.01095 = 1.31$. The corresponding "expected" values are given in parentheses in Table 19.3. Although the

TABLE 19.3   Frequencies of compatible $(+)$ and incompatible $(-)$ parental grafts to F_2 in mice

Graft survival $D \rightarrow R$	+	−	Total	X^2
Strain A $\rightarrow F_2$	2 (1.31)	118 (118.69)	120	0.3674
Strain CBA $\rightarrow F_2$	1 (1.69)	153 (152.31)	154	0.2848
Total	3	271	274	0.6522

X^2-test gives a good fit, we do not feel that it is meaningful since the frequencies in the classes of compatible grafts were very low.

19.6.2. Graftings $F_1 \rightarrow F_2$ and $F_2 \rightarrow F_2$

When grafting $F_1 \rightarrow F_2$, we notice that $p_{12} = p_{32} = \frac{1}{4}$ and $p_{22} = \frac{1}{2}$, with other p_{ij} equal to zero. We have

$$\Pr\{x = 0 \mid F_1 \rightarrow F_2\} = p_{22} = \tfrac{1}{2}, \quad \Pr\{x = 1 \mid F_1 \rightarrow F_2\} = \tfrac{1}{2}, \quad (19.15)$$

and thus

$$\Pr\{+ \mid F_1 \rightarrow F_2\} = (\tfrac{1}{2})^n, \quad \Pr\{- \mid F_1 \rightarrow F_2\} = 1 - (\tfrac{1}{2})^n. \quad (19.16)$$

When grafting $F_2 \rightarrow F_2$, the joint genotype distribution of pairs (D, R) will be the product of the genotype distributions of F_2 (see Table 19.4).

TABLE 19.4 Joint genotype distribution
of pairs (D, R) in F_2

R \ D	$\frac{1}{4}A_1A_1$	$\frac{2}{4}A_1A_2$	$\frac{1}{4}A_2A_2$
$\frac{1}{4}A_1A_1$	$\frac{1}{16}$	$\frac{2}{16}$	$\frac{1}{16}$
$\frac{2}{4}A_1A_2$	$\frac{2}{16}$	$\frac{4}{16}$	$\frac{2}{16}$
$\frac{1}{4}A_2A_2$	$\frac{1}{16}$	$\frac{2}{16}$	$\frac{1}{16}$

It should be noted that this table can be considered as a matrix $\mathbf{H}$ (see the definition in section 7.7). Therefore

$$\Pr\{x = 0 \mid F_2 \rightarrow F_2\} = \Pr\{x = 0 \mid \mathbf{H}\} = \tfrac{1}{16} + \tfrac{2}{16} + \tfrac{4}{16} + \tfrac{2}{16} + \tfrac{1}{16} = \tfrac{5}{8}, \quad (19.17)$$

$$\Pr\{x = 1 \mid F_2 \rightarrow F_2\} = \tfrac{3}{8},$$

and hence

$$\Pr\{+ \mid F_2 \rightarrow F_2\} = (\tfrac{5}{8})^n, \quad \Pr\{- \mid F_2 \rightarrow F_2\} = 1 - (\tfrac{5}{8})^n. \quad (19.18)$$

Some other populations, such as offspring of backcrosses, can be also considered for such experiments.

19.7. REQUIRED SAMPLE SIZE

We noticed in Example 19.1 that among the grafts performed only three were not rejected before the 100th postoperative day. One might ask this question: how large should the sample size be if we want to observe *at least one* graft which is not rejected, after a certain postoperative time, with probability not less than $1 - \alpha$ (where $1 - \alpha = 0.90$ or 0.95 or 0.99, say)?

This problem was discussed in section 11.10.3 (iii). Using formula (11.66) with N substituted for n, we find that the minimum sample size should not be less than

$$\frac{\log \alpha}{\log(1 - P)}. \tag{19.19}$$

In the mouse, 15 loci have already been identified by serological methods. Calculating P for $n = 15$, and taking $\alpha = 0.10$, we obtain for different types of experiments:

$$\begin{aligned}
\text{Pt} \to \text{F}_2: \quad & P = (\tfrac{3}{4})^{15} = 0.01336, & N &> 171; \\
\text{F}_1 \to \text{F}_2: \quad & P = (\tfrac{1}{2})^{15} = 0.00003052, & N &> 100{,}000; \\
\text{F}_2 \to \text{F}_2: \quad & P = (\tfrac{5}{8})^{15} = 0.0008674, & N &> 2{,}600.
\end{aligned} \tag{19.20}$$

From these figures we see that the experiments $\text{F}_1 \to \text{F}_2$ and $\text{F}_2 \to \text{F}_2$ are, in practice, impossible because they require enormous sample sizes. In planning experiments the necessary size of the sample should be taken into account.

19.8. GRAFTING AMONG RELATIVES

Grafts can also be exchanged among members of the same litter.

Let F_2 be the randomly mating population of the parents.

(i) *Parent → Child.* For a single locus, using the joint genotype distribution parent-child given in Table 7.4 with $p = q = \tfrac{1}{2}$ (matrix **M**), we find that the probabilities in Table 19.2 are as follows: $p_{13} = p_{31} = 0$, $p_{22} = \tfrac{1}{4}$, and all the rest are each equal to $\tfrac{1}{8}$.

$$\left.\begin{aligned}
\Pr\{x = 0 \mid \text{Parent} \to \text{Child}\} = \Pr\{x = 0 \mid \mathbf{M}\} = \tfrac{3}{4}, \\
\Pr\{x = 1 \mid \text{Parent} \to \text{Child}\} = \tfrac{1}{4},
\end{aligned}\right\} \tag{19.21}$$

so that

$$\Pr\{+ \mid \text{Parent} \to \text{Child}\} = (\tfrac{3}{4})^n, \quad \Pr\{- \mid \text{Parent} \to \text{Child}\} = 1 - (\tfrac{3}{4})^n. \tag{19.22}$$

Of course, the same results hold when grafting Child → Parent. It is worthwhile noting that these results are also identical with those when grafts from the parental strain are placed on F_2.

(ii) Sib → Sib. In this case, we may use the joint distribution of sib-sib pairs given in Table 7.12 (matrix **B**) or we may apply formula (7.78). Let us use (7.78). The diagonal matrix, **D**, when the arrangement of genotypes is as in Table 19.1, is

$$\mathbf{D} = \begin{bmatrix} \frac{1}{4} & 0 & 0 \\ 0 & \frac{1}{2} & 0 \\ 0 & 0 & \frac{1}{4} \end{bmatrix} \tag{19.23}$$

and we find

$$\Pr\{x = 0 \mid \mathbf{D}\} = 1, \quad \Pr\{x = 1 \mid \mathbf{D}\} = 0. \tag{19.24}$$

Recalling that $\Pr\{x = 0 \mid \mathbf{M}\} = \frac{3}{4}$ given by (19.21) and $\Pr\{x = 0 \mid \mathbf{H}\} = \frac{5}{8}$ given by (19.17), we obtain

$$\Pr\{x = 0 \mid \text{Sib} \to \text{Sib}\} = \Pr\{x = 0 \mid \mathbf{B}\}$$
$$= \tfrac{1}{4}\Pr\{x = 0 \mid \mathbf{D}\} + \tfrac{1}{2}\Pr\{x = 0 \mid \mathbf{M}\} + \tfrac{1}{4}\Pr\{x = 0 \mid \mathbf{H}\}$$
$$= \tfrac{1}{4} \cdot 1 + \tfrac{1}{2} \cdot \tfrac{3}{4} + \tfrac{1}{4} \cdot \tfrac{5}{8} = \tfrac{25}{32} = 0.7812. \tag{19.25}$$

Hence,

$$\Pr\{+ \mid \text{Sib} \to \text{Sib}\} = (\tfrac{25}{32})^n, \quad \text{and} \quad \Pr\{- \mid \text{Sib} \to \text{Sib}\} = 1 - (\tfrac{25}{32})^n. \tag{19.26}$$

Two remarks may be useful.

1. We notice that among the one-way single-grafting designs, discussed above, the experiments with sib-sib pairs seem to be most efficient since $\Pr\{x = 0\} = 0.7812$ is the largest value of this probability, so that the required minimal sample size is the smallest.

2. On the other hand, in view of what has been said about the sampling of related pairs, only *one pair* of siblings *per litter* should be used in the experiment (see section 12.9). This may not be economical in some situations and should be taken into consideration when planning experiments.

19.9. RECIPROCAL GRAFTING DESIGNS $(D_1 \rightleftharpoons D_2)$

In reciprocal grafting designs, a pair of donors, D_1 and D_2, say, is selected from a certain population and grafts are exchanged between the members of the pair. We denote this briefly as $D_1 \rightleftharpoons D_2$. Let x_1 and x_2 be random variables determining histoincompatibility when grafting $D_1 \to D_2$ and $D_2 \to D_1$, respectively, each taking values 0, 1 as defined in (19.3). Table 19.5 represents the joint sample space for the variables (x_1, x_2), at a single locus, for reciprocal grafting $D_1 \rightleftharpoons D_2$.

TABLE 19.5 Histocompatibility model for
reciprocal grafting $(D_1 \rightleftharpoons D_2)$. Single locus

D_2 \ D_1	A_1A_1	A_1A_2	A_2A_2
A_1A_1	00	10	11
A_1A_2	01	00	01
A_2A_2	11	10	00

Using again, the genotype distribution given in Table 19.2, we can evaluate the following probabilities:

(i) The probability, $\Pr\{0, 0 \mid D_1 \rightleftharpoons D_2\}$, that grafts in both directions are compatible, that is,

$$\Pr\{x_1 = 0, x_2 = 0 \mid D_1 \rightleftharpoons D_2\} = \Pr\{0, 0 \mid D_1 \rightleftharpoons D_2\} = p_{11} + p_{22} + p_{33}.$$
(19.27)

(ii) The probability, $\Pr\{0, 1 \mid D_1 \rightleftharpoons D_2\}$, that a graft from D_1 is compatible but that from D_2 incompatible, that is,

$$\Pr\{0, 1 \mid D_1 \rightleftharpoons D_2\} = p_{21} + p_{23}.$$
(19.28)

It should be noticed that

$$\Pr\{0, 1 \mid D_1 \leftrightharpoons D_2\} = \Pr\{0 \mid D_1 \rightarrow D_2\} - \Pr\{0, 0\},$$
(19.29)

where $\Pr\{0 \mid D_1 \rightarrow D_2\}$ corresponds to $\Pr\{x_1 = 0\}$ defined in (19.4).

(iii) Similarly, the probability, $\Pr\{1, 0\}$, that a graft from D_2 is compatible, but one from D_1 is incompatible, that is,

$$\Pr\{1, 0 \mid D_1 \leftrightharpoons D_2\} = \Pr\{0 \mid D_2 \rightarrow D_1\} - \Pr\{0, 0 \mid D_1 \leftrightharpoons D_2\} = p_{12} + p_{32}.$$
(19.30)

(iv) Finally, the probability, $\Pr\{1, 1 \mid D_1 \leftrightharpoons D_2\}$, that grafts in both directions are incompatible, that is,

$$\Pr\{1, 1 \mid D_1 \leftrightharpoons D_2\} = 1 - \Pr\{0, 1 \mid D_1 \leftrightharpoons D_2\} - \Pr\{1, 0 \mid D_1 \leftrightharpoons D_2\}$$
$$- \Pr\{0, 0 \mid D_1 \leftrightharpoons D_2\} = p_{13} + p_{33}. \quad (19.31)$$

Now let $\Pr\{+, + \mid D_1 \leftrightharpoons D_2\}$ be the probability that the grafts in both directions are compatible with respect to n histocompatibility *loci*. Assuming that the loci are *independent*, we have

$$\Pr\{+, + \mid D_1 \leftrightharpoons D_2\} = [\Pr\{0, 0 \mid D_1 \leftrightharpoons D_2\}]^n.$$
(19.32)

A graft from a donor is compatible if it is compatible with respect to all histocompatibility loci and is incompatible if it is incompatible with respect at least one locus. Therefore, the probability, $\Pr\{+, - \mid D_1 \leftrightarrows D_2\}$, that a graft from $D_1 \to D_2$ is not rejected whereas one from $D_2 \to D_1$ is rejected is

$$\Pr\{+, - \mid D_1 \leftrightarrows D_2\} = [\Pr\{0 \mid D_1 \to D_2\}]^n - [\Pr\{0, 0 \mid D_1 \leftrightarrows D_2\}]^n.$$
(19.33)

Similarly,

$$\Pr\{-, + \mid D_1 \leftrightarrows D_2\} = [\Pr\{0 \mid D_2 \to D_1\}]^n - [\Pr\{0, 0 \mid D_1 \leftrightarrows D_2\}]^n.$$
(19.34)

Finally,

$$\Pr\{-, - \mid D_1 \leftrightarrows D_2\} = 1 - \Pr\{+, - \mid D_1 \leftrightarrows D_2\}$$
$$- \Pr\{-, + \mid D_1 \leftrightarrows D_2\} - \Pr\{+, + \mid D_1 \leftrightarrows D_2\}. \quad (19.35)$$

The notation used in formulae (19.34) and (19.35) is obvious.

19.9.1. Reciprocal Grafting $F_2 \leftrightarrows F_2$

We can evaluate the probabilities given in (19.28)–(19.35) for different, related or unrelated pairs, (D_1, D_2). As an example we can consider an experiment in which each pair (D_1, D_2) is selected randomly from the F_2 generation. Using Table 19.4, we calculate

$$\left.\begin{array}{l}
\Pr\{0, 0 \mid F_2 \leftrightarrows F_2\} = \frac{6}{16} = \frac{3}{8}, \\[6pt]
\Pr\{0, 1 \mid F_2 \leftrightarrows F_2\} = \Pr\{1, 0 \mid F_2 \leftrightarrows F_2\} = \frac{4}{16} = \frac{1}{4}, \\[6pt]
\Pr\{1, 1 \mid F_2 \leftrightarrows F_2\} = \frac{1}{8}.
\end{array}\right\} \quad (19.36)$$

From (19.16) we have $\Pr\{0 \mid F_2 \to F_2\} = \frac{5}{8}$. Thus, for n independent histocompatibility loci the probabilities of compatible, $\Pr\{+, + \mid F_2 \leftrightarrows F_2\}$, incompatible, $\Pr\{+, - \mid F_2 \leftrightarrows F_2\}$, etc., grafts are as given in Table 19.6.

TABLE 19.6   Joint probabilities of compatible and incompatible grafts exchanged reciprocally ($F_2 \leftrightarrows F_2$). n independent loci

D_2 \ D_1	$+$	$-$	Total
$+$	$\left(\frac{3}{8}\right)^n$	$\left(\frac{5}{8}\right)^n - \left(\frac{3}{8}\right)^n$	$\left(\frac{5}{8}\right)^n$
$-$	$\left(\frac{5}{8}\right)^n - \left(\frac{3}{8}\right)^n$	$1 - 2 \times \left(\frac{5}{8}\right)^n + \left(\frac{3}{8}\right)^n$	$1 - \left(\frac{5}{8}\right)^n$
Total	$\left(\frac{5}{8}\right)^n$	$1 - \left(\frac{5}{8}\right)^n$	1

19.9.2. Estimation of the Number of Loci, n

In the sample of N grafts, one can observe N_{00} of $(+, +)$, N_{01} of $(+, -)$, N_{10} of $(-, +)$, and N_{11} of $(-, -)$, with $N_{00} + N_{01} + N_{10} + N_{11} = N$. Thus N_{00}, N_{01}, N_{10}, and N_{11}, considered as random variables, are multinomially distributed with parameters as given in Table 19.6.

We should notice that, when the two donors are unrelated or are a sib-sib pair, we usually do not distinguish which one is D_1 and which one D_2. Thus the classes $(+, -)$ and $(-, +)$ are pooled together, and we count the number of pairs in which one donor's graft is accepted and the other's graft is rejected, that is, $N_{01}' = N_{01} + N_{10}$. Then, there will be only three classes into which the joint responses can be classified. On the other hand, when the donors are, for instance, mother and child or uncle and nephew, the classes $(+, -)$ and $(-, +)$ are distinct.

The likelihood function for both situations is straightforward to construct. For instance, if the classes $(+, -)$ and $(-, +)$ are distinct, the likelihood function, L, is proportional to

$$[\Pr\{+, +\}]^{N_{00}}[\Pr\{+, -\}]^{N_{01}}[\Pr\{-, +\}]^{N_{10}}[\Pr\{-, -\}]^{N_{11}}.$$

19.10. DOUBLE-GRAFTING DESIGNS $(D_1 \to R \leftarrow D_2)$

In double-grafting designs, two grafts, each from a different donor (D_1 and D_2, say), are placed *simultaneously* on the same recipient, R. We denote this briefly as $D_1 \to R \leftarrow D_2$. Let y_1 and y_2 be random variables determining the histoincompatibility of D_1 and D_2 respectively, with R, each taking the values of 0 (if $D \to R$ compatible) and 1 (if $D \to R$ incompatible) at a single locus. Table 19.7 represents a sample space on which the joint variables (y_1, y_2) are defined.

TABLE 19.7 Histocompatibility model for double grafting $(D_1 \to R \leftarrow D_2)$. Single locus

D_1 $R \leftarrow$ D_2	A_1A_1			A_1A_2			A_2A_2		
	A_1A_1	A_1A_2	A_2A_2	A_1A_1	A_1A_2	A_2A_2	A_1A_1	A_1A_2	A_2A_2
A_1A_1	00	01	01	10	11	11	10	11	11
A_1A_2	00	00	00	00	00	00	00	00	00
A_2A_2	11	11	10	11	11	10	01	01	00

In the general case, the triplets $(D_1 \to R \leftarrow D_2)$ can be related as, for instance (mother $\to$ sib$_1$ $\leftarrow$ sib$_2$) or (sib$_2$ $\to$ sib$_1$ $\leftarrow$ sib$_3$). In these cases, the joint distributions of the corresponding triplets should be used (compare Problem 7.14). The probabilities $\Pr\{0, 0 \mid D_1 \to R \leftarrow D_2\}$, etc., and $\Pr\{+, + \mid D_1 \to R \leftarrow D_2\}$, etc., can be derived by using arguments similar to those for reciprocal grafting experiments.

Of particular interest are experiments in which the three marginal distributions of R, D_1, and D_2 are mutually independent. Examples are grafting from both parental strains to F_2 progeny $(Pt_1 \to F_2 \leftarrow Pt_2)$ (see section 19.10.1) and experiments in which recipient and two donors are randomly selected from F_2 $(F_2 \to F_2 \leftarrow F_2)$ (see section 19.10.2).

19.10.1. Double Grafting $(Pt_1 \to F_2 \leftarrow Pt_2)$

Table 19.8 represents the joint probabilities when the recipient R is randomly selected from F_2 and the grafts from parental strains are placed on R. The derivation of probabilities $\Pr\{+, +\}$, $\Pr\{+, -\}$, and so on is straightforward, using Table 19.7 and the fact that for $Pt_1 = D_1$ all genotypes are A_1A_1, so that we have $\Pr\{A_1A_1\} = 1$, similarly for $Pt_2 = D_2$ we have $\Pr\{A_2A_2\} = 1$, and for R the genotype distribution is $\Pr\{A_1A_1\} = \frac{1}{4}$, $\Pr\{A_1A_2\} = \frac{1}{2}$, and $\Pr\{A_2A_2\} = \frac{1}{4}$.

TABLE 19.8   Joint probabilities of compatible and incompatible grafts in double grafting, $Pt_1 \to F_2 \leftarrow Pt_2$. n independent loci

Pt$_1 \to$ F$_2$ ⟍ Pt$_2$	+	−	Total
+	$(\frac{1}{2})^n$	$(\frac{3}{4})^n - (\frac{1}{2})^n$	$(\frac{3}{4})^n$
−	$(\frac{3}{4})^n - (\frac{1}{2})^n$	$1 - 2 \times (\frac{3}{4})^n + (\frac{1}{2})^n$	$1 - (\frac{3}{4})^n$
Total	$(\frac{3}{4})^n$	$1 - (\frac{3}{4})^n$	1

19.10.2. Double Grafting $(F_2 \to F_2 \leftarrow F_2)$

In this case all marginal distributions are the same as those given in 19.10.1 for F_2. The corresponding probabilities are given in Table 19.9. The derivation is left to the reader.

TABLE 19.9   Joint probabilities of compatible and incompatible grafts in double grafting, $F_2 \rightarrow F_2 \leftarrow F_2$. n independent loci

$F_2 \rightarrow F_2$ \ F_2	+	−	Total
+	$(\frac{17}{32})^n$	$(\frac{5}{8})^n - (\frac{17}{32})^n$	$(\frac{5}{8})^n$
−	$(\frac{5}{8})^n - (\frac{17}{32})^n$	$1 - 2 \times (\frac{5}{8})^n + (\frac{17}{32})^n$	$1 - (\frac{5}{8})^n$
Total	$(\frac{5}{8})^n$	$1 - (\frac{5}{8})^n$	1

It appears that the experiments $Pt_1 \rightarrow F_2 \leftarrow Pt_2$, as well as $F_2 \rightarrow F_2 \leftarrow F_2$, are impractical to use for estimating the number of loci, n, since a rather large number of grafts are required. But they will be useful in the estimation of the number of some "major" loci, a topic discussed in the following sections.

19.11. STRENGTH OF IMMUNE RESPONSE TO TRANSPLANTATION ANTIGENS. THE H LOCI IN MICE

The immunogenetic properties of transplantation antigens were first studied in mice. At least 15 histocompatibility loci (briefly, the H loci) have been identified. Almost all of the earlier and more recent achievements and experimental results in the transplantation biology of mice are summarized in an excellent article by Snell and Stimpfling (1966). We shall confine ourselves here to a few key points.

The most important and best known is the H-2 locus. This happens to be both a blood group and a histocompatibility locus controlling the fate of tumor and tissue transplants. Therefore, this locus has been studied by serological techniques as well as by methods involving the transplantation of tumors and tissues, using highly inbred lines. At least 33 antigenic specificities, denoted by numbers 1–33, have been detected. Different sets of these specificities are transmitted as heredity units and are called "alleles." At present 20 allelic forms, H-2^a, H-2^b, etc., through H-2^w, have been assigned to this locus. The results are summarized on a special chart in Snell and Stimpfling (1966). The chart represents the specificities associated with each allele and also lists the inbred strains carrying a given allele.

The H-2 locus is called a "strong" locus, and alleles (more precisely, antigens associated with the corresponding alleles) are called "strong" alleles (antigens), because differences at this locus cause, in most cases, rapid and severe rejection. The terms *strong* and *weak* alleles (antigens) were, loosely, first introduced by Counce et al. (1956) to describe the rapidity of graft rejection. The "strength" of an allele is measured roughly in terms of median* survival time (MST). The longer the graft survives, the weaker must be the alleles (antigens) present in the donor and absent in the recipient. It is assumed that the immune response to a strong antigen cannot be suppressed by immunosuppressive therapy. It is also more difficult to induce tolerance against a strong allele. However, the "strength" of an allele *can be conditioned by the genetic relationship of* a donor-recipient pair. In some combinations $D \to R$' the allele present in D and absent in R may cause rapid rejection, whereas in others the rejection may be not so severe. It appears that almost every difference at the H-2 locus results in the rejection of incompatible grafts between the 9th and the 11th post-operative days. However, the experiments by Barnes and Krohn (1957) suggest that in some strains rejection can occur between the 10th and the 13th days.

The other H loci in mice, denoted as H-1, H-3, H-4, etc., are called "weak" loci. However, their role in transplantation immunity cannot be neglected since their cumulative effects can cause severe rejection.

The weak non-H-2 loci have been extensively studied by Graff et al. (1966a, 1966b), using congenic strains which differ with respect to one, two, three, and four loci. We summarize briefly their results and will utilize them in the next section for deriving our histocompatibility models.

(i) The strengths of the non-H-2 alleles are variable. For instance, in the H-1 locus the allelic combination in grafting H-$1^c \to H$-1^b yields a median survival time (briefly, MST) of about 15 days, whereas in grafting H-$1^b \to H$-1^c the MST is about 250 days. Similar discrepancies have been observed at the H-3, H-4, and H-11 loci.

(ii) A cumulative effect was observed if the single-allele effect was not very strong. A given host may have a maximum immune response, and the addition of other antigens, once this maximum response has been reached, cannot further accelerate graft rejection.

(iii) Weak alleles become insignificant or have only a small effect in the presence of a strong allele (e.g., a H-2 allele) on the median survival time.

(iv) The effects of weak antigens are cumulative but not always additive; interaction among some antigenic combinations may occur and cause more (or less) violent reaction than· if the effects were only additive.

---

* The *median* of the sample is the value which divides the observations in halves. If n is odd, this is the $\left(\dfrac{n}{2} + 1\right)$th observation; if n is even, the median is the average of the $\left(\dfrac{n}{2}\right)$th and the $\left(\dfrac{n}{2} + 1\right)$th observations.

(v) The median survival time for grafts placed on females are usually shorter than for those placed on males of the same strain.

19.12. SOME ARBITRARY MEASURE OF STRENGTH OF HISTOCOMPATIBILITY ALLELES (ANTIGENS)

"Strength" is, in fact, a property of an antigen. Since in our terminology each antigen is associated with an allele, we will speak about the strength of an allele or of an antigen interchangeably.

We now attempt to introduce some measure of the strength of a histocompatibility allele. It should be pointed out that this is an arbitrary measure, as yet not adopted. It is a sort of qualitative or "discrete" measure, whereas strength itself is rather a continuous character. But it may help to evaluate some rejection phenomena, at least with approximation.

We will call an allele *strong* if its presence in the donor and absence in the recipient is able, *on its own*, to cause rejection within a certain *determined time* after grafting. As we have mentioned, most of the grafts in mice were rejected before the 12th day after grafting, but in some strains the 14th day was the upper limit. We feel that it is reasonable to define an allele as strong if it is able to evoke rejection before the 14*th postoperative day*.

An allele will be called *weak* if it is not able, on its own, to reject the graft before the 14th postoperative day.

A locus will be called *absolutely strong* or *absolutely weak* if the alleles at this locus are strong or weak, respectively, in the whole species. The *H-2* locus in mice appears to be absolutely strong.

However, if we deal with two homogeneous strains only two different alleles are present at each locus. The locus will be called *conditionally strong* or *conditionally weak*, depending on whether both alleles are strong or weak, respectively.

In experimental work we usually have two parental strains and their F_2 progeny. Therefore, in the following text we will omit the word "conditionally" unless ambiguity is likely to arise.

A locus which possesses at least one strong and at least one weak allele will be called a *mixed* locus. Of course, a mixed locus defined for a particular combination of two strains is also called mixed for the whole species. The converse statement, however, is not necessarily true.

When the donor possesses at a single locus at least one allele which is absent in the recipient, we will say that there is *one locus difference* in histoincompatibility between the donor and the recipient.

Suppose that the recipient is A_1A_1. The donors A_1A_2 and A_2A_2 both show one locus difference with the recipient. The different allele is A_2; but in the donor A_1A_2 there is only a *single gene dose*, whereas in A_2A_2 there is a *double gene dose* of the antigen A_2. When an allele is strong, a single gene dose difference is sufficient to cause rejection before the 14th postoperative day. This does not need to be true, in general, when an allele is weak. In the models considered here, we assume that there is no difference in immune response to a single or double gene dose of an antigen.

Let d be the *maximum* number of loci differences which do *not cause rejection* before the 14th postoperative day. When $d = 0$, an allele is strong. We assign the strength, S, say, to this allele equal to 1. If $d = 1$, then one locus difference does not cause rejection before the 14th day, but two loci differences do cause rejection. We say that in this case $S = \frac{1}{2}$. In general, we define the strength as

$$S = \frac{1}{d + 1}, \qquad d = 0, 1, 2, \ldots . \qquad (19.37)$$

For instance, if three loci differences do not cause rejection, but four loci differences do cause rejection, $d = 3$ and $S = \frac{1}{4}$, and so on.

Several models, including cumulative, additive, and nonadditive effects of weak loci, are theoretically possible. Some of these have been discussed by Elandt-Johnson (1969a, 1969b). Some further immunogenetic experiments are necessary to confirm certain assumptions and to select the most reasonable model.

In the following two sections we confine ourselves to models including only alleles with $S = 1$ and $S = \frac{1}{2}$.

19.13. A MODEL INCLUDING STRONG, WEAK, AND MIXED LOCI FOR ONE-WAY SINGLE-GRAFTING DESIGNS

Let $D \rightarrow R$ denote a one-way single grafting design. Let C denote a "compatible" and I an "incompatible" graft at a single locus, and P and Q be the corresponding probabilities of the events C and I. Let us consider the generating function

$$(PC + QI)^r = (PC)^r + \binom{r}{1}(PC)^{r-1}(QI)$$
$$+ \binom{r}{2}(PC)^{r-2}(QI)^2 + \cdots + (QI)^r. \quad (19.38)$$

The coefficient of the first term P^r, gives the proportion of compatible grafts with respect to all r loci, the coefficient of the second term, $\binom{r}{1}P^{r-1}Q$, gives the proportion of grafts which are incompatible with respect to one locus, and so on.

Let $\Pr\{+ \mid d\}$ denote the probability of graft acceptance if *d or fewer loci differences* do not cause rejection, but $d + 1$ or more loci do cause rejection. Therefore

$$\Pr\{+ \mid d\} = \sum_{j=0}^{d} \binom{r}{j} P^{r-j} Q^{j}. \tag{19.39}$$

In particular, we have

$$\Pr\{+ \mid 0\} = P^{r}, \tag{19.40a}$$

$$\Pr\{+ \mid 1\} = P^{r} + \binom{r}{1} P^{r-1} Q = P^{r}\left(1 + \frac{rQ}{P}\right). \tag{19.40b}$$

We recall the probabilities $P = \Pr\{0\}$ defined for different one-way designs $D \to R$, discussed in sections 9.6.1 and 9.6.2. There we obtained: (a) Pt $\to F_2$, $P = \frac{3}{4}$; (b) $F_2 \to F_2$, $P = \frac{5}{8}$; (c) $F_1 \to F_2$, $P = \frac{1}{2}$. Table 19.10 gives the values of $\Pr\{+ \mid d\}$, $d = 0, 1$, for these designs.

We now consider the following model. Suppose that a parental strain, Pt, possesses m strong loci, (i.e., $d = 0$, or $S = 1$), and r weak loci, such that before the 14th postoperative day only two or more loci cause rejection (i.e., $d = 1$ or $S = \frac{1}{2}$). We will call these $m + r$ loci the *major* loci (again this definition is quite arbitrary). We assume that these $m + r$ loci are independent. Suppose that the remaining loci, which we will call the *minor* loci, are "inactive" in the presence of the major loci so that they are not included in our model.

The total probability, $\Pr\{+\}$, say, that the graft will not be rejected from Pt $\to F_2$ before the 14th day is

$$\Pr\{+\} = \left(\frac{3}{4}\right)^{m} \times \left[\left(\frac{3}{4}\right)^{r} + \binom{r}{1}\left(\frac{3}{4}\right)^{r-1}\frac{1}{4}\right] = \left(\frac{3}{4}\right)^{m+r} \cdot \frac{r + 3}{3}. \tag{19.41}$$

Example 19.2. The data in Table 19.11 represent the numbers of grafts surviving before the 14th day in two experiments, Pt $\to F_2$, on two strains of rats, B.N. (Brown Norway) and Lewis [Billingham et al. (1962)].

(a) Let us first estimate the number of "major" loci, n, in each strain under the hypothesis ($H_0^{(1)}$, say) that they are *strong*. The estimated number of loci, $\hat{n}$, will be found from the equation $\left(\frac{3}{4}\right)^{\hat{n}} = \hat{P}$, where $\hat{P}$ is the observed proportion of grafts not rejected. Thus for B.N. $\to F_2$ we have $\left(\frac{3}{4}\right)^{\hat{n}} = 0.4255$, hence $\hat{n} \doteq 3$; and for Lewis $\to F_2$ we have $\left(\frac{3}{4}\right)^{\hat{n}} = 0.3648$, so $3 < \hat{n} < 4$. From these experiments we cannot predict whether these are the same loci or whether, perhaps, some are mixed.

(b) It is known that the *Ag-B* locus in rats is *strong*, and our strains have different *Ag-B* alleles. Let us hypothesize that there is only *one strong* locus in both strains, and that r_1 loci in B.N. and r_2 loci in Lewis have alleles of strength $S = \frac{1}{2}$ ($H_0^{(2)}$, say). The expected proportions of grafts not rejected

TABLE 19.10 [a]Probabilities of grafts not being rejected, $\Pr\{+ \mid d\}$, for $d = 0, 1$ and different one-way designs ($D \to R$)

Design, $D \to R$	d	No. of loci, r											
		1	2	3	4	5	6	7	8	9	10	11	12
Pt → F₂ {	0	0.7500	0.5625	0.4219	0.3164	0.2373	0.1780	0.1335	0.1001	0.0751	0.0563	0.0422	0.0317
	1	1.0000	0.9375	0.8438	0.7383	0.6328	0.5339	0.4449	0.3671	0.3003	0.2440	0.1971	0.1584
F₂ → F₂ {	0	0.6250	0.3906	0.2441	0.1526	0.0954	0.0596	0.0373	0.0233	0.0146	0.0091	0.0057	0.0036
	1	1.0000	0.8594	0.6836	0.5188	0.3815	0.2742	0.1937	0.1350	0.0931	0.0637	0.0432	0.0291
F₁ → F₂ {	0	0.5000	0.2500	0.1250	0.0625	0.0312	0.0156	0.0078	0.0039	0.0020	0.0010	0.0005	0.0002
	1	1.0000	0.7500	0.5000	0.3125	0.1875	0.1094	0.0625	0.0352	0.0195	0.0107	0.0059	0.0032

[a]From Elandt-Johnson, *Biometrics 25* (1969b) p. 224.

TABLE 19.11   Grafts surviving on rats before the 14th day

| Experiment | Total number of grafts | Grafts surviving | |
		Number	Proportions
B.N. $\rightarrow$ F$_2$	329	140	0.4255
Lewis $\rightarrow$ F$_2$	318	116	0.3648

are given by formula (19.41), with $m = 1$. Thus, the maximum likelihood estimates of r_1 and r_2 are obtained by solving these equations:

for B.N. $\rightarrow$ F$_2$: $\dfrac{\hat{r}_1 + 3}{3}\left(\dfrac{3}{4}\right)^{\hat{r}_1+1} = \hat{P}_1 = 0.4255 \qquad (\hat{r}_1 \doteq 6);$

for Lewis $\rightarrow$ F$_2$: $\dfrac{\hat{r}_2 + 3}{3}\left(\dfrac{3}{4}\right)^{\hat{r}_2+1} = \hat{P}_2 = 0.3648 \qquad (\hat{r}_2 \doteq 6).$

[Note that the expected proportion for $m = 1$ and $r = 6$ is $3 \times (\frac{3}{4})^7 = 0.4005$.]

Again we cannot say whether these six alleles are assigned to the same or different loci in both strains.

It is easy to show that both hypotheses, $H_0^{(1)}$ and $H_0^{(2)}$, fit the data. To distinguish between them, some other experimental design(s) should be used. One such a possibility will be to use a double grafting experiment.

▼ 19.14. A MODEL INCLUDING STRONG AND WEAK LOCI FOR THE DOUBLE-GRAFTING DESIGN Pt$_1 \rightarrow$ F$_2 \leftarrow$ Pt$_2$

When grafts from both parental strains are placed simultaneously on F$_2$, we can calculate the probability, $\Pr\{+, +\}$, say, that both grafts are not rejected; the probability, $\Pr\{+, -\}$, that one graft is not rejected, and the other is rejected; and so on, assuming cumulative effects of some weak antigens. Several situations are possible. We will consider only one simple model based on the following assumptions:

(a) There are m strong loci.
(b) There are r weak loci with strength $S = \frac{1}{2}$, common to both strains, which can cumulate their effects and cause rejection before the 14th day. Cumulative effects of other weak loci may exist but occur after the 14th day.
(c) The $m + r$ loci are independent.

We notice that a graft will not be rejected if:

(i) Both donors are compatible with the recipient with respect to m strong loci. This can happen when the recipients are heterozygous with respect to these loci; the probability of this event is $(\frac{1}{2})^m$.

(ii) Each donor exhibits only one locus difference with respect to r weak loci. This can happen if:

either (a) the recipient is heterozygous with respect to all r or only with respect to $r - 1$ loci—the probability of this event is

$$\left(\frac{1}{2}\right)^r + \binom{r}{1}\left(\frac{1}{2}\right)^r = \left(\frac{1}{2}\right)^r (r + 1);$$

or (b) the recipient is heterozygous with respect to $r - 2$ loci *and* is simultaneously "mixed" homozygous with respect to the two remaining loci of this group. (If these two loci are A and B, say, the recipient must be $A_1A_1B_2B_2$ or $A_2A_2B_1B_1$, but not $A_1A_1B_1B_1$ or $A_2A_2B_2B_2$.) The probability of this event is

$$\frac{1}{2} \times \binom{r}{2}\left(\frac{1}{2}\right)^r = \left(\frac{1}{2}\right)^r \frac{r(r - 1)}{4}.$$

Thus, the probability that the grafts from both donors will not be rejected because there is a single locus difference between the recipient and each of the two donors is

$$\left(\frac{1}{2}\right)^r (r + 1) + \left(\frac{1}{2}\right)^r \frac{r(r - 1)}{4} = \left(\frac{1}{2}\right)^r \frac{r^2 + 3r + 4}{4}. \qquad (19.42)$$

Therefore, the total probability that both grafts will not be rejected before the 14th day is

$$\Pr\{+, +\} = \left(\frac{1}{2}\right)^{m+r} \frac{r^2 + 3r + 4}{4}. \qquad (19.43)$$

The probability, $\Pr\{+\}$, that the graft from Pt_1 will not be rejected is given by (19.41). Thus, the probability, $\Pr\{+, -\}$, that the graft from Pt_1 will not be rejected and the one from Pt_2 will be rejected is

$$\Pr\{+, -\} = \Pr\{+\} - \Pr\{+, +\}$$

$$= \left(\frac{3}{4}\right)^{m+r} \frac{r + 3}{3} - \left(\frac{1}{2}\right)^{m+r} \frac{r^2 + 3r + 4}{4}. \qquad (19.44)$$

Similarly, the probability $\Pr\{-, +\}$ yields the same value as (19.44).

And finally, the probability, $\Pr\{-, -\}$, that grafts from both strains will be rejected is

$$\Pr\{-, -\} = 1 - 2\Pr\{+, -\} - \Pr\{+, +\}$$
$$= 1 - 2\left(\frac{3}{4}\right)^{m+r}\frac{r+3}{3} + \left(\frac{1}{2}\right)^{m+r}\frac{r^2 + 3r + 4}{4}. \quad (19.45)$$

We notice that, for $r = 0$ probabilities (19.43)–(19.45) are identical with those given in Table 19.8, with m substituted for n.

Example 19.3. The data in Table 19.12 represent some results from a double-grafting experiment, $Pt_1 \rightarrow F_2 \leftarrow Pt_2$, in rats, where B.N. ($Pt_1$) and Lewis ($Pt_2$) were used [reconstructed from percentages given by Billingham et al. (1962)]. In Example 19.2 we found that two hypotheses—$H_0^{(1)}$: three strong loci are responsible for rejection before the 14th day; and $H_0^{(2)}$: one strong locus and six weak loci are responsible — both fit the data for one-way single-grafting experiments. We shall now fit the data from double grafting to these hypotheses, assuming additionally that the strong and weak loci are the same in both strains.

The test results are given in Table 19.12. As we can see, the goodness of fit X^2's are highly significant ($\chi^2_{0.999;3} = 16.27$), so neither of these hypotheses agrees with the data from double-grafting experiments, which are more informative than those from single grafting. Some of the reasons for this might be the following: (a) the weak loci may not be the same in both strains, and also the strong allele in each strain may belong to different (mixed) loci; (b) three or more loci differences can be responsible for early rejection; (c) double gene dose differences at a single locus might cause rejection (see, e.g., Problem 19.10).

Further studies of the strength of histocompatibility antigens in mice are in progress. If we were able to introduce more assumptions, we could, of course, make our probability models more specific.

TABLE 19.12   Double-grafting experiment in rats, B.N. $\rightarrow F_2 \leftarrow$ Lewis. Tests of goodness of fit

Response Frequency		$(+, +)$	$(+, -)$	$(-, +)$	$(-, -)$	Total	X^2
	Observed	56	78	59	123	316	
$H_0^{(1)}$	Probability	0.1250	0.2969	0.2969	0.2812	1.0000	35.54
	Expected	39.5	93.8	93.8	88.9	316	
$H_0^{(2)}$	Probability	0.1132	0.2872	0.2872	0.3123	1.0000	30.32
	Expected	35.8	90.8	90.8	9.87	316	

19.15. HISTOCOMPATIBILITY SYSTEMS IN MAN. THE *HL-A* SYSTEM

We have already seen how many complications arise in histocompatibility testing in genetically determined animals such as mice and rats. How many histocompatibility systems exist in man? As yet, the answer is unknown, and research in this field is aimed chiefly at finding the most important systems so that, in matching a donor with a recipient who needs an organ transplant, incompatibilities with respect to the major systems can be avoided.

There is no doubt that the *ABO* and the Rhesus blood group systems play an important part in graft rejection [Ceppellini et al. (1966), Dausset et al. (1970)]. Thus the donor and the recipient are usually first matched with respect to these systems.

From intense investigation on leucocyte antigens, it appears that the *HL-A* complex is the major histocompatibility system in man [Ceppellini et al. (1967), Dausset et al. (1970), Van Rood (1966), Walford (1969)]. The present status of information [Dausset et al. (1970), Walford (1969)] is as follows.

The *HL-A* system (region) consists of two closely linked loci. At the first locus, nine antigenic determinants, each transmitted as a unit to the offspring, have been detected and are denoted by the symbols HL-A1, A2, A3, A9, Da15, Da17, Da21, Da22, Da25. At the second locus, there are eleven antigenic determinants, denoted as HL-A5, A7, A8, Da4, Da6, Da9, Da18, Da20, Da23, Da24, and HN. There may possibly be other antigenic determinants at each locus which have not been detected as yet. These two closely linked loci resemble, in a certain way, the *Rh* system of three linked loci, *D, C, E.* (It should be mentioned that a few cross-overs between the two *HL-A* loci have been observed.) Recalling the discussion of the Rhesus system, we now associate the terms "genes" and "alleles" with antigenic determinants at each locus. Thus, for instance, *HL-A*1 is an "allele" at the first locus, while *HL-A*5 is an "allele" at the second locus.

Since these two loci are closely linked, two "genes" each from a different locus, appearing at the same chromosome, will be transmitted "in block." Ceppellini et al. (1967) suggested the term *haplotype* for each chromosomal combination. For instance, *HL-A*1, *A*8 or *HL-A*2, *Da*4 are such haplotypes, It should be noticed that this terminology differs from that for the *H*-2 system in mice, and from the definition of an "allele" in a system introduced in Chapters 1 and 14.

The number of possible haplotypes is $9 \cdot 11 = 99$. Several studies indicate that haplotypes, *HL-A*1, *A*8 and *HL-A*3, *A*7 occur more frequently than it is expected from gene frequencies calculated for each locus. Insufficient data are available to detect all $(99 \cdot 100)/2 = 4950$ possible combinations of pairs of haplotypes; nevertheless, it has been demonstrated that antigenic determinants exhibit widespread polymorphism. They are also of about the same strength. It is easy to conclude, in view of these facts, that the probability of finding a suitable unrelated donor is practically zero. Matching is reasonably likely only *within* families and mostly among siblings; in such cases the maximal number of *HL-A* haplotypes involved is four.

There are several different methods of matching a donor with the recipient, besides leucocyte typing. They can be divided into two groups: methods *in vivo* [see, e.g., Nelson and Russell (1968)] and methods *in vitro* [Hirschhorn (1968)]. Leucocyte typing is an *in vitro* method.

19.16. HISTOCOMPATIBILITY MODELS IN RANDOMLY MATING POPULATIONS

One of the major problems in human histocompatibility testing is that of working out in detail the role of leucocyte antigens. Several research projects have been devoted to estimating the "gene" frequencies corresponding to different antigenic determinants and finding the "hierarchy" in their strength. It appears that a double dose of an antigenic determinant is stronger than a single dose of the same determinant [Dausset et al. (1970)]. Methods of complex segregation analysis, designed especially for such problems, would be suitable tools for the analysis of family data collected in several research centers. Random samples from populations are also available; but, because the number of alleles appears to be very large, we would need very big samples for any kind of valid analysis.

Of course, there are some other histocompatibility systems. Complete leucocyte and blood group matching does not exclude rejection; it only indicates that graft survival can be prolonged.

Before the role of the *HL-A* system was established, several authors attempted to construct models appropriate for randomly mating populations. A single strong or mixed locus was postulated, and the expected proportions of early rejected grafts were evaluated. Single and double gene doses have been taken into account [Elandt-Johnson (1968*a*, 1968*b*)]. By the use of computer facilities, data from the kidney transplant registry have been fitted to some of these models [Kilpatrick and Gamble (1968), Serra and O'Mathuna (1966)]. As might be expected, the data fit several different models, and we cannot reach any firm conclusion from these speculations. Nevertheless, these models may be of some interest in studying histocompatibility systems in randomly mating populations other than man (e.g., rabbits or dogs). We do not present these models here but refer the reader who is interested in them to the papers mentioned above.

It should be noted that histocompatibility testing is a relatively new field, and not much mathematics or statistics on this problem has yet been worked out. A paper by Elandt-Johnson (1969*b*) reviews the mathematical work done on histocompatibility testing up to 1969.

19.17. GENERAL REMARKS

It should once more be emphasized that the models discussed in this chapter may be very rough approximations to the real situations in histocompatibility testing. They are based mostly on an analogy to the genetic systems which control the red cell antigens, and it may well be that they are oversimplified. Moreover, the models derived here are deterministic. The immune response which leads ultimately to graft destruction is a continuous process, and mathematical models which include time as a continuous variable might give a better fit to the biological phenomenon of histocompatibility.

More work needs to be done in designing experiments on laboratory animals to analyze the actions and interactions of weak alleles. It seems that more attention should also be paid to methods of designing experiments in which the responses to strong (and, perhaps, weak?) alleles can be studied.

SUMMARY

The fundamental assumptions regarding the immunogenetic mechanism of histocompatibility and the basic terminology in transplantation have been explained. Three main types of experiments for histocompatibility testing in laboratory animals have been discussed: one-way single-grafting, two-way (or reciprocal) grafting, and double-grafting designs. The concepts of strong and weak loci and an arbitrary definition of the strength of transplantation antigen have been introduced. Some histocompatibility models, including the cumulative effects of weak alleles, have been derived for different experimental designs. Using these models, we have attempted to estimate the number of strong and weak systems involved in early graft rejection and to test some hypotheses about the immune responses. The difficulty of obtaining a unique solution has been demonstrated, and the necessity for further studies in the biology of histocompatibility mechanisms has been indicated. The genetics of the *HL-A* (Human Leucocyte — locus *A*) system has been discussed, and the role of the *HL-A* complex in histocompatibility testing briefly described.

REFERENCES

Bailey, D. W., Histocompatibility associated with the X chromosome in mice, *Transplantation* 1 (1963), 70–74.

Barnes, A. D., and Krohn, P. L., The estimation of the number of histocompatibility genes controlling the successful transplantation of normal skin in mice, *Proc. Roy. Soc. Lond.* 146 (1957), 505–526.

Billingham, R. E., Hodge, B. A., and Silvers, W. K., An estimate of the number of histocompatibility loci in the rat, *Proc. Nat. Acad. Sci.* 48 (1962), 138–147.

Ceppellini, R., Curtoni, E. S., Mattiuz, P. L., Miggiano, V. C., and Visetti, M., An experimental approach to genetic analysis of histocompatibility in man. In: *Histocompatibility Testing*, Munksgaard, Copenhagen, 1965, p. 13.

Ceppellini, R., Curtoni, E. S., Mattiuz, P. L., Leigheb, G., Visetti, M., and Colombi, A., Survival of test skin grafts in man: effect of genetic relationship and blood groups incompatibility, *Ann. N.Y. Acad. Sci.* 129, Part I (1966), 421–425.

Ceppellini, R., Curtoni, E. S., Mattiuz, P. L., Miggiano, V., Scudeller, G., and Serra, A., Genetics of leukocyte antigens: a family study of segregation and linkage. In: *Histocompatibility Testing*, Munksgaard, Copenhagen, 1967, pp. 149–188.

Counce, S., Smith, P., Barth, R., and Snell, G. D., Strong and weak histocompatibility gene differences in mice and their role in rejection of homografts of tumor and skin, *Ann. Surg.* 144 (1956), 198–204.

Dausset, J., Colombani, J., Legrand, L., and Fellows, M. H., Genetics of the *HL-A* system. Deduction of 480 haplotypes. In: *Histocompatibility Testing*, Munksgaard, Copenhagen, 1970, pp. 53-77.

Davis, D. A. L., Manstone, A. J., Viza, D. C., Colombani, J., and Dausset, J., Human transplantation antigens: the *HL-A* (*Hu*-1) system and its homology with the mouse *H*-2 system, *Transplantation* 6 (1968), 571–586.

Eichwald, E. J., and Silmser, C. R., Communication, *Transplantation Bull.* 2 (1955), 148–149.

Elandt-Johnson, Regina C., General purpose probability models in histocompatibility testing, I. "Strong" and "weak" alleles at one locus, *Ann. Hum. Genet. Lond.* 31 (1968a), 293–308.

Elandt-Johnson, Regina C., General purpose probability models in histocompatibility testing, II. "Strong" and "weak" alleles at several loci, *Ann. Hum. Genet. Lond.* 32 (1968b), 81–88.

Elandt-Johnson, Regina, C., Estimation of the number of histocompatibility loci in laboratory animals. Experimental designs and statistical analysis, *Transplantation* 7 (1969a), 12–40.

Elandt-Johnson, Regina C., Survey of histocompatibility testing. Biological background, probabilistic and statistical models and problems, *Biometriċs* 25 (1969b), 207–283.

Graff, R. J., Hildeman, W. H. and Snell, G. D., Histocompatibility genes of mice. *Transplantation* 4, (1966a), 425–437.

Graff, R. J., Silvers, W. K., Billingham, R. E., Hildeman, W. H., and Snell, G. D., The cumulative effect of histocompatibility antigens, *Transplantation* 4 (1966b), 605–617.

Hirschhorn, K., *In vitro* histocompatibility testing. In: *Human Transplantation*, ed. by F. T. Rapaport and J. Dausset, Grune & Stratton, New York, 1968, pp. 406–422.

Humphrey, J. H., and White, R. G., *Immunology for Students of Medicine*, F. A. Davis Co., Philadelphia, 1963.

Kilpatrick, J. S., and Gamble, E. L., A model for incompatibility in kidney transplants, *Ann. Hum. Genet. Lond.* 31 (1968), 309–317.

McKhann, C. F., Transplantation studies of strong and weak histocompatibility barriers in mice, II. Tolerance, *Transplantation* **2** (1964), 620–626.

Murray, J. E., Gleason, R., and Bartholomay, A., Second report of registry in human kidney transplantation, *Transplantation* **2** (1964), 660–667.

Nelson, S. D., and Russell, P. S., *In vivo* histocompatibility testing. In: *Human Transplantation*, ed. by F. T. Rapaport and J. Dausset, Grune & Stratton, New York, 1968, pp. 394–405.

Owen, R. D., Genetic aspects of tissue transplantation and tolerance. *Amer. J. Hum. Genet.* **11** (1959), 366–383.

Rogers, B. O., Genetics of transplantation in humans, *Dis. Nervous System* **24**, Suppl. (April, 1957), 7–43.

Schwartz, R. S., Immunosuppressive drugs, *Prog. in Allergy* **9** (1965), 246–303.

Serra, A., and O'Mathuna, D., A theoretical approach to the study of genetic parameters of histocompatibility in man, *Ann. Hum. Genet. Lond.* **30** (1966), 96–118.

Snell, G. D., *Biology of the Laboratory Mouse*, Blakiston, Philadelphia, 1941.

Snell, G. D., Methods for the study of histocompatibility genes, *J. Genet.* **49** (1948), 87–103.

Snell, G. D., and Stimpfling, J. H., Genetics of tissue transplantation. In: *Biology of the Laboratory Mouse*, ed. by E. L. Green, McGraw-Hill Book Company, New York, 1966, pp. 457–491.

Van Rood, J. J., Leucocyte groups and histocompatibility, *Vox. Sang.* **11** (1966), 276–292.

Walford, R. L., *The Isoantigenic Systems of Human Leucocytes. Medical and Biological Significance*, Series Haematologica, Vol. II, 2, Munksgaard, Copenhagen, 1969, pp. 5–96.

Wilson, R. E. L., Henry, L., and Merill, J. P, A model system for determining histocompatibility in man, *J. Clin. Invest.* **42** (1963), 1497–1503.

PROBLEMS

19.1. (*a*) Using the data of Example 19.1, estimate the number of histocompatibility loci from each experiment separately.

(*b*) Show that the appropriate 95% confidence intervals for the number, n, of histocompatibility loci in each experiment are as below:

    (i) Strain A $\rightarrow$ F$_2$: 95% confidence interval, $10 \leqslant n \leqslant 21$;

    (ii) Strain CBA $\rightarrow$ F$_2$: 95% confidence interval, $12 \leqslant n \leqslant 31$;

    (iii) Pooled data: 95% confidence interval, $12 \leqslant n \leqslant 21$.

[*Hint:* Use formula (11.39).]

19.2. Find the proportions of compatible and incompatible grafts when a donor and a recipient are both members of backcrosses of two homogeneous strains which differ with respect to n independent histocompatibility loci for the following designs:

    (*a*) one-way single-grafting design;

    (*b*) reciprocal grafting design;

    (*c*) double-grafting design.

19.3. Derive the formulae for proportions of compatible and incompatible grafts in reciprocal grafting design for each of the following donor-recipient pairs: (*a*) parent-child, (*b*) sib-sib, when n independent histocompatibility loci, each with two alleles of equal gene frequencies, are involved.

19.4. Derive the likelihood function from which the number of n independent histocompatibility loci can be estimated, when the reciprocal grafting design between the sib-sib pairs is used.

19.5. Suppose that a double-grafting experiment is performed such that sibling 1 (S_1) is a recipient, and the mother (M) and sibling 2 (S_2) are the donors.

(a) Find the formulae for $\Pr\{+, + \mid M \rightarrow S_1 \leftarrow S_2\}$, $\Pr\{+, - \mid M \rightarrow S_1 \leftarrow S_2\}$, and $\Pr\{-, - \mid M \rightarrow S_1 \leftarrow S_2\}$, assuming n independent histocompatibility loci. Prepare a table of these probabilities, for different n. [*Hint:* Use the joint distribution of triplets (M, S_1, S_2) mentioned in Problem 7.14a.]

(b) Derive the likelihood equation for estimating the number of loci, n.

19.6. Using the joint genotype distribution of three siblings (S_1, S_2, S_3), (see Problem 7.14b) calculate the same probabilities as in Problem 19.5, for the double-grafting experiment $S_2 \rightarrow S_1 \leftarrow S_3$.

▼**19.7.** Derive a histocompatibility model for a one-way, Pt $\rightarrow F_2$, grafting design when there are m strong loci, w loci with the strength $S = \frac{1}{2}$, and r loci with $S = \frac{1}{3}$. In this model assume that two loci differences of type $S = \frac{1}{3}$ and one locus difference of type $S = \frac{1}{2}$ can cause rejection.

▼**19.8.** Derive a histocompatibility model, as in 19.7, for the double-grafting design $Pt_1 \rightarrow F_2 \leftarrow Pt_2$.

(a) Assume that w loci and r loci are common to both strains.

(b) Assume that w loci are common to both strains, and r loci are different in each strain.

(c) Assume that w and r, loci, all are different in each strain.

▼**19.9.** Derive a histocompatibility model for the double-grafting design $Pt_1 \rightarrow F_2 \leftarrow Pt_2$, including strong and weak alleles of strength $S = \frac{1}{2}$, for the following hypotheses:

(a) There are m strong loci.

(b) There are w weak loci common in both strains.

(c) There are r_1 mixed loci whose alleles are strong in Pt_1, and weak in Pt_2.

(d) There are r_2 mixed loci whose alleles are strong in Pt_2, and weak in Pt_1.

Answer:

$$\Pr\{+, + \mid Pt_1 \rightarrow F_2 \leftarrow Pt_2\}$$
$$= (\tfrac{1}{2})^{m+w+r_1+r_2} \left[\frac{r_1 + 2}{2} \cdot \frac{r_2 + 2}{2} + \frac{w(w + 3) + (r_1 + r_3)}{4} \right]$$

and

$$\Pr\{+ \mid Pt_1 \rightarrow F_2\} = (\tfrac{3}{4})^{m+w+r_1+r_2} \left(\frac{w + r_2 + 3}{3} \right)$$

[Elandt-Johnson (1969*b*)].

19.10. Interpret the situations when, in Problem 19.9, we have the following:

(a) $m = w = 0$;

(b) $m = 1$, $w = 0$;

(c) $r_1 = r_2 = 0$ (compare this situation with the model derived in section 19.14).

▼**19.11.** Suppose that the strength of a weak antigen is such that single gene dose differences between the donor and the recipient do not cause rejection at the early stage, but *double gene dose differences* do. In such cases, $A_1A_2 \rightarrow A_1A_1$ (or $A_1A_2 \rightarrow A_2A_2$) does not reject the graft, but $A_1A_1 \rightarrow A_2A_2$ (or $A_2A_2 \rightarrow A_1A_1$) does.

Using this assumption, find the probability that the graft will not be rejected, if there are s strong loci and r weak loci and these loci are independent, for the following experimental designs:

(a) $F_2 \rightarrow F_2$;

(b) $F_2 \leftrightarrows F_2$;

(c) $F_2 \rightarrow F_2 \leftarrow F_2$.

19.12. Let $A_1, A_2, \ldots, A_s$ be s alleles at a single histocompatibility locus with the frequencies $p_1, p_2, \ldots, p_s$, with $\sum p_i = 1$, in a population. We denote by

$$m_r = \sum_{i=1}^{s} p_i^{\,r}$$

the rth frequency moment; and by $\Pr\{0\}$ the probability that a donor is compatible with a recipient. Assume a population in equilibrium.

(a) Show that, when donors and recipients are unrelated,

$$\Pr\{0 \mid \text{Unrel.} \rightarrow \text{Unrel.}\} = 2m_2^{\,2} - m_4.$$

(b) Show that, when grafting is from parent to child,

$$\Pr\{0 \mid \text{Parent} \rightarrow \text{Child}\} = m_2.$$

(*Hint:* Use a table analogous to Table 7.3.)

(c) Show that, when grafting is from one sibling to another,

$$\Pr\{0 \mid \text{Sib} \rightarrow \text{Sib}\} = \tfrac{1}{4}(1 + 2m_2 + 2m_2^{\,2} - m_4).$$

[*Hint:* Use formula (7.78).]

▼**19.13.** Suppose that there is one strong histocompatibility locus with s codominant alleles. Suppose also that the parental genotypes are $A_iA_i \times A_iA_j$. The child genotypes can also be only A_iA_i or A_iA_j, each with probability $\tfrac{1}{2}$. In a *population* of the children the conditional probability that two siblings are histo-compatible, given the parents are $A_iA_i \times A_iA_j$, is $\Pr\{0 \mid A_iA_i \times A_iA_j\} = \tfrac{3}{4}$, as we can see from the accompanying table.

I Sibling ╲ II Sibling	A_iA_i	A_iA_j
A_iA_i	0	1
A_iA_j	0	0

However, this need not be so in any particular family. The proportion of compatible siblings depends on the family pattern and on which sibling in a pair is a donor and which a recipient. Suppose, for example, that in a family with $s = 3$ children the family pattern is A_iA_i, A_iA_i, A_iA_j. Let us order these children as 1, 2, 3, respectively. The *ordered* possible pairs of siblings are (1, 2), (1, 3), (2, 3), (2, 1),

(3, 1), (3, 2). If we assume that the ordered pair corresponds to the donor →
recipient pair, then only pairs (3, 1) (i.e., $A_iA_j → A_iA_i$) and (3, 2) (i.e., $A_iA_j →$
A_iA_i) are incompatible. Thus, among six possible pairs, four pairs are histocompatible, that is, the proportion of histocompatible pairs for this pattern is
$\frac{4}{6} = \frac{2}{3}$. In other words, given the parental combination $A_iA_i \times A_iA_j$ and the
family pattern A_iA_i, A_iA_i, A_iA_j, the conditional probability of compatible siblings
is $\frac{2}{3}$.

We can evaluate such conditional probabilities for each pattern for any family
size.

Suppose that the parental genotypes are $A_iA_i \times A_iA_j$ and there are $n = 5$
children in a family.

(a) Calculate the probabilities of occurrence of each pattern.

(b) Calculate the conditional probabilities of histocompatible pairs of siblings
for each family pattern.

(c) Thus calculate the average probability of histocompatible siblings over all
possible patterns in families of parents $A_iA_i \times A_iA_j$. (Hint: It should be equal to $\frac{3}{4}$.)

(d) Could you generalize the results for any family size, s?

▼19.14. Let the parental genotypes be $A_iA_j \times A_iA_k$.

(a) Show that in a population of children from such families the probability of
histocompatible siblings is $\frac{5}{8}$.

(b) Derive all possible patterns for families of size $s = 5$; calculate the probability
of occurrence of each pattern and the probabilities of histocompatible pairs for
each pattern, as in Problem 19.1.

(c) Extend the results to families of size s.

(d) Show that the average probability of histocompatible siblings in such families
is $\frac{5}{8}$.

Elements of Matrix Algebra

Vectors and matrices are very useful tools in algebraic manipulations. Just as mathematical formulae can stand for whole sentences, vector and matrix notations abbreviate whole sets of mathematical expressions.

There are many books, at different mathematical levels, on matrix algebra, and we do not attempt a formal development of this subject here. However, we feel that it may be useful for the readers of this book to have a brief summary of the essentials of matrix operations, with particular reference to ways in which they can be used in genetics. Therefore, we shall present the basic definitions and formulae which have been used in this book, especially in Chapters 4, 7, and 10.

VECTORS AND MATRICES. BASIC DEFINITIONS

Vectors. Let $p_1, p_2, \ldots, p_s$ be the frequencies of alleles $A_1, A_2, \ldots, A_s$, respectively. These frequencies can be written in an abbreviated form:

$$\mathbf{p} = \begin{pmatrix} p_1 \\ p_2 \\ \cdot \\ \cdot \\ \cdot \\ p_s \end{pmatrix} \qquad (\text{I.1a})$$

or

$$\mathbf{p}' = (p_1, p_2, \ldots, p_s). \qquad (\text{I.1b})$$

which are called *vectors*. The vector $\mathbf{p}$ in (I.1a) is represented by a column consisting of s ordered elements and is sometimes called a *column vector*. The vector $\mathbf{p}'$ is a *row vector* and is also called a *transpose* of the vector $\mathbf{p}$ ("prime" is used for transpose). It is customary to use boldface lower case letters for vectors, while boldface capital letters are reserved for matrices. However, in genetics, we may occasionally use capital letters to express in an abbreviated form an ordered set of alleles, that is,

$$
\mathbf{A} = \begin{pmatrix} A_1 \\ A_2 \\ \cdot \\ \cdot \\ \cdot \\ A_s \end{pmatrix}
\tag{I.2a}
$$

or

$$
\mathbf{A}' = (A_1, A_2, \ldots, A_s).
\tag{I.2b}
$$

In this book, the elements of a vector are usually real numbers. The "vectors of alleles" in (I.2) do not have numerical meaning; they are, rather, "formal" abstractions, but, as we shall see later, they give a convenient form of representation. It will always be obvious from the text whether $\mathbf{A}$ is a matrix or a "vector of alleles."

Of particular interest is the so-called *unit vector*, that is,

$$
\mathbf{1} = \begin{pmatrix} 1 \\ 1 \\ \cdot \\ \cdot \\ \cdot \\ 1 \end{pmatrix}
\tag{I.3a}
$$

or

$$
\mathbf{1}' = (1, 1, \ldots, 1).
\tag{I.3b}
$$

Matrices.   A rectangular arrangement of elements c_{ij} into s rows and m columns is called a matrix $\mathbf{C}$ of $s \times m$ *order*, that is,

$$
\mathbf{C} = \begin{bmatrix} c_{11} & c_{12} & \cdots & c_{1m} \\ c_{21} & c_{22} & \cdots & c_{2m} \\ \multicolumn{4}{c}{\cdots\cdots\cdots\cdots\cdots} \\ c_{s1} & c_{s2} & \cdots & c_{sm} \end{bmatrix}.
\tag{I.4}
$$

Another brief notation is

$$
\mathbf{C} = (c_{ij}).
\tag{I.5}
$$

To emphasize the order of the matrix $\mathbf{C}$, one may write $\mathbf{C}_{sm}$ or $\underset{s \times m}{\mathbf{C}}$.

Example I.1.   Let A, a and B, b be two alleles at each of two loci, A and B, respectively. There are nine distinct genotypes, $AABB$, $AABb$, etc. If

$h_1, h_2, \ldots, h_9$ are the frequencies of the genotypes arranged in a certain order, then $\mathbf{h}' = (h_1, h_2, \ldots, h_9)$ is a genotype frequency vector. However, the nine genotype frequencies can be represented in the form of a 3×3 matrix, $\mathbf{Z}$, say, as follows:

$$
\begin{array}{c}
 \begin{array}{ccc} BB & Bb & bb \end{array} \\
\begin{array}{c} AA \\ Aa \\ aa \end{array}
\begin{bmatrix}
z_{11} & z_{12} & z_{13} \\
z_{21} & z_{22} & z_{23} \\
z_{31} & z_{32} & z_{33}
\end{bmatrix} = \mathbf{Z} = (z_{ij}),
\end{array}
$$

where the first subscript indicates the genotype with regard to locus A and the second with regard to locus B, respectively (compare, for instance, Table 2.4b).

We notice that, for $m = 1$, a $s \times 1$ matrix will be reduced to a column vector. Thus all rules and properties which apply to matrices are also valid for vectors.

Interchanging the rows and columns of a $s \times m$ matrix $\mathbf{C}$, given in (I.4), we obtain a $m \times s$ matrix $\mathbf{C}'$ which is called the *transpose* of $\mathbf{C}$, that is,

$$
\mathbf{C}' =
\begin{bmatrix}
c_{11} & c_{21} & \cdots & c_{s1} \\
c_{12} & c_{22} & \cdots & c_{s2} \\
\multicolumn{4}{c}{\dotfill} \\
c_{1m} & c_{2m} & \cdots & c_{sm}
\end{bmatrix} = (c_{ji}). \tag{I.6}
$$

[Note that this is consistent with the definition of the vector in (I.1).]

If every element of a $s \times m$ matrix is 0, the matrix is called a *zero matrix* and is denoted by $\mathbf{O}$.

When $s = m$, $\mathbf{C}$ is called a *square matrix*. The elements $c_{11}, c_{22}, \ldots, c_{ss}$ of a $s \times s$ square matrix form the *principal diagonal* and are called *diagonal elements*. A square matrix with all its nondiagonal elements equal to 0 is called a *diagonal matrix*. In particular, a diagonal matrix with all its diagonal elements equal to 1 is called a *unit matrix* or an *identity matrix*. It is usually denoted by $\mathbf{I}$, that is,

$$
\mathbf{I} =
\begin{bmatrix}
1 & 0 & 0 & \cdots & 0 \\
0 & 1 & 0 & \cdots & 0 \\
\multicolumn{5}{c}{\dotfill} \\
0 & 0 & 0 & \cdots & 1
\end{bmatrix}. \tag{I.7}
$$

A $s \times s$ square matrix $\mathbf{C}$ which is symmetrical about the principal diagonal, that is, for which $c_{ij} = c_{ji}$ ($i, j = 1, 2, \ldots, s$), is called a *symmetric matrix*. For a symmetric matrix we have

$$
\mathbf{C} = \mathbf{C}'. \tag{I.8}
$$

OPERATIONS ON MATRICES

Sum. We can add two (or more) matrices only if they are of the same order.

Let $A = (a_{ij})$ and $B = (b_{ij})$ be two matrices, each of order $s \times m$. The *sum*

$$C = A + B \tag{I.9}$$

is defined by the sm equations

$$c_{ij} = a_{ij} + b_{ij} \qquad i = 1, 2, \ldots, s; \qquad j = 1, 2, \ldots, m. \tag{I.10}$$

Example I.2. Let

$$A = \begin{bmatrix} 2 & 1 & 0 \\ 1 & 1 & 2 \end{bmatrix}, \quad B = \begin{bmatrix} 1 & 1 & 2 \\ 0 & 1 & 0 \end{bmatrix}; \quad \text{then} \quad C = \begin{bmatrix} 3 & 2 & 2 \\ 1 & 2 & 2 \end{bmatrix}.$$

The following properties hold:

$$
\left.
\begin{aligned}
A + B = B + A; & \qquad (a) \\
(A + B) + C = A + (B + C) = A + B + C; & \qquad (b) \\
(A + B)' = A' + B'. & \qquad (c)
\end{aligned}
\right\} \tag{I.11}
$$

To *multiply* a matrix A by a real number, c, we have to multiply each element of A by c, so that

$$cA = (ca_{ij}) = Ac. \tag{I.12}$$

Product of Two Matrices. The matrix A can be multiplied by the matrix B if and only if the number of columns in A is the same as the number of rows in B.

Let A be a matrix of order $s \times k$ and B a matrix of order $k \times m$. The product matrix, C, say, of order $s \times m$, that is,

$$\underset{s \times k}{A} \; \underset{k \times m}{B} = \underset{s \times m}{C}, \tag{I.13}$$

is defined by the sm equations

$$c_{ij} = \sum_{t=1}^{k} a_{it} b_{tj}, \qquad i = 1, 2, \ldots, s; \qquad j = 1, 2, \ldots, m. \tag{I.14}$$

In other words, to obtain the (ij)th element of the product matrix C, we have to multiply the elements of the ith row in A by the corresponding elements of the jth column in B, and add the products.

In particular, when the number of columns in $\mathbf{B}$ is equal to the number of rows in $\mathbf{A}$, that is, $s = m$, the product $\mathbf{AB} = \mathbf{C}$ is a $s \times s$ square matrix, and we can also calculate $\mathbf{BA} = \mathbf{E}$, which is a $k \times k$ square matrix. It should be noted that matrix multiplication is *not commutative*, that is, in general

$$\mathbf{AB} \neq \mathbf{BA}. \tag{I.15}$$

We have to distinguish *premultiplication* and *postmultiplication* ($\mathbf{AB}$ means that $\mathbf{A}$ is postmultiplied by $\mathbf{B}$, and $\mathbf{B}$ is premultiplied by $\mathbf{A}$).

Example I.3.   Let

$$\mathbf{A}_{2\times 3} = \begin{bmatrix} 1 & 2 & 0 \\ 0 & 1 & 2 \end{bmatrix} \quad \text{and} \quad \mathbf{B}_{3\times 2} = \begin{bmatrix} 1 & 1 \\ 1 & 0 \\ 2 & 1 \end{bmatrix}.$$

We calculate $\mathbf{C} = \mathbf{AB}$.

We have $c_{11} = 1 \cdot 1 + 2 \cdot 1 + 0 \cdot 2 = 3$; $c_{12} = 1 \cdot 1 + 2 \cdot 0 + 0 \cdot 1 = 1$; $c_{21} = 0 \cdot 1 + 1 \cdot 1 + 2 \cdot 2 = 5$; $c_{22} = 0 \cdot 1 + 1 \cdot 0 + 2 \cdot 1 = 2$. Hence

$$\underset{2\times 3}{\mathbf{A}} \; \underset{3\times 2}{\mathbf{B}} = \underset{2\times 2}{\mathbf{C}} = \begin{bmatrix} 3 & 1 \\ 5 & 2 \end{bmatrix}.$$

Since here $s = m = 2$, we can also calculate $\mathbf{BA} = \mathbf{E}$. We have $e_{11} = 1 \cdot 1 + 1 \cdot 0 = 1$; $e_{12} = 1 \cdot 2 + 1 \cdot 1 = 3$; $e_{13} = 1 \cdot 0 + 1 \cdot 2 = 2$; $e_{21} = 1 \cdot 1 + 1 \cdot 0 = 1$; $e_{22} = 1 \cdot 2 + 0 \cdot 1 = 2$; $e_{23} = 1 \cdot 0 + 0 \cdot 2 = 0$; $e_{31} = 2 \cdot 1 + 1 \cdot 0 = 2$; $e_{32} = 2 \cdot 2 + 1 \cdot 1 = 5$; $e_{33} = 2 \cdot 0 + 1 \cdot 2 = 2$. Hence

$$\underset{3\times 2}{\mathbf{B}} \; \underset{2\times 3}{\mathbf{A}} = \underset{3\times 3}{\mathbf{E}} = \begin{bmatrix} 1 & 3 & 2 \\ 1 & 2 & 0 \\ 2 & 5 & 2 \end{bmatrix}.$$

It is easy to see that $\mathbf{AB} \neq \mathbf{BA}$.

We shall now consider the situation in which $k = 1$, that is, in which the matrix $\mathbf{A}$ is reduced to a $s \times 1$ vector, $\mathbf{a}$, say, and the matrix $\mathbf{B}$ is reduced to a $1 \times m$ vector, $\mathbf{b}'$, say, so that

$$\mathbf{a} = \begin{pmatrix} a_1 \\ a_2 \\ \cdot \\ \cdot \\ \cdot \\ a_s \end{pmatrix} \quad \text{and} \quad \mathbf{b}' = (b_1, b_2, \dots, b_m). \tag{I.16}$$

We now have

$$\mathbf{a} \ \mathbf{b}' = \begin{bmatrix} a_1b_1 & a_1b_2 & \cdots & a_1b_m \\ a_2b_1 & a_2b_2 & \cdots & a_2b_m \\ \cdots\cdots\cdots\cdots\cdots\cdots \\ a_sb_1 & a_sb_2 & \cdots & a_sb_m \end{bmatrix} = (a_ib_j). \qquad (\text{I.17})$$
$$\underset{s\times1\ 1\times m}{}$$

Thus, the product of a $s \times 1$ column vector by a $1 \times m$ row vector is a $s \times m$ matrix.

If $s = m$, we can also multiply $\mathbf{b}'\mathbf{a}$, that is,

$$\underset{1\times s\ s\times1}{\mathbf{b}'\ \mathbf{a}} = \underset{1\times s\ s\times1}{\mathbf{a}'\ \mathbf{b}} = \sum_{i=1}^{s} a_ib_i. \qquad (\text{I.18})$$

If $\mathbf{a}$ and $\mathbf{b}$ are vectors of real numbers, the product $\mathbf{b}'\mathbf{a}$ is also a number. But it is not essential that the elements of a vector represent numbers. Let, for instance, $\boldsymbol{\gamma}' = (\gamma_1, \gamma_2, \ldots, \gamma_v)$ be a $1 \times v$ vector of gametes, and $\mathbf{g}' = (g_1, g_2, \ldots, g_v)$ be a vector of their frequencies. The product

$$\mathbf{g}'\boldsymbol{\gamma} = \sum_{i=1}^{v} g_i\gamma_i$$

is called a *gametic array* [see formula (4.14)]. This is a convenient combined form or representation of gametic output with its frequencies.

Of particular interest is the case when $s = m$ and $\mathbf{b} = \mathbf{a}'$. We have

$$\underset{s\times1\ 1\times s}{\mathbf{a}\ \mathbf{a}'} = \begin{bmatrix} a_1^2 & a_1a_2 & \cdots & a_1a_s \\ a_2a_1 & a_2^2 & \cdots & a_2a_s \\ \cdots\cdots\cdots\cdots\cdots\cdots \\ a_sa_1 & a_sa_2 & \cdots & a_s^2 \end{bmatrix}. \qquad (\text{I.19})$$

The product $\mathbf{aa}'$ represents a $s \times s$ symmetric matrix. For instance, genotypes resulting from the random union of gametes γ_i and γ_j, $\Gamma_{ij} = \gamma_i\gamma_j$, say $(i, j = 1, 2, \ldots, v)$, can be represented in the form of a $v \times v$ matrix $\boldsymbol{\Gamma}$, that is,

$$\boldsymbol{\Gamma} = \boldsymbol{\gamma}'\boldsymbol{\gamma} = (\Gamma_{ij}) = (\gamma_i\gamma_j),$$

and their distribution in the form of matrix

$$\mathbf{G} = \mathbf{g}'\mathbf{g} = (g_{ij}) = (g_ig_j)$$

(see section 4.3). Of course, in this presentation not all genotypes are distinct.

The product $\mathbf{a}'\mathbf{a}$ is

$$\underset{1\times s\ s\times1}{\mathbf{a}'\ \mathbf{a}} = \sum_{i=1}^{s} a_i^2. \qquad (\text{I.20})$$

It is a number, if a's are real numbers.

Multiplication can be extended to more than two matrices, provided that the number of columns in each postmultiplying matrix is equal to the number of rows in the matrix being postmultiplied. For instance,

$$\underset{s \times k}{\mathbf{A}} \; \underset{k \times m}{\mathbf{B}} \; \underset{m \times n}{\mathbf{C}} = \underset{s \times n}{\mathbf{E}}. \tag{I.21}$$

PROPERTIES OF MATRIX MULTIPLICATION

Let $\underset{s \times k}{\mathbf{A}}$, $\underset{k \times m}{\mathbf{B}}$, $\underset{k \times m}{\mathbf{C}}$ and $\underset{m \times n}{\mathbf{E}}$ be matrices, and c be a number. It can easily be shown that

$$
\begin{aligned}
(\mathbf{AB})' &= \mathbf{B}'\mathbf{A}'; & (a) \\
\mathbf{A}(\mathbf{B} + \mathbf{C}) &= \mathbf{AB} + \mathbf{AC}; & (b) \\
(\mathbf{B} + \mathbf{C})\mathbf{E} &= \mathbf{BE} + \mathbf{CE}; & (c) \\
(\mathbf{AB})\mathbf{E} &= \mathbf{A}(\mathbf{BE}) = \mathbf{ABE}; & (d) \\
(c\mathbf{A})\mathbf{B} &= \mathbf{A}(c\mathbf{B}) = c(\mathbf{AB}) = c\mathbf{AB}. & (e)
\end{aligned}
\tag{I.22}
$$

QUADRATIC FORMS

Let $\mathbf{x}' = (x_1, x_2, \ldots, x_s)$ be a $1 \times s$ vector of variables, and $\mathbf{C}$ be a $s \times s$ symmetric matrix of real numbers. The expression

$$Q(\mathbf{x}) = Q = \underset{1 \times s}{\mathbf{x}'} \; \underset{s \times s}{\mathbf{C}} \; \underset{s \times 1}{\mathbf{x}} = \sum_{i=1}^{s} \sum_{j=1}^{s} c_{ij} x_i x_j \tag{I.23}$$

is called a *quadratic form* in the variables $x_1, x_2, \ldots, x_s$; $\mathbf{C}$ is the matrix of quadratic form Q. For instance, the genotype distribution in a population with genetic output $\boldsymbol{\gamma}' = (\gamma_1, \gamma_2, \ldots, \gamma_v)$ and corresponding frequencies $\mathbf{g}' = (g_1, g_2, \ldots, g_v)$ can be written as a quadratic form (in elements of $\boldsymbol{\gamma}$) and is called a *genotype array*, that is,

$$Q(\boldsymbol{\gamma}) = (\mathbf{g}\boldsymbol{\gamma})'(\mathbf{g}\boldsymbol{\gamma}) = \boldsymbol{\gamma}'(\mathbf{g}'\mathbf{g})\boldsymbol{\gamma} = \boldsymbol{\gamma}'\mathbf{G}\boldsymbol{\gamma} = \sum_{i=1}^{v} \sum_{j=1}^{v} g_i g_j \gamma_i \gamma_j = \sum_{i=1}^{v} \sum_{j=1}^{v} g_{ij} \gamma_i \gamma_j.$$

[Compare formula (4.15a). Note the application of properties (I.22) to the vector multiplication.]

We notice that, if the elements of $\mathbf{x}$ in (1.23) take some numerical values, Q becomes a number.

If, for all $\mathbf{x} \neq \mathbf{0}$, $\mathbf{x}'\mathbf{Cx} > 0$, the quadratic form is called *positive definite*, and if $\mathbf{x}'\mathbf{Cx} \geqslant 0$, for all $\mathbf{x}$, then the quadratic form is *positive semidefinite*. Similarly, if $\mathbf{x}'\mathbf{Cx} < 0$, the quadratic form is *negative definite*; if $\mathbf{x}'\mathbf{Cx} \leqslant 0$, it is *negative semidefinite*.

Many examples of the use of quadratic form can be found in Chapters 4, 7, and 10 of this book.

PARTITION OF MATRICES

Matrices can be subdivided (or partitioned) by horizontal and vertical lines into a certain number of blocks or nonoverlapping *submatrices*.

Let $\mathbf{A}$ and $\mathbf{B}$ be two matrices, each partitioned into submatrices. If $\underset{s \times k}{}$ $\underset{k \times m}{}$ submatrices are regarded as "elements" of a corresponding matrix, $\mathbf{A}$ can be multiplied by $\mathbf{B}$ in the ordinary fashion, provided that the partitioning of the *rows* in matrix $\mathbf{B}$ is the same as the partitioning of the columns in $\mathbf{A}$. The following example illustrates the multiplication of partitioned matrices.

Example I.4. Let us consider two matrices, $\mathbf{A}$ and $\mathbf{B}$, partitioned into 3×2 and 2×2 blocks, respectively, as shown below.

$$\mathbf{A} = \left[\begin{array}{ccc|cc} a_{11} & a_{12} & a_{13} & a_{14} & a_{15} \\ \hline a_{21} & a_{22} & a_{23} & a_{24} & a_{25} \\ a_{31} & a_{32} & a_{33} & a_{34} & a_{35} \\ \hline a_{41} & a_{42} & a_{43} & a_{44} & a_{45} \end{array}\right] = \begin{bmatrix} \mathbf{A}_{11} & \mathbf{A}_{12} \\ \mathbf{A}_{21} & \mathbf{A}_{22} \\ \mathbf{A}_{31} & \mathbf{A}_{32} \end{bmatrix}$$

and

$$\mathbf{B} = \left[\begin{array}{cc|c} b_{11} & b_{12} & b_{13} \\ b_{21} & b_{22} & b_{23} \\ b_{31} & b_{32} & b_{33} \\ \hline b_{41} & b_{42} & b_{43} \\ b_{51} & b_{52} & b_{53} \end{array}\right] = \begin{bmatrix} \mathbf{B}_{11} & \mathbf{B}_{12} \\ \mathbf{B}_{21} & \mathbf{B}_{22} \end{bmatrix}.$$

The columns in $\mathbf{A}$ are partitioned by the vertical line into two groups of three and two columns, respectively, and the rows in $\mathbf{B}$ are divided in the same fashion. It can then be shown that the submatrices in the product $\mathbf{AB} = \mathbf{C}$ satisfy the equation similar to that in (I.14), that is,

$$\mathbf{C}_{ij} = \sum_{t=1}^{r} \mathbf{A}_{it}\mathbf{B}_{tj}, \tag{I.24}$$

where r is the number of partitions of columns in $\mathbf{A}$ (or rows in $\mathbf{B}$). In the present example, $r = 2$.

In evaluation of a transpose $\mathbf{A}'$ we have to interchange rows and columns. Thus, in our example,

$$\mathbf{A}' = \begin{bmatrix} \mathbf{A}'_{11} & \mathbf{A}'_{21} & \mathbf{A}'_{31} \\ \mathbf{A}'_{12} & \mathbf{A}'_{22} & \mathbf{A}'_{32} \end{bmatrix}.$$

KRONECKER PRODUCT OF MATRICES

Let $\mathbf{A}$ and $\mathbf{B}$ be two matrices. If each element, a_{ij}, of the matrix $\mathbf{A}$
$_{s \times m}$ $_{k \times n}$
is multiplied by the matrix $\mathbf{B}$, the resulting matrix is called a *Kronecker*
or *direct product*. The symbol $\otimes$ is used to denote Kronecker's multiplication.
have

$$\mathbf{A} \otimes \mathbf{B} = \begin{bmatrix} a_{11}\mathbf{B} & a_{12}\mathbf{B} & \cdots & a_{1m}\mathbf{B} \\ a_{21}\mathbf{B} & a_{22}\mathbf{B} & \cdots & a_{2m}\mathbf{B} \\ \cdots\cdots\cdots\cdots\cdots\cdots \\ a_{s1}\mathbf{B} & a_{s2}\mathbf{B} & \cdots & a_{sm}\mathbf{B} \end{bmatrix}. \tag{I.25}$$

Note that this is a $sk \times mn$ matrix.

TERM-BY-TERM PRODUCT OF MATRICES

Another useful product of two matrices, $\mathbf{A}$ and $\mathbf{B}$, of the *same order* is
the matrix resulting from multiplication of the corresponding terms in the
two matrices. This is the *term-by-term product* of these two matrices. We
introduce the symbol $\square$ to denote this product. We have

$$\mathbf{A} \square \mathbf{B} = \begin{bmatrix} a_{11}b_{11} & a_{12}b_{12} & \cdots & a_{1m}b_{1m} \\ a_{21}b_{21} & a_{22}b_{22} & \cdots & a_{2m}b_{2m} \\ \cdots\cdots\cdots\cdots\cdots\cdots \\ a_{s1}b_{s1} & a_{s2}b_{s2} & \cdots & a_{sm}b_{sm} \end{bmatrix} = (a_{ij}b_{ij}). \tag{I.26}$$

This is a matrix of the same order as $\mathbf{A}$ and $\mathbf{B}$. This kind of multiplication
was introduced as early as 1903 by Hadamard and was used in 1911 by
Schur. Hence it is sometimes called the *Hadamard-Schur product*. [For
detail, see Halmos (1948).]

Some useful applications of term-by-term matrix multiplication are given
in Chapter 7 of this book.

DETERMINANTS

With every $s \times s$ square matrix $\mathbf{C}$ there is associated a number, usually
denoted by $|\mathbf{C}|$, which is called the *determinant* of the matrix $\mathbf{C}$. (Note that
here the two parallel lines $|\ |$ do not mean "absolute value," as in nonmatrix

algebra, but denote "determinant.") The number s is here called the *order* of the determinant.

The determinant of any square submatrix of **C** is called a *minor* of **C**. In other words, deleting $s - r$ rows and $s - r$ columns, we obtain a determinant of order r, which is a minor of **C**. A minor for which all the diagonal elements are also diagonal elements of **C** is called a *principal minor*. If the diagonal elements are from the first r rows (and columns), it is a *leading principal minor* of order r. It can be proved that a necessary and sufficient condition that the quadratic form $\mathbf{x}'\mathbf{C}\mathbf{x} > 0$ (be positive definite) is that all leading principal minors be positive. The condition for $\mathbf{x}'\mathbf{C}\mathbf{x} < 0$ (be negative definite) is that the leading principal minor of order r has the sign $(-1)^r$ $(r = 1, 2, \ldots, s)$. (For application, see Theorem 10.4 in section 10.6.)

The *cofactor*, $\mathscr{C}_{ij}$, of the element c_{ij} in the $s \times s$ matrix **C** is $(-1)^{i+j}$ times the determinant (minor) of order $s - 1$ obtained by deleting the ith row and the jth column in **C**. Thus cofactor is a minor with sign $+$ or $-$.

Example I.5. Let us consider the 3×3 matrix

$$\mathbf{C} = \begin{bmatrix} c_{11} & c_{12} & c_{13} \\ c_{21} & c_{22} & c_{23} \\ c_{31} & c_{32} & c_{33} \end{bmatrix}.$$

The cofactor of the element c_{32}, $\mathscr{C}_{32}$, is

$$\mathscr{C}_{32} = (-1)^{3+2} \begin{vmatrix} c_{11} & c_{13} \\ c_{21} & c_{23} \end{vmatrix} = - \begin{vmatrix} c_{11} & c_{13} \\ c_{21} & c_{23} \end{vmatrix}.$$

(It is obtained by crossing out the third row and the second column in **C**.)

The determinant $|\mathbf{C}|$ is evaluated either from the formula

$$|\mathbf{C}| = \sum_{j=1}^{s} c_{ij}\mathscr{C}_{ij}, \tag{I.27a}$$

which represents the expansion of the determinant by the elements of the ith row, or from the formula

$$|\mathbf{C}| = \sum_{i=1}^{s} c_{ij}\mathscr{C}_{ij}, \tag{I.27b}$$

which is the expansion by the elements of the jth column.

Note that for $s = 1$, $|\mathbf{C}| = c_{11}$, that is, the determinant is equal to the numerical value of the single element. Taking this into account, formula (I.27a) or [(I.27b)] can be used as a definition of the *value* of the determinant of order s.

Example I.6. (a) For the determinant of order $s = 2$ we have

$$|\mathbf{C}| = \begin{vmatrix} c_{11} & c_{12} \\ c_{21} & c_{22} \end{vmatrix} = c_{11}c_{22} - c_{12}c_{21}.$$

(b) Let us now consider a determinant of order $s = 3$ and expand it by the elements of the first row:

$$|\mathbf{C}| = \begin{vmatrix} c_{11} & c_{12} & c_{13} \\ c_{21} & c_{22} & c_{23} \\ c_{31} & c_{32} & c_{33} \end{vmatrix} = c_{11}\begin{vmatrix} c_{22} & c_{23} \\ c_{32} & c_{33} \end{vmatrix} - c_{12}\begin{vmatrix} c_{21} & c_{23} \\ c_{31} & c_{33} \end{vmatrix} + c_{13}\begin{vmatrix} c_{21} & c_{22} \\ c_{31} & c_{32} \end{vmatrix}$$

$$= c_{11}(c_{22}c_{33} - c_{23}c_{32}) - c_{12}(c_{21}c_{33} - c_{23}c_{31}) + c_{13}(c_{21}c_{32} - c_{22}c_{31})$$

$$= (c_{11}c_{22}c_{33} + c_{12}c_{23}c_{31} + c_{13}c_{21}c_{32}) - (c_{13}c_{22}c_{31} + c_{12}c_{21}c_{33} + c_{11}c_{23}c_{32}).$$

When $s > 3$, the calculations become laborious and the following properties (which are given without proof) can be helpful in the evaluation of determinants.

(i) If two rows or two columns in $\mathbf{C}$ are interchanged, the determinant changes its sign. Hence, if two rows or two columns in $\mathbf{C}$ are identical, the determinant $|\mathbf{C}|$ is equal to zero.

(ii) The value of the determinant remains the same if to any row we add any linear combination of other rows, and the same holds for columns. In particular, this is true if to any row we add (or subtract) another row, and the same for columns.

Example I.7. Evaluate the determinant

$$\begin{vmatrix} 0 & 1 & 0 & 4 \\ 4 & 1 & 1 & 2 \\ 3 & 1 & 1 & 0 \\ -3 & 0 & -1 & 3 \end{vmatrix} = \begin{vmatrix} 0 & 1 & 0 & 4 \\ 1 & 1 & 1 & 2 \\ 0 & 1 & 1 & 0 \\ 0 & 0 & -1 & 3 \end{vmatrix} = (-1)^3 \begin{vmatrix} 1 & 0 & 4 \\ 1 & 1 & 0 \\ 0 & -1 & 3 \end{vmatrix}$$

$$= -\begin{vmatrix} 1 & -1 & 2 \\ 1 & 0 & 0 \\ 0 & -1 & 3 \end{vmatrix} = \begin{vmatrix} -1 & 2 \\ -1 & 3 \end{vmatrix} = -1.$$

We multiply the third column by 3 and subtract from the first column. Now the first column has three elements equal to zero, so it is convenient to expand the determinant by the elements of this column yielding a determinant of order 3. In this determinant of order 3 we subtract the first column from the second, and then expand by the elements of the second row, yielding a determinant of order 2.

It is left as an exercise to the reader to prove that

$$\underset{n \text{ times}}{\begin{vmatrix} a & b & b & \cdots & b \\ b & a & b & \cdots & b \\ & & \cdots\cdots\cdots & \\ b & b & b & \cdots & a \end{vmatrix}} = [a + (n-1)b](a-b)^{n-1}.$$

n times

(*Hint:* Subtract the first row from each of the other rows, add to the first column the sum of the other columns, and expand the determinant by the elements of the first row.)

The further following properties of determinants can be proved:

$$|\mathbf{A}'| = |\mathbf{A}|; \qquad (a)$$
$$|\mathbf{AB}| = |\mathbf{A}||\mathbf{B}|; \quad (b) \qquad\qquad (\text{I}.28)$$
$$|c\mathbf{A}| = c^s |\mathbf{A}|, \quad (c)$$

if $|\mathbf{A}|$ is of order s.

RANK OF A MATRIX

The *rank* of a matrix $\mathbf{C}$ is the greatest integer r such that $\mathbf{C}$ contains at least one minor of order r which is not equal to zero.

Thus, the rank of a $s \times m$ matrix $\mathbf{C}$ can be at most equal to the smaller of the numbers s and m.

It can be shown that, if $s = m$ and $\mathbf{C}$ is symmetric of rank r, there exists at least one leading principal minor of order r in $\mathbf{C}$, which is not equal to zero.

INVERSE OF A MATRIX

A $s \times s$ square matrix $\mathbf{C}$ is called *singular* if its determinant is equal to zero ($|\mathbf{C}| = 0$). If $|\mathbf{C}| \neq 0$, the matrix $\mathbf{C}$ is *nonsingular*.

Let $\mathscr{C}_{ij}$ be the cofactor of the element c_{ij} in $\mathbf{C}$. We construct a matrix, $\mathbf{C}^*$, say, which is a transpose of a matrix with the cofactors as elements, that is,

$$\mathbf{C}^* = \begin{bmatrix} \mathscr{C}_{11} & \mathscr{C}_{21} & \cdots & \mathscr{C}_{s1} \\ \mathscr{C}_{12} & \mathscr{C}_{22} & \cdots & \mathscr{C}_{s2} \\ & & \cdots\cdots\cdots & \\ \mathscr{C}_{1s} & \mathscr{C}_{2s} & \cdots & \mathscr{C}_{ss} \end{bmatrix}. \qquad (\text{I}.29)$$

The matrix $\mathbf{C}^*$ is called the *adjugate matrix* of $\mathbf{C}$.

The *inverse* of the matrix C, denoted by C^{-1}, is defined as

$$C^{-1} = \frac{C^*}{|C|}. \tag{I.30}$$

If C is singular, that is, $|C| = 0$, then C^{-1} does not exist. It also should be noted that the inverse is defined only for square matrices.

It can be shown that

$$CC^{-1} = C^{-1}C = I. \tag{I.31}$$

Some other useful properties are as follows:

$$\left.\begin{aligned}
(C')^{-1} &= (C^{-1})'; \quad &(a)\\
(C^{-1})^{-1} &= C; \quad &(b)\\
(AB)^{-1} &= B^{-1}A^{-1}. \quad &(c)
\end{aligned}\right\} \tag{I.32}$$

Example I.8. Find the inverse, C^{-1}, of the matrix

$$C = \begin{bmatrix} 1 & 0 & 2 \\ 1 & 1 & 0 \\ 2 & 1 & 0 \end{bmatrix}.$$

Expanding by the elements of the third column, we find the determinant $|C| = 2(1 - 2) = -2$. We calculate the cofactors $\mathscr{C}_{ij}$. We have $\mathscr{C}_{11} = 0$; $\mathscr{C}_{12} = 0$; $\mathscr{C}_{13} = -1$; $\mathscr{C}_{21} = 2$; $\mathscr{C}_{22} = -4$; $\mathscr{C}_{23} = -1$; $\mathscr{C}_{31} = -2$; $\mathscr{C}_{32} = 2$; $\vartheta_{33} = 1$. The inverse is

$$C^{-1} = -\tfrac{1}{2}\begin{bmatrix} 0 & 2 & -2 \\ 0 & -4 & 2 \\ -1 & -1 & 1 \end{bmatrix} = \begin{bmatrix} 0 & -1 & 1 \\ 0 & 2 & -1 \\ \tfrac{1}{2} & \tfrac{1}{2} & -\tfrac{1}{2} \end{bmatrix}.$$

It is left to the reader to show that

$$CC^{-1} = \begin{bmatrix} 1 & 0 & 2 \\ 1 & 1 & 0 \\ 2 & 1 & 0 \end{bmatrix}\begin{bmatrix} 0 & -1 & 1 \\ 0 & 2 & -1 \\ \tfrac{1}{2} & \tfrac{1}{2} & -\tfrac{1}{2} \end{bmatrix} = \begin{bmatrix} 1 & 0 & 0 \\ 0 & 1 & 0 \\ 0 & 0 & 1 \end{bmatrix} = I.$$

SYSTEMS OF LINEAR EQUATIONS

Let us consider a nonhomogeneous system of linear equations:

$$\left.\begin{aligned}
c_{11}x_1 + c_{12}x_2 + \cdots + c_{1s}x_s &= y_1, \\
c_{21}x_1 + c_{22}x_2 + \cdots + c_{2s}x_s &= y_2, \\
\cdots\cdots\cdots\cdots\cdots\cdots\cdots\cdots & \\
c_{s1}x_1 + c_{s2}x_2 + \cdots + c_{ss}x_s &= y_s.
\end{aligned}\right\} \tag{I.33}$$

System (I.33) can be written in matrix form as

$$\underset{s \times s}{C} \; \underset{s \times 1}{x} = \underset{s \times 1}{y},$$ (I.34)

where C is called the matrix of the system. Premultiplying both sides of (I.34) by C^{-1}, we obtain

$$(C^{-1}C)x = C^{-1}y;$$

hence

$$Ix = C^{-1}y,$$

that is,

$$x = C^{-1}y$$ (I.35)

or

$$x = \frac{1}{|C|} C^*y.$$ (I.35a)

We notice that system (I.33) has nontrivial unique solutions if and only if the matrix C is nonsingular.

Some application of this theory is given, for instance, in section 10.6.

REFERENCES

Anderson, T. W., *Introduction to Multivariate Statistical•Analysis*, John Wiley & Sons, New York, 1958.

Cramér, H., *Mathematical Methods of Statistics*, Princeton University Press, Princeton, N.J., 1951.

Graybill, F. A., *An Introduction to Linear Statistical Models*, New York, 1961.

Halmos, P. R., *Finite Dimensional Vector Spaces*, Princeton University Press, Princeton, N.J., 1948.

Mostowski, A., and Stark, M., *Elementy Algebry Wyższej*, Państwowe Wydawnictwo Naukowe, Warszawa, 1958.

Scheffé, H., *The Analysis of Variance*, John Wiley & Sons, New York, 1959.

Minima and Maxima of Functions. Taylor's Expansions

We want to summarize here a few useful definitions and theorems, from differential calculus, which have been used in this book. It is assumed that the reader is familiar with the concepts of functions of real variables and their derivatives. The theorems will be given here without proofs, which can be found in any basic book on calculus.

MINIMA AND MAXIMA OF A FUNCTION OF ONE VARIABLE

The Derivative. Let $y = f(x)$ be a continous function of a real variable, x. Let Δx be a small increment (positive or negative) of x, and $\Delta y = f(x + \Delta x) - f(x)$ of the function corresponding to Δx (see Fig. II.1). The ratio $(\Delta y/\Delta x)$ is the slope of the chord through the points $(x, f(x))$ and $(x + \Delta x, f(x + \Delta x))$ (line l_1). The limit (if it exists),

$$\lim_{\Delta x \to 0} \frac{\Delta y}{\Delta x} = \frac{dy}{dx} = \frac{df}{dx} = f'(x), \tag{II.1}$$

is called the *derivative* of the function $f(x)$ at the point x. It represents the slope of the tangent to the curve $y = f(x)$ at the point $(x, f(x))$ (line l_0).

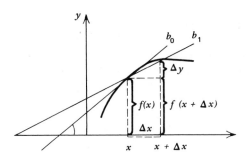

Figure II. 1

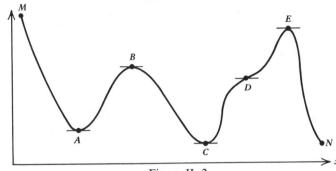

Figure II. 2

The first derivative may be again a (continuous) function of x and may be differentiated. The resulting derivative is called the *second* derivative. We have

$$\frac{d}{dx}\left(\frac{df}{dx}\right) = \frac{d^2f}{dx^2} = f''(x). \tag{II.2}$$

We can extend our definition to the nth derivative, that is,

$$\frac{d}{dx}\left(\frac{d^{n-1}f}{dx^{n-1}}\right) = \frac{d^nf}{dx^n} = f^{(n)}(x). \tag{II.3}$$

A *minimum* (more precisely, a *local minimum*) on the curve $y = f(x)$ is a point $(x, f(x))$, say, in the neighborhood of which the function decreases on the left-hand side and increases on the right-hand side of $f(x)$ (points A and C in Fig. II.2). It should be emphasized that a minimum is not necessarily the smallest value of the function (point N is not a minimum).

A *maximum* (more precisely, a *local maximum*) is a point on the curve in the neighborhood of which the function increases on the left-hand side and decreases on the right-hand side (points B and E). It is not necessarily the greatest value of $f(x)$ (e.g., point M is not a maximum).

Briefly, a minimum or maximum is called an *extremum* (more precisely, a *local extremum*).

We notice that at a local extremum the tangent is parallel to the x-axis. This is a necessary but not sufficient condition for an extremum. For instance, point D is a *point of inflection*.

Thus a *necessary* condition for $f(x)$ to attain an extremum at the point x is

$$\frac{df}{dx} = 0. \tag{II.4}$$

It can be shown that *sufficient* conditions are:

$$\frac{d^2f}{dx^2} > 0 \quad \text{for minimum,} \quad \frac{d^2f}{dx^2} < 0 \quad \text{for maximum,} \tag{II.5}$$

provided $(d^2f/dx^2) \neq 0$. (If $(d^2f/dx^2) = 0$ and $(d^3f/dx^3) \neq 0$, we have a point of inflection.)

MINIMA AND MAXIMA OF FUNCTIONS OF SEVERAL VARIABLES

Partial Derivatives. Let $z = f(x, y)$ be a continuous function of two variables, x and y. We may evaluate derivatives with respect to x and with respect to y, called *partial derivatives*, and denoted as

$$\frac{\partial z}{\partial x} = \frac{\partial f}{\partial x}, \quad \frac{\partial z}{\partial y} = \frac{\partial f}{\partial y}.$$

Each of these derivatives may be again a function of x and y, so we may be able to evaluate second derivatives:

$$\frac{\partial^2 f}{\partial x^2}, \quad \frac{\partial^2 f}{\partial x\,\partial y}, \quad \frac{\partial^2 f}{\partial y\,\partial x}, \quad \frac{\partial^2 f}{\partial y^2}.$$

If $f(x, y)$, $(\partial f/\partial x,)$ and $(\partial f/\partial y)$ are continuous, then for the mixed derivatives we have

$$\frac{\partial^2 f}{\partial x\cdot\partial y} = \frac{\partial^2 f}{\partial y\,\partial x}, \tag{II.6}$$

so that the order of differentiation is immaterial.

We can repeat differentiation procedures and obtain nth (pure and mixed) derivatives.

The definitions of partial derivatives can be extended to functions of k variables.

Minima and Maxima. Let $z = f(x_1, x_2, \ldots, x_k)$ be a continuous function of k variables. *Necessary* conditions for the function $z = f(x_1, x_2, \ldots, x_k)$ to attain an extremum at the point $(x_1, x_2, \ldots, x_k)$ are

$$\frac{\partial f}{\partial x_1} = 0, \quad \frac{\partial f}{\partial x_2} = 0, \quad \ldots, \quad \frac{\partial f}{\partial x_k} = 0. \tag{II.7}$$

Let

$$\mathbf{D}(x_1, x_2, \ldots, x_k) = \begin{bmatrix} \dfrac{\partial^2 f}{\partial x_1^2} & \dfrac{\partial^2 f}{\partial x_1\,\partial x_2} & \cdots & \dfrac{\partial^2 f}{\partial x_1\,\partial x_k} \\[2mm] \dfrac{\partial^2 f}{\partial x_2\,\partial x_1} & \dfrac{\partial^2 f}{\partial x_2^2} & \cdots & \dfrac{\partial^2 f}{\partial x_2\,\partial x_k} \\ \cdots\cdots\cdots\cdots\cdots\cdots\cdots\cdots \\ \dfrac{\partial^2 f}{\partial x_k\,\partial x_1} & \dfrac{\partial^2 f}{\partial x_k\,\partial x_2} & \cdots & \dfrac{\partial^2 f}{\partial x_k^2} \end{bmatrix} \tag{II.8}$$

be a $k \times k$ matrix of the second derivatives (called the Hessian matrix), evaluated at point $(x_1, \ldots, x_k)$ satisfying (II.7).

Sufficient conditions for the function $z = f(x_1, x_2, \ldots, x_k)$ to attain a minimum or maximum are as follows:

If all the leading principal minors of $D(x_1, x_2, \ldots, x_k)$ are positive, then the function $f(x_1, x_2, \ldots, x_k)$ has a minimum.

If the leading principal minor of order r has the sign $(-1)^r$ for $r = 1, 2, \ldots, k$, then the function $f(x_1, x_2, \ldots, x_k)$ has a maximum.

In particular, when $k = 2$ we have a function of *two* variables, $z = f(x, y)$. Necessary conditions for an extremum are

$$\frac{\partial f}{\partial x} = 0, \quad \frac{\partial f}{\partial y} = 0. \tag{II.9}$$

Matrix (II.8) takes the form

$$\mathbf{D}(x, y) = \begin{bmatrix} \dfrac{\partial^2 f}{\partial x^2} & \dfrac{\partial^2 f}{\partial x \, \partial y} \\[2mm] \dfrac{\partial^2 f}{\partial y \, \partial x} & \dfrac{\partial^2 f}{\partial y^2} \end{bmatrix}. \tag{II.10}$$

If the determinant $|\mathbf{D}(x, y)| > 0$ and $(\partial^2 f / \partial x^2) > 0$, the function $f(x, y)$ has a minimum; if $|\mathbf{D}(x, y)| > 0$ and $(\partial^2 f / \partial x^2)$ [or $(\partial^2 f / \partial y^2)$] < 0, $f(x, y)$ has a maximum. If $|\mathbf{D}(x, y)| < 0$ there is no minimum nor maximum.

Useful functions of k variables are the special polynomials of the second order, called *quadratic forms*

$$Q(x_1, x_2, \ldots, x_k) = \sum_{i=1}^{k} \sum_{j=1}^{k} c_{ij} x_i x_j = \mathbf{x}' \mathbf{C} \mathbf{x} \tag{II.11}$$

defined in Appendix I. It is easy to see that the matrix $\mathbf{C}$ of the quadratic form corresponds to the matrix $\mathbf{D}(x_1, x_2, \ldots, x_k)$ of the second derivatives. Thus, the conditions for quadratic forms are immediately obtained from the general theorem given above. Necessary conditions for an extremum are

$$\frac{\partial Q}{\partial x_1} = 0, \quad \frac{\partial Q}{\partial x_2} = 0, \ldots, \frac{\partial Q}{\partial x_k} = 0. \tag{II.12}$$

Necessary and sufficient conditions for an extremum to be a minimum or maximum are: if $Q(x_1, x_2, \ldots, x_k)$ is positive definite we have a minimum; if $Q(x_1, x_2, \ldots, x_k)$ is negative definite we have a maximum. (See Appendix I, "Quadratic forms" and "Determinants.") In fact, these extrema are local and overall extrema.

MINIMA AND MAXIMA WITH SIDE CONDITIONS. LAGRANGE MULTIPLIERS

Let $z = f(x, y)$ be a function of two variables, x and y, subject to the constraint

$$\psi(x, y) = 0. \tag{II.13}$$

To find an extremum we may apply two procedures.

(a) Suppose that it is possible to solve the equation $\psi(x, y) = 0$ in the form $y = \varphi(x)$. Substituting $\varphi(x)$ for y into $f(x, y)$ we obtain $z = f[x, \varphi(x)] = g(x)$. Thus $z = g(x)$ is a function of one variable, x, and the rules for extremum of a function of one variable can be used.

(b) It is sometimes difficult to obtain the explicit solution $y = \varphi(x)$. In such cases, a procedure for obtaining extremum is as follows.

We introduce a function

$$F(x, y) = f(x, y) + \lambda\psi(x, y). \tag{II.14}$$

The function $F(x, y)$ attains an extremum at the same points as $f(x, y)$. Thus, to obtain an extremum, we have to solve the system of three equations:

$$\frac{\partial F}{\partial x} = 0, \quad \frac{\partial F}{\partial y} = 0, \quad \psi(x, y) = 0. \tag{II.15}$$

The coefficient λ in (II.14) is called *Lagrange multiplier*. The sufficient conditions for two variables can now be applied to $F(x, y)$ in order to distinguish between minimum and maximum.

This method can be extended to functions of k variables subject to m ($m < k - 1$) constraints, introducing m Lagrange multipliers.

TAYLOR'S EXPANSION OF A FUNCTION OF ONE VARIABLE

Taylor Formula with Remainder.   Let $y = f(x)$ be a continuous function of x in the neighborhood of a certain point, x_0. Let $x - x_0 = \Delta x$ be a small increment (positive or negative) of x in the neighborhood of x_0. Suppose that all $(n + 1)$ derivatives of x exist and all are continuous in the neighborhood of x_0. It can be proved that $f(x)$ can be represented in the form

$$\begin{aligned}
f(x) &= f(x_0 + \Delta x) \\
&= f(x_0) + f'(x_0)\Delta x + \frac{1}{2!}f''(x_0)(\Delta x)^2 + \cdots \\
&\quad + \frac{1}{(n + 1)!}f^{(n+1)}(x_1)(\Delta x)^{n+1},
\end{aligned} \tag{II.16}$$

where x_1 is a certain value between x_0 and $x_0 + \Delta x$. Here

$$R_{n+1}(\Delta x) = \frac{1}{(n+1)!} f^{(n+1)}(x_1)(\Delta x)^{n+1} \tag{II.17}$$

is termed the *remainder*, and (II.16) is called *Taylor formula with remainder*.

Taylor (*Infinite*) Series. If all derivatives exist in the neighborhood of x_0, and $R_{n+1}(\Delta x) \to 0$ as $n \to \infty$, then the infinite power series

$$f(x) = f(x_0 + \Delta x)$$

$$= f(x_0) + f'(x_0)\Delta x + \frac{1}{2!} f''(x_0)(\Delta x)^i + \cdots + f^{(n)}(x_0)(\Delta x)^n + \cdots$$

$$= f(x_0) + \sum_{i=1}^{\infty} \frac{1}{i!} f^{(i)}(x_0)(\Delta x)^i$$

$$= f(x_0) + \sum_{i=1}^{\infty} \frac{1}{i!} \frac{d^i f}{dx^i}\bigg|_{x=x_0} (\Delta x)^i \tag{II.18}$$

converges and represents the function $f(x)$.

Formula (II.18) is sometimes called *Taylor series expansion* of $f(x)$ in the neighborhood of x_0.

In particular, when $x_0 = 0$, we obtain *Maclaurin series:*

$$f(x) = f(0) + f'(0)x + \tfrac{1}{2}f''(0)x^2 + \cdots + \frac{1}{n!} f^{(n)}(0)x^n + \cdots$$

$$= f(0) + \sum_{i=1}^{\infty} \frac{1}{i!} f^{(i)}(0)x^i. \tag{II.19}$$

For example, it can be shown that the function $f(x) = e^x$ can be expanded in the Maclaurin's series

$$e^x = 1 + x + \frac{1}{2!} x^2 + \frac{1}{3!} x^3 + \cdots + \frac{1}{n!} x^n + \cdots \tag{II.20}$$

for all x in the interval $(-\infty, +\infty)$.

On the other hand, the Maclaurin expansion

$$\frac{1}{1-x} = 1 + x + x^2 + \cdots + x^2 + \cdots \tag{II.21}$$

converges only for $-1 < x < 1$.

Formula (II.21) is called an *infinite power* series. It converges to $1/(1-x)$, provided $|x| < 1$.

The finite power series

$$1 + x + x^2 + \cdots + x^n \tag{II.22}$$

is called a *geometric series*. Its sum is $(1 - x^{n+1})/(1 - x)$.

Proof: Let S denote the sum of the geometric series. We have

$$S = 1 + x + x^2 + \cdots + x^n$$
$$Sx = x + x^2 + x^3 + \cdots + x^{n+1}$$

Subtracting the second row from the first, we have

$$S(1 - x) = 1 - x^{n+1}$$

Hence

$$S = \frac{1 - x^{n+1}}{1 - x}. \tag{II.23}$$

TAYLOR'S EXPANSION OF A FUNCTION OF TWO VARIABLES

Let $z = f(x, y)$ be a function of two variables. Assuming continuity of $f(x, y)$ in the neighborhood of (x_0, y_0) and the existence and continuity of all derivatives, we obtain Taylor's infinite multivariable expansion:

$$f(x, y) = f(x_0, y_0) + \left[\frac{\partial f}{\partial x}\bigg|_{\substack{x=x_0 \\ y=y_0}} \Delta x + \frac{\partial f}{\partial y}\bigg|_{\substack{x=x_0 \\ y=y_0}} \Delta y \right]$$

$$+ \frac{1}{2!} \left[\frac{\partial^2 f}{\partial x^2}\bigg|_{\substack{x=x_0 \\ y=y_0}} (\Delta x)^2 + 2 \frac{\partial^2 f}{\partial x\, \partial y}\bigg|_{\substack{x=x_0 \\ y=y_0}} \Delta x\, \Delta y + \frac{\partial^2 f}{\partial y^2}\bigg|_{\substack{x=x_0 \\ y=y_0}} (\Delta y)^2 \right] + \cdots$$

or, formally,

$$f(x, y) = f(x_0, y_0) + \sum_{i=1}^{\infty} \frac{1}{i!} \left[\frac{\partial}{\partial x}\bigg|_{\substack{x=x_0 \\ y=y_0}} \Delta x + \frac{\partial}{\partial y}\bigg|_{\substack{x=x_0 \\ y=y_0}} \Delta y \right]^i f, \tag{II.24}$$

provided that the remainder,

$$R_{n+1}(\Delta x, \Delta y) = \frac{1}{(n + 1)!} \left[\frac{\partial}{\partial x}\bigg|_{\substack{x=x_1 \\ y=y_1}} \Delta x + \frac{\partial}{\partial y}\bigg|_{\substack{x=x_1 \\ y=y_1}} \Delta y \right]^{n+1} f, \tag{II.25}$$

tends to zero as $n \to \infty$.

The theory can be extended to functions of k variables in a straightforward manner.

Taylor's expansions of functions of *random variables* are applied in the evaluation of approximate means and variances of these functions, assuming that the means, variances, and covariances of random variables are known (see section 8.20).

Index of Statistical and Mathematical Terms

Index of Genetic Terms

Index of Authors

Page numbers in parentheses signify that the name appears in references.